Abstract Algebra

Abstract Algebra: An Inquiry-Based Approach, Second Edition not only teaches abstract algebra, but also provides a deeper understanding of what mathematics is, how it is done, and how mathematicians think.

The second edition of this unique, flexible approach builds on the success of the first edition. The authors offer an emphasis on active learning, helping students learn algebra by gradually building both their intuition and their ability to write coherent proofs in context.

The goals for this text include:

- Allowing the flexibility to begin the course with either groups or rings.
- Introducing the ideas behind definitions and theorems to help students develop intuition.
- Helping students understand how mathematics is done. Students will experiment through examples, make conjectures, and then refine or prove their conjectures.
- Assisting students in developing their abilities to effectively communicate mathematical ideas.
- Actively involving students in realizing each of these goals through in-class and out-of-class activities, common in-class intellectual experiences, and challenging problem sets.

Changes in the Second Edition

- Streamlining of introductory material with a quicker transition to the material on rings and groups.
- New investigations on extensions of fields and Galois theory.
- New exercises added and some sections reworked for clarity.
- More online Special Topics investigations and additional Appendices, including new appendices on other methods of proof and complex roots of unity.

Encouraging students to do mathematics and be more than passive learners, this text shows students the way mathematics is developed is often different than how it is presented; definitions, theorems, and proofs do not simply appear fully formed; mathematical ideas are highly interconnected; and in abstract algebra, there is a considerable amount of intuition to be found.

Textbooks in Mathematics
Series editors:
Al Boggess, Kenneth H. Rosen

https://www.routledge.com/Textbooks-in-Mathematics/book-series/CANDHTEXBOOMTH

Abstract Algebra
An Inquiry-Based Approach
Second Edition

Jonathan K. Hodge, Steven Schlicker, and Ted Sundstrom

CRC Press
Taylor & Francis Group
Boca Raton London New York

CRC Press is an imprint of the
Taylor & Francis Group, an **informa** business

A CHAPMAN & HALL BOOK

Second edition published 2024
by CRC Press
2385 NW Executive Center Drive, Suite 320, Boca Raton FL 33431

and by CRC Press
4 Park Square, Milton Park, Abingdon, Oxon, OX14 4RN

CRC Press is an imprint of Taylor & Francis Group, LLC

© 2024 Taylor & Francis Group, LLC

First edition published by CRC Press 2013

Reasonable efforts have been made to publish reliable data and information, but the author and publisher cannot assume responsibility for the validity of all materials or the consequences of their use. The authors and publishers have attempted to trace the copyright holders of all material reproduced in this publication and apologize to copyright holders if permission to publish in this form has not been obtained. If any copyright material has not been acknowledged please write and let us know so we may rectify in any future reprint.

Library of Congress Control Number: 2023949885

ISBN: 978-0-367-55501-6 (hbk)
ISBN: 978-1-032-63492-0 (pbk)
ISBN: 978-1-032-63490-6 (ebk)

DOI: 10.1201/9781032634906

Typeset in Nimbus font
by KnowledgeWorks Global Ltd.

Publisher's note: This book has been prepared from camera-ready copy provided by the authors.

Access the Support Material: https://scholarworks.gvsu.edu/books/29/

Contents

II   Rings   55

4   Introduction to Rings   57

5   Integer Multiples and Exponents   69

6   Subrings, Extensions, and Direct Sums   83

Note to Students

This book may be unlike other mathematics textbooks you have read or used in previous courses. The investigations contained in it are designed to facilitate your learning by inviting you to be an active participant in the learning process. This is a book that is not meant to be simply *read*, but rather *engaged* in. It includes numerous activities within the text that are intended to motivate new material, illustrate definitions and theorems, and help you develop both the intuition and rigor that is necessary to understand and apply ideas from abstract algebra.

As professors of mathematics, we have found (and research confirms) that mathematics is not a spectator sport. To learn and understand mathematics, one must engage in the process of *doing* mathematics. This kind of engagement can be challenging and even frustrating at times. But if you are up to the challenge and willing to take responsibility for your own learning, you will indeed learn a great deal.

Obviously, this is a book about abstract algebra, and you will learn more about what that means as we begin our investigations. Our goal, however, is that you will not only learn about abstract algebra, but that you will also develop a deeper understanding of what mathematics is, how mathematics is done, and how mathematicians think. We hope that you will see that the way mathematics is developed is often different than how it is presented; that definitions, theorems, and proofs do not simply appear fully formed in the minds of mathematicians; that mathematical ideas are highly interconnected; and that even in a field like abstract algebra, there is a considerable amount of intuition to be found.

The objectives of this book and its inquiry-based format place the responsibility of learning the material where it belongs – on your shoulders. It is imperative that you engage the material by completing the preview activities and the in-class activities in order to develop your intuition and understanding of the material. Exercises are also important in developing your understanding of the material in the text. The exercises occur at a variety of levels of difficulty, and most will force you to extend your knowledge in different ways. While there are some standard, classic problems that are included in the exercises, many problems are open-ended and expect you to develop and then verify conjectures. Exercises that are highlighted with an asterisk (*) are referred to in the investigations and should be given special attention when working exercises.

The study of abstract algebra relies on some foundational ideas from other courses, including methods of proof, mathematical induction and the Well-Ordering Principle, properties of functions, and roots of unity. As you progress through the text, you may find it useful to review these ideas. To facilitate such a review, you can find supplemental investigations and appendices at `https://scholarworks.gvsu.edu/books/29/`. More information about the supplemental material is contained in the preface.

Thank you for joining us on this journey. We hope you enjoy both the challenges and the rewards that await you in these pages. Good luck!

Preface

The impetus for this book lies in our approach to teaching abstract algebra. We place an emphasis on active learning and on developing students' intuition through their investigation of examples. For us, active learning involves students—they are ***doing*** something instead of just being passive learners. What students are doing when they are actively learning might include discovering, processing, discussing, applying information, writing intensive assignments, and engaging in common intellectual in-class experiences or collaborative assignments and projects. We support all of these activities with peer review and substantial faculty mentoring. According to Meyers and Jones [2], active learning derives from the assumptions that learning is an active endeavor by nature and that different people learn in different ways. A number of reports and studies show that active learning has a positive impact on students. For example, active learning is described as a high-impact learning activity in the latest report from the Association of American Colleges and Universities' Liberal Education and America's Promise (LEAP) initiative [1]. Results of a study [3] testing the active learning findings in liberal arts education show, in part, that students who experience the type of instruction we describe as active learning show larger "value-added" gains on a variety of outcomes than their peers. Although it is difficult to capture the essence of active learning in a textbook, this book is our attempt to do just that.

Our goals for these materials are several:

- To carefully introduce the ideas behind definitions and theorems in order to help students develop intuition and understand the logic behind them.

- To help students understand that mathematics is not done as it is often presented. We expect students to experiment through examples, make conjectures, and then refine or prove their conjectures. We believe it is important for students to learn that definitions and theorems don't pop up completely formed in the minds of mathematicians, but are the result of much thought and work.

- To help students develop their communication skills in mathematics. We expect our students to read and complete activities before class and come prepared with questions. In-class group work, student presentations, and peer-evaluation are a regular part of our courses. Of course, students also individually write solutions (mostly proofs) to exercises and receive significant feedback. Communication skills are essential in any discipline, and we place a heavy emphasis on developing students' abilities to effectively communicate mathematical ideas and arguments.

- To have students actively involved in realizing each of these goals through in-class and out-of-class activities, common in-class intellectual experiences (which, for us, include student presentations and collaborative group work), and challenging problem sets.

Changes in the Second Edition

To prepare the second edition of this book, we adopted suggestions that were made by users of the text. The major changes to the second edition are the following:

Streamlining of introductory material: Investigations 1-6 in the first edition built the foundation for the study of rings and groups through an extensive exploration of the integers and other number systems. Some readers asked for a quicker transition to the material on rings and groups, and we addressed this feedback by streamlining the introductory activities, providing the necessary background for subsequent investigations, but in a more abbreviated manner.

New material: The second edition contains two new investigations, on extensions of fields and Galois theory. The investigation on field extensions provides the basics of single and multiple extensions and a discussion of automorphisms of field extensions, with the intent to build the foundation for the subsequent investigation on Galois theory.

The investigation of Galois theory describes Galois groups and delves into the fundamental Galois correspondence with the aim of tackling the problem of solvability of polynomials by radicals.

Material moved online: With the addition of the new investigations on field extensions and Galois theory, the text has become a bit long. To help keep the size (and cost) of the text manageable, several investigations have been moved online. These investigations include RSA Encryption (which has been updated with additional background material); Check Digits; Games: NIM and the 15 Puzzle; and Groups of Order 8 and 12: Semidirect Products of Groups. The Appendices on Functions and Mathematical Induction and the Well-Ordering Principle are also now accessible online at `https://scholarworks.gvsu.edu/books/29/`. More details about these sections is given later in this preface.

New appendices: In addition to the appendices on Functions and Mathematical Induction and the Well-Ordering Principle, we have added additional appendices on other methods of proof and complex roots of unity (to accompany the investigations on field extensions and Galois theory). We have also moved some material from the text (the proof that $R[x]$ is a ring, the cubic formula, and the Fundamental Theorem of Algebra) to the appendices, which are online, but accessible to users.

Of course, we addressed the known typos and errors in the first edition, added new exercises, and reworked a few sections. We hope you will find the changes to be worthwhile.

Layout

This text is formatted into investigations, each of which contains preview activities, in-class activities, concluding activities, exercises, and connections. The various types of activities serve different purposes.

- Preview activities are designed for students to complete before class to motivate the upcoming topic and prepare them with the background and information they need for the class activities and discussion.

- We generally use the regular activities to engage students in common in-class intellectual experiences. These activities provide motivation for new material, opportunities for students to prove substantial results on their own, and examples to help reinforce the meanings of definitions, theorems, and proofs. The ultimate goal is to help students build their intuition and develop a deep understanding of abstract algebra concepts. In our own practice, students often complete these activities—either during or before each class meeting—and then present their results to the entire class.

- Concluding activities are used to summarize, extend, or enhance the topics in a particular investigation. Concluding activities sometimes serve to foreshadow ideas that will be explored in more detail in subsequent investigations.

Each investigation contains a collection of exercises. The exercises occur at a variety of levels of difficulty, and most force students to extend their knowledge in different ways. While there are some standard, classic problems that are included in the exercises, many problems are open-ended and expect students to develop and then verify conjectures. Exercises that are highlighted with an asterisk (*) are referred to in the investigations and should be given special attention when assigning problems. Complete solutions to all activities and exercises are available to instructors – please contact the authors for access.

Most investigations conclude with a short discussion of the connections between the topics in that investigation and the corresponding topics in ring theory or group theory. These discussions are intended to help students see the relationships between the two main types of algebraic objects studied throughout the text.

Organization

At Grand Valley State University (GVSU), the first course we teach in modern algebra is focused on rings rather than the more simple structure of groups. Most of our majors intend to become elementary or secondary mathematics teachers, and the structure of the integers (and rings in general) is familiar to these students and therefore provides a comfortable entry point into the study of abstract algebra. Of course, a good argument can be made that groups, with their simpler structure, offer students an easier entrance to the subject. Both points are valid, and so we have designed this book so that, after completing some necessary background material, it is possible to begin with either rings or groups. One of the consequences of this flexibility is that investigations that treat similar topics for rings and groups have very similar formats. We feel that this is an asset in that students should naturally recognize the similarities and make connections between these topics in rings and groups.

A foundations course in reading and writing mathematical proofs is a prerequisite for modern algebra for all of our students, so these materials have been formatted with that in mind. Even with this background, we aim to help students learn the new algebra content by gradually building both their intuition and their ability to write coherent proofs in context. Early investigations include many situations where students are prompted to comment on or provide missing details in proofs to help them develop their proof-writing skills, while the activities help them develop their intuition. As the investigations proceed, it is expected that students will be able to better read and write proofs without this prompting, and so it is no longer provided.

As previously mentioned, this text is organized in such a way that it is possible to begin with either rings or groups.

Rings First: For a course that begins with ring theory, the organizational structure is linear. Investigations 1 – 3 provide background, specific examples, and motivation for the subsequent investigations. Investigations 4 – 7 contain the basics of ring theory, from the definitions of rings, integral domains, and fields to subrings, field extensions and direct sums, concluding with isomorphisms of rings. Many of our mathematics majors are aspiring elementary or secondary school teachers (for whom this class is required), and for them the study of polynomial rings develops a deeper understanding of an important subject that they will themselves teach. Investigations 8 – 11 deal in depth with polynomial rings and comprise an important and relevant conclusion to our first-semester course. Investigations 12 and 13 introduce the concepts of ideals, ring homomorphisms, and quotient rings for those who wish to have their students explore these topics. The ring theory portion of the text concludes with two additional investigations that require only some of the material preceding them.

- Investigation 14 treats divisibility and factorization in integral domains, proving in two different ways that every Euclidean domain is a unique factorization domain. The first approach relies primarily on the material from Investigations 1 – 4, with a few references to results about polynomials from Investigations 9 and 10. The second requires a more advanced understanding of ring theory, including results about ideals and principal ideal domains (from Investigation 13).

- Investigation 15 begins with the Peano axioms and then proceeds through the construction of $\mathbb{Q}$, $\mathbb{R}$, and $\mathbb{C}$. This investigation concludes with the characterization of the integers as the only ordered integral domain with a well-ordered set of positive elements. It requires an understanding of the material in Investigations 1 – 7.

Groups First: To begin a course with group theory, the background material needed is contained in Investigations 1 – 3. This material includes the Division Algorithm (Investigation 1); primes and prime factorizations (Investigation 1); equivalence relations, congruence, and $\mathbb{Z}_n$ (Investigation 2); and units and zero divisors in $\mathbb{Z}_n$ (Investigation 2). The instructor can choose from these investigations the material required for his/her students. We introduce groups with symmetries of planar objects (Investigation 16), and then the basic topics—groups, subgroups, cyclic groups, dihedral and symmetric groups, Lagrange's Theorem, normal subgroups and quotient groups, group isomorphisms and homomorphisms, the Fundamental Theorem of Finite Abelian Groups, and the Sylow theorems—follow (Investigations 17 – 30). This is an ambitious collection of investigations to complete in one semester.

Galois Theory: The printed text concludes with new investigations into fields extensions (Investigation 32) and Galois theory (Investigation 33). Investigation 31 provides some background, introducing splitting fields and minimal polynomials. The material on field extensions requires some knowledge of ring theory, especially ring homomorphisms, and this investigation provides the background for Galois theory. Galois theory makes connections between field extensions and groups, so a solid background in group theory is needed for Investigation 33.

Supplemental Investigations: There is an online supplement to this text at `https://scholarworks.gvsu.edu/books/29/`. This material includes applications of abstract algebra, along with appendices that provide additional information related to material presented in the text.

- **Investigation 34: RSA Encryption.** This investigation describes the RSA algorithm and assumes familiarity with modular congruence and prime numbers from Investigation 1. This investigation is referred to in Exercise 20 of Investigation 23 concerning Fermat's Little Theorem.

- **Investigation 35: Check Digits.** This investigation introduces the idea of check digits in several contexts and assumes familiarity with modular congruence (Investigation 1) and the dihedral groups (Investigation 21).

- **Investigation 36: Games: NIM and the 15 Puzzle.** This investigation applies group theory to develop a winning strategy in the game of NIM and to determine which 15 Puzzles are solvable. It assumes knowledge of groups (Investigation 17) and subgroups (Investigation 19), along with the symmetric groups (Investigation 22).

- **Investigation 37: Groups of Order 8 and 12: Semidirect Products of Groups.** In this investigation, we classify all groups of order 8, introduce semidirect products of groups, and then classify all groups of order 12. We assume familiarity with the earlier classification of groups of various orders (Investigation 26) and with products of groups (Investigation 25).

Appendices: The appendices from the first edition have been moved online and can be found at https://scholarworks.gvsu.edu/books/29/. In addition, some material from the text has been moved to the appendices, and we have added new appendices as reference material.

- **Appendix A: Functions.** This appendix appeared in the first edition and provides background information for students on properties of functions. This material is a helpful reference for the study of homomorphisms and isomorphisms in Investigations 7, 13, 26, symmetries in Investigation 17, as well as permutations in Investigation 22.

- **Appendix B: Mathematical Induction and the Well-Ordering Principle.** Mathematical induction and the Well-Ordering Principle are used throughout the book. This appendix can be used as a review of these important items, and also provides proofs of the equivalencies of the Well-Ordering Principle and the different flavors of mathematical induction for those who are interested.

- **Appendix C: Methods of Proof.** This is a new appendix that provides review material on different methods of proof other than induction, including direct proofs, using logical equivalencies in proofs, proof by contradiction, and proof using cases.

- **Appendix D: Proof that $R[x]$ is a Ring.** The formal proof that $R[x]$ is a ring when R is a ring is long and notationally complex. The details in the general case are omitted in the text in Investigation 8, but are included in this appendix for those who want to see a complete proof.

- **Appendix E: The Cubic Formula.** This material was in the first edition as a supplement to the investigation on irreducible polynomials. A complete derivation of the cubic formula is presented here. This formula is useful for Exercise 11 in Investigation 11 and is also referenced in Investigation 33 related to solvability by radicals.

- **Appendix F: The Fundamental Theorem of Algebra.** The Fundamental Theorem of Algebra is an important result regarding irreducible polynomials in Investigation 11. Since proofs of this theorem are not algebraic in nature, they don't usually appear in modern algebra texts. In this appendix we present what we believe is an accessible proof for the interested reader.

- **Appendix G: Complex Roots of Unity.** Complex roots of unity appear throughout Investigations 32 and 33 related to field extensions and Galois theory. Many students may already have a firm background in this topic. In this appendix we present a review of complex roots of unity for those who may benefit from one.

Acknowledgments

We wish to thank the Academy of Inquiry Based Learning and the Educational Advancement Foundation for their generous financial support of this project. We also wish to thank Grand Valley State University for providing the necessary time and resources to complete this project. We appreciate the feedback given by Audrey Malagon, Professor and Chair, Department of Mathematics, Virginia Wesleyan University, for comments and suggestions on the new investigations. Finally, we thank the many colleagues and students within the GVSU Mathematics Department who have inspired us to be better teachers and who have given us valuable feedback on this book.

References

[1] George D. Kuh. *High-impact educational practices: What they are, who has access to them, and why they matter.* Association of American Colleges and Universities, 2008.

[2] C. Meyers and T. Jones. *Promoting active learning: Strategies for the college classroom.* Jossey-Bass, 1993.

[3] Ernest T. Pascarella, Gregory C. Wolniak, Tricia A. D. Seifert, Ty M. Cruce, and Charles F. Blaich. Liberal arts colleges and liberal arts education: New evidence on impacts. *ASHE Higher Education Report* **31**(3), 2005.

Part I

Number Systems

Investigation 1

The Integers

Focus Questions

By the end of this investigation, you should be able to give precise and thorough answers to the questions listed below. You may want to keep these questions in mind to focus your thoughts as you complete the investigation.

- What arithmetic and ordering axioms are satisfied by the integers?

- How is divisibility defined within the integers, and what does the Division Algorithm say about division of integers?

- What does it mean for two integers to be congruent modulo n, and how is congruence useful?

- What does the Fundamental Theorem of Arithmetic say about prime factorization?

- What is the greatest common divisor of two integers, and how can the Euclidean Algorithm be used to find greatest common divisors?

- What is the relationship between greatest common divisors and linear combinations, and why does this relationship hold?

Preview Activity 1.1. We will begin by considering some simple mathematics problems that students in elementary or middle school might be asked to solve. For each problem, try to explain your solution as clearly as possible, describing not just *what* you are doing, but *why* you are able to do it. Once you've completed the rest of the investigation, you'll be asked to revisit your solutions to see how they are related to what you have learned.

(a) Find all real number solutions to the equation $3x^2 + x = 2$. Would your answer, or the methods you used to obtain it, be different if you were only interested in integer solutions?

(b) Is it possible for the product of two nonzero integers to be equal to zero? Can you think of any other contexts where it is possible for the product of two nonzero quantities to be zero?

(c) Find the quotient and remainder when 43 is divided by 5. Is your answer the only correct one?

(d) If it is 3 pm now, describe how you could find out what time it would be n hours from now.

(e) Write the fraction $\frac{28}{42}$ in lowest terms. Now do the same for the fraction $\frac{17161}{17947}$.

Introduction

This book is about abstract algebra. From your previous studies, you probably have at least an intuitive idea of what algebra is. When you think of algebra, you might think of solving equations, and that is an important part of algebra. But algebra also involves the structure and properties of number systems that make the techniques you learned in your previous algebra classes work. You may not have thought too much about these properties before, since – for the most part – they have just worked. Or course, most of us have made algebra mistakes because we did something we weren't allowed to do, like accidentally dividing both sides of an equation by an expression that might be equal to zero. With practice, we can learn to avoid these mistakes, but at some point in our mathematical education, it becomes important to take a closer and more systematic look at what's going on behind the scenes. For example, we might ask which rules are universal and which apply only in certain situations. These questions become especially important as our mathematical universe expands to include different types of mathematical objects (like matrices in linear algebra) and operations (like modular arithmetic). It turns out that, with some care and some important exceptions, much of what is familiar from our previous studies of algebra can be translated to these new settings as well. But without an understanding of why this is the case – and specifically, the underlying structure that makes algebra possible – we would have to constantly reinvent the wheel, building our algebraic toolkit from the ground up every time we encounter a new mathematical context.

Consider this analogy: Suppose we showed a friendly alien visiting Earth an apple. Having never seen such a thing before, our alien friend would likely study the various properties of the apple – its shape, color, texture, weight, etc. If we told them that the apple was food (assuming that, like us, their species obtained energy and nourishment from eating), they might even take a bite. Over time, they might learn that apples come in all sorts of varieties, but apples all share something in common – some concept of *appleness* that distinguishes them from rocks and tennis balls and even other fruits like pears. It might take our alien friend some time to figure all of this out. For example, if the first apple they saw was red, they might assume that a golden delicious apple is not an apple. Once they learned that apples can have different colors of skin, they might see a peach and think they had discovered yet another variety of apple. They would have to learn that apples are not fuzzy. And no, a nectarine is not an apple either. Over time, our alien visitor would have learned enough about appleness that they would be able to quickly identify what is – and what is not – an apple. If they saw a bowl of fruit, they would not need to carefully examine (or bite into) each piece to determine which were apples. And they would know with some level of certainty that if they did bite into an apple, it would have the taste and texture they had come to expect.

Our journey here will be similar, but instead of studying apples, we will be studying different number systems. Our "numbers" may be numbers in the familiar sense, or they may be different sorts of objects like matrices or sets. Operations like addition and multiplication might look very different as well. But we'll discover that all of these systems have certain features in common. Once we identify these features, and the properties that follow from them, we will better understand algebra in familiar contexts, and we will also be able to transfer our knowledge to realms of the mathematical universe that are new to us. We'll also learn what properties and techniques do *not* apply universally so that we can both avoid errors and choose the right tools for wherever our journey leads us.

Arithmetic and Ordering Axioms

In elementary school, you probably began your study of mathematics by learning to count with numbers like 1, 2, 3, and so on. In formal mathematics, these counting numbers are typically called the **natural numbers**, and the set of all such numbers is denoted $\mathbb{N}$. In other words,

$$\mathbb{N} = \{1, 2, 3, \ldots\}.$$

You may have also used the term **whole numbers**, which we'll denote $\mathbb{W}$ and define to be the natural numbers along with zero– i.e.,

$$\mathbb{W} = \{0, 1, 2, 3, \ldots\}.$$

Later on, when you learned about negative numbers, you began doing mathematics in the larger set of **integers**, denoted $\mathbb{Z}$ and defined as

$$\mathbb{Z} = \{\ldots, -3, -2, -1, 0, 1, 2, 3, \ldots\}.^*$$

After learning to count, the next step for most students is to learn how to add, subtract, and multiply integers. These arithmetic operations on the integers satisfy a number of important properties, or *axioms*, which are summarized below.

Axioms of Integer Arithmetic

- **The integers are closed under addition and multiplication,** meaning that for all integers a and b, both $a + b$ and ab are also integers.

- **Addition and multiplication are commutative,** meaning that for all integers a and b, $a + b = b + a$ and $ab = ba$.

- **Addition and multiplication are associative,** meaning that for all integers a, b, and c, $(a + b) + c = a + (b + c)$ and $(ab)c = a(bc)$.

- **Multiplication distributes over addition,** meaning that $a(b + c) = ab + ac$ for all integers a, b, and c.

- **The integer 0 is an additive identity,** meaning that $a + 0 = a$ for every integer a.

- **The integer 1 is a multiplicative identity,** meaning that $1a = a$ for every integer a.

- **Every integer a has an additive inverse,** typically denoted $-a$; in particular, $a + (-a) = 0$ for every integer a.

---

*The symbol $\mathbb{Z}$ for the integers is from the German word *Zahlen* for number. This symbol appeared in Bourbaki's *Algèbre*, Chapter 1. Nicolas Bourbaki was a name adopted by a group of mostly French mathematicians who wrote a series of books intended to thoroughly unify mathematics through set theory. In fact, the most common modern construction of the integers is based entirely on sets and set operations. Such a rigorous development of the integers is not necessary for our investigations, but it is good to be aware of the history, which spanned thousands of years and led to serious philosophical and even theological debates.

One thing you may notice in looking at this list of axioms is that it says nothing about subtraction or division. That's because subtraction is typically defined using addition and additive inverses, as follows:

$$a - b = a + (-b)$$

Likewise, division of integers can be thought of as repeated subtraction, and the Division Algorithm, which we will investigate shortly, formalizes this process.

You may also notice that several other properties of integer arithmetic – for example, that multiplication distributes over subtraction, or that any number multiplied by zero equals zero – are not included in the list above. That's because these properties can be *proved*, as *theorems*, from the *axioms*, which are *assumed* to be true without proof. Our goal is to develop algebra from the most basic set of assumptions possible, and the above list of axioms is robust enough to do the trick without containing any redundancies.

Activity 1.2. In this activity, we will explore a few of the familiar properties that can be proved from the axioms of integer arithmetic.

(a) Let a, b, and c be integers. Use the axioms of integer arithmetic to prove the property of *additive cancellation*, which states that if $a + c = b + c$, then $a = b$.

(b) Use additive cancellation to prove that for any integer a, $a \cdot 0 = 0$. (Hint: Start with the fact that $0 + 0 = 0$, and multiply both sides by a.)

(c) In general, an *additive inverse* of an integer x can be defined as an integer y such that $x + y = 0$. Can an integer have more than one additive inverse? If yes, give an example to illustrate. If no, prove that no such example exists. (Hint: For x to have more than one additive inverse, there would have to exist integers y and z such that both $x + y$ and $x + z$ are equal to 0.)

In addition to the arithmetic axioms, there are also axioms that describe the way the integers are ordered, and how this ordering interacts with addition and multiplication.

Ordering Axioms of the Integers

The "less than" relation on the integers, denoted $<$, satisfies the following properties:

- **Trichotomy**: For all integers a and b, exactly one of the following is true: $a < b$, $b < a$, or $a = b$.

- **Transitivity:** For all integers a, b, and c, if $a < b$ and $b < c$, then $a < c$.

- **Translation Invariance:** For all integers a, b, and c, if $a < b$, then $a + c < b + c$.

- **Scaling:** For all integers a, b, and c, if $a < b$ and $c > 0$, then $ac < bc$.

One of the results we can prove from the ordering axioms is that *multiplicative cancellation* works within the integers, provided the number being canceled is nonzero. Note that we cannot simply prove this result in the same way that we proved additive cancellation in Activity 1.2, as we have no axiom that tells us that every integer has a multiplicative inverse. In fact, if we want to work entirely within the integers (ignoring, for the time being, the fact that the integers are embedded within larger number systems like the rational and real numbers), then only 1 and -1 could be said to have multiplicative inverses. Instead, we will first show that the product of any two nonzero integers is a nonzero integer, and we will then use this result to establish the validity of multiplicative cancellation in $\mathbb{Z}$.

Activity 1.3. Consider the following theorem:

> **Theorem 1.4.** *There do not exist nonzero integers a and b such that $ab = 0$.*

(a) Explain why Theorem 1.4 is equivalent to each of the following:

- For all integers a and b, if $ab = 0$, then $a = 0$ or $b = 0$.
- For all integers a and b, if $ab = 0$ and $a \neq 0$, then $b = 0$.

(b) Use the ordering axioms of the integers to prove Theorem 1.4 or one of its equivalent forms. (Hint: Use the trichotomy axiom to set up cases.)

(c) Use Theorem 1.4 to prove the following result, which establishes the validity of *multiplicative cancellation* of a nonzero integer:

> **Theorem 1.5.** *For all integers a, b, and c, if $ac = bc$ and $c \neq 0$, then $a = b$.*

Recall again that "dividing by c" is not an option, as we have not yet defined division within the integers, and we cannot assume the existence of multiplicative inverses.

Divisibility in $\mathbb{Z}$

Once we understand how addition, subtraction, multiplication, and ordering work within the integers, the next step is to move on to division. We'll start by formalizing what it means for one integer to divide another.

Definition 1.6. An integer a **divides** an integer b, denoted $a \mid b$, if there is an integer q such that $b = aq$.

When a divides b, we often say that a is a *divisor* or *factor* of b, or that b is a *multiple* of a. Note also that the notation $a \mid b$ does not represent the rational number $\frac{b}{a}$. Rather, it expresses in shorthand a *relationship* between the integers a and b – namely, that a divides b. This distinction is important, especially when we consider the role of zero as a divisor.

Activity 1.7.

(a) Let b be a nonzero integer. Is there an integer q for which $b = 0q$?

(b) What does part (a) allow you to conclude about which integers 0 divides?

(c) Which integers q satisfy the equation $0 = 0q$?

(d) Does $0 \mid 0$? Is the quantity $\frac{0}{0}$ defined? Exactly which integers does 0 divide? Explain your answers.

To summarize, note that in order to say that $a \mid b$, there must *exist* an integer q for which $b = aq$. If this integer q is *unique* (which we will see is true except in one very special case), then we can use the notation $b \div a$ or $\frac{b}{a}$ to represent the integer q, which we call the *quotient*. It's important to remember that **the integers are not closed under division**. In particular, there are many integers a and b for which no such quotient exists. That said, you have probably completed many division problems that involved dividing an integer a into an integer b, even though a did not divide b (according to Definition 1.6).

As an example, consider the problem of dividing 43 by 5. Now 5 doesn't actually divide 43, but 5 does divide 40. So 5 goes into 43 eight times, with 3 left over. In other words, $43 = 5 \cdot 8 + 3$. More generally, to divide b by a (assuming, for the moment, that both a and b are natural numbers), we subtract multiples of a from b until there are no more multiples of a left to subtract. We can then write $b = aq + r$, where q is the number of times we subtracted a from b, and r is what's left over. The number q is called the *quotient*, and the number r is called the *remainder*. Note that when we are done with this process, the number r is always less than a; otherwise, we could subtract additional multiples of a from b. While it makes intuitive sense that this process works – and we've seen it work in lots of examples – the Division Algorithm, stated formally below, guarantees that we can always find such a quotient and remainder, and that they will be unique.

The Division Algorithm. *Let a and b be integers, with $a > 0$. Then there exist unique integers q and r such that*

$$b = aq + r \text{ and } 0 \leq r < a.$$

It's worth noting that the Division Algorithm is not really an algorithm at all, but rather a theorem that asserts the existence and uniqueness of quotients and remainders. This so-called "algorithm" doesn't really tell us anything about how to solve division problems. As we discussed above, long division will eventually lead us to the promised quotient and remainder, but the Division Algorithm is not concerned with *how* we find these values – just that we *can*, and that there is only one correct answer.

If the Division Algorithm seems obvious, it's probably because its proof relies on an equally obvious axiom called the *Well-Ordering Principle*, which states that every nonempty subset of the natural numbers has a smallest element. This axiom, along with the previously stated arithmetic and ordering axioms, is the key to making division of integers work the way we expect it to. A complete proof of the Division Algorithm is included at the end of this investigation.

Congruence

The Division Algorithm is also useful when studying the idea of congruence, which is used by mathematicians to describe cyclic phenomena in the world of the integers. For instance, time is a cyclic phenomenon in that the time of day repeats every 12 or 24 hours, depending on the clock we are using. The days of the week also cycle in this same fashion, with the same day occurring every 7 days. So if today is Tuesday, then it will be Friday in another 3 days, and then again in another 10 days, 17 days, 24 days, 31 days, and so on. Naturally, all of these numbers differ by multiples of 7 (the length of a week). Consequently, they all have the same remainder – namely, 3 – when divided by 7. The definition below, and the theorem that follows it, formalize these observations.

Definition 1.8. Let n be a natural number, and let a and b be integers. Then a **is congruent to** b **modulo** n, denoted $a \equiv b \pmod{n}$, provided that n divides $a - b$.

Theorem 1.9. *Let n be a natural number, and let a and b be integers. Then $a \equiv b \pmod{n}$ if and only if a and b yield the same remainder when divided by n.*

Note that, by Definition 1.8, all the numbers from our previous example $(3, 10, 17, 24, 31, \ldots)$ are congruent modulo 7. The proof of Theorem 1.9 is left as an exercise (see Exercise (15) on page 18), while Activity 1.10 presents several other results that allow us to treat congruence much like we treat equality of integers (at least for the purposes of doing arithmetic). Each of these results can be proved by first translating the given statement into one that involves divisibility. The first part is completed for you as an example.

Activity 1.10. Let n be a natural number, and let a, b, c, and d be integers. Prove each of the following results.

(a) If $a \equiv b \pmod{n}$ and $c \equiv d \pmod{n}$, then $(a + c) \equiv (b + d) \pmod{n}$.

> **Solution:** *Using the definition of congruence, the given result is equivalent to the following:*
>
> *If $n \mid (a - b)$ and $n \mid (c - d)$, then $n \mid [(a + c) - (b + d)]$.*
>
> *Thus, assume that $n \mid (a - b)$ and $n \mid (c - d)$. Then there exist integers j and k such that $a - b = nj$ and $c - d = nk$. Simple algebra (in particular, the associative and distributive axioms) then implies that*
>
> $$(a + c) - (b + d) = (a - b) + (c - d)$$
> $$= nj + nk$$
> $$= n(j + k).$$
>
> *Thus, $n \mid [(a + c) - (b + d)]$, as desired.*

(b) If $a \equiv b \pmod{n}$ and $c \equiv d \pmod{n}$, then $ac \equiv bd \pmod{n}$.

(c) If $a \equiv b \pmod{n}$ and $m \in \mathbb{N}$, then $a^m \equiv b^m \pmod{n}$.

(d) For every integer a, $a \equiv a \pmod{n}$. (This property is called the **reflexive** property of congruence.)

(e) If $a \equiv b \pmod{n}$, then $b \equiv a \pmod{n}$. (This property is called the **symmetric** property of congruence.)

(f) If $a \equiv b \pmod{n}$ and $b \equiv c \pmod{n}$, then $a \equiv c \pmod{n}$. (This property is called the **transitive** property of congruence.)

Factoring, Prime Numbers, and Greatest Common Divisors

Now that we understand how division works in the integers, let's move on to factoring, which involves breaking an integer down into a product of its divisors. Elementary school students learn how to factor integers, in part so that they can reduce fractions. To reduce a fraction means to write it in *lowest terms*, so that the numerator and denominator share no common divisors. For example, to reduce the fraction $\frac{28}{42}$, we simply divide the numerator and denominator by their common divisors until no common divisors remain. Dividing both the numerator and denominator first by 2 and then by 7, we obtain

$$\frac{28}{42} = \frac{14}{21} = \frac{2}{3}.$$

Note that if we wanted to complete this process in one step, we could have just divided both the numerator and denominator by 14, which is the largest integer that divides both 28 and 42. In other words, 14 is the *greatest common divisor* of 28 and 42, defined formally as follows:

Definition 1.11. Let a and b be integers, not both zero. A **common divisor** of a and b is any integer that divides both a and b. The largest integer that divides both a and b is called the **greatest common divisor** of a and b, denoted $\gcd(a, b)$.

Note that if a, b, and d are integers, and a and b are not both zero, then $d = \gcd(a, b)$ if and only if both of the following properties hold:

- $d \mid a$ and $d \mid b$. That is, d is a common divisor of a and b.

- If k is any other integer such that $k \mid a$ and $k \mid b$, then $k \leq d$. In other words, any other common divisor of a and b must be less than or equal to d.

One way to find the greatest common divisor of two integers is to first find the *prime factorization* of each integer. Recall that prime numbers are defined as follows:

Definition 1.12. A **prime number** is an integer $p > 1$ whose only positive divisors are 1 and p. A positive integer that is greater than 1 and not prime is said to be **composite**.

Moreover, the Fundamental Theorem of Arithmetic, stated formally below, guarantees both the existence and uniqueness of prime factorizations.

The Fundamental Theorem of Arithmetic. *Every integer greater than 1 is either prime or a product of primes. Furthermore, this factorization is unique up to the order of the factors.*

The phrase "unique up to the order of the factors" simply means that any two prime factorizations of the same integer will contain the same prime factors, with the only possible difference being the order in which the factors are listed. So, for example, $2 \cdot 3 \cdot 7$ and $3 \cdot 7 \cdot 2$ are considered the same prime factorization of the number 42, even though the factors are listed in different orders.

The proof of the Fundamental Theorem of Arithmetic, included at the end of this investigation, relies on an important lemma that is attributed to Euclid. Although his famous work, the *Elements*, is most commonly known for its contributions to geometry, it also contains many basic results in number theory. One of these results, which will be used often in subsequent investigations, is the following, which we will prove in Activity 1.26:

Euclid's Lemma. *Let a and b be integers, and let p be prime. If $p \mid ab$, then $p \mid a$ or $p \mid b$.*

One way to find the greatest common divisor of two integers utilizes a method elementary school students often use, called *factor trees*, to find prime factorizations. Figure 1.1 illustrates how this method can be used to find $\gcd(396, 780)$. Notice that $396 = 2^2 \cdot 3^2 \cdot 11$ and $780 = 2^2 \cdot 3 \cdot 5 \cdot 13$. Since the only prime factors common to both numbers are 2 (squared in each case) and 3, it follows that $\gcd(396, 780) = 2^2 \cdot 3 = 12$.

Of course, applying this process to larger numbers would be considerably more difficult. For instance, if we tried to use the factor tree method to find $\gcd(17947, 17161)$, we would likely get

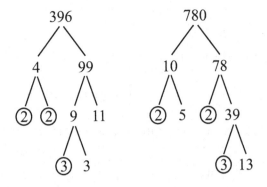

Figure 1.1
Factor trees for 396 and 780.

stuck on the first step, or we would at least spend a very long time trying to find the prime factors of each number. In fact, most secure communications over the internet use an encryption method that relies on how hard it is to find the prime factors of large numbers. (We explore this method in more detail in Investigation 15.)

Fortunately for us, even though finding prime factorizations is a difficult and time-consuming problem – even for computers – there is a simple and efficient algorithm for finding the greatest common divisor of two integers. This algorithm is called the *Euclidean Algorithm*, and it works by repeatedly applying the Division Algorithm. For example, to find $\gcd(17947, 17161)$ using the Euclidean Algorithm, we would carry out the following calculations:

$$17947 = 17161 \cdot 1 + 786$$
$$17161 = 786 \cdot 21 + 655$$
$$786 = 655 \cdot 1 + \boxed{131}$$
$$655 = 131 \cdot 5 + 0$$

Notice that each application of the Division Algorithm in this example uses the divisor and remainder from the previous iteration. The greatest common divisor is then equal to the last nonzero remainder we obtain through this iterative process (131 in this case). At first glance, it is not at all obvious why this works. In fact, the Euclidean Algorithm can be difficult to remember and apply if we don't understand what's going on behind the scenes – specifically, the following theorem:

Theorem 1.13. *Let a and b be integers, not both zero, and suppose that*

$$b = aq + r$$

for some integers q and r. Then $\gcd(b, a) = \gcd(a, r)$.

Activity 1.14. In this activity, we'll consider one way to prove Theorem 1.13. To begin, let a, b, q and r be as in Theorem 1.13, and let $d = \gcd(b, a)$. We need to show that (1) d divides both a and r, and (2) that d is the largest such common divisor.

(a) Let a and b be integers with $a > 0$, and let $b = aq + r$ for some integers q and r. Prove that if an integer d divides both a and b, then d also divides r. Explain why this implies that $d = \gcd(b, a)$ is a common divisor of a and r.

(b) Suppose that for some integer $k > d$, $k \mid a$ and $k \mid r$. Show that $k \mid b$ also. Deduce that k is a common divisor of b and a.

(c) Explain how part (a) contradicts the assumption that $d = \gcd(b, a)$.

Let's now revisit our previous example. Here it is again, but this time incorporating the conclusions of Theorem 1.13:

$$\begin{aligned}
\gcd(17947, 17161) &= \gcd(17161, 786) &\text{since} &&17947 &= 17161 \cdot 1 + 786 \\
&= \gcd(786, 655) &\text{since} &&17161 &= 786 \cdot 21 + 655 \\
&= \gcd(655, 131) &\text{since} &&786 &= 655 \cdot 1 + 131 \\
&= \gcd(131, 0) &\text{since} &&655 &= 131 \cdot 5 + 0 \\
&= 131.
\end{aligned}$$

What's really happening here is that each time we apply the Division Algorithm, we're reducing our greatest common divisor problem into one that involves smaller numbers. Eventually, we will arrive at a point where we are calculating the gcd of some positive number (the last nonzero remainder) and 0. And since $\gcd(r, 0) = r$ for every positive integer r (do you see why?), this will solve our original problem (in this case, finding the greatest common divisor of 17947 and 17161).

Linear Combinations

We'll conclude this investigation by using what we have learned about the Euclidean Algorithm to discover a useful result that relates greatest common divisors and linear combinations. To begin, consider the following definition:

Definition 1.15. Let a and b be integers. A **linear combination** of a and b is an integer that can be written as $ax + by$ for some integers x and y.

Activity 1.16.

(a) Find at least 10 different integers that are linear combinations of 60 and 95. Make sure some of these integers are positive and some are negative.

(b) Now find the greatest common divisor of 60 and 95. Do you notice a relationship between $\gcd(60, 95)$ and the linear combinations you found in part (a)?

Activity 1.16 suggests that there is a relationship between greatest common divisors and linear combinations. To explore this relationship in more detail, let's revisit the process we used to find $\gcd(17947, 17161)$ in the previous section.

Recall that we found $\gcd(17947, 17161) = 131$, and in the second-to-last step of the Euclidean Algorithm, we were able to write

$$786 = 655 \cdot 1 + 131.$$

Rearranging this equation, we obtain

$$131 = 786 \cdot 1 + 655 \cdot (-1). \tag{1.1}$$

In other words, we can write 131 as a linear combination of 786 and 655.

Notice also that in the previous step of the Euclidean Algorithm, we found that

$$17161 = 786 \cdot 21 + 655.$$

Rearranging this equation, we obtain

$$655 = 17161 \cdot 1 + 786 \cdot (-21). \qquad (1.2)$$

Combining Equations (1.1) and (1.2) yields

$$131 = 786 \cdot 1 + [17161 \cdot 1 + 786 \cdot (-21)] \cdot (-1)$$
$$= 17161 \cdot (-1) + 786 \cdot 22$$

Continuing in this fashion, we obtain

$$131 = 17161 \cdot (-1) + [17947 \cdot 1 + 17161 \cdot (-1)] \cdot 22$$
$$= 17947 \cdot 22 + 17161 \cdot (-23)$$

Thus, by solving for the remainders obtained in each step of the Euclidean Algorithm, we were able to find a way to write $\gcd(17947, 17161)$ as a linear combination of 17947 and 17161. This process can be applied whenever the Euclidean Algorithm is used to find the greatest common divisor of two integers (not both zero). Thus, we have the following theorem, which is sometimes called Bezout's Identity:

Theorem 1.17 (Bezout's Identity). *Let a and b be integers, not both zero. Then $\gcd(a, b)$ can be written as a linear combination of a and b. That is, there exist integers x and y such that*

$$\gcd(a, b) = ax + by.$$

Bezout's Identity is interesting in its own right, but we can actually say more. First, we can easily show that in addition to $\gcd(a, b)$ being equal to a linear combination of a and b, $\gcd(a, b)$ divides *every* linear combination of a and b.

Activity 1.18. Let a and b be integers, and let d be any common divisor of a and b. Prove that d is a divisor of $am + bn$ for all integers m and n.

Next, note that the greatest common divisor of two integers is always positive. This is because 1 divides every integer, and so the greatest common divisor of any two integers must be greater than or equal to 1. This means that:

- $\gcd(a, b)$ is a linear combination of a and b.

- $\gcd(a, b)$ divides every linear combination of a and b.

- $\gcd(a, b)$ is positive.

These three facts together imply that $\gcd(a, b)$ is not only a positive linear combination of a and b, but in fact the *smallest* positive linear combination of a and b. Theorem 1.19 formalizes this result, and Corollary 1.20 states a useful consequence.

Theorem 1.19. *Let a and b be integers, not both zero. Then $\gcd(a, b)$ is equal to the smallest positive linear combination of a and b.*

Corollary 1.20. *Let a and b be integers, not both zero. Then $\gcd(a, b) = 1$ if and only if there exist integers x and y such that*

$$ax + by = 1.$$

Activity 1.21. Explain why Corollary 1.20 is true. (Hint: The forward implication follows immediately from Theorem 1.17. The reverse implication requires the stronger Theorem 1.19.)

Corollary 1.20 provides a mechanism for showing that two integers a and b share no positive common divisors other than 1. Such pairs of integers are said to be *relatively prime*, defined formally below.

Definition 1.22. Let a and b be integers, not both zero. Then a and b are said to be **relatively prime** if and only if $\gcd(a, b) = 1$.

Returning to our earlier example of reducing fractions, we can say that a fraction is fully reduced when its numerator and demominator are relatively prime. The following theorem then implies that dividing the numerator and denominator of a fraction by their greatest common divisor does in fact reduce the fraction to lowest terms. The proof of this theorem is left as an exercise. (See Exercise (23) on page 19).

Theorem 1.23. *Let a and b be integers, not both zero, and let $d = \gcd(a, b)$. Then $\dfrac{a}{d}$ and $\dfrac{b}{d}$ are relatively prime integers.*

Proofs of the Division Algorithm and the Fundamental Theorem of Arithmetic

In this last section, we will consider proofs of both the Division Algorithm and the Fundamental Theorem of Arithmetic. In these proofs (and many others throughout the text), we will use the ⑦ symbol to denote places where more elaboration or justification may be desirable. When you encounter a ⑦ , you may want to pause and ask yourself, "Wait – why is that true?" If you can convince a classmate or peer that the statement or suggested technique is valid, then you are probably ready to continue reading. On the other hand, if you cannot provide a convincing explanation, then you may not fully understand the concepts behind the proof.

Proof of the Division Algorithm. Let a and b be integers, with $a > 0$. For the existence portion of the proof, define the set S as follows:

$$S = \{x \in \mathbb{Z} : x \geq 0 \text{ and } x = b - am \text{ for some } m \in \mathbb{Z}\}$$

We will use the Well-Ordering Principle to show that S has a smallest element. Since S is clearly a subset of the natural numbers, we need only to show that S is nonempty. ⑦ If $b \geq 0$, then $b \in S$. ⑦ Furthermore, if $b < 0$, then $b - ab = b(1 - a) \geq 0$, and so $b - ab \in S$. ⑦ In either case, S is nonempty and therefore has a smallest element, which we will call r. Because $r \in S$, it follows that $r = b - aq$ for some $q \in \mathbb{Z}$. Thus, we have found integers q and r such that $b = aq + r$.

To show that $0 \leq r < a$, we will assume, to the contrary, that $r \geq a$. (We already know that $0 \leq r$, since $r \in S$.) This implies, however, that $r - a \in S$, since $r - a \geq 0$, and

$$r - a = (b - aq) - a = b - a(q + 1).$$

Since $a > 0$, we also know that $r - a < r$. So $r - a \in S$ and $r - a < r$, a contradiction to the fact that r was defined to be the smallest element of S. Thus, it must be the case that $0 \leq r < a$.

To prove uniqueness, assume that there exist integers q' and r' such that

$$b = aq' + r' \text{ and } 0 \le r' < a.$$

Then since $aq + r = b = aq' + r'$, it follows that

$$a(q - q') = r' - r. \textcircled{?}$$

We know that $0 \le r' < a$, and it is also the case that $-a < -r \le 0.\textcircled{?}$ Therefore,

$$-a < r' - r < a. \textcircled{?}$$

It follows that $r' - r$ is both an integer multiple of a and strictly between $-a$ and a. As such, the only possibility is that $r' - r = 0$, which implies that $q - q' = 0$ as well. $\textcircled{?}$ Thus, the integers q and r determined by the Division Algorithm are unique, which completes the proof. ∎

Proof of the Fundamental Theorem of Arithmetic. We must show that every integer greater than 1 is either prime or a product of primes, and that all such prime factorizations are unique up to the order of the factors. For both the existence and uniqueness portions of the proof, we will proceed by induction.

For existence, first note that since 2 is prime, the base case is trivial. For our induction hypothesis, we will assume that, for some $n \ge 2$, every integer between 2 and n, inclusive, is either prime or a product of primes. We need to show that $n + 1$ is either prime or a product of primes. If $n + 1$ is prime, then this is true trivially. Therefore, assume that $n + 1$ is not prime. Then there exist integers x and y such that $n + 1 = xy$ and $2 \le x, y \le n.\textcircled{?}$ By the induction hypothesis, both x and y are either prime or a product of primes. Thus, $n + 1 = xy$ is a product of primes, $\textcircled{?}$ which completes the induction step.

For uniqueness, first note that 2 is prime and therefore cannot be factored in any nontrivial way. Thus, 2 (like any prime) has a unique – and trivial – prime factorization. Now assume that, for some $n \ge 2$, every integer between 2 and n, inclusive, has a factorization into primes that is unique up to the order of the factors. Reasoning by contradiction, suppose that $n + 1$ has two different prime factorizations. In other words, suppose that for some primes $p_1, p_2, \ldots p_j$, and $q_1, q_2, \ldots, q_k$,

$$p_1 p_2 \cdots p_j = n + 1 = q_1 q_2 \cdots q_k.$$

By Euclid's Lemma, $p_1 \mid q_i$ for some i with $1 \le i \le k.\textcircled{?}$ Without loss of generality, assume that $p_1 \mid q_1$. (If necessary, we can simply reorder and/or renumber the factors to make this true.) Since p_1 and q_1 are both prime and $p_1 \mid q_1$, it must be the case that $p_1 = q_1$, and so we can cancel this common factor, yielding

$$p_2 p_3 \cdots p_j = q_2 q_3 \cdots q_k \le n. \textcircled{?} \tag{1.3}$$

The induction hypothesis then implies that $j = k$ and the factors on each side of Equation (1.3) can be reordered and/or renumbered so that $p_i = q_i$ for all i with $2 \le i \le j = k.\textcircled{?}$ Thus, the factorization of $n + 1$ into primes is unique up to the order of the factors, as desired. ∎

Concluding Activities

Activity 1.24. Look back at your solutions to Preview Activity 1.1. For each part, explain the relationship between the problem you solved and what we learned in this investigation.

Activity 1.25. Identify all instances in this investigation where the concepts of *existence* and *uniqueness* were mentioned. Can you think of any other mathematical contexts where these concepts are important?

Activity 1.26. In this activity, we will prove Euclid's Lemma by first proving the following more general result:

> **Theorem 1.27.** *Let a, b, and c be integers. If $c \mid ab$ and $\gcd(c, a) = 1$, then $c \mid b$.*

To prove this theorem, let a, b, and c be integers, and suppose that $c \mid ab$ and $\gcd(c, a) = 1$.

(a) Use Bezout's Identity to translate the assumption that $\gcd(c, a) = 1$ into an equation involving a linear combination.

(b) Multiply both sides of your equation from part (a) by an appropriate quantity in order to obtain an equation of the form

$$b = \underline{\quad} + \underline{\quad}.$$

(c) Explain why each of the terms on the right-hand side of the equation from part (b) are divisible by c. Deduce that $c \mid b$. (Note that this conclusion establishes Theorem 1.27.)

(d) Now suppose that a, b, and p are integers with p prime. Assume further that $p \mid ab$ and $p \nmid a$. Explain why $\gcd(p, a) = 1$.

(e) Explain how Theorem 1.27 and the result from part (d) establish Euclid's Lemma.

Exercises

(1) **Addition and multiplication.** Let a and n be integers, with $n > 0$. Prove that the sum of n copies of a is equal to na. That is, prove that

$$\underbrace{a + a + \cdots + a}_{n \text{ terms}} = na.$$

(2) Let a, b, and c be integers. Is it always the case that $(a + b)c = ac + bc$? Prove your answer using only the axioms stated in this investigation.

(3) Find all integer solutions to the equation

$$x^3 + 3x^2 - 4x = 12.$$

Justify each step in your solution with one or more of the axioms or theorems from this investigation (possibly including Theorem 1.4).

(4) **Antisymmetry of the $\leq$ relation.** Prove that the $\leq$ relation is antisymmetric; that is, prove that for all integers a and b, if $a \leq b$ and $b \leq a$, then $a = b$.

(5) One of the properties of integer arithmetic is that the set of integers contains an additive inverse for each of its elements. The existence of additive inverses allows us to define an operation of subtraction on the set of integers. Although we have no operation of division on the integers, we can still ask if there are any integers that have a *multiplicative* inverse within the integers. We will call such integers *units*.

(a) State a formal definition of what it would mean for an integer a to have a multiplicative inverse within $\mathbb{Z}$.

(b) Determine all units in $\mathbb{Z}$. (Hint: There is more than one.) Use your definition from part (a) to verify your answer.

(c) Use the ordering axioms of the integers to prove that the units you found are the only integer units. (Warning: We have no operation of division in the integers, so you cannot "divide" in your proof.)

(6) Let a, b, and c be integers. Prove that if $a < b$ and $c < 0$, then $ac > bc$. Deduce that if $a < b$, then $-b < -a$.

(7) Let a, b, and c be integers. Prove that if $ac > bc$ and $c > 0$, then $a > b$.

(8) Let a, b, c, and d be integers.

(a) Prove that if $a < b$ and $c < d$, then $a + c < b + d$.

(b) Prove that the result from part (a) still holds if $a < b$ and $c \leq d$.

(9) In a popular seventh-grade mathematics textbook, students are asked to investigate the following conjecture:

The sum of any three consecutive whole numbers will always be divisible by 3.

(a) Is the conjecture true or false? Provide a proof or a counterexample to justify your answer.

(b) If the conjecture is true, can it be generalized in any way? If it is false, are there any special cases for which it does hold? Prove your answer.

(10) Let a, b, and c be integers. What conclusions, if any, can be drawn in each of the following situations? Prove your answers.

(a) $a \mid c$ and $b \mid c$

(b) $a \mid b$ and $b \mid c$

(c) $a \mid b$ and $a \mid c$

(11) Let a and b be integers. Prove that if $a \mid b$ and $b \mid a$, then $|a| = |b|$.

(12) Let a and b be positive integers, and suppose that $a \mid b$. Prove that $(a + 1) \mid (b + \frac{b}{a})$.

(13) Let a, $b \in \mathbb{N}$. Use the arithmetic and ordering axioms of the integers to prove that if $a \mid b$, then $a \leq b$.

(14) A nonempty subset S of $\mathbb{R}$ is said to be **well-ordered** if every nonempty subset of S contains a least element.

(a) Use this definition to concisely restate the Well-Ordering Principle. (Hint: You should be able to do so in no more than six words.)

(b) Is $\mathbb{R}$ well-ordered? Why or why not?

(c) Is the set $\mathbb{R}^* = \{x \in \mathbb{R} : x \geq 0\}$ well-ordered? Why or why not?

(d) Is $\{-9, -7, -5, \ldots\}$ well-ordered? Why or why not?

(e) Prove or disprove: If a set S is well-ordered, then S contains a least element.

(f) Prove or disprove: If a set S contains a least element, then S is well-ordered.

(15) In this exercise, we will prove Theorem 1.9. To begin, let n be a natural number, and let a and b be integers.

    (a) Use the Division Algorithm to write equations (together with the appropriate inequalities) that represent the result of dividing each of a and b by n. For convenience, use q_1, q_2, r_1, r_2 to denote the resulting quotients and remainders.

    (b) If you haven't already done so, write your equations from part (a) so that they are in the form $a = \ldots$ and $b = \ldots$. Then use subtraction to obtain a new equation of the form $a - b = \ldots$.

    (c) Now assume that $n \mid (a-b)$. Use your equation from part (b) to argue that $n \mid (r_1 - r_2)$ as well.

    (d) Use the result you proved in part (c) to deduce that $r_1 = r_2$. (Hint: Both r_1 and r_2 satisfy a certain inequality. Use these inequalities to argue that $r_1 - r_2$ is a multiple of n and is strictly between $-n$ and n.)

    (e) Which direction of the biconditional statement from Theorem 1.9 did you prove in parts (c) and (d)? What remains to be shown?

    (f) Use your equation from part (b) to prove that if $r_1 = r_2$, then $n \mid (a - b)$. Explain how this argument finishes the proof of Theorem 1.9.

(16) Prove or disprove: For every integer a, if $a \not\equiv 0 \pmod 3$, then $a^2 \equiv 1 \pmod 3$. (Hint: Consider two cases.)

(17) (a) Is the following theorem true or false?

        For every integer n, if n is odd, then $8 \mid (n^2 - 1)$.

        Give a proof or a counterexample to justify your answer.

    (b) Translate the statement from part (a) into a corresponding statement dealing with congruence modulo 8.

(18) Prove or disprove: Let $a, b \in \mathbb{Z}$. If 3 divides $(a^2 + b^2)$, then 3 divides a and 3 divides b.

(19) Let a be an integer. After looking at several examples, make a general conjecture about the value of $\gcd(a - 1, a + 1)$. Then prove your conjecture.

(20) Fill in the blank, and prove your answer: For every integer a,

$$\gcd(a, a + 1) = \underline{}.$$

(21) For each of the following values of a and b, use the Euclidean algorithm to determine $\gcd(a, b)$. Then find integers x and y such that $ax + by = \gcd(a, b)$.

    (a) $a = 525, b = 252$

    (b) $a = 54321, b = 12345$

    (c) $a = 27182, b = -3141$

(d) $a = -61880, b = -60678$

(e) $a = 12906, b = 42905$

(22) Decide whether each of the following statements is true or false. For those that are true, explain why. For those that are false, give a counterexample and then change **one word or symbol** in the statement to make it true. For each statement, assume that a, b, and d are positive integers.

(a) If $ax + by = 1$ for some integers x and y, then $\gcd(a, b) = 1$.

(b) If $ax + by \neq 1$ for some integers x and y, then $\gcd(a, b) \neq 1$.

(c) If $ax + by = d$ for some integers x and y, then $\gcd(a, b) = d$.

(23) Use Bezout's Identity and Corollary 1.20 to prove Theorem 1.23.

(24) Determine all values of n for which the following statement is true:

There exist integers x and y such that $63x + 147y = n$.

Give a convincing argument to justify your answer.

(25) (a) Prove or disprove: For all nonzero integers a, b, and c, $\gcd(a, bc) = 1$ if and only if $\gcd(a, b) = 1$ and $\gcd(a, c) = 1$.

(b) Now take this a step further. Let n be a positive integer and let $a, b_1, b_2, \ldots, b_n$ be nonzero integers. Prove or disprove: $\gcd(a, b_1 b_2 \cdots b_n) = 1$ if and only if $\gcd(a, b_i) = 1$ for all $1 \leq i \leq n$.

(26) Let a and b be integers, not both zero. Prove that if $\gcd(a, b) = 1$, then $\gcd(a^2, b^2) = 1$. Is the converse true? Verify your answer.

(27) Let a be any integer. What is $\gcd(a, a + 2)$? Prove your answer.

(28) Let a and x be integers, with $x > 0$. Prove that $\gcd(a, a + x) = \gcd(a, x)$.

(29) Let a and b be integers, not both zero. Prove that if $\gcd(a, b) = 1$, then $\gcd(a + b, ab) = 1$. Is the converse true? Verify your answer.

(30) Our definition of greatest common divisors (Definition 1.11) uses the standard "less than" ordering relation on the integers ($<$) to define what is meant by *greatest*. However, some number systems cannot be ordered in the same way. (For example, consider sets of matrices, polynomials, or functions.) The theorem below establishes an equivalent definition that defines *greatest* using only divisibility. The questions that follow guide you through a proof of the theorem.

> **Theorem 1.28.** *Let a, b, and d be integers, with a and b not both zero. Then $d = \gcd(a, b)$ if and only if all the following conditions hold:*
>
> (i) *$d \mid a$ and $d \mid b$.*
>
> (ii) *If k is an integer such that $k \mid a$ and $k \mid b$, then $k \mid d$ also.*
>
> (iii) *d is positive.*

(a) Suppose $d = \gcd(a, b)$. Explain why conditions (i) and (iii) from Theorem 1.28 are automatically satisfied. Then use Bezout's Identity to prove condition (ii).

(b) Now suppose d is an integer that satisfies all three of the conditions from Theorem 1.28. Explain why there cannot exist an integer $k > d$ such that $k \mid a$ and $k \mid b$.

* (31) Use induction to prove the strong form of Euclid's Lemma, stated below.

> **Euclid's Lemma** (Strong Form). *Let $a_1, a_2, \ldots, a_n$ be integers, and let p be prime. If $p \mid a_1 a_2 \cdots a_n$, then $p \mid a_i$ for some i with $1 \le i \le n$.*

(32) Recall that an irrational number is one that cannot be written as a ratio $\frac{a}{b}$, where a and b are integers and $b \ne 0$. Use Euclid's Lemma to prove that for all positive integers n and q, if $\sqrt[q]{n}$ is not an integer, then $\sqrt[q]{n}$ is irrational.

(33) Prove or disprove: For all integers a and b, $a \mid b$ if and only if $a^2 \mid b^2$. (Hint: Exercise (32) may be helpful.)

(34) Let p_i denote the ith prime integer (so that $p_1 = 2$, $p_2 = 3$, $p_3 = 5$, and so on). Prove or disprove: For all $n \in \mathbb{Z}^+$, $p_1 p_2 p_3 \cdots p_n + 1$ is prime.

(35) **The Infinitude of the Primes.** Prove that there are infinitely many prime integers. (Hint: Suppose there are only a finite number of primes, say $p_1, p_2, \ldots p_n$. Show that $p_1 p_2 p_3 \cdots p_n + 1$ can be neither prime nor a product of primes.)

(36) (a) Let $n \in \mathbb{N}$. Prove that for each integer k with $2 \le k \le n+1$, k divides $[(n+1)! + k]$.

   (b) Deduce from part (a) that for each positive integer n, there exist at least n consecutive composite numbers.

(37) Prove or disprove: If n is an odd integer and 3 does not divide n, then $24 \mid (n^2 - 1)$.

(38) Let $y \in \mathbb{N}$. Use the Fundamental Theorem of Arithmetic to prove that there exists an odd natural number x and a nonnegative integer k such that $y = 2^k x$.

(39) **Goldbach's Conjecture.** *Goldbach's Conjecture*, which was made by Christian Goldbach in a letter to Leonhard Euler in 1742, states the following:

> Every even integer greater than 2 can be expressed as the sum of two (not necessarily distinct) prime numbers.

As of this printing, it is not known whether Goldbach's Conjecture is true or false, although most mathematicians believe it to be true.

   (a) Write each of 78, 90, and 138 as a sum of two primes.

   (b) Is there an even integer that can be written as a sum of two primes in more than one way? If so, find the smallest such integer.

   (c) Prove that Goldbach's Conjecture implies that every integer greater than 5 can be written as a sum of three primes.

   (d) Prove that Goldbach's Conjecture implies that every odd integer greater than 7 can be written as a sum of three odd primes.

(40) **The Twin Prime Conjecture.** A quick look at the first dozen or so prime numbers reveals several cases in which consecutive prime numbers differ by 2 (for instance, 3 and 5, 11 and 13, etc.) Such pairs of primes are called *twin primes*, and the *Twin Prime Conjecture*, which as of this printing has been neither proved nor disproved, states that there are infinitely many

twin primes. Answer each of the following questions related to the Twin Prime Conjecture:

(a) Find the first 10 pairs of twin primes.

(b) How many pairs of primes p and q satisfy $q - p = 3$? (Note that twin primes satisfy $q - p = 2$.)

(c) How many triplets of primes of the form $p, p + 2, p + 4$ are there? That is, how many triplets of primes are there where each prime is 2 more than the preceding prime? Prove your answer. (Hint: Set up cases using congruence modulo 3.)

Investigation 2

Equivalence Relations and $\mathbb{Z}_n$

Focus Questions

By the end of this investigation, you should be able to give precise and thorough answers to the questions listed below. You may want to keep these questions in mind to focus your thoughts as you complete the investigation.

- What is a congruence class, and what are some properties of congruence classes?

- What is an equivalence relation, and what are some strategies for proving the properties that characterize equivalence relations?

- What are equivalence classes, and how do the equivalence classes corresponding to an equivalence relation divide the underlying set into subsets?

- What is $\mathbb{Z}_n$, and what arithmetic axioms hold within $\mathbb{Z}_n$?

- What is a binary operation, and what does it mean for a binary operation to be well-defined?

- What are zero divisors and units in $\mathbb{Z}_n$? How are zero divisors and units related to solving linear equations?

Preview Activity 2.1. When working with large sets of objects, it is often useful to group these objects according to some common attribute or property. For instance, in a cooler containing 100 cans of soft drinks, there may be 30 cans of Coke, 30 cans of Pepsi, and 40 cans of 7 Up. If someone wanted to drink a can of Coke, they probably would not care exactly which can of Coke they pulled out of the cooler. In other words, they would probably consider all the different cans of Coke to be indistinguishable, or equivalent. This same kind of grouping can be applied to a set of mathematical objects by defining an *equivalence relation*. In this preview activity, we will investigate how congruence can be used to define such a relation on the integers.

(a) For every integer a, let $[a]_3$ denote the set of all integers that are congruent to a modulo 3. Using the roster method,* list the elements in $[0]_3$.

(b) Repeat part (a) for $[1]_3$, $[2]_3$, $[3]_3$, $[4]_3$, and $[5]_3$. Do you notice anything about the relationships between these sets?

(c) What is the remainder when 734 is divided by 3? Which, if any, of the sets $[0]_3$, $[1]_3$, and $[2]_3$ contain 734?

---

*To specify a set using the roster method, we simply list the elements of the set between braces, as we did when we defined the natural and whole numbers on page 5.

(d) Repeat part (c), replacing 734 with another integer of your choosing.

(e) Which elements belong to $[0]_3 \cap [1]_3$? What about $[0]_3 \cap [2]_3$, or $[1]_3 \cap [2]_3$?

(f) What familiar set is $[0]_3 \cup [1]_3 \cup [2]_3$ equal to, and why?

(g) Based on your answers to parts (c) – (f), make as many conjectures as you can about the sets $[0]_3$, $[1]_3$, and $[2]_3$.

Congruence Classes

In Preview Activity 2.1, we investigated several sets of integers called *congruence classes*, which we can define formally as follows:

Definition 2.2. Let n be a natural number, and let a be an integer. The **congruence class of** a **modulo** n, denoted $[a]_n$, is the set of all integers congruent to a modulo n. In other words,

$$[a]_n = \{x \in \mathbb{Z} : x \equiv a \pmod{n}\}.$$

Based on our work in Preview Activity 2.1, we can make several observations and conjectures about congruence classes. In particular:

- For $0 \leq a \leq n-1$, $[a]_n$ contains all integers x for which x divided by n yields a remainder of a.

- It is possible for the congruence classes of two distinct integers to be equal to each other. That is, it is possible for $[a]_n = [b]_n$ even when $a \neq b$. This occurs exactly when $a \equiv b \pmod{n}$. In other words,

$$[a]_n = [b]_n \text{ if and only if } a \equiv b \pmod{n}.$$

- If two congruence classes are not equal, then they are disjoint. That is, $[a]_n \cap [b]_n = \emptyset$ for all integers a and b such that $[a]_n \neq [b]_n$.

- There are exactly n distinct congruence classes modulo n, which can be represented by $[0]_n$, $[1]_n$, $[2]_n$, ..., $[n-1]_n$. All other congruence classes modulo n (such as $[n+3]_n$, $[-1]_n$, etc.) are equal to one of these n classes.

- For every natural number n, the union of the n distinct congruence classes modulo n is equal to $\mathbb{Z}$. That is,

$$[0]_n \cup [1]_n \cup [2]_n \cup \cdots \cup [n-1]_n = \mathbb{Z}.$$

- Every integer x belongs to exactly one of the n distinct congruence classes modulo n. In other words, for every natural number n and every integer x, there exists a unique integer a with $0 \leq a \leq n-1$ such that $x \in [a]_n$.

As you may have noticed, some of these properties are related to each other or even implied by each other. Most of them, in fact, are consequences of the more general theory of equivalence relations that we will study in the next section. Before we move on, however, the last three observations on our list merit some special attention. Together, they imply that congruence modulo n divides

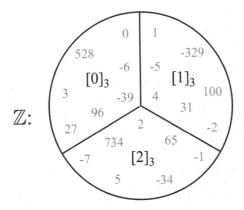

Figure 2.1
Dividing $\mathbb{Z}$ into classes based on congruence modulo 3.

the set of integers into n distinct congruence classes, and that each integer belongs to exactly one of these classes. In the context of our original example of congruence modulo 3, this allows us to picture the integers as shown in Figure 2.1.

This nice division of the integers into congruence classes is made possible by the fact that congruence satisfies three very important properties. These three properties characterize *equivalence relations*, which we will study in the next section.

Equivalence Relations

Throughout mathematics, *binary relations* (or just *relations* for short) are often used to specify certain associations or relationships between the elements of a set of mathematical objects. For instance, the equals relation ($=$), the less than or equal to relation ($\leq$), and congruence modulo n are all examples of relations on the set of integers. In geometry, similarity and congruence (of shapes—not to be confused with congruence modulo n) both define relations on the set of all triangles. In everyday life, we use the same kind of language; we talk about two people being related if they are, for lack of a better word, relatives—either by blood or by another familial relationship such as marriage.

Technically, a *binary relation* on a set S is just a set of ordered pairs, where both coordinates of each pair in the relation are elements of S. So, for example, all the following pairs (along with many others) would belong to the congruence modulo 3 relation:

$$(0, 3), (1, 4), (2, -1), (3, 99)$$

For our purposes, we will usually describe a relation not by listing the ordered pairs that belong to it, but rather by specifying some kind of rule, as we did when we defined congruence modulo n (i.e., $a \equiv b \pmod{n}$ provided that $n \mid (a - b)$). We will often use the $\sim$ symbol to denote a relation on a set. For instance, if we wanted to define two integers a and b to be related whenever b was the square of a, we might write something like this:

Let $\sim$ be the relation on $\mathbb{Z}$ defined by $a \sim b$ if and only if $a^2 = b$.

We could then use this notation to say that $3 \sim 9$ (since $3^2 = 9$), but $2 \not\sim 5$ (since $2^2 \neq 5$). If we had used a different symbol to define our relation, say $\propto$ instead of $\sim$, then we would simply adjust our notation accordingly, writing $3 \propto 9$ and $2 \not\propto 5$.

When we use a rule to define a relation, we are specifying exactly what it means for two elements to be related to each other. So, in the context of the preceding example, if we wanted to prove that $a \sim b$ for two integers a and b, we would need to show that $a^2 = b$. Likewise, if we were assuming that $a \sim b$, then this assumption would allow us to conclude that $a^2 = b$. Stated another way, the rules that define relations allow us to translate generic statements, such as $a \sim b$, into more specific statements, such as $a^2 = b$, and vice versa.

There are many different kinds of relations, but for now we will be mainly interested in the type of relations suggested by Preview Activity 2.1—that is, those that identify certain objects as being equivalent in some way. Recall that such relations are called *equivalence relations*. We will use equivalence relations throughout many of our later investigations. The formal definition of an equivalence relation is as follows:

Definition 2.3. Let S be a set, and let $\sim$ be a binary relation on S. Then $\sim$ is called an **equivalence relation** on S provided that $\sim$ satisfies the following properties:

- **Reflexivity:** For all $a \in S$, $a \sim a$.

- **Symmetry:** For all $a, b \in S$, if $a \sim b$, then $b \sim a$.

- **Transitivity:** For all $a, b, c \in S$, if $a \sim b$ and $b \sim c$, then $a \sim c$.

In other words, an equivalence relation is a binary relation that is reflexive, symmetric, and transitive.

If the properties from Definition 2.3 sound familiar, it's probably because we proved all of them for congruence modulo n back in Activity 1.10. (See page 9.) In other words, we proved that congruence modulo n is an equivalence relation. This result is very important, and we state it formally in Theorem 2.4 below. The proof that follows ties together the ideas and arguments from parts (d) – (f) of Activity 1.10. You should read this proof carefully, inserting additional details or explanations wherever you see the ⑦ symbol.

Theorem 2.4. *Let n be any natural number. Then congruence modulo n is an equivalence relation on* $\mathbb{Z}$. *In other words, the relation $\sim$ defined on* $\mathbb{Z}$ *by the rule*

$$a \sim b \text{ if and only if } a \equiv b \pmod{n}$$

is an equivalence relation.

Proof. We must show that the congruence modulo n relation $\sim$, as defined in the statement of the theorem, is reflexive, symmetric, and transitive.

For reflexivity, let $a \in \mathbb{Z}$. Since $a - a = 0$, it follows that $n \mid (a - a)$.⑦ Thus, $a \equiv a \pmod{n}$, and congruence modulo n is reflexive.

For symmetry, suppose $a \equiv b \pmod{n}$ for some $a, b \in \mathbb{Z}$. Then $n \mid (a - b)$, which implies that $n \mid (b - a)$.⑦ Thus, $b \equiv a \pmod{n}$, and congruence modulo n is symmetric.

For transitivity, suppose $a \equiv b \pmod{n}$ and $b \equiv c \pmod{n}$ for some $a, b, c \in \mathbb{Z}$. Then $n \mid (a - b)$ and $n \mid (b - c)$. Thus, $n \mid [(a - b) + (b - c)]$,⑦ which implies that $n \mid (a - c)$. It follows that $a \equiv c \pmod{n}$, and so congruence modulo n is transitive.

By showing that the congruence modulo n relation is reflexive, symmetric, and transitive, we have established that congruence modulo n is an equivalence relation on $\mathbb{Z}$. ∎

Equivalence Classes

Recall that congruence modulo n naturally divides the integers into n congruence classes. Now we have also shown that congruence modulo n is an equivalence relation on $\mathbb{Z}$. As you might suspect, these two results are closely related. In fact, we will show that *every* equivalence relation divides the set on which it is defined into subsets called *equivalence classes*. The definition of an equivalence class, stated formally below, generalizes Definition 2.2.

Definition 2.5. Let $\sim$ be an equivalence relation on a nonempty set S, and let $a \in S$. The **equivalence class of** a (with respect to $\sim$), denoted $[a]_\sim$, is the set of all elements of S that are related to a by $\sim$. More precisely,

$$[a]_\sim = \{x \in S : x \sim a\}.$$

Note that Definition 2.5 implies that if $x \in S$, then $x \in [a]_\sim$ if and only if $x \sim a$. In other words, for all elements $x \in S$, the symbolic expressions $x \in [a]_\sim$ and $x \sim a$ are interchangeable.

When the context is clear (that is, when we have defined a particular equivalence relation and are interested only in that relation), we will often omit the subscript on the $[a]_\sim$ notation, simply writing $[a]$ instead. The same convention applies to the notation for congruence classes that we studied earlier. For instance, if we had already stated that we were working with congruence modulo 6, then statements like $[3] \neq [4]$ would make perfect sense even without the usual subscripts. On the other hand, if we were trying to compare congruence classes with respect to two different moduli, then the subscripts would be absolutely essential. To illustrate, note that the true statement $[3]_{12} \neq [3]_6$ becomes ambiguous and potentially confusing when written as $[3] \neq [3]$. Thus, while there are times when it is convenient to simplify our notation, we must be careful to do so only when the context is clear.

Now that we have defined equivalence classes, let's turn our attention to generalizing some of the observations about congruence classes that we noted earlier in the investigation. The key result that we will prove is the following:

Theorem 2.6. *Let S be a nonempty set, and let $\sim$ be an equivalence relation on S. Then S can be written as the disjoint union of the distinct equivalence classes corresponding to $\sim$. That is, the equivalence classes corresponding to $\sim$ are pairwise disjoint, and every element of S belongs to exactly one equivalence class. In particular:*

(i) *For all $a, b \in S$, if $[a] \neq [b]$, then $[a] \cap [b] = \emptyset$.*

(ii) *For all $a \in S$, $a \in [a]$.*

(iii) *For all $a \in S$, if $a \in [b]$ for some $b \in S$, then $[a] = [b]$.*

Note that, like our earlier observation about congruence classes, Theorem 2.6 acknowledges the fact that the equivalence classes of two elements of set can be equal to each other, even if the elements themselves are not. The next lemma characterizes exactly when this equality can occur; it will be particularly helpful in our proof of Theorem 2.6.

Lemma 2.7. *Let S be a nonempty set, and let $\sim$ be an equivalence relation on S. Then for all $a, b \in S$, $[a] = [b]$ if and only if $a \sim b$.*

To prove Lemma 2.7, we will need to remember exactly what it means for two sets to be equal. In particular, we will need to make use of the fact that $[a] = [b]$ if and only if $[a] \subseteq [b]$ and $[b] \subseteq [a]$. Activity 2.8 below will help us work out the rest of the details.

Activity 2.8. Suppose, as in Lemma 2.7, that $\sim$ is an equivalence relation on a nonempty set S. Let $a, b \in S$.

(a) Suppose $[a] = [b]$. Using the fact that $[a] \subseteq [b]$, argue that $a \in [b]$. Then deduce that $a \sim b$. (Hint: Use one of the three properties that define an equivalence relation.)

(b) For the converse, suppose that $a \sim b$, and let $x \in [a]$. Show that $x \in [b]$. (Hint: Use one of the properties that you didn't use in part (a).)

(c) Now suppose that $x \in [b]$. Still assuming that $a \sim b$, show that $x \in [a]$. Is there a difference between your argument here and the one you made in part (b)?

Having established Lemma 2.7, we are now ready to move on to the proof of Theorem 2.6. As you read this proof, you should try to fill in the missing details, treating the $\textcircled{?}$ symbol as we have in previous proofs.

Proof of Theorem 2.6. Let $\sim$ be an equivalence relation on a nonempty set S. For part (i), we will prove the _____. So let $a, b \in S$, and assume that $[a] \cap [b] \neq \emptyset.\textcircled{?}$ Then there exists an element $x \in [a] \cap [b]$, which implies that $x \sim a$ and $x \sim b.\textcircled{?}$ By the _____ property, it follows that $a \sim x$. But _____ then implies that $a \sim b.\textcircled{?}$ Since $a \sim b$, Lemma 2.7 now lets us conclude that _____, as desired.

For part (ii), let $a \in S$, and note that by the _____ property, $a \sim a$. Thus, by the definition of equivalence class, $a \in [a]$.

For part (iii), let $a, b \in S$, and suppose that $a \in [b]$. Then $a \sim b,\textcircled{?}$ which implies by _____ that $[a] = [b]$.   ∎

Activity 2.9. Finish the proof of Theorem 2.6 by explaining why conditions (i) – (iii) imply that S can be written as the disjoint union of the distinct equivalence classes corresponding to $\sim$.

The Number System $\mathbb{Z}_n$

So far, we have shown that equivalence relations, such as the congruence modulo n relation on $\mathbb{Z}$, always divide the sets on which they are defined into pairwise disjoint subsets called equivalence classes. In this section, we will use what we have learned to investigate a family of number systems whose elements are not numbers at all, but rather congruence classes. We define this family of number systems as follows:

Definition 2.10. For every integer $n \geq 2$, the **integers modulo** n, denoted $\mathbb{Z}_n$, is the set of the n distinct congruence classes of $\mathbb{Z}$ modulo n, i.e.,

$$\mathbb{Z}_n = \{[0]_n, [1]_n, [2]_n, \ldots, [n-1]_n\}.$$

We can make $\mathbb{Z}_n$ into a number system by defining an addition and multiplication on the set. There is a seemingly natural way to do this:

$$[a] + [b] = [a+b] \text{ and } [a] \cdot [b] = [a \cdot b]$$

for all $[a], [b] \in \mathbb{Z}_n$.

A few observations are worth noting. First, we haven't yet defined exactly what a *number system* is. For now, we will think of a number system as a set of mathematical objects with one or more operations, like addition and multiplication, defined on it. Later on, we will add more precision and clarity to this informal definition.

Second, we must keep in mind that the elements of $\mathbb{Z}_n$ are not actually numbers, but rather *sets* of numbers, and *infinite* sets at that. It is for this reason that we must formally define how exactly addition and multiplication should work in $\mathbb{Z}_n$. Our definitions are quite natural, so much so that they may not seem like definitions at all, but rather statements of fact. Be assured that this is not the case, for while it may seem natural to write $[3] + [2] = [5]$ when working in $\mathbb{Z}_7$, what we are really defining with this notation is the set operation that specifies

$$\{\ldots, -11, -4, 3, 10, \ldots\} + \{\ldots, -12, -5, 2, 9, \ldots\} = \{\ldots, -9, -2, 5, 12, \ldots\}.$$

If we saw an expression like this outside of the context of $\mathbb{Z}_7$, we would probably be quite puzzled, since the notions of addition and multiplication of sets are not universally defined. Also note that the addition and multiplication operations we have defined for $\mathbb{Z}_n$ are quite different from some of the more familiar set operations, such as unions and intersections. In Investigation 3, we will see an example of a number system whose elements are sets and for which addition and multiplication are defined in terms of unions, intersections, and relative complements.

Finally, note that in our definition of $\mathbb{Z}_n$, we have used the same notation, namely the $+$ symbol, to represent both addition in $\mathbb{Z}_n$ and addition in $\mathbb{Z}$. The same could be said for multiplication, and we will rely on other notation (such as brackets) to make the context of our operations clear. So, for instance, if we write

$$2 + 4 = 6,$$

we will assume that the $+$ sign is indicating addition in $\mathbb{Z}$. However, if we write

$$[2]_5 + [4]_5 = [6]_5 = [1]_5,$$

then the bracket notation should indicate clearly to us that our addition is being performed within $\mathbb{Z}_5$. Note that in the latter case, we took the extra step of reducing our final answer so that the representative a chosen to denote $[a]_5$ satisfied $0 \le a < 5$. Adopting this standard reducing convention, we can construct the addition and multiplication tables for $\mathbb{Z}_n$. For example, the addition and multiplication tables for $\mathbb{Z}_3$ are as follows:

$+$	$[0]$	$[1]$	$[2]$
$[0]$	$[0]$	$[1]$	$[2]$
$[1]$	$[1]$	$[2]$	$[0]$
$[2]$	$[2]$	$[0]$	$[1]$

$\cdot$	$[0]$	$[1]$	$[2]$
$[0]$	$[0]$	$[0]$	$[0]$
$[1]$	$[0]$	$[1]$	$[2]$
$[2]$	$[0]$	$[2]$	$[1]$

Activity 2.11.

(a) Make the addition and multiplication tables for $\mathbb{Z}_4$ and $\mathbb{Z}_5$.

(b) Do you notice any patterns or symmetries in the tables for $\mathbb{Z}_3$, $\mathbb{Z}_4$, and $\mathbb{Z}_5$? If so, what do your observations allow you to conclude about the way arithmetic works in $\mathbb{Z}_n$? State your conclusions specifically and precisely, as we did when we stated the axioms for $\mathbb{Z}$ in Investigation 1.

(c) Are there any differences between the way arithmetic works in $\mathbb{Z}$ and the way it works in either $\mathbb{Z}_3$, $\mathbb{Z}_4$, or $\mathbb{Z}_5$? If so, state these differences precisely.

(d) Do addition and multiplication behave the same in $\mathbb{Z}_3$, $\mathbb{Z}_4$, and $\mathbb{Z}_5$, or are there differences between these three number systems? Give specific examples to justify your answer.

As you may have noticed from Activity 2.11, $\mathbb{Z}_n$ and $\mathbb{Z}$ are similar in many ways, especially with regard to the axioms of addition and multiplication. In fact, for at least some of the arithmetic axioms from Investigation 1, there are analogous properties that hold in $\mathbb{Z}_n$. For instance, addition in $\mathbb{Z}_n$ is associative, just as it is in $\mathbb{Z}$. If we wanted to prove this result, our argument might look something like this:

Proof that addition in $\mathbb{Z}_n$ *is associative.* Let $n \in \mathbb{N}$, and let $[a], [b], [c] \in \mathbb{Z}_n$. Then

$$([a] + [b]) + [c] = [a + b] + [c] \;^{\textcircled{?}}$$
$$= [(a + b) + c] \;^{\textcircled{?}}$$
$$= [a + (b + c)] \;^{\textcircled{?}}$$
$$= [a] + [b + c] \;^{\textcircled{?}}$$
$$= [a] + ([b] + [c]), \;^{\textcircled{?}}$$

as desired. ∎

As you read this proof, you should have been able to provide an explanation for each step (as indicated by the $\textcircled{?}$ symbol). In particular, the first two and the last two steps were simply applications of the definition of addition in $\mathbb{Z}_n$, and the middle step used the fact that addition in $\mathbb{Z}$ is associative. In other words, the proof used associativity of addition in one number system, $\mathbb{Z}$, to prove a similar property for a related number system, $\mathbb{Z}_n$. The same strategy can be employed to prove other properties in $\mathbb{Z}_n$, and Activity 2.12 asks us to do exactly that.

Activity 2.12.

(a) For each of the arithmetic axioms listed on page 5, either prove a corresponding property for $\mathbb{Z}_n$ or give a counterexample to show that no such property holds in $\mathbb{Z}_n$.

(b) Do the ordering axioms from page 6 hold in $\mathbb{Z}_n$? Why or why not?

(c) Does an analogous version of Theorem 1.4 hold in $\mathbb{Z}_n$? (Hint: Consider the multiplication tables you made in Activity 2.11. Does your answer depend on n?)

Binary Operations

What we did in the previous section may seem very natural and may not concern you at all, but it will all be total nonsense if addition and multiplication on $\mathbb{Z}_n$ are not *well-defined*. The next activity illustrates why it is important for operations to be well-defined by showing what can happen if an operation is *not* well-defined.

Activity 2.13. Let $\diamond$ be the operation on $\mathbb{Z}_n$ defined as follows:

$$[a] \diamond [b] = \begin{cases} [1] & \text{if } a \text{ and } b \text{ have the same parity} \\ [0] & \text{if } a \text{ and } b \text{ have opposite parity} \end{cases}$$

Note that *parity* refers to whether an integer is even or odd. Thus, two integers have the same parity if they are both even or both odd. Likewise, two integers have opposite parity if one is even and the other is odd.

(a) Use the above definition to calculate each of the following quantities in $\mathbb{Z}_5$:

- $[1] \diamond [3]$

- $[1] \diamond [8]$

- $[6] \diamond [13]$

- $[11] \diamond [13]$

(b) What is the relationship between $[1]$, $[6]$, and $[11]$ in $\mathbb{Z}_5$?

(c) What is the relationship between $[3]$, $[8]$, and $[13]$ in $\mathbb{Z}_5$?

(d) In light of your answers to parts (b) and (c), does anything seem strange or unusual about your answers to part (a)? Explain.

You should have observed something unusual in Activity 2.11. In particular, the result of the operation depended on which representative we used for our input. This is a critically important observation to make in sets like $\mathbb{Z}_n$ where each of the equivalence classes can be represented in infinitely many different ways. If the result of an operation depends on which way we choose to represent one of the inputs, then two people working the same problem could, without making any mistakes in computation, arrive at different answers. This would result in chaos! What defines these equivalence classes, however, is not the way we choose to represent them, but rather the elements that they contain. Thus, if two equivalence classes contain exactly the same elements, and are therefore equal, we would expect any reasonable operation to treat them exactly the same.

This, however, is not what happened in Activity 2.13. There, in spite of the fact that $[1] = [6]$ and $[3] = [13]$, we saw that $[1] \diamond [3] = [1]$, but $[6] \diamond [13] = [0]$. In other words, the output of the $\diamond$ operation depended not only on the elements that we used (they were the same in each case), but also on the way we chose to name, or represent, those elements. Because of this, we might say that $\diamond$ is *ill-defined*. Not surprisingly, the opposite of an ill-defined operation is a *well-defined* operation.

Although we have already seen and used several different binary operations (the most familiar being standard addition and multiplication in $\mathbb{Z}$), we have not yet formally defined exactly what a binary operation is. The next definition fills in this gap. It formalizes the idea that a binary operation on a set S maps every ordered pair of elements of S to single element of S. So, for instance, the operation of addition on the integers maps the pair $(1, 4)$ to the integer 5 (since $1 + 4 = 5$). This same idea forms the basis of the definition of a binary operation on any set.

Definition 2.14. Let S be a set. The **Cartesian product** $S \times S$ is the set of all ordered pairs of elements from S; that is,

$$S \times S = \{(x, y) : x \in S \text{ and } y \in S\}.$$

A **binary operation** on S is a function $f : S \times S \to S$.

Activity 2.15. For each part below, find the image of the given pair of elements under the given binary operation. The first part is completed for you as an example.

(a) $(3, -5)$; addition in $\mathbb{Z}$

   Solution: *Using function notation,* $+(3, -5) = 3 + (-5) = -2.$

(b) $([2], [5])$; addition in $\mathbb{Z}_8$

(c) $([4], [3])$; multiplication in $\mathbb{Z}_5$

You may be wondering at this point why it is important to define binary operations so formally. It is certainly not standard practice to use function notation when doing arithmetic. For instance, we would rarely if ever write

$$+(2, 3) = 5$$

in $\mathbb{Z}$. Instead, we would use the more natural notation of

$$2 + 3 = 5.$$

So, on the surface, it seems that our formal definition only makes matters more complicated. While this may be the case from a notational perspective, defining a binary operation as a function does help to make explicit a very important property that we should expect all binary operations to satisfy. This property, which we will define next, is one that you have used many times, perhaps without even realizing it. As we saw in Activity 2.11, the result of an operation must not be dependent on the particular representation or name we choose for the input, but must be independent of such superficial distinctions. That is, if we have two elements that may "look" different ([1] and [4] in $\mathbb{Z}_3$, for example) but are actually equal, any operation must treat them the same way. Activity 2.11 shows that this is a property of operations that cannot be taken for granted. We formalize what we mean by a well-defined operation in the next definition.

Definition 2.16. Let $\star$ denote a binary operation on a set S. Then $\star$ is said to be **well-defined** provided that whenever $a = x$ and $b = y$ in S we have $a \star b = x \star y$.

Definition 2.16 suggests that in order to prove that an operation $\star$ on S is well-defined, we should assume that $a = x$ and $b = y$ for some $a, b, x, y \in S$, and then try to show that $a \star b = x \star y$. To illustrate this technique, consider the following proof that addition in $\mathbb{Z}_n$ is well-defined:

> *Proof.* Let $[a], [b], [x], [y] \in \mathbb{Z}_n$, and suppose that $[a] = [x]$ and $[b] = [y]$. Then $a \equiv x$ $(\bmod\ n)^{①}$ and $b \equiv y$ $(\bmod\ n)$,^{②} which implies that $n \mid (a - x)^{③}$ and $n \mid (b - y)$.^{④} From this it follows that $n \mid [(a - x) + (b - y)]$,^{⑤} or $n \mid [(a + b) - (x + y)]$. Thus,
>
> $$[a] + [b] \stackrel{⑥}{=} [a + b] \stackrel{⑦}{=} [x + y] \stackrel{⑧}{=} [x] + [y],$$
>
> as desired.   ∎

You will show that multiplication is well-defined on $\mathbb{Z}_n$ in Activity 2.25.

A few observations are in order before we move on. The first is that, after all of this discussion, it turns out in many cases, we don't have to worry too much about whether our operations are well-defined or not. This is because many of the number systems we have studied (and will study) do not allow for multiple representations of the same element. For such number systems, the definition of well-defined is trivially satisfied. Thus, the systems that merit more attention are those whose elements are equivalence classes, like $\mathbb{Z}_n$. In general, whenever the elements of a number system can be represented in multiple ways, we will need to verify that the operations within the number system do not depend upon these varying representations.

A second observation is that our definition of well-defined is actually redundant and even unnecessary in some sense. This is due to the fact that we defined a binary operation on S to be a *function* $f : S \times S \rightarrow S$, which implies that equal inputs must be mapped to equal outputs, regardless of their representation. Thus, for all $a, b, x, y \in S$, if $a = b$ and $x = y$, then it must be the case that $(a, b) = (x, y)$, which implies that $f(a, b) = f(x, y)$. In other words, in order for a binary operation to really be a binary operation, it must first pass the test of being well-defined. Along these same lines, because the codomain of a binary operation on S is always S itself, closure is also guaranteed by definition.

Throughout subsequent investigations, we will be referencing $\mathbb{Z}_n$ frequently. We will also be considering several other number systems, all of which satisfy at least some of the same axioms that the integers do. Our goal in these investigations will be to identify a certain common structure that is shared by a variety of number systems. Once we have done so, we will be able to prove general results about the way arithmetic works in various contexts. Before we move on, however, we will consider some important ideas related to solving linear equations in $\mathbb{Z}_n$.

Zero Divisors and Units in $\mathbb{Z}_n$

Solving equations is an important part of algebra, and the simplest type of equation is the linear equation of the form $ax = b$. These types of equations are among the first studied in beginning algebra courses, but here we will consider them from a more general perspective.

Activity 2.17. If $a \neq 0$ in the real number system, then the equation $ax = b$ always has a unique solution.

(a) Does the equation $2x = 1$ have a solution in $\mathbb{Z}$? Explain.

(b) Find all solutions to $[2]x = [4]$ in $\mathbb{Z}_6$, and then reflect on your answer. Do you notice anything interesting or surprising?

(c) Find all solutions to $[2]x = [3]$ in $\mathbb{Z}_6$, and then reflect on your answer. Do you notice anything interesting or surprising?

Activity 2.17 shows that it is possible for the equation $ax = b$ to have no solutions or more than one solution even if $a \neq 0$, depending on the number system in question. The element $[2]$ in $\mathbb{Z}_6$ has a special property that leads to the behavior in Activity 2.17.

Activity 2.18. In Investigation 1, we saw that there are no nonzero integers a and b such that $ab = 0$.

(a) Show that this is not true in general in $\mathbb{Z}_n$ by exhibiting a particular value of n and specific nonzero classes $[a]$ and $[b]$ so that $[a][b] = [0]$ in $\mathbb{Z}_n$.

(b) Is there a multiplicative cancellation law in $\mathbb{Z}_n$ as there is in $\mathbb{Z}$? That is, if $[c] \neq [0]$ and $[a][c] = [b][c]$ in $\mathbb{Z}_n$, must it be true that $[a] = [b]$? Verify your conjecture.

Elements that behave like those described in Activity 2.18 are given a special name.

Definition 2.19. A nonzero element $[a] \in \mathbb{Z}_n$ is a **zero divisor** in $\mathbb{Z}_n$ if there is a nonzero element $[b] \in \mathbb{Z}_n$ so that $[a][b] = [0]$.

So if a is a zero divisor, then the equation $ax = b$ may have no solutions or more than one solution. But when can we be sure that the equation $ax = b$ has exactly one solution?

Activity 2.20. Just as we did in $\mathbb{Z}$ (see the axioms on page 5), we can define the additive inverse of an element $[x] \in \mathbb{Z}_n$ to be an element $[y] \in \mathbb{Z}_n$ such that $[x] + [y] = [0]$.

(a) Write a similar definition of the multiplicative inverse of an element $[x] \in \mathbb{Z}_n$.

(b) Find all elements of $\mathbb{Z}_3$ that have a multiplicative inverse in $\mathbb{Z}_3$.

(c) Find all elements of $\mathbb{Z}_4$ that have a multiplicative inverse in $\mathbb{Z}_4$.

(d) Exercise (5) on page 16 tells us that there are only two integers that have a multiplicative inverse in $\mathbb{Z}$, namely 1 and -1. Is the same result true in $\mathbb{Z}_n$ in general? Explain. (Hint: What element of $\mathbb{Z}_n$ is analogous to -1 in $\mathbb{Z}$?)

(e) In the set of real numbers, we know that every nonzero element has a real multiplicative inverse. Is the comparable statement true in $\mathbb{Z}_n$? Explain.

Elements that behave like those described in Activity 2.20 are also given a special name.

Definition 2.21. An element $[x] \in \mathbb{Z}_n$ is a **unit** in $\mathbb{Z}_n$ if there is an element $[y] \in \mathbb{Z}_n$ such that $[x][y] = [1]$. In this case, the element $[y]$ is called a **multiplicative inverse** of $[x]$.

It can be shown that if $[a]$ is a unit in $\mathbb{Z}_n$, then the equation $[a]x = [b]$ has exactly one solution for every $[b] \in \mathbb{Z}_n$. (See Exercise (19).) There is also a pattern to the zero divisors and units in $\mathbb{Z}_n$, as we will see in the next activity.

Activity 2.22. Construct the multiplication tables for $\mathbb{Z}_3$, $\mathbb{Z}_4$, $\mathbb{Z}_5$, $\mathbb{Z}_6$, and $\mathbb{Z}_8$.

(a) Identify the zero divisors in each set.

(b) Identify the elements in each set that have a multiplicative inverse in the set.

(c) Make a conjecture in the form of a biconditional statement, such as:

Let $n \in \mathbb{N}$, and let $[a] \in \mathbb{Z}_n$. Then $[a] \neq [0]$ is a zero divisor in $\mathbb{Z}_n$ if and only if
...

(d) Make a conjecture in the form of a biconditional statement, such as:

Let $n \in \mathbb{N}$, and let $[a] \in \mathbb{Z}_n$. Then $[a]$ has a multiplicative inverse in $\mathbb{Z}_n$ if and only if ...

Proofs of correct conjectures for parts (c) and (d) are outlined in Exercises 16 and 17.

We will now conclude our explorations of $\mathbb{Z}_n$ with a short discussion of how to solve an equation of the form $[a]x = [b]$ in $\mathbb{Z}_n$ if $[a]$ is a unit.

Activity 2.23.

(a) What tool can we use to quickly determine if 231 and 4210 are relatively prime?

(b) Bezout's Identity states that we can write

$$231x + 4210y = 1 \tag{2.1}$$

for some integers x and y. Find integers x and y that satisfy Equation (2.1). (Hint: Think back to an algorithm we learned in a previous investigation.)

(c) Reduce both sides of Equation (2.1) modulo 4210 and find a multiplicative inverse of $[231]$ in $\mathbb{Z}_{4210}$.

(d) Summarize this process to explain how to find a multiplicative inverse of $[a]$ in $\mathbb{Z}_n$. Then explain how such an inverse could be used to solve an equation of the form $[a]x = [b]$.

Concluding Activities

Activity 2.24. Consider the following proof of Lemma 2.7:

> *Proof.* Let $a, b \in S$, and suppose that $[a] = [b]$. By the reflexive property, $a \sim a$, which implies that $a \in [a]$. But since $[a] = [b]$, it then follows that $a \in [b]$, which implies that $a \sim b$.
>
> For the converse, suppose that $a \sim b$. We must show that $[a] \subseteq [b]$ and $[b] \subseteq [a]$. For the former, let $x \in [a]$. Then $x \sim a$. But since $a \sim b$ as well, transitivity implies that $x \sim b$. Thus, $x \in [b]$, and $[a] \subseteq [b]$, as desired. A similar argument establishes that $[b] \subseteq [a]$, which completes the proof. ∎

(a) Is this proof correct? If so, are there any places in the proof where more detail should be provided, or where the argument could be made clearer? If not, where are the errors in the proof? Give a thorough and precise answer.

(b) As you may have noticed, the proof above made no reference to the symmetric property. Does this mean that Lemma 2.7 holds for all relations that are reflexive and transitive (instead of only those that are equivalence relations)? Why or why not?

(c) Consider the "less than or equal to" ($\leq$) relation on $\mathbb{Z}$. Is $\leq$ both reflexive and transitive? Give a proof or counterexample to justify your answer for each property.

(d) For each $a \in \mathbb{Z}$, define $[a]_\leq$ as in Definition 2.5. That is, let

$$[a]_\leq = \{x \in \mathbb{Z} : x \leq a\}.$$

Use roster notation to list the elements of $[1]_\leq$ and $[2]_\leq$. Is $1 \leq 2$? Does $[1]_\leq = [2]_\leq$? How do your answers to these questions relate to your answers to parts (a) – (c) above?

Activity 2.25. We have already shown that addition is well-defined in $\mathbb{Z}_n$. In this activity, we will show that multiplication is also well-defined.

(a) Let $n \in \mathbb{N}$, and let $[a], [b], [x], [y] \in \mathbb{Z}_n$ such that $[a] = [x]$ and $[b] = [y]$. Prove that $[a][b] = [x][y]$. (There **is** something to prove here. If the result seems trivial, then you should go back and re-read the section on well-defined operations.)

(b) Why is it important for the operations on $\mathbb{Z}_n$ to be well-defined?

Exercises

(1) Let $S = \{1, 2, 3, \ldots, 999, 1000\}$, and define the relation $\sim$ on S as follows:

   For all $x, y \in S$, $x \sim y$ if and only if y has the same number of digits as x.

   Prove that $\sim$ is an equivalence relation on S, and find all distinct equivalence classes corresponding to $\sim$.

(2) Let $\sim$ be the relation on $\mathbb{Z}$ defined by $a \sim b$ if and only if $3a + 4b = 7n$ for some integer n. Prove that $\sim$ is an equivalence relation, and give a precise description of the equivalence classes of $\sim$.

(3) Let $\sim$ be the relation on $\mathbb{Z}$ defined by $a \sim b$ if and only if $a + b$ is even.

   (a) Is $\sim$ an equivalence relation? Verify your answer.

   (b) Suppose the definition of $\sim$ is changed so that $a \sim b$ if and only if $a + b$ is odd. Would this change your answer to part (a)? Why or why not?

(4) Let $\sim$ be the relation on $\mathbb{Z}$ defined by $a \sim b$ if and only if $ab \geq 0$.

   (a) Is $\sim$ an equivalence relation? Verify your answer.

   (b) Suppose the definition of $\sim$ is changed so that $a \sim b$ if and only if $ab > 0$. Would this change your answer to part (a)? Why or why not?

(5) Let $S = \{(x, y) \in \mathbb{Z} \times \mathbb{Z} : y \neq 0\}$ and let $\sim$ be the relation on S defined by $(a, b) \sim (c, d)$ if and only if $ad = bc$. Prove that $\sim$ is an equivalence relation on S, and describe the equivalence classes corresponding to $\sim$. Then find and explain a connection between $\sim$ and the rational numbers.

(6) For each part below, find a binary relation on the set $S = \{1, 2, 3, 4\}$ that satisfies the given combination of properties.

   (a) reflexive, symmetric, and transitive

   (b) not reflexive, but symmetric and transitive

   (c) not symmetric, but reflexive and transitive

   (d) not transitive, but reflexive and symmetric

   (e) neither reflexive nor symmetric, but transitive

   (f) neither reflexive nor transitive, but symmetric

   (g) neither symmetric nor transitive, but reflexive

   (h) not reflexive, not symmetric, and not transitive

(7) Which of the properties of reflexive, symmetric, and transitive are satisfied by each of the following relations on the set of natural numbers? Give a proof or counterexample to justify each of your answers.

   (a) The relation $\sim$ defined by $a \sim b$ if and only if $a \mid b$

(b) The relation $\sim$ defined by $a \sim b$ if and only if $a \mid b$ or $b \mid a$

(c) The relation $\sim$ defined by $a \sim b$ if and only if $a + b = 10$

(8) Which of the following sets could be the set of all equivalence classes for an equivalence relation on $\{a, b, c, d, e\}$? Cite a theorem from this investigation to justify each of your answers.

(a) $\{\{a, b, c, d, e\}\}$

(b) $\{\{a\}, \{b, c\}, \{d\}\}$

(c) $\{\{a\}, \{b, c\}, \{d, e\}\}$

(d) $\{\{a\}, \{b\}, \{c\}, \{d\}, \{e\}\}$

(e) $\{\{a, b\}, \{b, c\}, \{d, e\}\}$

(f) $\{\{a, d, e\}, \{b, c\}\}$

(9) Let $*$ be the binary operation defined on $\mathbb{R}$ by:

$$\text{For all } x, y \in \mathbb{R}, x * y = x + y - xy.$$

(a) Is the binary operation $*$ commutative? In other words, is $x * y = y * x$ for all x and y in $\mathbb{R}$? Justify your conclusion.

(b) Is the binary operation $*$ associative? In other words, is $(x * y) * z = x * (y * z)$ for all x, y, and z in $\mathbb{R}$? Justify your conclusion.

(c) Does the binary operation $*$ have an identity? In other words, is there an element $e \in \mathbb{R}$ such that $x * e = e * x = x$ for all $x \in \mathbb{R}$? Justify your conclusion.

(d) Determine all real numbers x for which there exists a real number y such that $x * y = 0$.

(10) A relation $\sim$ on a set S is said to be **circular** provided that for all $a, b, c \in S$, if $a \sim b$ and $b \sim c$, then $c \sim a$.

(a) Suppose a relation $\sim$ on a set S is both reflexive and circular. Prove that S must also be symmetric.

(b) Use part (a) to show that if a relation $\sim$ on a set S is both reflexive and circular, then S is also transitive.

(c) Parts (a) and (b) establish one direction of the biconditional statement in the following theorem:

Theorem. A relation $\sim$ on a set S is an equivalence relation if and only if $\sim$ is reflexive and circular.

Complete the proof of this theorem by proving the converse of the statement you proved in parts (a) and (b).

(11) Decide whether each of the following statements is true or false. Prove each of your answers.

(a) $572 \in [11]_{17}$

(b) $-37 \in [7]_{10}$

(c) $[5]_7 \subseteq [10]_{14}$

(d) $[3]_8 \subseteq [3]_4$

(e) $[3]_4 \subseteq [3]_8$

(12)   (a) Determine $\gcd(112, 475)$ and find integers x and y such that $112x + 475y = \gcd(112, 475)$.

(b) In $\mathbb{Z}_{475}$, does the element $[112]$ have a multiplicative inverse? That is, does an element $[k]$ exist in $\mathbb{Z}_{145}$ such that $[112][k] = [1]$? If so, what is the multiplicative inverse of $[112]$ in $\mathbb{Z}_{475}$. Justify all conclusions.

(c) Find all solutions of the equation $[112]x + [125] = [60]$ in $\mathbb{Z}_{475}$. Explain how you know that you have found all the solutions of this equation.

(d) Find all solutions of the equation $[112]x + [125] = [60]$ in $\mathbb{Z}_{200}$. (Note: This problem is independent of the work in Parts (a), (b), and (c).)

(13) Let a and b be integers. Suppose $a \equiv 7 \pmod 9$ and $b \equiv 1 \pmod 6$. What is the remainder when $a^2 + 2b$ is divided by 3? Explain.

(14) Let $m, n \in \mathbb{N}$, and let $a \in \mathbb{Z}$. Find and prove a necessary and sufficient condition for $[a]_m \subseteq [a]_n$.

(15) Let $m, n \in \mathbb{N}$, and let $a \in \mathbb{Z}$. Show that $[a]_m \cap [a]_n$ contains infinitely many elements.

* (16) **Zero divisors in $\mathbb{Z}_n$.** Let n be a positive integer.

(a) Prove that if $\gcd(a, n) > 1$ and $n \nmid a$, then $[a]$ is a zero divisor in $\mathbb{Z}_n$.

(b) Prove that if $[a]$ is a zero divisor in $\mathbb{Z}_n$, then $\gcd(a, n) > 1$.

(c) Correctly complete the biconditional statement:

Let $n \in \mathbb{N}$, and let $[a] \in \mathbb{Z}_n$. Then $[a] \neq [0]$ is a zero divisor in $\mathbb{Z}_n$ if and only if ...

* (17) **Units in $\mathbb{Z}_n$.** Let n be a positive integer. Let $[a] \in \mathbb{Z}_n$.

(a) Let $[a] \in \mathbb{Z}_n$ with $\gcd(a, n) = 1$. Using Bezout's Identity we can find integers x and y so that $ax + ny = 1$. Explain how $[a]$ and $[x]$ are related in $\mathbb{Z}_n$.

(b) Prove that if $\gcd(a, n) = 1$, then $[a]$ is a unit in $\mathbb{Z}_n$.

(c) Prove that if $[a]$ is a unit in $\mathbb{Z}_n$, then $\gcd(a, n) = 1$. (Hint: Is the converse of Bezout's Identity ever true?)

(d) Correctly complete the biconditional statement:

Let $n \in \mathbb{N}$, and let $[a] \in \mathbb{Z}_n$. Then $[a]$ is a unit in $\mathbb{Z}_n$ if and only if ...

(18) For the given value of n and the given $[a] \in \mathbb{Z}_n$, show that $[a]$ is a unit in $\mathbb{Z}_n$ and find an element $[x]$ so that $[a][x] = [1]$.

(a) $n = 5$, $[a] = [2]$

(b) $n = 15$, $[a] = [7]$

(c) $n = 24672$, $[a] = [443]$

* (19) Let n be a positive integer. Prove that $[a]$ is a unit in $\mathbb{Z}_n$ if and only if the equation $[a]x = [b]$ has a unique solution for each $[b] \in \mathbb{Z}_n$.

(20) **Units and zero divisors.** Let $[a] \in \mathbb{Z}_n$. Determine whether each of the following statements are true or false. Verify your answers.

   (a) If $[a]$ is a zero divisor, then $[a]$ is not a unit.

   (b) If $[a]$ is not a unit, then $[a]$ is a zero divisor.

   (c) If $[a]$ is a unit, then $[a]$ is not a zero divisor.

   (d) If $[a]$ is not a zero divisor, then $[a]$ is a unit.

(21) In this exercise, we will create a different number system from $\mathbb{Z}$ using absolute value as our relation. In other words, let $\sim$ be the relation on $\mathbb{Z}$ defined by $a \sim b$ for $a, b \in \mathbb{Z}$ if $|a| = |b|$.

   (a) List all integers that are related to 0 under the relation $\sim$.

   (b) List all integers that are related to 1 under the relation $\sim$.

   (c) Show that $\sim$ is an equivalence relation.

   (d) What is the equivalence class of -3, under the relation $\sim$?

We will now attempt to define a number system based on this equivalence relation. Let S be the set of all distinct equivalence classes under the absolute value relation $\sim$. Define addition and multiplication on S the same way that we did in $\mathbb{Z}_n$:

$$[x] + [y] = [x + y] \quad \text{and} \quad [x][y] = [xy].$$

Note here that $[x]$ denotes the set of all elements equivalent to x, according to the absolute value relation $\sim$. Now let's see if these operations are well-defined.

   (e) Calculate each of the following quantities in S:
      (i) $[1] + [1]$
      (ii) $[-1] + [1]$
      (iii) $[1] + [-1]$
      (iv) $[-1] + [-1]$

   (f) If addition in S were well-defined, what would have to be true about all the calculations you just performed? What can you conclude about addition in S?

   (g) Is multiplication in S well-defined? Give a convincing argument or counterexample to justify your answer.

(22) Define the following operation on $\mathbb{Z}_n$:

$$[a] \star [b] = \begin{cases} [1] & \text{if } a \equiv b \ (\mathrm{mod}\ 5) \\ [0] & \text{if } a \not\equiv b \ (\mathrm{mod}\ 5) \end{cases}$$

   (a) Is $\star$ a well-defined operation on $\mathbb{Z}_4$?

   (b) Is $\star$ a well-defined operation on $\mathbb{Z}_5$?

Investigation 3

Algebra in Other Number Systems

Focus Questions

By the end of this investigation, you should be able to give precise and thorough answers to the questions listed below. You may want to keep these questions in mind to focus your thoughts as you complete the investigation.

- What are some familiar subsets of the real numbers, and what algebraic properties are satisfied by these various subsets?

- What are the complex numbers, and what algebraic properties do they satisfy?

- What algebraic properties are satisfied by matrix addition and multiplication? What are some common number systems that involve matrices?

- How can set operations be used to define addition and multiplication on collections of sets? What algebraic properties are satisfied by the resulting number systems?

Preview Activity 3.1. Up to this point, our study of number systems has focused mainly on systems that are related in some way to the integers. Let's now, however, shift our focus to some number systems that aren't quite as familiar. To begin, let $S = \{\gamma, \alpha, \delta, \beta\}$, and define addition and multiplication on S as shown in Table 3.1.

(a) Is addition commutative in S? Is multiplication commutative in S?

(b) Does S contain an additive identity? If so, which elements of S have an additive inverse? Find the additive inverse of each such element.

(c) Does S contain a multiplicative identity? If so, which elements of S have a multiplicative inverse? Find the multiplicative inverse of each such element.

(d) Do you think that addition is associative in S? Do you think that multiplication is associative in S?

(e) Does multiplication distribute over addition in S? Does addition distribute over multiplication? (Hint: Exactly one of these potential distributive laws holds.)

(f) What makes the questions in parts (d) and (e) harder to answer than those in parts (a) – (c)?

(g) Solve each of the following equations for x. That is, find all elements $x \in S$ for which the equation holds. Justify each of your steps by citing one or more of the axioms from parts (a) – (e) (including associativity from part (d) and the appropriate distributive law from part (e), both of which you may assume to be true).

41

Table 3.1
An unfamiliar number system.

+	γ	α	δ	β
γ	α	γ	β	δ
α	γ	α	δ	β
δ	β	δ	γ	α
β	δ	β	α	γ

$\cdot$	γ	α	δ	β
γ	α	α	γ	γ
α	α	α	α	α
δ	γ	α	β	δ
β	γ	α	δ	β

(i) $\alpha \cdot x = \alpha$

(ii) $\delta = \beta + x$

(iii) $\delta + x = x \cdot (\delta + \delta)$

(iv) $\delta \cdot x = \gamma \cdot (\beta + x)$

(v) $\gamma + x = \gamma + \beta$

(vi) $\gamma \cdot x = \gamma \cdot \beta$

Introduction

In Preview Activity 3.1, we investigated a number system that, at least on the surface, didn't seem to involve numbers at all. Instead of starting with the integers or even congruence classes of integers, we simply picked four letters from the Greek alphabet and defined, using tables, how addition and multiplication of these letters should work. Our definition of addition and multiplication was exhaustive in the sense that it specified exactly what the sum and product of every pair of elements of $S = \{\gamma, \alpha, \delta, \beta\}$ should be. This information allowed us to determine, or at least attempt to determine, which arithmetic axioms were satisfied by S.

Some of these axioms were easier to observe than others. For instance, the fact that S is closed under addition and multiplication can be ascertained by simply noticing that every entry in the addition and multiplication tables is an element of S. The fact that addition and multiplication are commutative in S is implied by the diagonal symmetry exhibited by each table.

Associativity, on the other hand, requires more than a simple observation. This is primarily due to the fact that the definition of associativity involves three elements instead of just two. The same could be said of the fact that multiplication distributes over addition. Each of these properties would require numerous cases to prove; in particular, every ordered triple of elements of S (for instance, (α, β, γ)) would need to be considered. This work is tedious, but not difficult. Thus, we will skip over the details and simply state the conclusion—that both addition and multiplication are associative in S, and multiplication does indeed distribute over addition.

These properties are important and very useful, just as they were in the integers. For instance, to solve the equation

$$\delta \cdot (x + \beta) = x \cdot \gamma,$$

which is similar to one from part (g) of Preview Activity 3.1, we must take advantage of many of the axioms that we were able to use in the integers. In the solution that follows, see if you can identify the axiom or axioms being used in each of the steps marked with the ⑦ symbol:

$$\delta \cdot (x + \beta) = x \cdot \gamma$$
$$\delta \cdot x + \delta \cdot \beta = x \cdot \gamma \;⑦$$
$$\delta \cdot x + \delta = \gamma \cdot x \;⑦$$
$$(\delta \cdot x + \delta) + \beta = \gamma \cdot x + \beta$$
$$\delta \cdot x + (\delta + \beta) = \gamma \cdot x + \beta \;⑦$$
$$\delta \cdot x + \alpha = \gamma \cdot x + \beta$$
$$\delta \cdot x = \gamma \cdot x + \beta \;⑦$$
$$\gamma \cdot x + \delta \cdot x = \gamma \cdot x + (\gamma \cdot x + \beta)$$
$$(\gamma + \delta) \cdot x = (\gamma \cdot x + \gamma \cdot x) + \beta \;⑦$$
$$\beta \cdot x = (\gamma + \gamma) \cdot x + \beta \;⑦$$
$$x = \alpha \cdot x + \beta \;⑦$$
$$x = \alpha + \beta$$
$$x = \beta$$

A few comments about this solution are in order. First, note that we obtained the solution by using the standard technique that is taught in almost every high school algebra course. In particular, we started with the equation we were trying to solve, and we carried out a sequence of simplifications, stopping when we had isolated x on one side of the equation. When we use this technique, we are essentially working backward. That is, we are *assuming* that there is some solution to our equation, and then finding out what that solution would have to be. What holds this kind of argument together is the fact that each of our steps is reversible. In other words, if we had wanted to, we could have started with the fact that $x = \beta$ and shown that $\delta \cdot (x + \beta) = x \cdot \gamma$ by simply carrying out the above steps in reverse order. Of course, if we knew, or even suspected, that $x = \beta$ was a solution, then we wouldn't have needed all of those steps in the first place. Instead, we could have simply substituted $x = \beta$ into the original equation and verified that the equation was indeed satisfied.

It is almost always easier to verify a solution than it is to find that solution in the first place. This is why mathematicians have developed algebraic techniques, such as the one illustrated above, for systematically reducing equations into simpler, but logically equivalent, forms. For number systems with only a few elements, these techniques may be more trouble than they are worth. In fact, in small number systems, it may be easier to solve equations by simply checking all possible solutions. For instance, to solve our equation $\delta \cdot (x + \beta) = x \cdot \gamma$ in S, we could have just made a table to compare both sides of the equation for each value of $x \in S$:

x	$\delta \cdot (x + \beta)$	$x \cdot \gamma$
γ	β	α
α	δ	α
δ	α	γ
β	γ	γ

This table clearly demonstrates that $x = \beta$ is the only solution to our equation. Of course, for number systems with more elements, such a table would be difficult, if not impossible, to make. This fact underscores the limitations of "guess-and-check" methods, and it suggests that the algebraic techniques we used earlier are worth further investigation.

With that said, our main goal in this investigation will be to compare and contrast the way algebra works in a variety of different number systems, some familiar and some not. Our work will be focused on the axioms we have studied in previous investigations and the theorems that follow from them. As we will see, these axioms are the glue that holds algebra together, not only in the integers, but in many other contexts as well. When certain axioms are satisfied, algebra works exactly the way we would expect it to. But when these axioms fail to hold, even algebraic manipulations that seem trivial may not be valid.

Subsets of the Real Numbers

The first number system we will consider is the set of real numbers, denoted $\mathbb{R}$, with addition and multiplication defined in the usual way. We will refrain from giving a formal definition of $\mathbb{R}$, mainly because it is surprisingly difficult to do so without developing a more sophisticated framework that is better suited for a course on advanced calculus or real analysis. * We can, however, define certain subsets of the real numbers in a slightly more precise fashion. For instance, we have already given definitions of the natural numbers ($\mathbb{N}$), the whole numbers ($\mathbb{W}$), the integers ($\mathbb{Z}$), and the even integers ($\mathbb{E}$). The *rational numbers* can be defined as follows:

Definition 3.2. A **rational number** is a real number that can be expressed as the quotient of two integers a and b, with $b \neq 0$.

Thus, the set of all rational numbers, denoted $\mathbb{Q}$,[†] is defined to be

$$\mathbb{Q} = \left\{ \frac{a}{b} : a \in \mathbb{Z}, b \in \mathbb{Z}, \text{ and } b \neq 0 \right\}.$$

Two rational numbers $\frac{a}{b}$ and $\frac{c}{d}$ are considered equal if and only if

$$ad = bc.$$

Furthermore, addition and multiplication within $\mathbb{Q}$ are defined by

$$\frac{a}{b} + \frac{c}{d} = \frac{ad + bc}{bd}$$

and

$$\frac{a}{b} \cdot \frac{c}{d} = \frac{ac}{bd}$$

for all $\frac{a}{b}, \frac{c}{d} \in \mathbb{Q}$.

The rationals are a familiar number system and you are probably comfortable with the fact that a rational number can be represented in many different ways—for example, $\frac{1}{2} = \frac{2}{4} = \frac{-5}{-10}$. This

---

*There are two main approaches to formally defining the real numbers: one involving sequences of rational numbers called *Cauchy sequences*, and one involving sets of rational numbers called *Dedekind cuts*.

[†]The symbol $\mathbb{Q}$ for the rational numbers comes from the German word *Quotient*, as rational numbers are quotients of integers. This symbol appeared in Bourbaki's *Algèbre*, Chapter 1.

should prompt you to ask whether the operations we have defined on $\mathbb{Q}$ are well-defined. You will show that they are in Activity 3.10.

Using the definitions of addition and multiplication in $\mathbb{Q}$, it is easy to show that the rational numbers satisfy many of the same axioms and properties as the integers. For instance, consider the following proof that the rational numbers are closed under addition:

Proof. Let $\frac{a}{b}$, $\frac{c}{d} \in \mathbb{Q}$. Then a, b, c, and d are all integers, with $b \neq 0$ and $d \neq 0$. Because the integers are closed under addition and multiplication, both $ad + bc$ and bd are integers. Furthermore, Theorem 1.4 implies that $bd \neq 0$. Thus, it follows from Definition 3.2 that

$$\frac{a}{b} + \frac{c}{d} = \frac{ad + bc}{bd} \in \mathbb{Q},$$

as desired. ∎

Notice that this proof uses facts that we have already assumed or proved for $\mathbb{Z}$ in order to establish a related result for $\mathbb{Q}$. Similar arguments can be used to prove several other properties of addition and multiplication in $\mathbb{Q}$, including closure under multiplication, associativity of addition and multiplication, commutativity of addition and multiplication, and distribution of multiplication over addition.

Activity 3.3.

(a) Does $\mathbb{Q}$ contain an additive identity? If so, what is it?

(b) Does every element of $\mathbb{Q}$ have an additive inverse?

(c) Does $\mathbb{Q}$ contain a multiplicative identity? If so, what is it?

(d) Which elements of $\mathbb{Q}$ are units—that is, which elements of $\mathbb{Q}$ have a multiplicative inverse in $\mathbb{Q}$? Are there elements of $\mathbb{Q}$ (besides 0) that do not have a multiplicative inverse in $\mathbb{Q}$?

(e) Does $x \cdot 0 = 0$ for all $x \in \mathbb{Q}$? (This property is sometimes called the *zero property of multiplication*.)

(f) Does $\mathbb{Q}$ contain any *zero divisors*? That is, do there exist nonzero elements x, $y \in \mathbb{Q}$ such that $xy = 0$?

(g) Does *additive cancellation* hold in $\mathbb{Q}$? That is, if

$$x + z = y + z$$

for some x, y, $z \in \mathbb{Q}$, does it follow that $x = y$?

(h) Does *multiplicative cancellation* hold in $\mathbb{Q}$? That is, if

$$x \cdot z = y \cdot z$$

for some x, y, $z \in \mathbb{Q}$, does it follow that $x = y$? Would your answer change if z was assumed to be nonzero?

(i) Is there a natural ordering defined on the elements of $\mathbb{Q}$? If so, does this ordering satisfy the axioms listed on page 6?

In Activity 3.3, we proved a variety of important properties of the rational numbers. But what about the *irrational numbers*—that is, the real numbers that are not rational?

There is no standard symbol for denoting the set of irrational numbers, but for convenience here, we will use the symbol $\mathbb{J}$, so that $\mathbb{R} = \mathbb{Q} \cup \mathbb{J}$. If we were to go through the same list of properties that we proved for $\mathbb{Q}$ and try to prove them for $\mathbb{J}$ instead, we wouldn't get very far before we ran into trouble. In fact, the very first property we considered, closure under addition, does not hold in $\mathbb{J}$. To see this, note that if $x \in \mathbb{J}$, then $-x \in \mathbb{J}$ also. (The proof of this simple fact is left to you in Exercise (1).) But then $x + (-x) = 0 \notin \mathbb{J}$. Note that the same issue would arise if we investigated the odd integers, say $\mathbb{O}$, in more detail.

It's hard to imagine doing any kind of meaningful arithmetic or algebra in a number system that isn't even closed under its operations. For this reason, we will restrict our attention to number systems that are closed under both addition and multiplication. As we just saw, this restriction rules out both $\mathbb{J}$ and $\mathbb{O}$. Note that if we also required closure under subtraction, then we would have to eliminate both $\mathbb{N}$ and $\mathbb{W}$ from consideration as well.

The Complex Numbers

Having considered several of the most familiar subsets of the real numbers, let's now turn our attention to the complex numbers, defined formally below.

Definition 3.4. A **complex number** is any number of the form $a + bi$, where a and b are real numbers and i is an imaginary number with the property that $i^2 = -1$. For a complex number $x = a + bi$, the real number a is called the **real part** of x, and the real number b is called the **imaginary part** of x.

The set of all complex numbers is denoted $\mathbb{C}$, so that

$$\mathbb{C} = \{a + bi : a \in \mathbb{R}, b \in \mathbb{R}, \text{ and } i^2 = -1\}.$$

Two complex numbers, $a + bi$ and $c + di$, are considered equal if and only if both their real and imaginary parts are equal. In other words,

$$a + bi = c + di \text{ if and only if } a = c \text{ and } b = d.$$

Furthermore, addition and multiplication within $\mathbb{C}$ are defined by

$$(a + bi) + (c + di) = (a + c) + (b + d)i$$

and

$$(a + bi) \cdot (c + di) = (ac - bd) + (ad + bc)i$$

for all $a + bi, c + di \in \mathbb{C}$.

Note that since every real number x can be written in the form $x + 0i$, it follows that the real numbers are a subset of the complex numbers. Not surprisingly, the complex numbers satisfy many of the same properties as the real numbers, and most of these properties are fairly easy to prove. For instance, to show that addition is commutative in $\mathbb{C}$, it suffices to observe that

$$
\begin{aligned}
(a + bi) + (c + di) &= (a + c) + (b + d)i \; ^{①} \\
&= (c + a) + (d + b)i \; ^{②} \\
&= (c + di) + (a + bi) \; ^{③}
\end{aligned}
$$

for all $a + bi$, $c + di \in \mathbb{C}$. Notice that, like some of the other arguments we have seen, this one used the fact that addition is commutative in $\mathbb{R}$ to establish an analogous result for $\mathbb{C}$. Many other properties of $\mathbb{C}$ can be proved in a similar manner, but some require more work.

Take, for example, the existence of multiplicative inverses within $\mathbb{C}$. How can we determine which elements of $\mathbb{C}$ are units? Although it is easy to show that $1 + 0i$ is a multiplicative identity for $\mathbb{C}$, it is more difficult to find the multiplicative inverse of an arbitrary nonzero element of $\mathbb{C}$. To illustrate, let $a + bi \in \mathbb{C}$. At first glance, we may be tempted to say that since

$$(a + bi) \cdot \frac{1}{a + bi} = 1 = 1 + 0i,$$

the multiplicative inverse of $a + bi$ is just $\frac{1}{a+bi}$. But is $\frac{1}{a+bi}$ even an element of $\mathbb{C}$? It certainly isn't written in the form $c + di$ for some real numbers c and d. In fact, it's not even clear what exactly $\frac{1}{a+bi}$ would represent in light of Definition 3.4.

The real question then is this: Can we find real numbers c and d such that

$$(a + bi) \cdot (c + di) = 1 + 0i?$$

In other words, can we find real numbers c and d such that

$$ac - bd = 1$$

and

$$ad + bc = 0?$$

Activity 3.5.

(a) Viewing a and b as constants, solve the following system of equations for c and d:

$$ac - bd = 1$$
$$ad + bc = 0$$

(b) Verify that, for the values of c and d you found in part (a),

$$(a + bi) \cdot (c + di) = 1 + 0i.$$

Activity 3.5 demonstrates that every nonzero element of $\mathbb{C}$ has a multiplicative inverse. We will have a chance to consider several other important properties of $\mathbb{C}$ later on in the investigation. For now, however, let's shift gears and briefly consider two other important types of number systems.

Matrices

Given a number system, say $\mathbb{R}$, we can define a new number system, denoted $\mathcal{M}_{n \times n}(\mathbb{R})$, that consists of all $n \times n$ matrices whose entries are elements of $\mathbb{R}$, with addition and multiplication of matrices defined in the usual way. As with the complex numbers, some properties of $\mathcal{M}_{n \times n}(\mathbb{R})$ are easy to establish, and others are more difficult, mainly because the calculations involved can become fairly tedious. As an example of one of the more straightforward properties, let's try to find an additive inverse of an arbitrary element of $\mathcal{M}_{2 \times 2}(\mathbb{R})$. Note that for any $a, b, c, d \in R$,

$$\begin{bmatrix} a & b \\ c & d \end{bmatrix} + \begin{bmatrix} -a & -b \\ -c & -d \end{bmatrix} = \begin{bmatrix} a + (-a) & b + (-b) \\ c + (-c) & d + (-d) \end{bmatrix} = \begin{bmatrix} 0 & 0 \\ 0 & 0 \end{bmatrix}.$$

Since $\begin{bmatrix} 0 & 0 \\ 0 & 0 \end{bmatrix}$ is the additive identity in $\mathcal{M}_{2\times2}(\mathbb{R})$, it follows that $\begin{bmatrix} -a & -b \\ -c & -d \end{bmatrix}$ is an additive inverse of $\begin{bmatrix} a & b \\ c & d \end{bmatrix}$.

Like $\mathbb{Q}$ and $\mathbb{C}$, the set $\mathcal{M}_{n\times n}(\mathbb{R})$ has a multiplicative identity (the familiar identity matrix). Unlike $\mathbb{Q}$ and $\mathbb{C}$, however, not every element of $\mathcal{M}_{n\times n}(\mathbb{R})$ is a unit—a fact that we should recall from linear algebra. In fact, the units in $\mathcal{M}_{n\times n}(\mathbb{R})$—that is, the elements of $\mathcal{M}_{n\times n}(\mathbb{R})$ that have a multiplicative inverse within $\mathcal{M}_{n\times n}(\mathbb{R})$—are just the invertible matrices.

Collections of Sets

Most of the number systems we have looked at so far have involved operations that are at least in some way related to addition and multiplication of real numbers. The last type of number system we will consider is like $\mathbb{Z}_n$ in the sense that its elements are sets. Unlike $\mathbb{Z}_n$, however, addition and multiplication in this new number system are defined in terms of standard set operations such as unions and intersections. We define these operations, and the corresponding family of number systems, as follows:

Definition 3.6.

- Let S be a set. The **power set** of S, denoted $\mathcal{P}(S)$, is the collection of all subsets of S. That is,
$$\mathcal{P}(S) = \{T : T \subseteq S\}.$$

- For any sets A and B, the **symmetric difference** of A and B, denoted $A \triangle B$, is the set of all elements that belong to either A or B, but not both. That is,
$$A \triangle B = \{x : x \in A \cup B \text{ and } x \notin A \cap B\}.$$

- For any natural number n, the number system $\mathcal{P}_n$ is the power set of the set $\{1, 2, \ldots, n\}$, with addition defined as symmetric difference, and multiplication defined as intersection. In other words,
$$\mathcal{P}_n = \mathcal{P}(\{1, 2, \ldots, n\}),$$
with
$$A + B = A \triangle B \quad \text{and} \quad A \cdot B = A \cap B$$
for all $A, B \in \mathcal{P}_n$.

Note that we have used symmetric differences, rather than just unions, to define addition in $\mathcal{P}_n$. This is because defining addition as set union turns out to be problematic. (Exercise (7) explores why this is the case.)

To become more familiar with $\mathcal{P}_n$, consider the following addition and multiplication tables for $\mathcal{P}_2$.

+	$\emptyset$	$\{1\}$	$\{2\}$	$\{1,2\}$
$\emptyset$	$\emptyset$	$\{1\}$	$\{2\}$	$\{1,2\}$
$\{1\}$	$\{1\}$	$\emptyset$	$\{1,2\}$	$\{2\}$
$\{2\}$	$\{2\}$	$\{1,2\}$	$\emptyset$	$\{1\}$
$\{1,2\}$	$\{1,2\}$	$\{2\}$	$\{1\}$	$\emptyset$

$\cdot$	$\emptyset$	$\{1\}$	$\{2\}$	$\{1,2\}$
$\emptyset$	$\emptyset$	$\emptyset$	$\emptyset$	$\emptyset$
$\{1\}$	$\emptyset$	$\{1\}$	$\emptyset$	$\{1\}$
$\{2\}$	$\emptyset$	$\emptyset$	$\{2\}$	$\{2\}$
$\{1,2\}$	$\emptyset$	$\{1\}$	$\{2\}$	$\{1,2\}$

From these tables, we can observe many important properties of $\mathcal{P}_2$. For instance, it is easy to see that both addition and multiplication are commutative in $\mathcal{P}_2$. To prove a property like this in general—that is, for $\mathcal{P}_n$ and not just for $\mathcal{P}_2$—we would need to construct a more general set theoretic argument. To illustrate what such an argument might look like, let's consider a property that is somewhat more involved than commutativity. In particular, we will show that addition is associative in $\mathcal{P}_n$. Throughout the proof, we have left out some details for you to fill in as you read.

Proof. Let $A, B, C \in \mathcal{P}_n$. We will argue that $(A \triangle B) \triangle C \subseteq A \triangle (B \triangle C)$ and $A \triangle (B \triangle C) \subseteq (A \triangle B) \triangle C$.[1]

For the forward inclusion, let $x \in (A \triangle B) \triangle C$. Then either $x \in A \triangle B$ and $x \notin C$, or _____ and _____. Consider these two cases:

Case 1: $x \in A \triangle B$ *and* $x \notin C$. In this case, $x \in A$ and $x \notin B$, or _____ and _____. If $x \in A$ and $x \notin B$, then since _____ also, it follows that $x \notin B \triangle C$. Thus, $x \in A \triangle (B \triangle C)$.[2] In the case that $x \in B$ and $x \notin A$, it must be that $x \in B \triangle C$,[3] which again implies that $x \in A \triangle (B \triangle C)$,[4] as desired.

Case 2: $x \in C$ *and* $x \notin A \triangle B$. Since $x \notin A \triangle B$, it follows that either _____ and _____, or _____ and _____. For the former, since $x \notin B$ and $x \in C$, it must be that $x \in B \triangle C$. But _____, and so it follows that $x \in A \triangle (B \triangle C)$. For the latter, note that since $x \in B$ and $x \in C$, it is again the case $x \notin B \triangle C$.[5] Thus, $x \in A \triangle (B \triangle C)$ in this case as well.

The above two cases establish that $(A \triangle B) \triangle C \subseteq A \triangle (B \triangle C)$. A similar argument establishes the reverse inclusion. Thus, $(A \triangle B) \triangle C = A \triangle (B \triangle C)$, as desired. ∎

While formal proofs like the one above are necessary in order to prove conjectures involving set operations, making those conjectures in the first place often requires the use of less formal exploratory tools, such as Venn diagrams. For instance, the Venn diagrams in Figure 3.1 illustrate the associative property that we just proved. You may find Venn diagrams such as these to be helpful as you explore other properties of $\mathcal{P}_n$ in the next section.

Putting It All Together

In this investigation, we have introduced several new number systems, and we have identified some of the algebraic properties that each of these systems satisfy. Table 3.2 lists the number systems and

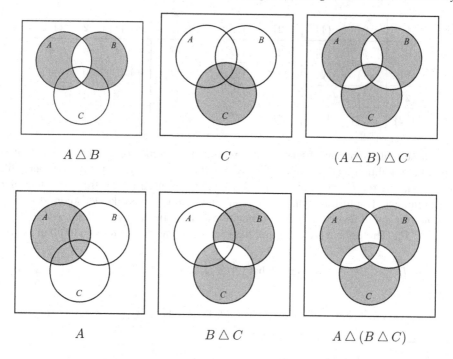

Figure 3.1
Venn diagrams illustrating $(A \triangle B) \triangle C = A \triangle (B \triangle C)$.

properties we have considered. To summarize our work thus far, fill in this table, using the following codes to indicate whether each number system satisfies the given properties:

- **Y**, to indicate that you have proved the property or are confident that it holds.

- **N**, to indicate that the property does not hold, and you have found a counterexample to illustrate its failure.

- **Y/N**, to indicate, for $\mathbb{Z}_n$, $\mathcal{M}_{n \times n}(\mathbb{R})$, or $\mathcal{P}_n$, that the property holds for some values of n but not for others. In this case, you should be able to give at least one value of n for which the property holds and at least one for which it does not. For an extra challenge, try to give a precise description of all values of n for which the property does hold.

- **?**, to indicate that you have doubts about whether the property holds, but you have not been able to find a counterexample to disprove it.

- **N/A**, to indicate that the property does not apply. (For instance, it wouldn't make sense to talk about multiplicative inverses in a number system that does not have a multiplicative identity.)

For the purposes of this activity, it is not necessary to prove every property that you suspect is true, but you should be able to find counterexamples for those that are not true. When the variable n is used, you may assume that $n > 1$. Furthermore, you may assume that the variable p (as in $\mathbb{Z}_p$) represents a prime number.

Table 3.2

Properties of various number systems.

Property	$\mathbb{Z}$	$\mathbb{E}$	$\mathbb{Q}$	$\mathbb{R}$	$\mathbb{C}$	$\mathbb{Z}_n$	$\mathbb{Z}_p$	$\mathcal{M}_{n \times n}(\mathbb{R})$	$\mathcal{P}_n$
The system is closed under addition.									
The system is closed under multiplication.									
Addition is associative.									
Multiplication is associative.									
Addition is commutative.									
Multiplication is commutative.									
Multiplication distributes over addition.									
There is an additive identity.									
Every element has an additive inverse.									
There is a multiplicative identity.									
Every nonzero element has a multiplicative inverse.									
The zero property of multiplication holds.									
There are no zero divisors.									
Additive cancellation holds.									
Multiplicative cancellation (of nonzero elements) holds.									
The ordering axioms hold.									

Concluding Activities

Activity 3.7. Suppose that you were asked to solve the equation

$$x^2 + [3]x = [4]$$

in $\mathbb{Z}_6$.

(a) Would the following solution be correct?

$$x^2 + [3]x = [4]$$
$$x^2 + [3]x + [2] = [0]$$

$$(x + [1])(x + [2]) = [0]$$

Therefore, $x = [5]$ or $x = [4]$.

If so, explain the logic behind each step. Otherwise, explain why the solution is not correct.

(b) Evaluate $x^2 + [3]x$ for every $x \in \mathbb{Z}_6$. Are your findings consistent with your answer for part (a)?

(c) Look back to part (g) of Preview Activity 3.1. Does $\gamma \cdot x = \gamma \cdot \beta$ imply $x = \beta$? Why or why not?

Activity 3.8. When we discussed number systems whose elements were matrices, we considered only square ($n \times n$) matrices. Why is this? Would it be possible to define a meaningful number system using nonsquare matrices? Why or why not?

Activity 3.9. Prove that for all $n \in \mathbb{N}$ and $a \in \mathbb{Z}$, the following statements are equivalent:

(i) $\gcd(a, n) = 1$.

(ii) $[a]$ has a multiplicative inverse in $\mathbb{Z}_n$.

(iii) The function $g : \mathbb{Z}_n \to \mathbb{Z}_n$ defined by $g([x]) = [ax]$ is injective.

Activity 3.10. The rational numbers can be defined formally as equivalence classes of ordered pairs of integers. In particular, let

$$Q = \mathbb{Z} \times (\mathbb{Z} - \{0\}) = \{(a, b) \in \mathbb{Z} \times \mathbb{Z} : b \neq 0\},$$

and define the following equivalence relation on Q:

$$(a, b) \sim (x, y) \text{ if and only if } ay = bx$$

Let $\mathbb{Q}$ be the set of all equivalence classes of this relation, and define addition and multiplication on $\mathbb{Q}$ as follows:

$$[(a, b)] + [(c, d)] = [(ad + bc, bd)] \text{ and } [(a, b)][(c, d)] = [(ac, bd)]$$

Using only these definitions and the axioms of the integers, prove that addition and multiplication are well-defined binary operations on $\mathbb{Q}$. Then give two compelling mathematical reasons why we do NOT define the sum of $\frac{a}{b}$ and $\frac{c}{d}$ in $\mathbb{Q}$ to be $\frac{a+c}{b+d}$.

Exercises

* (1) Let $\mathbb{J}$ denote the set of all irrational numbers. Prove that if $x \in \mathbb{J}$, then $-x \in \mathbb{J}$ also.

* (2) A **Gaussian integer** is a complex number whose real and imaginary parts are both integers. The set of Gaussian integers is usually denoted $\mathbb{Z}[i]$, so that

$$\mathbb{Z}[i] = \{a + bi : a \in \mathbb{Z}, b \in \mathbb{Z}, \text{ and } i^2 = -1\}.$$

Which of the properties in Table 3.2 are satisfied by $\mathbb{Z}[i]$? Use what you know about $\mathbb{C}$ to give a proof or counterexample for each property.

* (3) Let $n \geq 2$ be an integer. Which properties from Table 3.2 are satisfied by each of the following number systems? Give a proof or counterexample for each property, possibly using properties that we have already stated or proved for $\mathcal{M}_{n \times n}(\mathbb{R})$.

  (a) $\mathcal{M}_{n \times n}(\{0, 1\})$

  (b) $\mathcal{M}_{n \times n}(\mathbb{Z})$

  (c) $\mathcal{M}_{n \times n}(\mathbb{E})$

  (d) $\mathcal{M}_{n \times n}(\mathbb{Z}_5)$

(4) Let $m, n \in \mathbb{N}$, and let $A, B \in \mathcal{M}_{m \times n}(\mathbb{R})$. The **Hadamard product** of A and B is the matrix $A \cdot B$ whose (i, j) entry is equal to the product of the (i, j) entries of A and B. That is,

$$(A \cdot B)_{i,j} = A_{i,j} \cdot B_{i,j}.$$

With multiplication defined by the Hadamard product, and addition defined as usual, which of the properties from Table 3.2 are satisfied by $\mathcal{M}_{m \times n}(\mathbb{R})$? Give a proof or counterexample for each property.

(5) For any subset S of $\mathbb{R}$, let S^c denote the complement of S in $\mathbb{R}$; that is,

$$S^c = \{x \in \mathbb{R} : x \notin S\}.$$

Assuming that the ordering axioms hold in $\mathbb{R}$ just as they do in $\mathbb{Z}$, answer each of the following questions:

  (a) Does there exist a nonempty, proper subset S of $\mathbb{R}$ such that both S and S^c are closed under addition? Prove your answer.

  (b) Does there exist a nonempty, proper subset S of $\mathbb{R}$ such that both S and S^c are closed under multiplication? Prove your answer.

(6) Find all the units in $\mathcal{P}_n$. Verify your answer.

* (7) Suppose that addition in $\mathcal{P}_n$ was defined to be set union instead of symmetric difference. Which properties from Table 3.2 would be satisfied in this case?

(8) Let n be a positive integer

  (a) Is the identity matrix I_n a unit in $\mathcal{M}_{n \times n}(\mathbb{R})$? Explain.

  (b) Assume A and B are units in $\mathcal{M}_{n \times n}(\mathbb{R})$. Is the product AB a unit in $\mathcal{M}_{n \times n}(\mathbb{R})$? Prove your answer. (Be careful; is matrix multiplication commutative?)

(9) Assume that $*$ is a binary operation on a set S. Suppose that a, b, and c are elements of S with $a = b$. Explain why $a * c = b * c$. (Hint: Think of $*$ as a function.)

(10) Which of the following operations are well-defined? Verify your answers.

  (a) The operation $\odot$ defined on $\mathbb{Q}$ by $\dfrac{a}{b} \odot \dfrac{c}{d} = \dfrac{adbc}{b+d}$.

  (b) The operation $*$ defined on $\mathbb{Z}_7$ by $[a] * [b] = [2a + 3b]$.

  (c) The operation $\star$ defined on $\mathcal{P}_n$ by

$$\{a_1, a_2, \ldots, a_j\} \star \{b_1, b_2, \ldots, b_k\} = \{a_1, b_1\}.$$

(11) Define a **blip** to be a pair of integers, denoted $\langle a, b \rangle$, and define two blips $\langle a, b \rangle$ and $\langle x, y \rangle$ to be equal whenever $a + b = x + y$ (so that, for instance, $\langle 3, 5 \rangle$ and $\langle 10, -2 \rangle$ would be considered equal since $3 + 5 = 8 = 10 + (-2)$). Using these definitions, decide whether each of the following are well-defined operations on the set B of all blips. Give a proof or counterexample to justify each of your answers.

(a) The operation $*$ defined by $\langle a, b \rangle * \langle c, d \rangle = \langle a + c, b + d \rangle$.

(b) The operation $\bullet$ defined by $\langle a, b \rangle \bullet \langle c, d \rangle = \langle a^2 + c^2, b^2 + d^2 \rangle$.

Part II

Rings

Investigation 4

Introduction to Rings

Focus Questions

By the end of this investigation, you should be able to give precise and thorough answers to the questions listed below. You may want to keep these questions in mind to focus your thoughts as you complete the investigation.

- What is a ring, and what are some special types of rings?

- What uniqueness properties are satisfied by identities and inverses in rings, and why do these properties hold?

- What is a field? What is an integral domain? What is the relationship between fields and integral domains, and why does this relationship hold?

Preview Activity 4.1. In Investigation 3, we considered a variety of different number systems, some familiar and some not. While there were significant differences between these number systems, there were also common features that seemed to be shared by all of them. In this investigation, we will focus on these common features and their implications. With that in mind, look back at Table 3.2 (on page 51), and make a list of the properties that were satisfied by *all* number systems from Investigation 3. Include in your list the properties that you were able to prove as well as those that you suspected were true but were not able to prove.

Introduction

In Preview Activity 4.1, we identified a set of properties that seemed to be satisfied by a variety of different number systems, such as $\mathbb{Z}$, $\mathbb{E}$, $\mathbb{Q}$, $\mathbb{R}$, $\mathbb{C}$, $\mathbb{Z}_n$, $\mathcal{M}_{n \times n}(\mathbb{R})$, and $\mathcal{P}_n$. All of these number systems are examples of a special type of algebraic structure known as a *ring*.[*] The properties that they have in common can be used to define rings in general, as follows:

---

[*]The phrase *number ring* (from the German *Zahlring*), shortened to *ring*, is due to Hilbert in his *Zahlbericht*. Hilbert used this term to refer to certain collections of algebraic integers. Some speculate that Hilbert used the word ring because of the cyclical (ring-shaped) behavior of powers of certain algebraic integers.

Definition 4.2. A **ring** is a set R together with two binary operations, called addition $(+)$ and multiplication $(\cdot)$, such that the following axioms hold:

The Ring Axioms

- **The set R is closed under addition and multiplication**, meaning that for all $x, y \in R$, $x + y \in R$ and $x \cdot y \in R$.

- **Addition is associative**, meaning that for all $x, y, z \in R$, $(x + y) + z = x + (y + z)$.

- **Addition is commutative**, meaning that for all $x, y \in R$, $x + y = y + x$.

- **The set R contains an additive identity**, also called a **zero element**, meaning that there exists some element $0_R \in R$ such that $x + 0_R = x$ for all $x \in R$.

- **Every element of R has an additive inverse within R**, meaning that for every $x \in R$, there exists $y \in R$ such that $x + y = 0_R$.

- **Multiplication is associative**, meaning that for all $x, y, z \in R$, $(x \cdot y) \cdot z = x \cdot (y \cdot z)$.

- **Multiplication distributes over addition**, meaning that for all $x, y, z \in R$, $x \cdot (y + z) = x \cdot y + x \cdot z$ and $(x + y) \cdot z = x \cdot z + y \cdot z$.

Consistent with the usual convention, we will often omit the symbol for multiplication $(\cdot)$, writing xy instead of $x \cdot y$. When the context is clear, we will also omit the subscript from our notation for the zero element, writing 0 instead of 0_R. There are some situations in which we would want to include this subscript—for instance, to differentiate between the zero elements in two different rings, or to distinguish the zero element in a ring from the integer 0. Most of the time, however, we will be able to use 0 in lieu of 0_R without any confusion or ambiguity.

Definition 4.2 raises a number of questions that will form the basis of our work in this investigation. These questions include:

- Why are the ring axioms listed the only ones included in the definition of a ring? What other properties are satisfied by all rings, and how can these properties be proved from the ring axioms?

- What are the differences between addition and multiplication in rings, and what special types of rings can be defined in light of these differences?

Before we try to answer these questions, a quick reminder is in order. Recall that in Investigation 2, we defined a binary operation on a set S to be a function from $S \times S$ to S. Because the codomain of a binary operation on a set S is always S itself, closure is guaranteed whenever we have a well-defined operation. Thus, our first ring axiom is not entirely necessary; we include it in the definition to remind us more explicitly that every ring must be closed under its operations. Indeed, if this were not the case, then the ring operations themselves would not be true binary operations in the sense of Definition 2.14. (See page 31.)

Basic Properties of Rings

The ring axioms themselves should seem quite familiar, since we have used them both implicitly and explicitly throughout previous investigations. But what about the other properties we have studied? You may have noticed that at least a few of these properties (including some that were satisfied by all the number systems from Investigation 3) were not included in Definition 4.2. Do these properties in fact hold for all rings, and if so, why are they not included in our list of ring axioms?

To answer this question, we must think back to our discussion in Investigation 1 regarding the difference between axioms and theorems. There are certain properties, such as additive cancellation and the zero property of multiplication, that are in fact satisfied by all rings. These properties, however, can be proved from the ring axioms. Thus, to include them would cause our axiom system to be redundant. The primary benefit of our definition of a ring is that it provides a minimal set of axioms from which numerous other algebraic properties and theorems can be proved. Moreover, any property that we can prove using only the ring axioms must necessarily hold in *every* number system that satisfies these axioms. Thus, the theory of rings gives us a way to study algebra more abstractly, instead of just within the context of specific number systems. In fact, the entire field of abstract algebra revolves around the study of general algebraic structures, such as rings, and their applications.

With that said, let's now formally state and prove the two properties we just mentioned (additive cancellation and the zero property of multiplication). The integer versions of each of these properties were part of Investigation 1—as Activity 1.2 (see page 6). As it turns out, the arguments we used there generalize easily to the context of arbitrary rings.

Theorem 4.3. *Let R be a ring. For all $x, y, z \in R$, if $x + z = y + z$, then $x = y$.*

Activity 4.4. Prove Theorem 4.3.

Theorem 4.5. *Let R be a ring. Then $0x = 0 = x0$ for all $x \in R$.*

Activity 4.6. Prove Theorem 4.5. (Hint: Start with the fact that $0_R + 0_R = 0_R$, and multiply both sides of this equation by an arbitrary element $x \in R$.)

Commutative Rings and Rings with Identity

In the definition of a ring, you may have noticed similarities between the axioms for addition and the axioms for multiplication. For instance, both addition and multiplication satisfy the closure axiom, and both are associative. With that said, there are also some significant differences between the two operations. One of these differences is that multiplication is not required to be commutative. This observation leads to the following definition:

Definition 4.7. Let R be a ring. Then R is said to be **commutative** if multiplication in R is commutative—that is, if $xy = yx$ for all $x, y \in R$.

Notice that when we call a ring *commutative*, we are always referring to the multiplication operation. This is because addition is guaranteed to be commutative in *every* ring. Thus, the important distinguishing feature with regard to commutativity is whether or not multiplication commutes.

A similar distinction can be made with regard to additive and multiplicative identities. In particular, we call an additive identity (which every ring must have) a *zero element*, and we refer to a multiplicative identity (which a ring may or may not have) as simply an identity. The next definition formalizes this language.

Definition 4.8. Let R be a ring. An **identity** for R is an element $1_R \in R$ such that $1_R \neq 0_R$ and $1_R \cdot x = x = x \cdot 1_R$ for all $x \in R$. If such an element exists, then R is said to be a **ring with identity**.

Note that, unlike the definition of a zero element, it is necessary to specify in the definition of an identity that both $1_R \cdot x$ and $x \cdot 1_R$ are equal to x. This is again due to the fact that addition is commutative in every ring, but multiplication may not be. Note also that, as with zero elements, we will often omit the subscript from our notation for an identity, writing 1 instead of 1_R whenever the context is clear.

We have seen many examples of commutative rings, including $\mathbb{Z}$, $\mathbb{E}$, $\mathbb{Q}$, $\mathbb{R}$, $\mathbb{C}$, $\mathbb{Z}_n$, and $\mathcal{P}_n$. In fact, the only family of noncommutative rings we have studied is $\mathcal{M}_{n \times n}(\mathbb{R})$. Of the commutative rings mentioned above, all but $\mathbb{E}$ have identity. The noncommutative ring $\mathcal{M}_{n \times n}(\mathbb{R})$ also has identity. Thus, the rings we have studied so far can be divided into the following categories:

- **Commutative rings with identity:** $\mathbb{Z}, \mathbb{Q}, \mathbb{R}, \mathbb{C}, \mathbb{Z}_n, \mathcal{P}_n$

- **Commutative rings without identity:** $\mathbb{E}$

- **Noncommutative rings with identity:** $\mathcal{M}_{n \times n}(\mathbb{R})$

Activity 4.9. Find a noncommutative ring without identity, or prove that no such ring exists.

Uniqueness of Identities and Inverses

As we suggested in Investigation 1, nothing in the definition of a zero element or a multiplicative identity requires uniqueness. That is, the definitions alone do not rule out the possibility of a ring having more than one zero element or identity. Fortunately, however, uniqueness does hold for both zero elements and multiplicative identities, as stated in the theorems below.

Theorem 4.10. *Let R be a ring, and suppose that both 0 and $0'$ are zero elements for R. Then $0 = 0'$.*

Theorem 4.11. *Let R be a ring, and suppose that both 1 and $1'$ are identities for R. Then $1 = 1'$.*

Activity 4.12. Let R be a ring, and suppose that 0 and $0'$ are zero elements for R.

(a) Let a be any element of R. What must $a + 0$ and $a + 0'$ equal, and why?

(b) Use your answer to part (a) to equate $a + 0$ and $a + 0'$.

(c) What axiom or theorem, along with your answer to part (b), allows you to conclude that $0 = 0'$?

(d) Combine your work in parts (a) – (c) to write a proof of Theorem 4.10.

(e) Explain why the strategy from parts (a) – (c) would be invalid for a proof of Theorem 4.11. (Hint: See part (c).)

(f) Prove Theorem 4.11 by evaluating $1 \cdot 1'$ in two different ways.

Because of Theorems 4.10 and 4.11, we can now refer to *the* zero element and *the* identity (when it exists) of a ring. Similar uniqueness results can be established for both additive and multiplicative inverses, the latter of which is defined formally below.

Definition 4.13. Let R be a ring with identity, and let $x \in R$. An element $y \in R$ is said to be a **multiplicative inverse** of x provided that $xy = 1 = yx$.

It is important to note that the definition of a ring does not require any particular ring element to actually have a multiplicative inverse. In fact, the very notion of a multiplicative inverse makes sense only in rings with identity, as suggested by Definition 4.13. Furthermore, we have seen several examples of rings with identity—for instance, $\mathcal{M}_{n \times n}(\mathbb{R})$, $\mathcal{P}_n$, and $\mathbb{Z}_n$ for certain values of n—in which numerous ring elements do not have a multiplicative inverse. In Investigation 2, we called elements of $\mathbb{Z}_n$ that do have a multiplicative inverse *units*. This definition generalizes to arbitrary rings as follows:

Definition 4.14. Let R be a ring with identity. An element $x \in R$ is said to be a **unit** provided that R contains a multiplicative inverse for x. In other words, $x \in R$ is a unit if and only if there exists $y \in R$ such that $xy = 1 = yx$.

The next two theorems show that both additive and multiplicative inverses (when they exist) must be unique. Their proofs use ideas similar to those in Activity 4.12. As you read these proofs, you should be able to fill in the missing details and provide additional explanation or justification wherever it is needed.

Theorem 4.15. *Let R be a ring, and let $x \in R$. Suppose that both y and y' are additive inverses for x. Then $y = y'$.*

Proof. Let R be a ring, let $x \in R$, and let y and y' be additive inverses for x. Then $x + y =$ ___ and $x + y' =$ ___. Thus, $x + y = x + y'$, and _____ implies that $y = y'$, as desired. ∎

Theorem 4.16. *Let R be a ring with identity, and let $x \in R$ be a unit. Suppose that both y and y' are multiplicative inverses for x. Then $y = y'$.*

Proof. Let R be a ring with identity, let $x \in R$ be a unit, and let y and y' be multiplicative inverses for x. Then $xy = 1 = xy'$, which implies that

$$y(xy) = y(xy').^{①}$$

But then

$$(yx)y = (yx)y',^{②}$$

which implies that $y = y',^{③}$ as desired. ∎

Now that we have shown that additive and multiplicative inverses must be unique, we can use the standard notations of $-x$ (for the additive inverse of a ring element x) and x^{-1} (for the multiplicative inverse, if one exists) without any risk of ambiguity. We will use these notations regularly throughout subsequent investigations. We will also define subtraction within an arbitrary ring R just as we did in $\mathbb{Z}$—that is,

$$x - y = x + (-y)$$

for all $x, y \in R$.

Zero Divisors and Multiplicative Cancellation

In Investigation 3, you should have noticed that in some number systems, multiplicative cancellation, even of nonzero elements, does not hold. For instance, consider the following equation in $\mathbb{Z}_6$:

$$[2]x = [2]y$$

If we were to naively apply the same cancellation rules that apply in $\mathbb{Z}$ (and many other rings), we might conclude that $x = y$. This conclusion, however, would not be valid, since

$$[2][2] = [4] = [10] = [2][5],$$

but clearly $[2] \neq [5]$.

So under what circumstances does multiplicative cancellation hold? One obvious answer to this question is when the element being canceled is a unit. In this case, the proof of Theorem 4.3 generalizes easily to multiplication, yielding the following result:

Theorem 4.17. *Let R be a ring with identity, and let z be a unit in R. For all $x, y \in R$, if $xz = yz$, then $x = y$. Similarly, if $zx = zy$, then $x = y$.*

Theorem 4.17 provides a sufficient condition for multiplicative cancellation to hold, but this condition is far from being necessary. For instance, the only units in $\mathbb{Z}$ are 1 and -1, but multiplicative cancellation works in $\mathbb{Z}$ as long as the integer being canceled is nonzero. The same can be said of $\mathbb{E}$, which has no identity and therefore no units. In these rings, the validity of multiplicative cancellation rests on the fact that the elements being canceled are not *zero divisors*, defined formally as follows:

Definition 4.18. *Let R be a ring. An element $x \in R$ is said to be a **zero divisor** if $x \neq 0$, and $xy = 0$ or $yx = 0$ for some nonzero $y \in R$.*

We first studied zero divisors, and their relation to multiplicative cancellation, all the way back in Investigation 1. (See Activity 1.3 on page 7.) We revisited these ideas again in Investigations 5 and 6 (Activities 2.18, 2.22, and 3.7 on pages 33, 34, and 51, respectively), where we investigated zero divisors within the context of $\mathbb{Z}_n$ and another number system. The ideas from these activities can be generalized to arbitrary rings, as stated in the next two theorems. The proof of the first theorem is left as an exercise, and the proof of the second should mirror your work in part (c) of Activity 1.3.

Theorem 4.19. *Let R be a ring. The following statements are equivalent:*

- *R contains no zero divisors.*

- *For all $x, y \in R$, if $xy = 0$, then $x = 0$ or $y = 0$.*

- *For all $x, y \in R$, if $xy = 0$ and $x \neq 0$, then $y = 0$.*

- *For all x, $y \in R$, if $yx = 0$ and $x \neq 0$, then $y = 0$.*

Theorem 4.20. *Let R be a ring, and let z be a nonzero element of R that is not a zero divisor. For all x, $y \in R$, if $xz = yz$, then $x = y$. Similarly, if $zx = zy$, then $x = y$.*

Proof. Let R be a ring, let z be a nonzero element of R that is not a zero divisor, and let x, $y \in R$. Suppose that $xz = yz$. Then

$$xz + (-y)z = yz + (-y)z,$$

which implies that

$$(x + (-y))z = (y + (-y))z \,^{\oslash}$$
$$= 0z \,^{\oslash}$$
$$= 0. \,^{\oslash}$$

Since $(x + (-y))z = 0$, and since z is neither zero nor a zero divisor, it must be the case that $x + (-y) = 0.\,^{\oslash}$ Thus,

$$(x + (-y)) + y = 0 + y,$$

which implies that

$$x + (y + (-y)) = y. \,^{\oslash}$$

It then follows that

$$x + 0 = y, \,^{\oslash}$$

and so $x = y,\,^{\oslash}$ as desired. A similar argument establishes the result when $zx = zy$. ∎

As it turns out, Theorem 4.17 is actually implied by Theorem 4.20. This is because of the following result, which states that a unit can never be a zero divisor:

Theorem 4.21. *Let R be a ring with identity, and let $x \in R$ be a unit. Then x is not a zero divisor. That is, if $xy = 0$ or $yx = 0$ for some $y \in R$, then $y = 0$.*

Proof. Let R be a ring with identity, and let $x \in R$ be a unit. Then $x \neq 0$. Thus, suppose that $xy = 0$. Then

$$x^{-1}(xy) = x^{-1}(0),$$

which implies that

$$(x^{-1}x)y = 0, \,^{\oslash}$$

so $y = 0$. A similar argument shows that $y = 0$ in the case that $yx = 0$. Thus, x is not a zero divisor. ∎

The statement of Theorem 4.21, and the subsequent proof, relies on a particularly useful form of the negation of the definition of zero divisor. In particular, $x \in R$ is *not* a zero divisor if and only if either (i) $x = 0$, or (ii) $xy \neq 0$ and $yx \neq 0$ for all nonzero $y \in R$. Equivalently, x is not a zero divisor if and only if the following statement is true:

If $x \neq 0$, and $xy = 0$ or $yx = 0$ for some $y \in R$, then $y = 0$.

Thus, to show that a nonzero element $x \in R$ is *not* a zero divisor, we can assume that $xy = 0$ for some $y \in R$, and show that $y = 0$. If R is noncommutative, then we must also show that $y = 0$ whenever $yx = 0$. This is exactly the strategy we used in the proof above, and the equivalence it relies upon is quite similar to the ones asserted by Theorem 4.19.

Fields and Integral Domains

Having investigated the relationships between zero divisors, units, and the property of multiplicative cancellation, we are now ready to define and give examples of two very important types of rings.

Definition 4.22. An **integral domain** is a commutative ring with identity that contains no zero divisors.

Definition 4.23. A **field** is a commutative ring with identity in which every nonzero element has a multiplicative inverse. [†]

Notice that fields and integral domains are similar in many ways. Both are commutative. Both have identity. And, by Theorems 4.17 and 4.20, both satisfy the property of multiplicative cancellation. The following result, which follows from Theorem 4.21 and some of the examples we have already considered, establishes the relationship between fields and integral domains.

Corollary 4.24. *Every field is an integral domain, but not every integral domain is a field.*

We have already investigated several examples of fields (for instance, $\mathbb{Q}$, $\mathbb{R}$, and $\mathbb{C}$) and integral domains (for instance, $\mathbb{Z}$, which is an integral domain but not a field). We will consider numerous other examples in future investigations.

Concluding Activities

Activity 4.25. As noted earlier, subtraction in arbitrary rings can be defined in the same way we defined it for $\mathbb{Z}$. Using such a definition when needed, prove the following theorem:

> **Theorem 4.26.** *Let R be a ring, and let x, y, $z \in R$. Then:*
>
> (i) $-(-x) = x$
>
> (ii) $x(-y) = -(xy) = (-x)y$
>
> (iii) $(-x)(-y) = xy$
>
> (iv) $x(y - z) = xy - xz$
>
> (v) $(y - z)x = yx - zx$
>
> (vi) $-(x + y) = -x - y$

Activity 4.27. Let $\mathbb{Z}^\star$ be the number system consisting of the set of all integers, with addition ($\oplus$) and multiplication ($\otimes$) defined as follows:

$$x \oplus y = x + y - 1 \quad \text{and} \quad x \otimes y = x + y - x \cdot y$$

Note that $+$ and $\cdot$ denote the normal operations of addition and multiplication in $\mathbb{Z}$. Which of the ring axioms are satisfied by $\mathbb{Z}^\star$, and which are not? Is $\mathbb{Z}^\star$ a ring? If so, is $\mathbb{Z}^\star$ commutative? Does $\mathbb{Z}^\star$ have an identity? Prove your answers.

[†]Dedekind introduced the term *field* (from the German *Köper*, or body).

Activity 4.28.

(a) Find all values of n for which $\mathbb{Z}_n$ is a field. State (and prove) your answer in the form of an if and only if statement. (Hint: See Activity 2.20 on page 34.)

(b) How, if at all, would your answer to part (a) change if you were instead asked to find the values of n for which $\mathbb{Z}_n$ is an integral domain?

Exercises

(1) Let R be a ring. Suppose that, due to a printer error, the addition and multiplication tables for R were printed with several entries missing, as shown below:

+	a	b	c
a	a	b	c
b		c	
c		a	

×	a	b	c
a			
b		a	
c			

Using only the ring axioms, complete the tables. Explain how each entry you add can be determined from the ring axioms and the entries already completed.

(2) Let R be a ring. Suppose that, due to a printer error, the addition and multiplication tables for R were printed with several entries missing, as shown below:

+	a	b	c
a	a	b	c
b		c	
c		a	

×	a	b	c
a			
b			
c			b

Using only the ring axioms, complete the tables. Explain how each entry you add can be determined from the ring axioms and the entries already completed.

(3) Let R be the number system consisting of the set of all integers, with addition ($\oplus$) and multiplication ($\otimes$) defined as follows:

$$x \oplus y = x \cdot y \quad \text{and} \quad x \otimes y = x + y$$

Note that $+$ and $\cdot$ denote the normal operations of addition and multiplication in $\mathbb{Z}$. Which of the ring axioms are satisfied by R, and which are not? Is R a ring? If so, is R commutative? Does R have an identity? Prove your answers.

(4) Let $\mathbb{R}^+$ denote the set of all positive real numbers. For all $x, y \in \mathbb{R}^+$, define

$$x \oplus y = xy \quad \text{and} \quad x \otimes y = x^{\log y}.$$

(a) With these operations, does $\mathbb{R}^+$ have an additive identity? If so, what is it?

(b) Does $\mathbb{R}^+$ have a multiplicative identity? If so, what is it?

(c) Is $\mathbb{R}^+$ a ring with the operations defined above? Prove your answer.

(5) For this problem, we will be working with equations of the form $[a]x = [b]$ in the ring of integers modulo n, $\mathbb{Z}_n$ with $[a] \neq [0]$.

(a) In $\mathbb{Z}_{12}$, find all solutions of each of the following equations:

(i) $[6]x = [9]$ (iii) $[11]x = [5]$ (v) $[8]x = [3]$
(ii) $[6]x = [10]$ (iv) $[8]x = [4]$ (vi) $[7]x = [1]$

(b) Let n be a natural number and let a and b be integers with $[a] \neq [0]$ in $\mathbb{Z}_n$, and let $d = \gcd(a, n)$. Assume that d does not divide b. Determine the number of solutions of the equation $[a]x = [b]$ in $\mathbb{Z}_n$. State a proposition and prove it.

(6) (a) If $[a] \in \mathbb{Z}_8$, then what are the possible values for $[a]^2$ in $\mathbb{Z}_8$? State a proposition and prove it.

(b) An integer a is said to be the **sum of two squares** provided that there exist positive integers b and c such that $a = b^2 + c^2$. For example,

- The integer 0 is the sum of two squares since $0 = 0^2 + 0^2$, and the integer 5 is the sum of two squares since $5 = 1^2 + 2^2$.

- The integer 937 is the sum of two squares since $937 = 19^2 + 24^2$.

  (i) Explain why 11 is not a sum of squares.

  (ii) If $[b]$ and $[c]$ are in $\mathbb{Z}_8$, then what are the possible values of $\left([b]^2 + [c]^2\right)$ in $\mathbb{Z}_8$? State a proposition and prove it.

(c) Write a proposition (and prove it) about integers that are the sum of two squares. Your proposition must be written in the form

> For each integer a, if a is the sum of two squares, then *"your conclusion."*

The conclusion that you use should state something about what the integer a can be congruent to modulo 8.

(d) Is the integer $513\,281\,799\,218\,322\,423$ the sum of two squares? Justify your conclusion.

★ (7) Let n be a nonnegative integer, and let $n\mathbb{Z} = \{nx : x \in \mathbb{Z}\}$, with addition and multiplication defined as in $\mathbb{Z}$. Is $n\mathbb{Z}$ a ring? If so, is $n\mathbb{Z}$ commutative? Does $n\mathbb{Z}$ have an identity? Does your answer depend on the value of n? Explain.

(8) Let $\mathbb{Z}(\sqrt{2}) = \{a + b\sqrt{2} : a, b \in \mathbb{Z}\}$. Is $\mathbb{Z}(\sqrt{2})$ a ring? If so, is $\mathbb{Z}(\sqrt{2})$ commutative? Does $\mathbb{Z}(\sqrt{2})$ have an identity? Verify your answers.

(9) Let n and k be natural numbers, both greater than 1.

(a) How many elements does $M_{n \times n}(\mathbb{Z}_k)$ have?

(b) Is $M_{n \times n}(\mathbb{Z}_k)$ a ring? If so, is $M_{n \times n}(\mathbb{Z}_k)$ commutative? Does $M_{n \times n}(\mathbb{Z}_k)$ have an identity? Verify your answers.

(10) For all elements $x_1, x_2, \ldots, x_n$ in a ring R, we can define $x_1 + x_2 + \cdots + x_n$ recursively as follows:

$$x_1 + x_2 + \cdots + x_n = (x_1 + x_2 + \cdots + x_{n-1}) + x_n$$

Use this definition, along with mathematical induction, to prove the following **generalized distributive laws**:

Theorem 4.29 (Generalized Distributive Laws). *Let R be a ring, and let a, b_1, b_2, $\ldots$, $b_n \in R$. Then*

(i) $a(b_1 + b_2 + \cdots + b_n) = ab_1 + ab_2 + \cdots ab_n$.

(ii) $(b_1 + b_2 + \cdots + b_n)a = b_1a + b_2a + \cdots b_na$.

(11) Let R be a ring with at least two elements. Prove that $\mathcal{M}_{2\times 2}(R)$ is always a ring (with addition and multiplication of matrices defined as usual).

(12) **Rings of functions.** Let $\mathcal{F}(\mathbb{R})$ denote the set of all functions from $\mathbb{R}$ to $\mathbb{R}$. Define addition and multiplication on $\mathcal{F}(\mathbb{R})$ as follows:

  - For all $f, g \in \mathcal{F}(\mathbb{R})$, $(f + g) : \mathbb{R} \to \mathbb{R}$ is the function defined by
    $$(f + g)(x) = f(x) + g(x)$$
    for all $x \in \mathbb{R}$.
  - For all $f, g \in \mathcal{F}(\mathbb{R})$, $(fg) : \mathbb{R} \to \mathbb{R}$ is the function defined by
    $$(fg)(x) = f(x)g(x)$$
    for all $x \in \mathbb{R}$.

  Prove that $\mathcal{F}(\mathbb{R})$ is a commutative ring with identity.

(13) A **Boolean ring** R is one in which $x^2 = x$ for all $x \in R$.

  (a) Prove that in a Boolean ring, every element is its own additive inverse. Deduce that in a Boolean ring, addition and subtraction are the same. (Hint: Square a convenient element of R.)

  (b) Prove that every Boolean ring is commutative. (Hint: Square *another* convenient element of R. You may want to eventually use part (a).)

(14) Let R be a ring, and suppose there exists a positive even integer n such that $x^n = x$ for all $x \in R$. Prove that $-x = x$ for all $x \in R$.

(15) Let n and k be positive integers, both greater than 1. State and prove a necessary and sufficient condition for a matrix A to be a unit in $M_{n\times n}(\mathbb{Z}_k)$.

(16) Let R be a ring with identity, and let x and y be units in R. Prove or disprove each of the following statements:

  (a) $x + y$ is a unit in R.

  (b) xy is a unit in R.

  (c) Let z be any element of R such that $xz = 1$. Then z is a unit in R.

(17) Let R be a ring with identity, and let $x, y \in R$. Prove or disprove: If xy is a unit in R, then both x and y are units in R.

(18) (a) For which values of n is $\mathcal{M}_{n\times n}(\mathbb{R})$ a field? An integral domain? Find all such values, and prove your answer.

  (b) For which values of n is $\mathcal{P}_n$ a field? An integral domain? Find all such values, and prove your answer.

(19) **Finite integral domains and fields.** Prove that if R is a finite ring with identity, then every nonzero element of R is either a zero divisor or a unit. Deduce that every finite integral domain is a field. (Hint: Let x be a nonzero element of R that is not a zero divisor. Show that $x^n = 1$ for some $n \in \mathbb{N}$, and deduce from this that x must be a unit.)

(20) Prove that if R is a finite ring that contains at least one nonzero element that is not a zero divisor, then R has identity. Use this result to state a stronger version of the result from Exercise (19).

Connections

This investigation introduced the concept of a ring. Rings are algebraic objects that share the same basic structure as the integers that we saw in Investigation 1 and the different number systems discussed in Investigation 3. There is a great deal of power to be had in recognizing the features these number systems have in common and then creating a larger category (rings) that encapsulates all of these features. By doing so, we can learn about all of these number systems at one time by studying arbitrary rings.

If you studied group theory before ring theory, you should notice connections between the topics in this investigation and those in Investigation 17. Rings and groups are both algebraic objects—that is, sets on which an operation or operations are defined, yielding a particular algebraic structure. The main difference between a ring and a group is that a ring comes with two binary operations and a group comes with only one. In fact, every ring is a group under its operation of addition, but not every group is a ring.

From a structural standpoint, rings may be more familiar to you than groups in that many of the sets with which you have worked in your mathematical past (e.g., $\mathbb{Z}$, $\mathbb{Q}$, $\mathbb{R}$, and sets of polynomials) are all rings. For this reason, starting our exploration of modern algebra with rings is a reasonable choice. On the other hand, since there is only one operation in a group, groups are simpler objects than rings and for that reason a good argument can be made that the study of modern algebra should begin with groups. In either case, many of the concepts we will encounter in these investigations will apply to both rings and groups.

Investigation 5

Integer Multiples and Exponents

Focus Questions

By the end of this investigation, you should be able to give precise and thorough answers to the questions listed below. You may want to keep these questions in mind to focus your thoughts as you complete the investigation.

- How can integer multiplication be defined in an arbitrary ring, and what properties are satisfied by integer multiplication in rings?

- How can integer exponentiation be defined in an arbitrary ring? What properties are satisfied by integer exponentiation in rings, and what special considerations must be taken into account when using nonpositive exponents?

- What is the characteristic of a ring? What are some examples of rings with characteristic zero? What are some examples of rings with nonzero characteristic?

Preview Activity 5.1. Our study of rings began in Investigation 1, where we learned about the integers and their various axioms. Our next example of a ring was $\mathbb{Z}_n$, a set of equivalence classes of integers. As you might suspect from these two examples, the integers play an important role in the general theory of rings. In fact, even in rings whose elements are not integers, it is possible to define notions of integer multiplication and integer exponentiation. In other words, it is possible to multiply and exponentiate ring elements by integers, even though the ring elements themselves may not be integers. In fact, it turns out that integer multiplication and exponentiation work exactly the way we would expect them to. To see this, use your intuition to calculate as many of the quantities listed below as you can. For those that you are not able to calculate, explain why. Throughout your calculations, you will be applying the definitions that we will formally develop in this investigation.

In $\mathbb{Z}_6$:	In $\mathcal{M}_{2\times2}(\mathbb{R})$:	In $\mathcal{P}_3$:
$0[4]$	$0\begin{bmatrix} 1 & \pi \\ -\pi & 3 \end{bmatrix}$	$0\{1,2\}$
$3[5]$	$3\begin{bmatrix} e^2 & e^3 \\ e^3 & e^4 \end{bmatrix}$	$3\{2\}$
$(-4)[2]$	$(-4)\begin{bmatrix} 1 & 1+\sqrt{2} \\ 1-\sqrt{2} & -1 \end{bmatrix}$	$(-4)\{2,3\}$
$[3]^4$	$\begin{bmatrix} \pi & 0 \\ 0 & -\pi \end{bmatrix}^4$	$\{3\}^4$
$[2]^0$	$\begin{bmatrix} 1 & 1+\sqrt{2} \\ 1-\sqrt{2} & -1 \end{bmatrix}^0$	$\{1,3\}^0$
$[3]^{-5}$	$\begin{bmatrix} e^2 & e^3 \\ e^3 & e^4 \end{bmatrix}^{-3}$	$\{1,2,3\}^{-3}$
$[5]^{-3}$	$\begin{bmatrix} \pi & 0 \\ 0 & -\pi \end{bmatrix}^{-5}$	$\{1\}^{-5}$

Introduction

In Preview Activity 5.1, we began to intuitively develop the notions of integer multiplication and exponentiation for arbitrary rings. You may have performed the requested calculations by simply thinking of multiplication as repeated addition, and exponentiation as repeated multiplication. For instance, in $\mathcal{P}_3$, we can calculate $3\{1,2\}$ as follows:

$$3\{1,2\} = \{1,2\} + \{1,2\} + \{1,2\}$$
$$= (\{1,2\} + \{1,2\}) + \{1,2\}$$
$$= \emptyset + \{1,2\}$$
$$= \{1,2\}.$$

Likewise,

$$\{1,2\}^3 = \{1,2\}\{1,2\}\{1,2\}$$
$$= (\{1,2\}\{1,2\})\{1,2\}$$
$$= \{1,2\}\{1,2\}$$
$$= \{1,2\}.$$

This intuitive formulation of integer multiplication and exponentiation makes sense as long as we are multiplying or exponentiating by a positive integer. For nonpositive integers, however, we will need to be a bit more careful. Furthermore, in order to prove that integer multiplication and exponentiation work the way we would expect them to, we will need to make use of a more formal definition.

We will develop such a definition in the next section, and we will use this definition to prove several fundamental properties of integer multiplication and exponentiation. We will then use integer multiplication to define the *characteristic* of a ring, which, intuitively speaking, measures the extent to which the addition operation exhibits cyclic behavior, as in $\mathbb{Z}_n$.

Integer Multiplication and Exponentiation

In the previous section, we observed that multiplication and exponentiation by a positive integer could be defined in terms of repeated addition or multiplication, respectively. Note, however, that in order to actually carry out such a repeated operation, we must first parenthesize the corresponding expression so that we will only be dealing with two ring elements at a time. Addition and multiplication are, after all, *binary* operations. Thus, expressions of the form

$$x_1 + x_2 + \cdots + x_n$$

or

$$x_1 x_2 \cdots x_n$$

only make sense if we define them in terms of a sequence of binary additions or multiplications.

As an example, suppose you were asked to calculate $37 + 63 + 29$. You might begin by adding 37 and 63 to get 100, and then add 29 to 100 to yield a final answer of 129. In other words, you would perform a sequence of two binary additions, effectively carrying out the computation as follows:

$$37 + 63 + 29 = (37 + 63) + 29 = 100 + 29 = 129.$$

The next definition formalizes this type of process for calculating sums of three or more ring elements.

Definition 5.2. Let R be a ring, let $n \geq 3$ be an integer, and let $x_1, x_2, \ldots, x_n \in R$. Then we define

$$x_1 + x_2 + \cdots + x_n = (x_1 + x_2 + \cdots + x_{n-1}) + x_n$$

and

$$x_1 x_2 \cdots x_n = (x_1 x_2 \cdots x_{n-1}) x_n.$$

Note that Definition 5.2 is what mathematicians typically call a *recursive* definition. In essence, this means that the definition refers to a simpler case of itself and would typically need to be applied repeatedly in order to actually yield a final answer. For instance, to calculate

$$x_1 + x_2 + x_3 + x_4 + x_5,$$

one must first calculate

$$x_1 + x_2 + x_3 + x_4,$$

which requires

$$x_1 + x_2 + x_3,$$

and so on.

We can formally define multiplication and exponentiation by a positive integer in a similar manner:

Definition 5.3. Let R be a ring, and let $x \in R$. Then the expressions $1x$ and x^1 are both defined to be equal to x; that is,

$$1x = x \text{ and } x^1 = x.$$

Furthermore, for every integer $n \geq 2$, we define the expressions nx and x^n recursively as follows:

- $nx = \underbrace{x + x + \cdots + x}_{n \text{ terms}} = \underbrace{(x + x + \cdots + x)}_{n-1 \text{ terms}} + x = (n-1)x + x$

- $x^n = \underbrace{x \cdot x \cdots x}_{n \text{ factors}} = \underbrace{(x \cdot x \cdots x)}_{n-1 \text{ factors}} x = x^{n-1}x$

Nonpositive Multiples and Exponents

Definition 5.3 is quite natural, but unfortunately it applies only to multiplication or exponentiation by a *positive* integer. To extend the definition to nonpositive multiples and exponents, let's begin by making a few observations, some of which you may have noted in your answers to Preview Activity 5.1.

First, we would like multiplication by an integer to possess the same properties as other notions of multiplication (such as multiplication of ring elements). For instance, if n is a positive integer and x is an element of a ring R, then we would expect

$$-(nx) = (-n)x = n(-x),$$

since we stated a similar property for multiplication of ring elements in Theorem 4.26. (See page 64.) Likewise, it would make sense for the integer 0 times x to be the same as the zero element of R (0_R) times x. In other words, it should be the case that

$$0x = 0_R x = 0_R.$$

We would also like exponentiation by an integer in an arbitrary ring to behave similarly to exponentiation in more familiar number systems. For instance, we know that $x^0 = 1$ (the multiplicative identity) for every nonzero element x of $\mathbb{Z}$, $\mathbb{Q}$, $\mathbb{R}$, or $\mathbb{C}$. It would therefore be natural to define x^0 to be equal to 1_R in an arbitrary ring R. Of course, in order for such a definition to make sense, R would need to be a ring with identity. Likewise, if $x \in R$ and n is a positive integer, it would seem natural to define

$$x^{-n} = (x^{-1})^n.$$

This definition, however, requires R to be a ring with identity *and* x to be a unit in R.

So, to summarize, there are natural ways to extend Definition 5.3 to nonpositive multiples and exponents, provided that certain conditions are met. The next definition incorporates these conditions, which are based on the observations that we noted above.

Definition 5.4. Let R be a ring, and let n be a positive integer.

- For all $x \in R$, we define $0x = 0_R$ and $(-n)x = n(-x)$.

- If R is a ring with identity, then for each nonzero $x \in R$, we define $x^0 = 1_R$. If R does not have identity, then x^0 remains undefined.

- If R is a ring with identity, then for each unit $x \in R$, we define $x^{-n} = (x^{-1})^n$, where x^{-1} denotes the multiplicative inverse of x. If R does not have identity, or if x is not a unit in R, then x^{-n} remains undefined.

Note that if n is a negative integer, then Definition 5.4 implies that $nx = (-n)(-x)$ and $x^n = (x^{-1})^{-n}$, where $-n > 0$. This alternative definition will be particularly useful in the proofs in the next section.

Properties of Integer Multiplication and Exponentiation

Now that we have precisely defined integer multiplication and exponentiation for both positive and nonpositive integers, we are ready to state and prove the familiar properties that we commonly associate with these operations. We begin with the following theorem:

Theorem 5.5. *Let R be a ring, let x, $y \in R$, and let m and n be integers. Then*

(i) $m(x + y) = mx + my$

(ii) $-(mx) = m(-x) = (-m)x$

(iii) $(m + n)x = mx + nx$

(iv) $m(nx) = (mn)x$

(v) $m(xy) = (mx)y = x(my)$

(vi) $(mx)(ny) = (mn)(xy).$

Because the proof of Theorem 5.5 is fairly long, we will only consider parts of it here. In particular, we will explore the inductive argument behind part (i) in Activity 5.6, and we will then consider a more complete proof of part (ii). The proofs of parts (iii) – (vi) will be left as exercises for you to complete at the conclusion of the investigation.

Activity 5.6.

(a) Parts (i) and (iii) look very similar to the distributive axiom in the definition of a ring. Explain why, in spite of these similarities, these properties still must be proved. (Hint: What is the fundamental difference between these properties and the ring axioms?)

(b) We will prove part (i) of Theorem 5.5 by considering three separate cases: $m = 0$, $m > 0$, and $m < 0$. Apply Definition 5.4 to prove the first case (that is, when $m = 0$).

(c) Now let $m > 0$, and let $P(m)$ be the predicate, "$m(x + y) = mx + my$." Prove that $P(1)$ is true.

(d) Let k be a positive integer, and assume that $P(k)$ (as defined in part (c)) is true. Use this assumption, along with Definition 5.3 and one or more of the ring axioms, to show that

$$(k + 1)(x + y) = (k + 1)x + (k + 1)y.$$

(e) Deduce from parts (b) – (d) that the statement from part (i) holds for all $m \geq 0$.

(f) Now assume that m is a negative integer. Give a justification for each step in the following argument:

$$
\begin{aligned}
m(x + y) &= (-m)[-(x + y)]\;^\oslash \\
&= (-m)[(-x) + (-y)]\;^\oslash \\
&= (-m)(-x) + (-m)(-y)\;^\oslash \\
&= mx + my.\;^\oslash
\end{aligned}
$$

(g) Use your work from parts (b) – (f) to write a clear and convincing proof of part (i) of Theorem 5.5.

Now that we have proved part (i) of Theorem 5.5, we can use this result in the proofs of the remaining parts. To illustrate, try to fill in the missing details in the following proof of part (ii). In particular, make sure that you are able to identify all instances in which the proof relies on part (i).

Proof of Theorem 5.5, part (ii). Let m be an integer. We will first prove that $m0_R = 0_R$. By _____, we know that

$$
m0_R = m(0_R + 0_R) = m0_R + m0_R,
$$

which implies that $m0_R = 0_R.\;^\oslash$ Now let $x \in R$. Then

$$
mx + m(-x) \overset{\oslash}{=} m(x + (-x)) \overset{\oslash}{=} m0_R \overset{\oslash}{=} 0_R.
$$

This proves that $m(-x)$ is the additive inverse of mx; that is, $-(mx) = m(-x)$.

To complete the proof, we will now show that $m(-x) = (-m)x$. Note that if $m \geq 0$, then $-m \leq 0$, and so $(-m)x = m(-x)$ by Definition _____. If $m < 0$, then

$$
m(-x) \overset{\oslash}{=} (-m)[-(-x)] \overset{\oslash}{=} (-m)x.
$$

Thus, we have shown that for every integer m, $-(mx) = m(-x) = (-m)x$.   ∎

Theorem 5.5 asserted several familiar properties for integer multiplication within an arbitrary ring. The next theorem deals with integer exponentiation.

Theorem 5.7. *Let R be a ring, let x, $y \in R$, and let m and n be positive integers. Then:*

(i) $x^{m+n} = x^m x^n$

(ii) $(x^m)^n = x^{mn}$

If R is a ring with identity, then the above properties hold for all nonnegative integers m and n (provided that $x \neq 0$, as 0^0 remains undefined). Furthermore, if R is a ring with identity and x is a unit in R, then the above properties hold for all integers m and n.

We will prove part (i) of Theorem 5.7 and leave part (ii) as an exercise. For convenience, we will begin with the case where both m and n are positive. Our proof will employ an induction argument, but instead of using induction on m or n, we will instead induct on the sum $m + n$.

Proof of Theorem 5.7, part (i), for m, n positive. Let m and n be positive integers, and let $x \in R$. We will proceed by induction on $m + n$. For the base case, suppose $m + n = 2$. Then $m = 1$ and $n = 1,$ [⊙] and so

$$x^{m+n} = x^2 \overset{⊙}{=} x^1 x \overset{⊙}{=} x^1 x^1 = x^m x^n,$$

as desired. Now suppose that for some integer $k \geq 2$,

$$x^{m+n} = x^m x^n$$

whenever $m + n = k$. Let m' and n' be positive integers such that $m' + n' = k + 1$. Then

$$x^{m'+n'} = x^{(m'+n'-1)+1} \; ⊙$$
$$= x^{(m'+n'-1)} x^1 \; ⊙$$
$$= (x^{m'} x^{n'-1}) x \; ⊙$$
$$= x^{m'} (x^{n'-1} x) \; ⊙$$
$$= x^{m'} x^{n'}. \; ⊙$$

This completes the induction argument, and so the result holds for all positive integers m and n.   ■

The preceding proof established part (i) of Theorem 5.7 for the case in which m and n are positive. It is easy to show that, in a ring with identity, this result also holds when m or n is zero. (See Exercise (2).) To show that it is true for negative values of m and/or n, however, we will need the following lemma:

Lemma 5.8. *Let R be a ring with identity, and let x be a unit in R. Then:*

(i) *The element x^{-1} is a unit in R, and $(x^{-1})^{-1} = x$.*

(ii) *For every integer n, $x^n = x \cdot x^{n-1}$.*

(iii) *For every integer n, x^n is a unit and $(x^n)^{-1} = (x^{-1})^n$.*

Although the statements in Lemma 5.8 may seem obvious, their proofs are not trivial (with the possible exception of (i)). The next activity suggests some strategies for proving these results.

Activity 5.9. Let R be a ring with identity, and let x be a unit in R.

(a) Apply the definition of unit to prove part (i) of Lemma 5.8.

(b) Explain why part (ii) is satisfied by definition for $n = 0$, $n = 1$, and $n = 2$.

(c) The following argument can be used, along with part (b), to establish part (ii) of the lemma for every nonnegative integer n. Explain what proof technique is being used and what assumptions would need to be made for this technique to be valid. Also provide a justification for each step in the argument.

$$x^n \overset{⊙}{=} x^{n-1} x \overset{⊙}{=} (x \cdot x^{n-2}) x \overset{⊙}{=} x(x^{n-1})$$

(d) The following argument can be used to prove part (ii) of Lemma 5.8 for negative values of n. Provide a justification for each step in the argument, being sure to identify explicitly where the assumption that $n < 0$ would be used.

$$x \cdot x^{n-1} = x \cdot (x^{-1})^{-n+1} \;\text{⑦}$$
$$= x \cdot (x^{-1})^{1+(-n)} \;\text{⑦}$$
$$= x \cdot [(x^{-1})^1 (x^{-1})^{-n}] \;\text{⑦}$$
$$= (x \cdot x^{-1})(x^{-1})^{-n} \;\text{⑦}$$
$$= 1_R \cdot (x^{-1})^{-n} \;\text{⑦}$$
$$= x^n \;\text{⑦}$$

(e) To prove part (iii) of Lemma 5.8, it suffices to show that

$$x^n \cdot (x^{-1})^n = (x^{-1})^n \cdot x^n = 1_R.$$

Explain why this equality holds for $n = 0$ and $n = 1$.

(f) Assume that the equality from part (e) holds for some integer $n \geq 1$. Complete the following inductive argument to establish that

$$x^n \cdot (x^{-1})^n = 1_R$$

for every positive integer n:

$$x^n \cdot (x^{-1})^n = (x^{n-1} \cdot x) \cdot [x^{-1} \cdot (x^{-1})^{n-1}] \;\text{⑦}$$
$$= \cdots$$
$$= 1_R$$

(g) An argument similar to the one you used in part (f) can be used to show that $(x^{-1})^n \cdot x^n = 1_R$ for every nonnegative integer n. Thus, all that remains to show is that

$$x^n \cdot (x^{-1})^n = (x^{-1})^n \cdot x^n = 1_R$$

for $n < 0$. Use what you have proved in the previous parts of this activity to establish this equality. You may want to begin your proof by noting that

$$x^n \cdot (x^{-1})^n = (x^{-1})^{-n} \cdot [(x^{-1})^{-1}]^{-n},$$

with $-n > 0$ whenever $n < 0$.

(h) Combine your work in parts (a) – (g) to write a clear and convincing proof of Lemma 5.8.

With Lemma 5.8 in hand, we are now ready to finish our proof of part (i) of Theorem 5.7, which entails proving the result for negative values of m and/or n. Fortunately, we can use what we have proved in the case where m and n are both positive to accomplish this goal.

Proof of Theorem 5.7, part (i), for m and/or n negative. Without loss of generality, assume that $m < 0$. We will consider three cases.

Case 1: $m + n > 0$. In this case, $-m > 0$, $n > 0$, and $-m < n$.⑦ Thus,

$$x^{m+n} = [(x^{-1})^{-m} x^{-m}] x^{m+n} \;\text{⑦}$$
$$= (x^{-1})^{-m} (x^{-m} x^{m+n}) \;\text{⑦}$$
$$= x^m x^n. \;\text{⑦}$$

Case 2: $m + n = 0$. In this case, $n = -m > 0$, and so

$$
\begin{aligned}
x^{m+n} &= 1_R \\
&= (x^{-1})^n x^n \\
&= (x^{-1})^{-m} x^n \\
&= x^m x^n.
\end{aligned}
$$

Case 3: $m + n < 0$. In this case, either $n < 0$, or $-m > n \geq 0$. For the former, note that

$$
\begin{aligned}
x^{m+n} &= (x^{-1})^{-(m+n)} \\
&= (x^{-1})^{(-m)+(-n)} \\
&= (x^{-1})^{-m}(x^{-1})^{-n} \\
&= x^m x^n.
\end{aligned}
$$

For the latter, assume that $n > 0$. (The case where $n = 0$ is treated in Exercise (2).) Then

$$
\begin{aligned}
x^{m+n} &= (x^{-1})^{-(m+n)} \\
&= (x^{-1})^{-(m+n)}[(x^{-1})^n x^n] \\
&= [(x^{-1})^{-(m+n)}(x^{-1})^n]x^n \\
&= (x^{-1})^{-m} x^n \\
&= x^m x^n.
\end{aligned}
$$

Since we have established the desired result in all three cases, our proof is complete. ∎

The Characteristic of a Ring

We will conclude this investigation by introducing a way to classify rings like $\mathbb{Z}_n$ whose addition operations are cyclic in some sense. Preview Activity 5.10 demonstrates what we mean by this and provides the motivation for the more formal treatment that follows.

Preview Activity 5.10.

(a) Suppose that a standard 12-hour clock reads 9:00. What time will it be in 6 hours?

(b) Write an equation in $\mathbb{Z}_{12}$ that represents the question from part (a) and your answer to it.

(c) Repeat parts (a) and (b), but this time add 12 hours, 18 hours, and 24 hours.

(d) Again using a 12-hour clock, describe all positive integers k that make the following statement true:

No matter what time it is currently, it will be the same time k hours from now.

(e) Write an equation in $\mathbb{Z}_{12}$ that represents the statement from part (d).

(f) For which positive integers k does $k[3] = [0]$ in $\mathbb{Z}_{12}$?

(g) For which positive integers k does $k[4] = [0]$ in $\mathbb{Z}_{12}$?

(h) For which positive integers k does $k[5] = [0]$ in $\mathbb{Z}_{12}$?

(i) For which positive integers k does $k[x] = [0]$ for *all* $[x] \in \mathbb{Z}_{12}$? What is the smallest of these values?

(j) What is the relationship between your answers to parts (d) and (i)? Why does this relationship hold?

(k) Is there a positive integer k such that $kx = 0$ for all $x \in \mathbb{Z}$? Why or why not?

The questions in Preview Activity 5.10 begin to formalize an important feature of rings like $\mathbb{Z}_n$—namely, that they behave cyclically when it comes to addition and integer multiplication. Just like on a clock, if we add the right element in $\mathbb{Z}_n$, or if we add the same element the right number of times, we'll always end up right back where we started. Figure 5.1 illustrates this cyclic behavior for $\mathbb{Z}_6$:

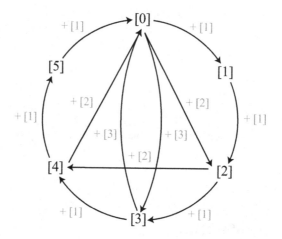

Figure 5.1
Additive cycles in $\mathbb{Z}_6$.

Notice that, for any element $[x]$ in $\mathbb{Z}_6$, if we add $[1]$ to $[x]$ exactly 6 times, the resulting sum will be equal to $[x]$. In other words,

$$[x] + [1] + [1] + [1] + [1] + [1] + [1] = [x] + 6[1] = [x].$$

This, of course, implies that $6[1] = [0]$ which should come as no surprise to us, since we know that $6[x] = [0]$ for *all* $[x] \in \mathbb{Z}_6$. In fact, the integer 6 is special in this sense. Notice that 6 is the first positive integer that yields a result of $[0]$ when multiplied by *any* element of $\mathbb{Z}_6$. Other integers may yield a similar result in some cases (for instance, $3[x] = [0]$ for $[x] = [0], [2], [4]$, but not for any other $[x] \in \mathbb{Z}_6$), but 6 is the first positive integer that does so universally. Furthermore, it is easy to show that any other integer k for which

$$k[x] = [0] \text{ for all } [x] \in \mathbb{Z}_6$$

must be an integer multiple of 6.

Putting all of these observations together, we might say that the integer 6 in some way *charac-terizes* the cyclic nature of the ring $\mathbb{Z}_6$. The next definition uses similar language to describe cyclic behavior in arbitrary rings.

Definition 5.11. Let R be a ring. The **characteristic** of R, denoted char(R), is the smallest positive integer k such that
$$kx = 0_R \text{ for all } x \in R.$$
If no such integer exists, then R is said to have **characteristic zero**.

Using the language and notation from Definition 5.11, we can now state, as we argued earlier, that char$(\mathbb{Z}_6) = 6$. In fact, our prior observations can be generalized to establish that char$(\mathbb{Z}_n) = n$ for every integer $n \geq 2$. We state this result and several others in the next theorem. The proof that follows uses a simple lemma that we have not yet stated or proved. This lemma plays an essential role in at least three different locations marked by the $?$ symbol. Pay careful attention, as you will be asked to state and prove the missing lemma in Activity 5.13.

Theorem 5.12.

- *For every integer* $n \geq 2$, char$(\mathbb{Z}_n) = n$.

- *The rings* $\mathbb{Z}$, $\mathbb{Q}$, $\mathbb{R}$, *and* $\mathbb{C}$ *all have characteristic zero.*

Proof. Let $n \geq 2$ be given. To show that char$(\mathbb{Z}_n) = n$, we must show three things:

- that n is a positive integer;

- that $n[x] = [0]$ for all $[x] \in \mathbb{Z}_n$; and

- that there is no positive integer $k < n$ for which $k[x] = [0]$ for all $[x] \in \mathbb{Z}_n$.

The first is immediate. For the second, note that for all $[x] \in \mathbb{Z}_n$,
$$n[x] \overset{?}{=} [nx] \overset{?}{=} [n][x] \overset{?}{=} [0][x] = [0].$$
Now let k be any positive integer with $k < n$. It follows that $n \nmid k$,$^?$ and so
$$k[1] = [k] \neq [0].^?$$
Therefore, char$(\mathbb{Z}_n) = n$, as desired.

To show that $\mathbb{Z}$, $\mathbb{Q}$, $\mathbb{R}$, and $\mathbb{C}$ all have characteristic zero, it suffices to show that there does not exist a positive integer k such that $k \cdot 1 = 0$.$^?$ This, however, follows immediately from the ordering axioms of the integers. $^?$ ∎

Concluding Activities

Activity 5.13. State and prove the missing lemma that is used in the proof of Theorem 5.12.

Activity 5.14. A classmate of yours claims that char$(\mathcal{P}_n) = 2$ and offers the following argument in support of their claim:

Let $S \in \mathcal{P}_n$. Then

$$2S = S + S = S \triangle S = \emptyset = 0_{\mathcal{P}_n}.$$

Thus, $\mathrm{char}(\mathcal{P}_n) = 2$.

Is your classmate's claim correct? Is their argument correct? Why or why not?

Activity 5.15. Consider the following alternative definition of the characteristic of a ring with identity.

> **Definition 5.16.** Let R be a ring with identity. Then $\mathrm{char}(R)$ is the smallest positive integer k such that $k \cdot 1_R = 0_R$. If no such integer exists, then $\mathrm{char}(R) = 0$.

Is this definition equivalent to Definition 5.11? If so, prove the equivalence. Otherwise, give an example to show that the definitions are not equivalent.

Activity 5.17. Is the following statement true for all rings, some rings, or no rings?

> Let R be a ring, and let $x, y \in R$. For every positive integer m,
> $$(xy)^m = x^m y^m.$$

Prove your answer, including a necessary and sufficient condition if you think that the result holds in only some rings.

Activity 5.18. Find all rings that have characteristic 1. Use Definition 5.11 to justify your answer.

Exercises

(1) **Cancellation of integer multiples.**

    (a) For which types of rings is the following statement true?

        If $x, y \in R$ and m is a nonzero integer, then $mx = my$ implies $x = y$.

        Find a sufficient condition for the statement to be true, and prove the resulting theorem.

    (b) Is your sufficient condition from part (a) also necessary? Explain.

* (2) Prove Theorem 5.7 for the case where at least one of m or n is zero. (You may assume that R is a ring with identity.)

(3) Prove part (ii) of Theorem 5.7.

(4) Let R be a ring with the following addition table:

+	a	b	c	d
a	a	b	c	d
b	b	a	d	c
c	c	d	a	b
d	d	c	b	a

What is the characteristic of R? Prove your answer.

(5) Let m and n be integers, both greater than 1. What is the characteristic of $\mathcal{M}_{n \times n}(\mathbb{Z}_m)$? Prove your answer.

(6) A **Boolean ring** R is one in which $x^2 = x$ for all $x \in R$. Prove that every nontrivial (i.e., not $\{0\}$) Boolean ring has characteristic 2. (Hint: See Exercise (13) from Investigation 4.)

$\star$ (7) **The characteristic of an integral domain.** Prove that the characteristic of an integral domain is either zero or prime. (Hint: Reason by contradiction.)

(8) **The characteristic of a finite ring.** Prove that a finite ring R cannot have characteristic zero. (Hint: Begin by showing that for each element $x \in R$, there must exist some positive integer k_x such that $k_x x = 0_R$.) Deduce from this and Exercise (7) that every finite integral domain (and consequently, every finite field) has prime characteristic.

(9) (a) Give an example of a commutative ring of characteristic 4 and give an example of two elements a and b in a commutative ring of characteristic 4 for which $(a+b)^4 \neq a^4 + b^4$.

   (b) Let R be a commutative ring of characteristic 4. For $a, b \in R$, use the distributive property to expand $(a+b)^4$ into a sum of terms and find a new expression that is equal to $(a+b)^4$. State a proposition and follow it with a proof.

$\star$ (10) **The Binomial Theorem.** If you have ever expanded an expression of the form $(x+y)^n$, then you probably made use of the *Binomial Theorem*. The Binomial Theorem is often stated for the real numbers, but here we will prove the following generalization for commutative rings.

> **Theorem 5.19** (Binomial Theorem for Commutative Rings). *Let R be a commutative ring, and let n be a positive integer. Then for all $x, y \in R$,*
>
> $$(x+y)^n = \sum_{k=0}^{n} \binom{n}{k} x^{n-k} y^k,$$
>
> *where*
>
> $$\binom{n}{k} = \frac{n!}{k!(n-k)!}.$$

   (a) For several values of n and k, calculate $\binom{n}{k}$, $\binom{n}{k-1}$ and $\binom{n+1}{k}$. Use your calculations to state a conjecture about the relationship between these three quantities.

   (b) Prove the conjecture you made in part (a).

   (c) Use induction to prove the Binomial Theorem for commutative rings. (Hint: You will need to use your conjecture from part (a).)

   (d) Where in your proof did you make use of the fact that R was commutative? Would your proof still be valid if R was not commutative?

$\star$ (11) **The Freshman's Dream.** How often have you been tempted to expand $(x+y)^n$ as $x^n + y^n$? Wouldn't life be so much simpler if this were true in general? In this exercise, we will state and prove a related theorem that is sometimes amusingly called the "Freshman's Dream." The theorem requires x and y to be elements of a ring with prime characteristic. (So, if you ever mistakenly simplify $(x+y)^n$ as $x^n + y^n$, you can recover from your mistake by simply saying, "Oh, I thought we were working in a ring with prime characteristic.")

(a) Let R be a commutative ring with characteristic 3, and let x, $y \in R$. Expand and simplify the expression $(x+y)^3$. Do you notice anything interesting? (Hint: Remember that 3 times any element of R is equal to 0_R.)

(b) Let R be a commutative ring with characteristic 4, and let x, $y \in R$. Expand and simplify the expression $(x+y)^4$. Compare and contrast your results to those from part (a).

(c) Prove that if p is a prime number, then $p \mid \binom{p}{k}$ for every integer k such that $1 \le k \le p - 1$.

(d) Explain why the assumption that p is prime is essential to the result in part (c). Use this insight to explain the difference between the simplifications in parts (a) and (b).

(e) Use the result from part (c), along with the Binomial Theorem (see Exercise (10)) to prove the following general version of the Freshman's Dream:

> **Theorem 5.20** (Freshman's Dream). *Let p be a prime number, and let R be a commutative ring with characteristic p. Then for all $x, y \in R$,*
>
> $$(x + y)^p = x^p + y^p.$$

(f) Explain why the assumption that p is prime is an essential part of the statement of the Freshman's Dream. Give a specific example (that is, an example from a particular ring with nonprime characteristic) to illustrate the failure of the Freshman's Dream when p is not prime.

Connections

In this investigation, we studied integer multiples and integer powers of elements in rings. If you studied group theory before ring theory, you should notice connections between the topics in this investigation and those in Investigation 18. The major difference between rings and groups in this context is that there are two operations in a ring and only one in a group. As a result, we need to understand both integer multiples (under addition) and integer powers (under multiplication) of ring elements. With only one operation in a group, we only need one of these ideas. However, we use integer multiples when we represent a group operation as addition, and we use integer powers when we write a group operation multiplicatively. Consequently, we need to understand both notations even when working with algebraic structures—such as groups—that have only one operation.

Investigation 6

Subrings, Extensions, and Direct Sums

Focus Questions

By the end of this investigation, you should be able to give precise and thorough answers to the questions listed below. You may want to keep these questions in mind to focus your thoughts as you complete the investigation.

- What is a subring? What conditions must be verified in order to show that a subset of a ring is a subring?

- In what ways are a ring and all of its subrings guaranteed to be similar? In what ways can a ring and its subrings be different?

- What is a field extension, and how can field extensions be used to construct larger rings from smaller ones?

- What is a direct sum, and how can direct sums be used to construct larger rings from smaller ones?

- How are the properties of field extensions and direct sums related to the properties of the individual rings used to construct them?

Preview Activity 6.1. Throughout mathematics, the relationship between mathematical objects and their sub-objects is of central importance. For instance, in linear algebra, we study vector spaces and their subspaces. In discrete mathematics, many graph theory problems can be solved by finding a subgraph that is optimal in some sense. Furthermore, many other applied optimization problems involve minimizing or maximizing a certain function subject to certain constraints. These constraints define what is known as a *feasible region*, which is nothing more than a subset of the space of all possible solutions.

In light of these examples and our recent investigations of rings, it seems natural that we would be interested in defining and characterizing *subrings*. To begin thinking along these lines, we will consider a set of numbers that is larger than $\mathbb{Q}$ but smaller than $\mathbb{R}$. The set, denoted $\mathbb{Q}(\sqrt{2})$, is defined as follows:

$$\mathbb{Q}(\sqrt{2}) = \{a + b\sqrt{2} : a, b \in \mathbb{Q}\}$$

(a) Show that $\mathbb{Q} \subset \mathbb{Q}(\sqrt{2}) \subset \mathbb{R}$ (that is, $\mathbb{Q}$ is a proper subset of $\mathbb{Q}(\sqrt{2})$, which is a proper subset of $\mathbb{R}$).

(b) With addition and multiplication defined as in $\mathbb{R}$, which of the ring axioms does $\mathbb{Q}(\sqrt{2})$ satisfy? Give a brief explanation to justify your answer for each axiom.

(c) Critique the following proof that multiplication distributes over addition in $\mathbb{Q}(\sqrt{2})$. Is the proof correct? If so, could it be improved in any way? If not, what is the main error in the argument?

> *Proof.* Let x, y, $z \in \mathbb{Q}(\sqrt{2})$. Since $\mathbb{Q}(\sqrt{2}) \subseteq \mathbb{R}$, it follows that x, y, $z \in \mathbb{R}$ as well. Thus,
> $$x(y+z) = xy + xz$$
> and
> $$(x+y)z = xz + yz,$$
> since multiplication distributes over addition in $\mathbb{R}$. This, however, proves that multiplication distributes over addition in $\mathbb{Q}(\sqrt{2})$. ∎

(d) Critique the following proof that $\mathbb{Q}(\sqrt{2})$ is closed under addition. Is the proof correct? If so, could it be improved in any way? If not, what is the main error in the argument?

> *Proof.* Let x, $y \in \mathbb{Q}(\sqrt{2})$. Since $\mathbb{Q}(\sqrt{2}) \subseteq \mathbb{R}$, it follows that x, $y \in \mathbb{R}$ as well. But $\mathbb{R}$ is closed under addition, so $x + y \in \mathbb{R}$. This shows that $\mathbb{Q}(\sqrt{2})$ is closed under addition. ∎

(e) Which of the ring axioms were easiest to establish for $\mathbb{Q}(\sqrt{2})$? Which required the most work?

Introduction

In Preview Activity 6.1, we considered $\mathbb{Q}(\sqrt{2})$, a subset of the real numbers that was formed, loosely speaking, by "adding" $\sqrt{2}$ (an irrational number) to $\mathbb{Q}$. As it turns out, this special set is not only a subset of $\mathbb{R}$, but a *subring* as well. The next definition formalizes this terminology.

Definition 6.2. Let R be a ring, and let S be a subset of R. Then S is said to be a **subring** of R provided that S itself is a ring with the operations of addition and multiplication defined the same as in R.

Definition 6.2 is not terribly surprising. Nevertheless, there is one important caveat to note—namely, the condition that both R and S have the same operations. Although this condition is often satisfied trivially, it is still important to verify. For instance, Activity 4.27 (see page 64) describes an alternative way to define addition and multiplication on the integers. Let $\mathbb{Z}^\star$ denote the set of integers with these new operations. Although $\mathbb{Z}^\star$ is both a ring and a subset of $\mathbb{R}$, $\mathbb{Z}^\star$ is not a subring of $\mathbb{R}$. This is because the operations of addition and multiplication are defined differently in $\mathbb{Z}^\star$ than in $\mathbb{R}$.

Putting strange examples like the previous one aside, let's assume that we are given a ring R and a subset S of R, with the operations in S defined the same way as in R. What must we verify in order to show that S is a subring of R? The most obvious answer to this question is exactly what is suggested by Definition 6.2: we must show that S satisfies all ring axioms. But is there any information we can use to simplify this task? As it turns out, there is. If we are careful, we can use the fact that S is a subset of R to automatically establish more than half of the ring axioms. This leaves only a handful of axioms to prove, and these axioms form the basis of a result known as the *Subring Test*, which we will investigate in the next section.

After we have proved the Subring Test, we will turn the question around and consider ways of constructing larger rings from smaller ones. In particular, we will see how *field extensions* and *direct sums* can be used to construct rings that contain one or more prespecified subrings.

The Subring Test

As you may have observed in Preview Activity 6.1, some of the ring axioms are satisfied almost trivially by *any* subset of a ring R. These include associativity of addition and multiplication, commutativity of addition, and distribution of multiplication over addition. The proof of the latter is given in part (c) of Preview Activity 6.1, and the others follow in a similar fashion. Notice that all four of these axioms share a common form or structure. In particular, they all assert that a particular equality holds for *all* elements in a particular set. When an axiom of this form is *assumed* for a set R, it can be automatically *inferred* for every subset of R.

But what about the remaining three axioms: closure under addition and multiplication, the existence of a zero element, and the existence of additive inverses? What do they have in common, and why do they require more attention than the others?

The answer to this question lies in one small phrase: "there exists." Notice that two of the three axioms mentioned above contain this phrase, which, as you may remember from previous courses, is called an *existential quantifier*. Note also that the condition

$$x + y \in R,$$

which appears in the closure axiom, is equivalent to the following:

There exists $z \in R$ such that $x + y = z$.

A similar equivalence holds for the statement $xy \in R$. Thus, the closure axiom can be rephrased in an equivalent form as follows:

For all $x, y \in R$, there exists $z_1, z_2 \in R$ such that $x + y = z_1$ and $xy = z_2$.

So, to summarize, of the seven ring axioms, four are satisfied automatically by any subset of a ring R. The statements of the remaining three all contain an existential quantifier, and these existential axioms (as we might call them) are what need to be established in order to show that a subset of R is in fact a subring. The following theorem states these observations more formally:

Theorem 6.3. *Let R be a ring, and let S be a subset of R. Then S is a subring of R if*

(i) *S is closed under addition;*

(ii) *S is closed under multiplication;*

(iii) *S contains 0_R (that is, $0_R \in S$); and*

(iv) *S is closed under additive inverses (that is, for every $x \in S$, $-x \in S$).*

Theorem 6.3 significantly simplifies the task of showing that a particular set is a subring. That being said, the next theorem (commonly called the Subring Test) makes the task even easier by combining closure under addition and closure under additive inverses into a single condition: closure under subtraction.

Theorem 6.4 (Subring Test). *Let R be a ring, and let S be a subset of R. Then S is a subring of R if and only if*

  (i) *S is nonempty;*

  (ii) *S is closed under multiplication; and*

  (iii) *S is closed under subtraction.*

The Subring Test allows us to prove that a subset S of a ring R is a subring of R by verifying only three essential conditions: that S is nonempty, that S is closed under multiplication, and that S is closed under subtraction. To illustrate, consider the set $\mathbb{Q}(\sqrt{2})$ introduced in Preview Activity 6.1. This set is clearly nonempty. Furthermore, for all $x = a + b\sqrt{2}$ and $y = c + d\sqrt{2}$ in $\mathbb{Q}(\sqrt{2})$, notice that

$$x - y = (a + b\sqrt{2}) - (c + d\sqrt{2}) = (a - c) + (b - d)\sqrt{2},$$

and

$$xy = (a + b\sqrt{2})(c + d\sqrt{2}) = (ac + 2bd) + (ad + bc)\sqrt{2}.$$

Since a, b, c, and d are all elements of $\mathbb{Q}$, and since $\mathbb{Q}$ is closed under addition, subtraction, and multiplication, it follows that both $x - y$ and xy are elements of $\mathbb{Q}(\sqrt{2})$. Thus, by the Subring Test, $\mathbb{Q}(\sqrt{2})$ is a subring of $\mathbb{R}$.

Now that we have seen how the Subring Test can be used, we are ready to investigate its proof, which we will present shortly. You may notice that, unlike previous proofs we have considered, we have not included any instances of the ⑦ symbol. This is because you have now gained enough experience reading proofs to be able to decide for yourself where additional details or explanations are necessary. Although we will still occasionally use the ⑦ symbol throughout the remainder of our investigations on ring theory, we will do so less frequently than in past investigations. You should still try to fill in missing details and add clarifying information to the proofs we consider, even when you are not explicitly prompted to do so. You may even want to use the ⑦ symbol as we have in the past to remind yourself where these additional details are necessary.

Proof of the Subring Test. Let R be a ring, and let S be a subset of R. If S is a subring of R, then S clearly satisfies conditions (i) – (iii) of the theorem. For the converse, suppose that S is nonempty and closed under both multiplication and subtraction. By Theorem 6.3, it suffices to show that S contains 0_R, S is closed under additive inverses, and S is closed under addition. Since S is nonempty, S contains some element, say z. But then, since S is closed under subtraction, we know that

$$0_R = z - z \in S.$$

Now let x be any element of S. Since $0_R \in S$ and S is closed under subtraction, it follows that

$$-x = 0_R - x \in S.$$

Thus, S is closed under additive inverses. To show closure under addition, let x, $y \in S$. Since we have assumed that S is closed under subtraction and proved that S is closed under additive inverses, it follows that

$$x + y = x - (-y) \in S.$$

Thus, S is closed under addition, and the proof is complete. ∎

Now that we have proved the Subring Test, we will conclude our study of subrings with a series of three activities. The first provides an opportunity to practice using the Subring Test, while the second and third explore the relationship between rings and their subrings, in particular with regard to the properties satisfied by each.

Activity 6.5. Use the Subring Test to prove or disprove each of the following statements:

(a) The set $\mathbb{E}$ of even integers is a subring of $\mathbb{Z}$.

(b) The set

$$K = \left\{ \begin{bmatrix} a & b \\ c & 0 \end{bmatrix} : a, b, c \in \mathbb{R} \right\}$$

is a subring of $\mathcal{M}_{2 \times 2}(\mathbb{R})$.

(c) The set $2\mathbb{Z}_6 = \{[0], [2], [4]\}$ (with addition and multiplication defined as in $\mathbb{Z}_6$) is a subring of $\mathbb{Z}_6$.

Activity 6.6.

(a) Does $\mathbb{Z}_6$ have a multiplicative identity? If so, what is it?

(b) Does $2\mathbb{Z}_6$ (defined in part (c) of Activity 6.5) have a multiplicative identity? If so, what is it?

(c) Is $\mathbb{Z}_6$ a field? Why or why not?

(d) Is $2\mathbb{Z}_6$ a field? Why or why not?

Activity 6.7. Let R be a ring, and let S be a subring of R. Which of the following conjectures do you think are true, and which do you think are false? Whenever possible, provide brief arguments or examples to justify your answers.

(a) If R has a multiplicative identity, then S has a multiplicative identity.

(b) If S has a multiplicative identity, then R has a multiplicative identity.

(c) If both R and S have a multiplicative identity, then $1_R = 1_S$.

(d) If R is commutative, then S is commutative.

(e) If S is commutative, then R is commutative.

(f) If R and S both have identity and x is a unit in R, then x is a unit in S. (Does your answer depend on whether $1_R = 1_S$?)

(g) If R and S both have identity and x is a unit in S, then x is a unit in R. (Does your answer depend on whether $1_R = 1_S$?)

(h) If R is a field, then S is a field.

(i) If S is a field, then R is a field.

(j) If R is an integral domain, then S is an integral domain.

(k) If S is an integral domain, then R is an integral domain.

Activities 6.6 and 6.7 demonstrate that the relationships between rings and subrings are not always as clear-cut as we might like them to be. For instance, a ring and its subrings can have different multiplicative identities. A ring that is not a field can contain a subring that is a field. And the set of units in a ring may be completely disjoint from the set of units in one of its subrings. These results are somewhat surprising, and there are numerous others like them. The bottom line is that we must be very careful not to make unwarranted assumptions when dealing with rings and subrings. A ring may satisfy a property that its subrings do not, and vice versa.

Subfields and Field Extensions

Preview Activity 6.8. In Preview Activity 6.1, we considered the set

$$\mathbb{Q}(\sqrt{2}) = \{a + b\sqrt{2} : a, b \in \mathbb{Q}\},$$

with addition and multiplication defined in the usual way (i.e., as in $\mathbb{R}$). On the one hand, we can
(and did) show that $\mathbb{Q}(\sqrt{2})$ is a subring of $\mathbb{R}$. On the other hand, since $\mathbb{Q}$ is a subring of $\mathbb{Q}(\sqrt{2})$, we
can also think of $\mathbb{Q}(\sqrt{2})$ as being an *extension* of $\mathbb{Q}$. (We will define this term more precisely in just
a bit.) In either case, since both $\mathbb{Q}$ and $\mathbb{R}$ are fields, it makes sense to ask whether $\mathbb{Q}(\sqrt{2})$ is also a
field. The questions below do exactly that, and they also explore in more detail the general methods
used to construct $\mathbb{Q}(\sqrt{2})$.

(a) We have already shown that $\mathbb{Q}(\sqrt{2})$ is a subring of $\mathbb{R}$. What would we still need to prove in
order to establish that $\mathbb{Q}(\sqrt{2})$ is a *subfield* of $\mathbb{R}$ (that is, a subring of $\mathbb{R}$ that is also a field)?

(b) Prove each of the properties you identified in part (a). (Hint: For one of these properties, you
will need to show that for all nonzero $a + b\sqrt{2} \in \mathbb{Q}(\sqrt{2})$, there exists $c + d\sqrt{2} \in \mathbb{Q}(\sqrt{2})$
such that

$$(a + b\sqrt{2})(c + d\sqrt{2}) = 1 + 0\sqrt{2}.$$

It may be helpful to look back at Activity 3.5 and the discussion that precedes it on page 47.)

(c) Is there a proper subfield of $\mathbb{Q}(\sqrt{2})$ that contains all rational numbers and the real number
$\sqrt{2}$? That is, can we find a field that is smaller than $\mathbb{Q}(\sqrt{2})$ and yet contains both $\mathbb{Q}$ (as a
subfield) and $\sqrt{2}$? Give a convincing argument to justify your answer.

(d) Suppose we defined $\mathbb{Q}(\sqrt[3]{2})$ analogously to $\mathbb{Q}(\sqrt{2})$—that is,

$$\mathbb{Q}(\sqrt[3]{2}) = \{a + b\sqrt[3]{2} : a, b \in \mathbb{Q}\}.$$

With this definition, is $\mathbb{Q}(\sqrt[3]{2})$ a field? Why or why not?

(e) How is the definition of the complex numbers, $\mathbb{C}$, similar to that of $\mathbb{Q}(\sqrt{2})$? In what ways do
these definitions differ?

The topic of field extensions is of central importance to much of abstract algebra. Many of the
questions pertaining to field extensions have to do with finding solutions to polynomial equations.
For instance, there are numerous polynomials with rational coefficients that fail to have any roots in
the rational numbers. One such polynomial is $f(x) = x^2 - 2$. This polynomial does have two roots
in $\mathbb{R}$ (that is, two values of $x \in \mathbb{R}$ for which $f(x) = 0$): $\sqrt{2}$ and $-\sqrt{2}$. But there are no values of
$x \in \mathbb{Q}$ for which $x^2 - 2 = 0$.

Now suppose that we wanted to "enlarge" $\mathbb{Q}$ to create a new ring that also contains both of these
irrational roots. One possibility would be to just use the set $\mathbb{Q} \cup \{-\sqrt{2}, \sqrt{2}\}$. Unfortunately, this set
is not even closed under addition, so we would have a hard time doing even basic algebra within it.
What we really want is to construct a set that has the same algebraic properties as $\mathbb{Q}$, but that also
contains $\pm\sqrt{2}$. Since $\mathbb{Q}$ is a field, this means that we are looking for a larger field that contains both
$\mathbb{Q}$ (as a subset) and $\pm\sqrt{2}$. Of course, there are many fields, such as $\mathbb{R}$, that satisfy these properties.
But some of these fields contain other elements that we might not be interested in. So the question
becomes this:

Given a field F and a polynomial[*] $p(x)$ with coefficients from F, what is the smallest field that contains every element of F and one or more solutions to the equation $p(x) = 0$?

In general, this is a surprisingly difficult question to answer. First off, it is not always immediately obvious that such a field even exists. For example, in the history of mathematics, imaginary numbers got their name for a reason. For a long time, it seemed impossible for mathematicians to conceive of any kind of number x that would satisfy the equation $x^2 + 1 = 0$. But the complex numbers are formed exactly by "adding" such a solution (we call it i) to $\mathbb{R}$, just as $\mathbb{Q}(\sqrt{2})$ is formed by "adding" a solution to $x^2 - 2 = 0$ to $\mathbb{Q}$. In each case, the resulting number system is what is known as a *field extension*, which we define formally as follows:

Definition 6.9. Let F be a field.

- A **subfield** of F is a subring of F that is also a field.

- If F is a subfield of another field E, then E is said to be a **field extension** (or simply an **extension**) of F.

- If E is an extension of F, and S is a subset of E, then the set $F(S)$, called the extension of F **generated by** S, is defined to be the smallest subfield of E that contains all elements of both F and S. In the case that S contains a single element $\alpha \in E$, then $F(\alpha)$ is called a **simple extension**.

You may have noticed that the notation used in Definition 6.9 to denote a simple extension is the same as the notation used in Preview Activities 6.1 and 6.8, where we defined $\mathbb{Q}(\sqrt{2})$. The next theorem explains the reason for this similarity in the case that α is the root of a quadratic polynomial. You should read its proof very carefully, adding additional details and explanations as you see fit.

Theorem 6.10. *Let F be a field, and let $p(x)$ be a quadratic polynomial with coefficients from F such that $p(x)$ has no roots in F. Suppose also that $p(x)$ does have a root α in some extension E of F. Then the simple extension of F generated by α (that is, the smallest field containing both F and α) can be described as follows:*

$$F(\alpha) = \{u + v\alpha : u, v \in F\}.$$

Proof. Let $p(x) = ax^2 + bx + c$ for some $a, b, c \in F$. Without loss of generality, we may assume that $a = 1$, so that $p(x) = x^2 + bx + c$. Since α is a root of $p(x)$, it follows that

$$\alpha^2 + b\alpha + c = 0,$$

or equivalently,

$$\alpha^2 = -b\alpha - c.$$

To prove the theorem, we must show that the set $S = \{u + v\alpha : u, v \in F\}$ is the smallest subfield of E that contains both α and all elements of F. We will first show that S is a subfield of E. We will do so by showing that S is a subring of E and S is a field. By definition, S is a

---

[*]We have not yet formally defined what a polynomial is, but for the purposes of this investigation, your intuitive understanding should suffice. If you are interested in a formal definition, we define a polynomial with coefficients from a field F to be an expression of the form

$$p(x) = c_0 + c_1 x + c_2 x^2 + c_3 x^3 + \cdots c_n x^n,$$

where each c_i belongs to F. We will study polynomials in much more detail in Investigations $8 - 12$.

nonempty subset of E. Thus, to establish that S is a subring of E, we must show that S is closed under subtraction and multiplication.

Let $u + v\alpha$ and $w + z\alpha$ be elements of S. Then

$$(u + v\alpha) - (w + z\alpha) = (u - w) + (v - z)\alpha \in S.$$

Thus, S is closed under subtraction. For closure under multiplication, note that

$$
\begin{aligned}
(u + v\alpha)(w + z\alpha) &= uw + vw\alpha + uz\alpha + vz\alpha^2 \\
&= uw + vw\alpha + uz\alpha + vz(-b\alpha - c) \\
&= (uw - vzc) + (vw + uz - vzb)\alpha,
\end{aligned}
$$

which is an element of S. Thus, we have shown that S is a subring of E.

Since E is a field, it follows that S is commutative. Furthermore, S clearly contains 1. Thus, to show that S is a subfield of E, we need only to show that every nonzero element of S is a unit. To this end, consider an arbitrary element $u + v\alpha \in S$, with u and v not both zero. We must show that there exists an element $w + z\alpha \in S$ such that

$$(u + v\alpha)(w + z\alpha) = 1.$$

First, we will argue that $\beta = -u^2 + buv - cv^2$ is a nonzero element of F. If $v = 0$, then $u \neq 0$, which implies that $\beta = -u^2 \neq 0$. Now suppose $v \neq 0$, and suppose also that $\beta = -u^2 + buv - cv^2 = 0$. Then

$$\left(-\tfrac{u}{v}\right)^2 + b\left(-\tfrac{u}{v}\right) + c = 0.$$

This, however, is a contradiction to our assumption that $p(x)$ has no roots in F. Thus, β is nonzero, which implies that β is a unit in F.

Now let $w = (-u + vb)\beta^{-1}$ and $z = v\beta^{-1}$. Then

$$
\begin{aligned}
(u + v\alpha)(w + z\alpha) &= (u + v\alpha)[(-u + bv)\beta^{-1} + v\beta^{-1}\alpha] \\
&= (u + v\alpha)(-u + bv + v\alpha)\beta^{-1} \\
&= (-u^2 + buv + uv\alpha - uv\alpha + bv^2\alpha + v^2\alpha^2)\beta^{-1} \\
&= [-u^2 + buv + bv^2\alpha + v^2(-b\alpha - c)]\beta^{-1} \\
&= (-u^2 + buv + bv^2\alpha - bv^2\alpha - cv^2)\beta^{-1} \\
&= (-u^2 + buv - cv^2)\beta^{-1} \\
&= \beta\beta^{-1} \\
&= 1
\end{aligned}
$$

So far, we have shown that S is a subfield of E. To finish the proof, we must argue that S is the smallest subfield of E that contains both α and all elements of F. To do so, it suffices to note that if a field K contains both α and F, then K must also contain $u + v\alpha$ for all $u, v \in F$. Thus, S is a subfield of any such K, which is what we needed to prove.

Since we have shown that S is the smallest subfield of E that contains both α and F, it follows by Definition 6.9 that

$$F(\alpha) = S = \{u + v\alpha : u, v \in F\}. \qquad \blacksquare$$

Activity 6.11. The proof of Theorem 6.10 was long and somewhat involved. In this activity, we will take a closer look at two of the important details within this proof.

(a) At the very beginning of the proof, we made the assumption that the leading coefficient of $p(x)$ was equal to 1. (Incidentally, such polynomials are said to be *monic*.) To see why this assumption is valid, show that if α is a root of any polynomial $p(x)$ with coefficients from F, then α is also a root of a monic polynomial $\tilde{p}(x)$ with coefficients from F.

(b) When we argued that every nonzero element $(u + v\alpha) \in S$ was a unit, it may have seemed as if we pulled our choice for $(u + v\alpha)^{-1}$ out of thin air. This was intentional, and it was done to illustrate an unfortunate feature of many of the proofs that you will read throughout your mathematical career. In particular, while the method we used was entirely correct, and we certainly proved that $u + v\alpha$ was a unit, the way we presented our argument provided virtually no insight into the way we actually found out what $(u + v\alpha)^{-1}$ should be. Perhaps we just guessed and got lucky. Or perhaps we remembered a similar technique from another proof. While either of these is a possibility, it seems more likely that we would have worked backward, starting with the *assumption* that $(u + v\alpha)(w + z\alpha) = 1$ and then determining from this assumption what w and z would need to be.

Try this technique; that is, assuming that $(u + v\alpha)(w + z\alpha) = 1$, set up and solve an appropriate system of equations, writing w and z in terms of u and v. How does your solution compare to that presented in our proof?

(c) Is your work from part (b) an acceptable alternative to our original argument? Does it establish that every nonzero element of S is a unit, or is the original proof still necessary? Explain.

(d) Are there any changes you would make to the proof of Theorem 6.10 to make it easier to understand or more insightful? If so, what would these changes be?

Before moving on, it is worth noting that Theorem 6.10 is actually a very special case of a much more general result on field extensions—in particular, extensions generated by roots of polynomials. The method we used to prove it is adequate for *quadratic extensions* (that is, extensions generated by the roots of quadratic polynomials), but it is not easy to generalize to broader contexts. For this, we will need to develop some more sophisticated tools, which we will do in Investigation 31.

Finally, the statement of Theorem 6.10 *assumed* the existence of an extension field E containing a root of $p(x)$. As it turns out, this assumption is always valid, thanks to the following result:

Theorem 6.12 (Kronecker's Theorem). *Let F be a field, and let $p(x)$ be a nonconstant polynomial with coefficients from F. Then there exists an extension E of F and an element $\alpha \in E$ such that $p(\alpha) = 0$.*

Like the more general version of Theorem 6.10, the proof of Kronecker's Theorem requires a more thorough treatment of polynomials and field extensions than we have considered up to this point. Thus, we will omit the proof for now and return to it (in Investigation 12) after we have developed more fully the necessary theoretical foundations.

Direct Sums

Preview Activity 6.13. In the previous section, we considered a way to enlarge a given field so that the resulting extension would contain an additional element of interest, such as the root of a polynomial. We will now consider a different, but related, problem.

Table 6.1

The addition table for $\mathbb{Z}_2 \oplus \mathbb{Z}_3$.

+	$([0]_2, [0]_3)$	$([0]_2, [1]_3)$	$([0]_2, [2]_3)$	$([1]_2, [0]_3)$	$([1]_2, [1]_3)$	$([1]_2, [2]_3)$
$([0]_2, [0]_3)$	$([0]_2, [0]_3)$	$([0]_2, [1]_3)$	$([0]_2, [2]_3)$	$([1]_2, [0]_3)$	$([1]_2, [1]_3)$	$([1]_2, [2]_3)$
$([0]_2, [1]_3)$	$([0]_2, [1]_3)$	$([0]_2, [2]_3)$	$([0]_2, [0]_3)$	$([1]_2, [1]_3)$	$([1]_2, [2]_3)$	$([1]_2, [0]_3)$
$([0]_2, [2]_3)$	$([0]_2, [2]_3)$	$([0]_2, [0]_3)$	$([0]_2, [1]_3)$	$([1]_2, [2]_3)$	$([1]_2, [0]_3)$	$([1]_2, [1]_3)$
$([1]_2, [0]_3)$	$([1]_2, [0]_3)$	$([1]_2, [1]_3)$	$([1]_2, [2]_3)$	$([0]_2, [0]_3)$	$([0]_2, [1]_3)$	$([0]_2, [2]_3)$
$([1]_2, [1]_3)$	$([1]_2, [1]_3)$	$([1]_2, [2]_3)$	$([1]_2, [0]_3)$	$([0]_2, [1]_3)$	$([0]_2, [2]_3)$	$([0]_2, [0]_3)$
$([1]_2, [2]_3)$	$([1]_2, [2]_3)$	$([1]_2, [0]_3)$	$([1]_2, [1]_3)$	$([0]_2, [2]_3)$	$([0]_2, [0]_3)$	$([0]_2, [1]_3)$

Suppose that we have two rings R and S, and we want to construct a new ring that contains both R and S as subrings. One way to do so is by using what is known as a *direct sum*. We will soon define direct sums formally, but before doing so, let's take a look at an example. The addition table for the direct sum of $\mathbb{Z}_2$ and $\mathbb{Z}_3$, denoted $\mathbb{Z}_2 \oplus \mathbb{Z}_3$, is shown in Table 6.1.

(a) Describe precisely how the elements of $\mathbb{Z}_2 \oplus \mathbb{Z}_3$ are related to the elements of $\mathbb{Z}_2$ and $\mathbb{Z}_3$.

(b) How does addition in $\mathbb{Z}_2 \oplus \mathbb{Z}_3$ seem to be related to the addition operations in $\mathbb{Z}_2$ and $\mathbb{Z}_3$?

(c) Defining multiplication analogous to the way addition was defined in Table 6.1, make the multiplication table for $\mathbb{Z}_2 \oplus \mathbb{Z}_3$.

(d) With the given addition table and the multiplication table you made in part (c), does $\mathbb{Z}_2 \oplus \mathbb{Z}_3$ seem to be a ring? If so, is $\mathbb{Z}_2 \oplus \mathbb{Z}_3$ an integral domain and/or a field? Why or why not?

(e) Consider the set S defined by

$$S = \{([0]_2, [x]_3) \in \mathbb{Z}_2 \oplus \mathbb{Z}_3\}.$$

Is S a subring of $\mathbb{Z}_2 \oplus \mathbb{Z}_3$? Why or why not?

(f) What is the relationship between $\mathbb{Z}_3$ and the set S defined in part (e)? Explain.

Preview Activity 6.13 introduces a new type of number system called a *direct sum*, which can be defined formally as follows:

Definition 6.14. Let R and S be rings. The **Cartesian product** of R and S is the set

$$R \times S = \{(r, s) : r \in R, s \in S\}.$$

The **direct sum** of R and S, denoted $R \oplus S$, is the set $R \times S$, with addition and multiplication defined componentwise—that is,

$$(r_1, s_1) + (r_2, s_2) = (r_1 + r_2, s_1 + s_2)$$

and

$$(r_1, s_1)(r_2, s_2) = (r_1 r_2, s_1 s_2).$$

It is important to note in Definition 6.14 that the $+$ sign really denotes three distinct operations: addition in R, addition in S, and addition in the new number system $R \oplus S$. The same could be said for multiplication; although the notation does not explicitly indicate this fact, there are really three different multiplication operations being used. In either case, the context in which the notation is used should alleviate any potential ambiguities. For instance, when the $+$ symbol is used *between* two ordered pairs of elements, we know that it is referring to addition in $R \oplus S$. Likewise, when $+$ is used *within* an ordered pair of elements, we know that it is referring to addition either within R or within S, depending on the coordinate in which it is used.

As you may have observed in Preview Activity 6.13, direct sums satisfy a number of important properties. The most fundamental of these is the following:

Theorem 6.15. *Let R and S be rings. Then $R \oplus S$ is also a ring.*

Theorem 6.15 is not difficult to prove. In fact, each of the ring axioms for $R \oplus S$ can be established by simply invoking the corresponding axioms within R and S. To illustrate, consider the following argument that addition in $R \oplus S$ is commutative:

Let $x = (r_1, s_1)$, $y = (r_2, s_2) \in R \oplus S$. Then

$$
\begin{aligned}
x + y &= (r_1, s_1) + (r_2, s_2) \\
&= (r_1 + r_2, s_1 + s_2) \quad \text{(by Definition 6.14)} \\
&= (r_2 + r_1, s_2 + s_1) \quad \text{(by commutativity of $+$ in R and S)} \\
&= (r_2, s_2) + (r_1, s_1) \quad \text{(again, by Definition 6.14)} \\
&= y + x.
\end{aligned}
$$

The other ring axioms for $R \oplus S$ can be established in a similar manner. Therefore, we will leave the rest of the proof of Theorem 6.15 as an exercise. (See Exercise (16).)

Finally, it is worth noting that $R \oplus S$ always contains both R and S (or, more accurately, rings that look and behave just like R and S) as subrings. The next theorem formalizes this fact. Its proof is left as an exercise. (See Exercise (17).)

Theorem 6.16. *Let R and S be rings. Then*

$$
R \oplus \{0_S\} = \{(r, 0_S) : r \in R\}
$$

and

$$
\{0_R\} \oplus S = \{(0_R, s) : s \in S\}
$$

are both subrings of $R \oplus S$.

You may have noticed that $R \oplus \{0_S\}$ contains the elements of R, each juxtaposed with the zero element for S. Thus, while $R \oplus \{0_S\}$ is not technically the same ring as R, the two are virtually identical, both in their makeup and in the way they behave with respect to addition and multiplication. We might say that R and $R \oplus \{0_S\}$ are essentially the same, a notion that we will make more precise in the next investigation when we study ring isomorphism. Likewise, S and $\{0_R\} \oplus S$ can also be considered to be essentially the same ring, which implies that $R \oplus S$ in some sense contains a copy of both R and S. Thus, direct sums provide a way to construct a larger ring that contains each of two smaller rings. This type of construction is useful in numerous examples, and it also plays an essential role in the classification of algebraic objects called *finite groups*, which we will study in Investigation 28.

Concluding Activities

Activity 6.17. Fill in the blank to make the following statement true, and prove the resulting theorem.

> **Theorem 6.18.** *Every nontrivial ring contains at least _____ subrings.*

Activity 6.19. Theorem 6.3 states that every subring S of a ring R must contain the zero element of R, 0_R. However, this statement seems to be stronger than what is required by Definition 6.2. In particular, the definition requires S to be a ring with the same operations as R, which implies that S must contain *some* zero element, say 0_S, which could conceivably be different than 0_R. You may recall from Activity 6.7 that it is possible for a ring and one or more of its subrings to have different multiplicative identities. Could this same behavior occur with respect to additive identities? In other words, does Theorem 6.3 need to be modified to account for the possibility that $0_S \neq 0_R$? Why or why not?

Activity 6.20. Under what conditions is $R \oplus S$ commutative? Under what conditions does $R \oplus S$ have an identity? Prove your answers.

Activity 6.21.

(a) Compare the addition and multiplication tables for $\mathbb{Z}_2 \oplus \mathbb{Z}_3$ and for $\mathbb{Z}_6$. How are these rings similar? How are they different?

(b) Make the addition and multiplication tables for $\mathbb{Z}_2 \oplus \mathbb{Z}_2$, and compare these tables to the tables for $\mathbb{Z}_4$. How are $\mathbb{Z}_2 \oplus \mathbb{Z}_2$ and $\mathbb{Z}_4$ similar? How are they different?

Exercises

(1) Let S denote the set of all 2×2 matrices of the form

$$\begin{bmatrix} x & 0 \\ y & 0 \end{bmatrix},$$

where $x, y \in \mathbb{R}$. Is S a subring of $\mathcal{M}_{2\times 2}(\mathbb{R})$? Prove your answer.

(2) Let S denote the set of all 2×2 matrices of the form

$$\begin{bmatrix} x & y \\ -y & x \end{bmatrix},$$

where $x, y \in \mathbb{R}$. Is S a subring of $\mathcal{M}_{2\times 2}(\mathbb{R})$? Prove your answer.

(3) Let $S = \{[0], [2], [4], [6], [8]\}$ be a subset of $\mathbb{Z}_{10}$.

(a) Construct addition and multiplication tables for S using addition and multiplication modulo 10.

(b) Is S a subring of $\mathbb{Z}_{10}$? Justify your conclusion.

(c) If S has a multiplicative identity, do the elements of S have multiplicative inverses? Specify the multiplicative inverse for each element that has one.

(4) (a) Let R be a ring and consider the following proposition:

$$\text{For all } a, b \in R, (a + b)(a - b) = a^2 - b^2. \tag{6.1}$$

Write the negation of Statement (1).

(b) Give an example of a ring R in which the negation of the Statement (1) is true (and hence, Statement (1) is false) or explain why no such example exists. If you give an example, you must justify why the negation of Statement (1) is true for your example.

(c) Give an example of a noncommutative ring R and a subring of R that is commutative, or explain why no such example exists.

(5) An element r in a ring R is called an **idempotent element** or an **idempotent** provided that $r^2 = r$.

(a) Determine all the idempotent elements in $\mathbb{Z}_6$.

(b) Determine all the idempotent elements in $\mathbb{Z}_{20}$.

(c) Let $S = \left\{ \begin{bmatrix} a & b \\ 0 & c \end{bmatrix} \mid a, b, c \in \mathbb{R} \right\}$. Prove that S is a subring of $\mathcal{M}_{2 \times 2}(\mathbb{R})$ and determine all idempotent elements in S.

(6) See Exercise (5) for the definition of an idempotent element, and for idempotents in $\mathbb{Z}_6$ and $\mathbb{Z}_{20}$. Prove that if $m, n \in \mathbb{N}$ with $m \geq 2$ and $n \geq 2$ and $\gcd(m, n) = 1$, then the ring $\mathbb{Z}_{mn}$ has at least four different idempotent elements. (Hint: There exist integers x and y such that $mx + ny = 1$.)

(7) The set H of **quaternions** is defined as follows:

$$H = \left\{ \begin{bmatrix} a + bi & c + di \\ -c + di & a - bi \end{bmatrix} : a, b, c, d \in \mathbb{R} \right\}.$$

Prove that H is a subring of $\mathcal{M}_{2 \times 2}(\mathbb{C})$. Then prove that H is a noncommutative ring with identity in which every nonzero element has a multiplicative inverse. (Such rings are called *division rings* or *skew fields*. They differ from fields only in the fact that their multiplication operation is not required to be commutative.)

(8) **Subrings and set operations.** Let S and T be subrings of a ring R. For each of the questions below, give a proof or a pair of examples (whichever is most appropriate) to justify your answer.

(a) Is $S \cup T$ always, sometimes, or never a subring of R?

(b) Is $S \cap T$ always, sometimes, or never a subring of R?

(c) Is $S \triangle T$ always, sometimes, or never a subring of R?

(9) **Subrings and commutativity.** Is every subring of a commutative ring necessarily commutative? Give a proof or counterexample to justify your answer.

(10) Let R be a ring. The **center** of R is defined to be the set of all elements $x \in R$ such that $xr = rx$ for all $r \in R$. Prove that the center of R is always a commutative subring of R.

(11) Let R be a ring, and let $r \in R$. The **centralizer** of r is defined to be the set of all $x \in R$ such that $xr = rx$.

    (a) Prove that the centralizer of r is a subring of R.

    (b) Is the centralizer of r necessarily commutative? Explain.

(12) Let R be a ring, and let $r \in R$. Prove that $S = \{x \in R : rx = 0\}$ is a subring of R.

(13) Let R be a ring, and let $n \in \mathbb{Z}$. Prove that $S = \{x \in R : nx = 0\}$ is a subring of R.

(14) **Subrings and units.** Let R be a ring with identity, and let S be the set of all units of R. Is S always, sometimes, or never a subring of R? Give a proof or a pair of examples (whichever is most appropriate) to justify your answer.

(15) **Subrings of $\mathbb{Z}$.** For every nonnegative integer n, let $n\mathbb{Z} = \{nx : x \in \mathbb{Z}\}$, with addition and multiplication defined as in $\mathbb{Z}$. (See Exercise (7) on page 66.)

    (a) Prove that $n\mathbb{Z}$ is always a subring of $\mathbb{Z}$.

    (b) Prove that every subring of $\mathbb{Z}$ is equal to $n\mathbb{Z}$ for some nonnegative integer n.

* (16) Prove Theorem 6.15.

* (17) Prove Theorem 6.16.

(18) **Direct sums of integral domains and fields.**

    (a) Suppose that R and S are both integral domains. Is $R \oplus S$ always, sometimes, or never an integral domain? Give a proof or a pair of examples (whichever is most appropriate) to justify your answer.

    (b) Would your answer to part (a) change if the words *integral domain* were replaced by *field*? Why or why not?

(19) **Direct sums and characteristic.**

    (a) What is the characteristic of $\mathbb{Z}_m \oplus \mathbb{Z}_n$? Prove your answer.

    (b) Let R and S be rings with characteristic m and n, respectively, where $m, n > 0$. Generalize your work in part (a) to determine (with proof) the characteristic of $R \oplus S$.

(20) **Direct sums and rings with characteristic zero.** Let R and S be rings. Decide whether each of the following statements is true or false. Verify your answers.

    (a) If both R and S have nonzero characteristic, then $R \oplus S$ has nonzero characteristic.

    (b) If $R \oplus S$ has nonzero characteristic, then both R and S have nonzero characteristic.

    (c) If both R and S have characteristic zero, then $R \oplus S$ has characteristic zero.

    (d) If $R \oplus S$ has characteristic zero, then both R and S have characteristic zero.

(21) For each natural number i, let R_i be a ring. Define the **infinite direct sum**

$$\bigoplus_{i=1}^{\infty} R_i = R_1 \oplus R_2 \oplus R_3 \oplus \cdots$$

to be the set of all sequences of the form

$$x = (x_1, x_2, x_3, \ldots),$$

where $x_i \in R_i$ for all i, and $x_i = 0_{R_i}$ for all but finitely many values of i. To illustrate, let $R_i = \mathbb{Z}_{i+1}$ for every natural number i, and define

$$R = \bigoplus_{i=1}^{\infty} R_i = \bigoplus_{i=1}^{\infty} \mathbb{Z}_{i+1} = \mathbb{Z}_2 \oplus \mathbb{Z}_3 \oplus \mathbb{Z}_4 \oplus \cdots$$

Then

$$x = ([1]_2, [0]_3, [0]_4, [4]_5, [3]_6, [0]_7, [0]_8, [0]_9, \ldots) \in R,$$

since x_i is the zero element in $\mathbb{Z}_{i+1}$ for all but three values of i. In contrast,

$$y = ([1]_2, [1]_3, [1]_4, [1]_5, \ldots) \notin R,$$

since y_i is nonzero for infinitely many values of i.

(a) With addition and multiplication defined componentwise (as in the definition of a finite direct sum), prove that every infinite direct sum of rings is also a ring.

(b) An **infinite direct product** is defined similarly to an infinite direct sum, but without the restriction that $x_i = 0_{R_i}$ for all but finitely many values of i. Is every infinite direct product of rings also a ring? Prove your answer.

(22) **The characteristic of an infinite direct sum.** Let R be the infinite direct sum

$$R = \bigoplus_{i=1}^{\infty} \mathbb{Z}_{i+1} = \mathbb{Z}_2 \oplus \mathbb{Z}_3 \oplus \mathbb{Z}_4 \oplus \cdots,$$

as defined in Exercise (21).

(a) Show that for every $x = (x_1, x_2, x_3, \ldots) \in R$, there exists a positive integer k such that

$$kx = 0_R = ([0]_2, [0]_3, [0]_4, \ldots).$$

(Hint: Recall that x_i is nonzero for only finitely many values of i.)

(b) Show that, in spite of your conclusion in part (a), the characteristic of R is zero.

(c) Explain the apparent contradiction between your answers to parts (a) and (b).

Connections

In this investigation, we studied subrings and direct sums of rings. If you studied group theory before ring theory, you should notice connections between the topics in this investigation and those in Investigations 19 and 25. The idea of a subgroup is the same as that of a subring; in particular, a subgroup of a group G is just a subset of G that is also a group using the operation from G. The only significant difference is that a ring comes with two operations and a group comes with only one, so it is a bit easier to determine if a subset of a group is a subgroup than if a subset of a ring is a subring.

The constructions of direct sums of rings and groups are also analogous. To create a direct sum of a pair of objects (either rings or groups), we make the Cartesian product of the sets into a ring or group using operations defined componentwise. Because of the simpler structure of groups, we can define more than one type of direct sum. In addition, we often use multiplicative notation to denote the operation in a group. For this reason, the group theoretic analog of a direct sum of rings is often called an *external direct product* to distinguish it from other direct sums.

Investigation 7

Isomorphism and Invariants

Focus Questions

By the end of this investigation, you should be able to give precise and thorough answers to the questions listed below. You may want to keep these questions in mind to focus your thoughts as you complete the investigation.

- Intuitively, what does it mean for two rings to be "essentially the same"?

- What does it mean for two rings to be isomorphic? How does the definition of isomorphism reflect the informal definition of "essentially the same"?

- What strategies can be used to prove that two rings are isomorphic? How are these strategies motivated by the definition of isomorphism?

- What is an invariant, and how does one prove that a property is an invariant?

- How can invariants be used to prove that two rings are not isomorphic? Can invariants be used to prove that two rings are isomorphic?

Preview Activity 7.1. In Investigation 2, we saw how equivalence relations can be used to identify mathematical objects that are the same (or equivalent) in one way or another. The notion of "sameness" is very important in mathematics, for it allows us to identify when two objects should be considered indistinguishable, and thus treated identically. Identifying sameness also makes our analysis more efficient, since it allows us to consider entire classes of objects at the same time, instead of dealing with each object individually.

In this investigation, we will define precisely what it means for two rings to be the same, or *isomorphic*. Before we do so, however, let's apply our intuitive ideas about sameness to a few examples.

The addition and multiplication tables for four rings are shown below. Which of these rings would you consider to be essentially the same, and which would you consider to be different? Consider each possible pair of rings, and give a convincing argument to justify your answer for each.

R_1 :

+	γ	α	δ	β
γ	α	γ	β	δ
α	γ	α	δ	β
δ	β	δ	γ	α
β	δ	β	α	γ

$\cdot$	γ	α	δ	β
γ	α	α	γ	γ
α	α	α	α	α
δ	γ	α	β	δ
β	γ	α	δ	β

R_2 :

+	w	x	y	z
w	w	x	y	z
x	x	w	z	y
y	y	z	w	x
z	z	y	x	w

$\cdot$	w	x	y	z
w	w	w	w	w
x	w	x	w	x
y	w	y	w	y
z	w	z	w	z

R_3 :

+	q	r	s	t
q	q	r	s	t
r	r	q	t	s
s	s	t	q	r
t	t	s	r	q

$\cdot$	q	r	s	t
q	q	q	q	q
r	q	r	q	r
s	q	q	s	s
t	q	r	s	t

R_4 :

+	i	j	k	l
i	i	j	k	l
j	j	k	l	i
k	k	l	i	j
l	l	i	j	k

$\cdot$	i	j	k	l
i	i	i	i	i
j	i	j	k	l
k	i	k	i	k
l	i	l	k	j

Introduction

In Preview Activity 7.1, you were asked to decide which of the four rings shown were essentially the same, and which were not. At first glance, it would be easy to think that all the rings are different. After all, their elements are certainly different. But is this enough to conclude that the rings themselves are different?

To answer this question, let's consider the addition and multiplication tables for $\mathbb{Z}_4$:

$\mathbb{Z}_4$:

+	[0]	[1]	[2]	[3]
[0]	[0]	[1]	[2]	[3]
[1]	[1]	[2]	[3]	[0]
[2]	[2]	[3]	[0]	[1]
[3]	[3]	[0]	[1]	[2]

$\cdot$	[0]	[1]	[2]	[3]
[0]	[0]	[0]	[0]	[0]
[1]	[0]	[1]	[2]	[3]
[2]	[0]	[2]	[0]	[2]
[3]	[0]	[3]	[2]	[1]

Let's suppose also that we decided to abbreviate the names of the equivalence classes in $\mathbb{Z}_4$ by assigning a variable to each one. In particular, we'll let $i = [0]$, $j = [1]$, $k = [2]$, and $l = [3]$.

Activity 7.2. Substitute i, j, k, and l for $[0]$, $[1]$, $[2]$, and $[3]$ in the tables for $\mathbb{Z}_4$. That is, each time $[0]$ appears in the tables, replace it with i. Do the same for the other classes as well, replacing $[1]$ with j, $[2]$ with k, and $[3]$ with l. What do you notice?

If you completed Activity 7.2 correctly, you probably observed that the addition and multiplication tables for $\mathbb{Z}_4$ can be made to look exactly like those of R_4 in Preview Activity 7.1, simply by renaming the elements. In other words, the only differences between $\mathbb{Z}_4$ and R_4 are the names of the elements. We might even say that $\mathbb{Z}_4$ and R_4 are essentially the same ring.

As it turns out, we can carry out a similar renaming to show that $\mathbb{Z}_4$ and R_1 are also essentially the same. Here we have to be a bit more careful, however, since the elements of R_1 seem to be arranged in a different order than those of $\mathbb{Z}_4$ and R_4. To illustrate, let's see what would happen if we replaced the names of the elements of R_1 with the names of the elements of $\mathbb{Z}_4$, keeping the elements in the same order as they are listed in the tables. We would replace γ with $[0]$, α with $[1]$, δ with $[2]$, and β with $[3]$, which would yield the following addition and multiplication tables:

+	[0]	[1]	[2]	[3]
[0]	[1]	[0]	[3]	[2]
[1]	[0]	[1]	[2]	[3]
[2]	[3]	[2]	[0]	[1]
[3]	[2]	[3]	[1]	[0]

·	[0]	[1]	[2]	[3]
[0]	[1]	[1]	[0]	[0]
[1]	[1]	[1]	[1]	[1]
[2]	[0]	[1]	[3]	[2]
[3]	[0]	[1]	[2]	[3]

A quick look at these tables reveals that they certainly do not appear to be the addition and multiplication tables for $\mathbb{Z}_4$. In this case, simply renaming the elements of R_1 did not yield a ring that looked essentially the same as $\mathbb{Z}_4$. The ring that resulted from such a renaming had the same elements as $\mathbb{Z}_4$, but its operations seem different from those of $\mathbb{Z}_4$.

So what can we conclude from this? Does our failed attempt at renaming the elements of R_1 necessarily imply that R_1 and $\mathbb{Z}_4$ are different rings? In fact, it does not. All that we know at this point is that the particular renaming that we used yields a ring that appears to behave differently than $\mathbb{Z}_4$. But what if we used a different renaming? For instance, what if we replaced γ with $[2]$, α with $[0]$, δ with $[3]$, and β with $[1]$? Doing so would yield the following addition and multiplication tables:

+	[2]	[0]	[3]	[1]
[2]	[0]	[2]	[1]	[3]
[0]	[2]	[0]	[3]	[1]
[3]	[1]	[3]	[2]	[0]
[1]	[3]	[1]	[0]	[2]

·	[2]	[0]	[3]	[1]
[2]	[0]	[0]	[2]	[2]
[0]	[0]	[0]	[0]	[0]
[3]	[2]	[0]	[1]	[3]
[1]	[2]	[0]	[3]	[1]

Notice that these tables are nearly identical to the addition and multiplication tables for $\mathbb{Z}_4$. In fact, the only difference is the order in which the elements have been assigned to the rows and columns of the tables. This minor detail affects only the way the tables are displayed, and not the information that they contain. A simple rearrangement of the rows and columns would put the tables in their more standard form. Thus, we can see that by renaming the elements of R_1 and possibly reordering the rows and columns of the resulting tables, we are able to produce the addition and multiplication tables for $\mathbb{Z}_4$. Because of this, we might say that R_1 and $\mathbb{Z}_4$ are essentially the same ring.

Now that we have seen a few examples, we are ready to state an informal definition, which we will use to formally define the notion of isomorphism in the next section.

Informal Definition 7.3. Let R and S be finite rings. Then R is said to be **essentially the same** as S if the addition and multiplication tables for R can be transformed into the addition and multiplication tables for S by doing nothing more than renaming the elements of R and/or reordering the rows and columns of R's addition and multiplication tables.

Note that this informal definition can easily be used to show that two rings are *not* essentially the same. For example, each of the rings from Preview Activity 7.1 contains an element that appears more often than any other element in the ring's multiplication table. For R_1, that element is α, which appears 8 times in the multiplication table for R_1. For R_2, it is w (10 times). For R_3, it is q (9 times). And for R_4, it is i (8 times).

From these observations, it is clear that R_1 and R_2 are not the same ring. This is because no matter how we rename and/or reorder the elements of R_1, we will still end up with a multiplication table whose most common element appears 8, not 10, times. Thus, no matter how we rename or reorder the elements of R_1, we will never get a ring whose addition and multiplication tables look just like those of R_2. A similar argument can be made for every other pair of rings from Preview Activity 7.1, with the exception of R_1 and R_4, both of which are the same as $\mathbb{Z}_4$ (and thus the same as each other) by our previous arguments.

To summarize, note that in order to show that two rings are the same, we must find a way to rename and/or reorder the elements of one ring so that its addition and multiplication tables are identical to those of the other ring. In order to show that two rings are different, however, it often suffices to identify a property that is different between the two rings—in particular, a property that could not possibly be different if one ring had been obtained from the other by simply renaming and reordering elements.

Activity 7.4. Consider each possible pair of rings from Preview Activity 7.1, with the exception of R_1 and R_4. For each such pair, make a list of properties that are different between the two rings. Then explain why each difference you listed would contradict Informal Definition 7.3. Include in your list at least three properties for each pair of rings.

Isomorphisms of Rings

Informal Definition 7.3 provides a helpful and intuitive way of thinking about what it means for two rings to be the same. This informal definition, however, has some significant limitations. First, it only works for finite rings. This is because it would be impossible to actually create the addition and multiplication tables for a ring with infinitely many elements. Second, even for finite rings, the definition can be extremely cumbersome to work with, especially if the rings in question have more than a few elements. Can you imagine trying to create a multiplication table for a ring with 50 or 1000 or even 50,000 elements? The task would be daunting at best, and practically impossible at worst.

To deal with these difficulties, we will adopt a formal definition that captures the idea behind Informal Definition 7.3, but does so in a more precise manner. In order to motivate this definition, let's consider again the two main parts of Informal Definition 7.3.

Renaming Elements

When we argued that $\mathbb{Z}_4$ and R_4 were essentially the same ring, we found a way to rename the elements of $\mathbb{Z}_4$ using the same names as the elements of R_4. This renaming was really just a bijective function (that is, a function that is both one-to-one and onto) from $\mathbb{Z}_4$ to R_4. Denoting this function by $\varphi : \mathbb{Z}_4 \to R_4$, we could write:

$$\varphi([0]) = i$$
$$\varphi([1]) = j$$
$$\varphi([2]) = k$$
$$\varphi([3]) = l$$

Note that any function that actually corresponds to a valid renaming would have to be bijective. This is because it wouldn't make sense to give two ring elements the same name, or to leave a ring element out. If two rings are truly the same, then the elements of one should be able to be matched in a one-to-one correspondence with the elements of the other. As such, our formal definition of "sameness" will begin with a bijective function.

Preserving Operations

As we saw in our earlier example, just having a bijective function from one ring to another is not enough to say that the two rings are the same. Indeed, this bijective function must also transform the addition and multiplication tables of the first ring into those of the second. To see exactly what this means, let's look at an example.

Consider two rings R and S, each having three elements. Suppose also that we have defined a bijective "renaming" function $\varphi : R \to S$. Let's consider the multiplication table for R, which we can write generically as follows (using a, b, and c to denote the elements of R):

$\cdot$	a	b	c
a	aa	ab	ac
b	ba	bb	bc
c	ca	cb	cc

If we simply replace each entry in this table with its new name (as given by φ), we obtain the following table:

$\cdot$	$\varphi(a)$	$\varphi(b)$	$\varphi(c)$
$\varphi(a)$	$\varphi(aa)$	$\varphi(ab)$	$\varphi(ac)$
$\varphi(b)$	$\varphi(ba)$	$\varphi(bb)$	$\varphi(bc)$
$\varphi(c)$	$\varphi(ca)$	$\varphi(cb)$	$\varphi(cc)$

Is this table the multiplication table for S? Its entries are certainly elements of S, since φ maps from R to S. But the actual multiplication table for S would be defined as follows:

·	$\varphi(a)$	$\varphi(b)$	$\varphi(c)$
$\varphi(a)$	$\varphi(a)\varphi(a)$	$\varphi(a)\varphi(b)$	$\varphi(a)\varphi(c)$
$\varphi(b)$	$\varphi(b)\varphi(a)$	$\varphi(b)\varphi(b)$	$\varphi(b)\varphi(c)$
$\varphi(c)$	$\varphi(c)\varphi(a)$	$\varphi(c)\varphi(b)$	$\varphi(c)\varphi(c)$

If the notation seems confusing here, just keep in mind that $\varphi(a)$, $\varphi(b)$, and $\varphi(c)$ are the elements of S, and we have formed the multiplication table for S in the usual way. In particular, each entry in the table is the product of the corresponding row and column headers (so, for instance, the entry in the $\varphi(a)$ row and $\varphi(b)$ column is just $\varphi(a)\varphi(b)$).

So what can we conclude? Recall that we wanted the renamed R table to be equal to the S table. In order for this to happen, each entry of the renamed R table must be equal to the corresponding entry of the S table. Thus, it must be the case that $\varphi(aa) = \varphi(a)\varphi(a)$, $\varphi(ab) = \varphi(a)\varphi(b)$, $\varphi(ac) = \varphi(a)\varphi(c)$, $\varphi(ba) = \varphi(b)\varphi(a)$, and so on.

In other words, we want φ to *preserve multiplication*, which means:

For all x, $y \in R$, $\varphi(xy) = \varphi(x)\varphi(y)$.

And, of course, since we want the renamed addition table for R to be the same as the addition table for S, we would like φ to *preserve addition* as well:

For all x, $y \in R$, $\varphi(x + y) = \varphi(x) + \varphi(y)$.

These two conditions help us to state in a more precise way exactly what it means for two rings to have the same addition and multiplication tables—or, in other words, the same algebraic structure. Any bijective function φ that satisfies both of these conditions is called an *isomorphism*, defined formally below.

Definition 7.5. Let R and S be rings. An **isomorphism** is a bijective function $\varphi : R \to S$ such that for all x, $y \in R$,
$$\varphi(x + y) = \varphi(x) + \varphi(y) \quad \text{and} \quad \varphi(xy) = \varphi(x)\varphi(y).$$
If there exists an isomorphism from R to S, then R is said to be *isomorphic* to S.

The word *isomorphic* comes from two Greek words: *isos*, which means *equal* or *same*, and *morphe*, which means *form* or *structure*. Thus, when we say that two rings are isomorphic, we mean that they have the same structure. Since *structure* in abstract algebra is really defined by the way a number system's operations work, our specific definition of isomorphism requires φ to preserve both the addition and multiplication operations. Thus, we can think of an isomorphism as being an *operation-preserving bijection* or a *structure-preserving bijection*.

It is worth noting that the isomorphism relation is symmetric. (In fact, the isomorphism relation is an equivalence relation, as we will show in Activity 7.17.) In particular, if $\varphi : R \to S$ is an isomorphism, then $\varphi^{-1} : S \to R$ is also an isomorphism. For this reason, we will often simply say that two rings R and S are isomorphic, rather than saying R is isomorphic to S, or S is isomorphic to R. When R and S are isomorphic, we denote this relationship by writing $R \cong S$. So $R \cong S$ means that there is an isomorphism from R to S (or, equivalently, an isomorphism from S to R). It's also important to note that the *function* in Definition 7.5 is called an *isomorphism*, while the *rings* R and S are said to be *isomorphic*.

Activity 7.6.

(a) Use Definition 7.5 to explain why the function $\varphi : R_1 \to R_4$ (from Preview Activity 7.1) defined by

$$\varphi(\gamma) = k, \quad \varphi(\alpha) = i, \quad \varphi(\delta) = l, \quad \varphi(\beta) = j$$

is an isomorphism.

(b) Use Definition 7.5 to explain why the function $\varphi : R_1 \to R_4$ defined by

$$\varphi(\gamma) = i, \quad \varphi(\alpha) = j, \quad \varphi(\delta) = k, \quad \varphi(\beta) = l$$

is **not** an isomorphism.

(c) Is the function $\varphi : R_3 \to R_2$ defined by

$$\varphi(q) = w, \quad \varphi(r) = x, \quad \varphi(s) = y, \quad \varphi(t) = z$$

an isomorphism? Use Definition 7.5 to justify your answer.

(d) What does your answer to part (c) allow you to conclude about whether R_2 and R_3 are isomorphic?

(e) Are R_2 and R_3 isomorphic? Use Definition 7.5 to justify your answer.

Proving Isomorphism

Now that we have precisely defined what it means for two rings to be isomorphic, let's consider how we might use this definition in the context of rings that have more than just a few elements. In particular, we will consider the ring M defined as follows:

$$M = \left\{ \begin{bmatrix} x & 0 \\ -x & 0 \end{bmatrix} : x \in \mathbb{R} \right\}.$$

Activity 7.7.

(a) Notice that M (as defined above) is a subset of $\mathcal{M}_{2 \times 2}(\mathbb{R})$. Show that M is actually a *subring* of $\mathcal{M}_{2 \times 2}(\mathbb{R})$.

(b) To which familiar ring do you think M is isomorphic? You don't have to prove your answer now, but you should make a reasonable conjecture with some solid reasoning to back it up.

Looking again at the definition of M, it appears that each element of M corresponds to a unique real number (and vice versa). Thus, it seems reasonable that we would try to prove that M is isomorphic to $\mathbb{R}$. Since both M and $\mathbb{R}$ have infinitely many elements, we will not be able to simply work with their addition and multiplication tables. Instead, we must use Definition 7.5, which suggests the following steps:

(1) We must define an appropriate function $\varphi : M \to \mathbb{R}$.

(2) We must show that φ is bijective; that is, we must show that φ is both *injective* and *surjective*.

(3) We must show that φ preserves addition.

(4) We must show that φ preserves multiplication.

Activity 7.8. Carefully read the following proof that M is isomorphic to $\mathbb{R}$, filling in all the missing details and providing additional explanations where appropriate. As you read the proof, try to identify where each of the four steps outlined above is taking place.

Theorem. *Let* $M = \left\{ \begin{bmatrix} x & 0 \\ -x & 0 \end{bmatrix} : x \in \mathbb{R} \right\}$. *Then* M *is isomorphic to* $\mathbb{R}$.

Proof. Let $\varphi : M \to \mathbb{R}$ be defined by

$$\varphi\left(\begin{bmatrix} x & 0 \\ -x & 0 \end{bmatrix} \right) = x.$$

To see that φ is injective, suppose that

$$\varphi\left(\begin{bmatrix} x & 0 \\ -x & 0 \end{bmatrix} \right) = \varphi\left(\begin{bmatrix} y & 0 \\ -y & 0 \end{bmatrix} \right)$$

for some $\begin{bmatrix} x & 0 \\ -x & 0 \end{bmatrix}, \begin{bmatrix} y & 0 \\ -y & 0 \end{bmatrix} \in M$. Then $x = y$, which implies that

$$\begin{bmatrix} x & 0 \\ -x & 0 \end{bmatrix} = \begin{bmatrix} y & 0 \\ -y & 0 \end{bmatrix}.$$

Thus, φ is injective.

To see that φ is surjective, observe that for all $x \in \mathbb{R}$,

$$\varphi\left(\begin{bmatrix} x & 0 \\ -x & 0 \end{bmatrix} \right) = x.$$

We must now show that φ preserves both addition and multiplication. To this end, note that for all $\begin{bmatrix} x & 0 \\ -x & 0 \end{bmatrix}, \begin{bmatrix} y & 0 \\ -y & 0 \end{bmatrix} \in M$,

$$\varphi\left(\begin{bmatrix} x & 0 \\ -x & 0 \end{bmatrix} + \begin{bmatrix} y & 0 \\ -y & 0 \end{bmatrix} \right) = \varphi\left(\begin{bmatrix} x+y & 0 \\ (-x)+(-y) & 0 \end{bmatrix} \right)$$
$$= \varphi\left(\begin{bmatrix} x+y & 0 \\ -(x+y) & 0 \end{bmatrix} \right)$$
$$= x+y$$
$$= \varphi\left(\begin{bmatrix} x & 0 \\ -x & 0 \end{bmatrix} \right) + \varphi\left(\begin{bmatrix} y & 0 \\ -y & 0 \end{bmatrix} \right),$$

and

$$\varphi\left(\begin{bmatrix} x & 0 \\ -x & 0 \end{bmatrix}\begin{bmatrix} y & 0 \\ -y & 0 \end{bmatrix}\right) = \varphi\left(\begin{bmatrix} xy & 0 \\ (-x)y & 0 \end{bmatrix}\right)$$

$$= \varphi\left(\begin{bmatrix} xy & 0 \\ -(xy) & 0 \end{bmatrix}\right)$$

$$= xy$$

$$= \varphi\left(\begin{bmatrix} x & 0 \\ -x & 0 \end{bmatrix}\right)\varphi\left(\begin{bmatrix} y & 0 \\ -y & 0 \end{bmatrix}\right).$$

Since we have shown that there exists a bijective function $\varphi : M \to \mathbb{R}$ that preserves both addition and multiplication, it follows that M and $\mathbb{R}$ are isomorphic. ∎

Well-Defined Functions

Preview Activity 7.9. Let f be a mapping that assigns to each element $[a]_3$ in $\mathbb{Z}_3$ the element $[a]_6$ in $\mathbb{Z}_6$—that is, $f([a]_3) = [a]_6$ for all $[a]_3 \in \mathbb{Z}_3$.

(a) Does f preserve the operations in $\mathbb{Z}_3$? Explain.

(b) Consider the following proof that f is an injection:

Let $[a]_3$ and $[b]_3$ be in $\mathbb{Z}_3$, and assume $f([a]_3) = f([b]_3)$. Then $[a]_6 = [b]_6$, and so 6 divides $b - a$. Thus, 3 divides $b - a$, which implies that $[a]_3 = [b]_3$.

This proof might seem to imply that $\mathbb{Z}_3$ is isomorphic to the set $f(\mathbb{Z}_3) = \{f([a]_3) : [a]_3 \in \mathbb{Z}_3\} = \{[0]_6, [1]_6, [2]_6\}$. What do you think about this conclusion? Explain your answer in detail.

There is one additional important piece of information we need to consider when proving isomorphism. Activity 7.9 shows that we can define a map that preserves a ring's structure and seems to behave like an isomorphism, but if the map treats equal elements with different representations in different ways, then whatever conclusions we might draw will make little sense. We saw this same idea in Investigation 2 when we discussed well-defined operations. To emphasize this point, any time we have multiple ways to represent the elements in a set (like in $\mathbb{Z}_n$ or $\mathbb{Q}$), we need to be sure that anything that acts on the elements of that set (like an operation or a function) is well-defined. We formalize this idea for functions in the next definition.

Definition 7.10. Let S and T be sets. A mapping $f : S \to T$ is **well-defined** if $f(a) = f(b)$ whenever $a = b$ in S.

When we use the word *function*, we always mean a well-defined mapping. Well-defined mappings or functions are also called *single-valued*. In many cases we do not need to worry about a function begin well-defined. In particular, if there is only one way to represent each element in the domain, then there is nothing to show. If, however, there are multiple ways to represent elements in the domain (like in $\mathbb{Z}_n$ or $\mathbb{Q}$), then we need to know whether our mapping is well-defined before we worry about any other properties the mapping might possess.

Activity 7.11. Let f be a map from $\mathbb{Q}$ to $\mathbb{Z}$ defined by $f\left(\frac{a}{b}\right) = ab$. Is f well-defined? Verify your answer.

Disproving Isomorphism

In a previous section, we saw an example of how Definition 7.5 can be used to show that two rings are isomorphic. Although other examples may require more sophisticated arguments, the basic structure will often be the same: first, define a particular function (well-defined), and then show that this function is bijective and preserves both addition and multiplication.

What should we do, however, if we want to prove that two rings are *not* isomorphic? For instance, consider the rings (fields, actually) $\mathbb{C}$ and $\mathbb{R}$. From an algebraic standpoint, these two rings seem to be quite different. After all, $\mathbb{C}$ contains an element i such that $i^2 = -1$, while $\mathbb{R}$ does not.

If we wanted to use the definition of isomorphism to prove that $\mathbb{C}$ and $\mathbb{R}$ are not isomorphic, we would have to show that there does not exist an isomorphism $\varphi : \mathbb{C} \to \mathbb{R}$. In other words, we would have to show that every function that we could possibly define from $\mathbb{C}$ to $\mathbb{R}$ would violate at least one of the conditions that define isomorphisms. To show this directly seems daunting, if not impossible.

Let us consider, therefore, a proof by contradiction. Perhaps we could begin by assuming that there does exist a function $\varphi : \mathbb{C} \to \mathbb{R}$ that is both bijective and operation-preserving. The next activity suggests how this assumption naturally leads to a contradiction.

Activity 7.12. Assume that $\varphi : \mathbb{C} \to \mathbb{R}$ is an isomorphism.

(a) Show that $\varphi(0) = 0$. (Hint: It suffices to show that $x + \varphi(0) = x$ for all $x \in \mathbb{R}$. You may need to use the fact that φ is surjective.)

(b) Using a similar argument as in part (a), show that $\varphi(1) = 1$.

(c) Use Definition 7.5 and your answers to part (a) and (b) to argue that $\varphi(i)$ is an element of $\mathbb{R}$, and that $\varphi(i)^2 = -1$. (Hint: Apply φ to both sides of the equation $i^2 + 1 = 0$.)

(d) Explain why the result you proved in part (c) is a contradiction. Deduce that $\mathbb{C}$ cannot be isomorphic to $\mathbb{R}$.

Invariants

In Activity 7.12, we identified a property (namely, the existence of element i such that $i^2 = -1$) that $\mathbb{C}$ satisfied and $\mathbb{R}$ did not. We then argued that this difference between $\mathbb{C}$ and $\mathbb{R}$ was incompatible with the definition of isomorphism, which allowed us to conclude that $\mathbb{C}$ and $\mathbb{R}$ could not be isomorphic.

The property mentioned above is known as an *invariant*, or more specifically, an invariant of ring isomorphism. Invariants are properties that must be preserved by isomorphism. Thus, if P is an invariant and a ring R satisfies P, then every ring that is isomorphic to R must also satisfy P. Consequently, if two rings differ with respect to an established invariant, then they cannot be isomorphic.

There are many different invariants, and we provide a list of the more common ones in Table 7.1. But how does one prove that a particular property is an invariant, and how can one use this fact once it has been established? To answer these questions, consider again the result we proved in part (b) of Activity 7.12. We can state a slight generalization of this result as follows:

Theorem 7.13. *Let R and S be rings, and let $\varphi : R \to S$ be an isomorphism. If R has an identity, say 1_R, then S also has an identity. Specifically, $\varphi(1_R)$ is an identity for S.*

Activity 7.14. Fill in the missing details in the following proof of Theorem 7.13:

Proof. To show that $\varphi(1_R)$ is an identity for S, we must show that $\varphi(1_R) \cdot x = x = x \cdot \varphi(1_R)$ for all $x \in S$. Let $x \in S$. Since φ is _____, there exists $r \in R$ such that $\varphi(r) = x$. Thus,

$$\varphi(1_R) \cdot x = \varphi(1_R) \cdot \varphi(r) \,^{①}$$
$$= \varphi(1_R \cdot r) \,^{②}$$
$$= \varphi(r) \,^{③}$$
$$= x.$$

A similar argument establishes that $x \cdot \varphi(1_R) = x$. Thus, $\varphi(1_R)$ is an identity for S. ∎

The following corollary of Theorem 7.13 is immediate:

Corollary 7.15. *The existence of an identity is an invariant. Specifically, if R and S are isomorphic rings and R has identity, then S has identity also.*

Corollary 7.15 can be used to show, for example, that $\mathbb{Z}$ and $\mathbb{E}$ are not isomorphic, since $\mathbb{Z}$ has an identity and $\mathbb{E}$ does not. But what about two rings that are both the same with respect to the existence of an identity? For instance, $\mathbb{Z}_4$ and $\mathbb{Z}_2 \oplus \mathbb{Z}_2$ both have an identity. Are they isomorphic?

The answer to this question is a resounding "NO!" Although $\mathbb{Z}_4$ and $\mathbb{Z}_2 \oplus \mathbb{Z}_2$ have the same number of elements, are both commutative, and both have identity, they differ with regard to a number of other invariants. For instance, $\mathbb{Z}_4$ has exactly one zero divisor ([2]), while $\mathbb{Z}_2 \oplus \mathbb{Z}_2$ has two (([0], [1]) and ([1], [0])). Similarly, $\mathbb{Z}_4$ has exactly two units ([1] and [3]), while $\mathbb{Z}_2 \oplus \mathbb{Z}_2$ has only one (([1], [1])). Finally $\mathbb{Z}_4$ has characteristic 4, while $\mathbb{Z}_2 \oplus \mathbb{Z}_2$ has characteristic 2. Any one of these properties (i.e., invariants) would be sufficient to establish that $\mathbb{Z}_4$ and $\mathbb{Z}_2 \oplus \mathbb{Z}_2$ are not isomorphic. However, even if two rings agree with respect to every invariant we can think of, this does not prove that they are isomorphic. **Although invariants can be used to prove that two rings are different, they cannot be used to prove that two rings are the same. In order to prove that two rings are isomorphic, we must find an appropriate function from one ring to the other, and prove that this function is in fact an isomorphism.**

Table 10.1 lists some of the more common and useful invariants. This list, however, is far from complete, and we will add to it as needed throughout the remainder of our study of rings.

Concluding Activities

Activity 7.16. To which familiar ring is R_3 (from Preview Activity 7.1) isomorphic? Prove your answer.

Table 7.1
Some common invariants.

A Partial List of Invariants of Ring Isomorphism

- Number of elements

- Commutativity

- The existence of an identity

- The existence or number of zero divisors

- The existence or number of units (in a ring with identity)

- Being a field

- Being an integral domain

- Characteristic

Activity 7.17.

(a) Prove that the identity function is always an isomorphism, and thus every ring is isomorphic to itself.

(b) Let R and S be rings. Prove that if a function $\varphi : R \to S$ is a bijection, then $\varphi^{-1} : S \to R$ is also a bijection. (Hint: Use the definition of bijection to guide you. Remember that you need to show two things: that φ^{-1} is injective and that φ^{-1} is surjective.)

(c) Let R and S be rings. Prove that if $\varphi : R \to S$ is an isomorphism, then φ has an inverse and $\varphi^{-1} : S \to R$ is also an isomorphism. Deduce that if R is isomorphic to S, then S is isomorphic to R. (Hint: For each $s \in S$, use the fact that φ is surjective to write $s = \varphi(r)$ for some $r \in R$. Then note that $\varphi^{-1}(s) = r$. Use this kind of reasoning to prove that φ^{-1} preserves addition and multiplication, and then use part (b) to complete the proof of isomorphism.)

(d) Let R, S, and T be rings. Prove that if two functions $\varphi : R \to S$ and $\psi : S \to T$ are both bijections, then the composition $\psi \circ \varphi : R \to T$ is also a bijection.

(e) Let R, S, and T be rings. Prove that if $\varphi : R \to S$ and $\psi : S \to T$ are both isomorphisms, then $\psi \circ \varphi : R \to T$ is also an isomorphism. Deduce that if R and S are isomorphic, and S and T are isomorphic, then R and T are isomorphic. (Hint: Part (d) establishes part of the result. What else do you need to show?)

(f) What have you proved about the isomorphism relation in parts (a), (c), and (e)?

Activity 7.18. Prove that commutativity is an invariant. That is, prove that if R and S are isomorphic rings and R is commutative, then S must also be commutative.

Exercises

(1) Consider the ring R_2^* defined by the addition and multiplication tables below.

R_2^* :

+	w	x	y	z
w	w	x	y	z
x	x	w	z	y
y	y	z	w	x
z	z	y	x	w

·	w	x	y	z
w	w	w	w	w
x	w	x	y	z
y	w	w	w	w
z	w	x	y	z

Is R_2^* isomorphic to R_2 (from Preview Activity 7.1)? Why or why not?

(2) **Subrings of direct sums.** Let R and S be rings. Prove that $R \oplus S$ contains a subring isomorphic to R and a subring isomorphic to S.

(3) In $\mathbb{Z}_{10}$, let $S = \{[0], [2], [4], [6], [8]\}$. The fact that S is a subring of $\mathbb{Z}_{10}$ is the result of Exercise (3) in Investigation 6. Now define $f : \mathbb{Z}_5 \to S$ as follows:

$$\text{For each } [x]_5 \in \mathbb{Z}_5, \, f([x]_5) = [6x]_{10}.$$

   (a) Prove that the function f is well-defined.

   (b) Is the function f an isomorphism? Justify your conclusion.

(4)   (a) Let S and T be subrings of the real numbers, $\mathbb{R}$, and assume that $1 \in S$ and $1 \in T$. Prove that $2 \in S$ and $2 \in T$.

   (b) Prove that if $S \cong T$ and $f : S \to T$ is a (ring) isomorphism, then if S contains an element x such that $x^2 = 2$, then $[f(x)]^2 = 2$ in T.

   (c) Let $\mathbb{Z}(\sqrt{2}) = \{a + b\sqrt{2} \mid a, b \in \mathbb{Z}\}$ and $\mathbb{Z}(\sqrt{3}) = \{a + b\sqrt{3} \mid a, b \in \mathbb{Z}\}$.

   (i) Show that $\mathbb{Z}(\sqrt{2})$ and $\mathbb{Z}(\sqrt{3})$ are subrings of $\mathbb{R}$.

   (ii) Is the ring $\mathbb{Z}(\sqrt{2})$ isomorphic to the ring $\mathbb{Z}(\sqrt{3})$? Justify your conclusion.

(5) Let R and S be rings, and let $f : R \to S$ be an isomorphism. Prove that for any integer m and any $x \in R$, $f(mx) = mf(x)$.

(6) Prove that characteristic is an invariant. (Hint: Use Exercise (5).)

(7) Prove that being a field is an invariant.

(8) Prove that being an integral domain is an invariant.

(9) For any ring R, let R_0 be the subset of R containing the zero divisors of R. Show that if R and S are isomorphic rings, then there is a one-to-one correspondence between the elements of R_0 and the elements of S_0. Conclude that the number of zero divisors in a ring is an invariant.

(10) For any ring R, let R^* be the subset of R containing the units of R. Show that if R and S are isomorphic rings, then there is a one-to-one correspondence between the elements of R^* and the elements of S^*. Conclude that the number of units in a ring is an invariant.

(11) Is it possible for two fields with the same characteristic to not be isomorphic to each other? Verify your answer.

(12) Is it possible for a ring to be isomorphic to one of its proper subrings? Prove your answer. (Hint: Use Exercise (21) on page 96 of Investigation 6 with a direct sum of copies of some familiar ring.)

(13) **Rings of order 3.** Show that there are only two nonisomorphic rings of order 3 (that is, with exactly three elements).

(14) (a) Prove that every integral domain with characteristic zero contains a subring isomorphic to $\mathbb{Z}$. (Hint: Use multiples of the identity to define the desired subring.)

   (b) One of the hypotheses in the statement from part (a) can be weakened, and the statement will still be true. State this improved result, and explain why your proof from part (a) is sufficient to establish it.

(15) **Matrix representations of complex numbers.** Find a subring of $M_{2\times 2}(\mathbb{R})$ that is isomorphic to $\mathbb{C}$. Prove your answer.

(16) **Subsets of $\mathbb{C}$.** We have already seen that $\mathbb{C}$ and $\mathbb{R}$ are not isomorphic, but let's now consider other subsets of $\mathbb{C}$.

   (a) Are $\mathbb{R}$ and $\mathbb{Q}$ isomorphic? Prove your answer.

   (b) Are $\mathbb{Q}$ and $\mathbb{Z}$ isomorphic? Prove your answer.

(17) Which of the following functions are well-defined? Prove each of your answers.

   (a) $f : \mathbb{Q} \to \mathbb{Q}$ defined by $f\left(\frac{a}{b}\right) = \frac{a+b}{b}$

   (b) $f : \mathbb{Z}_4 \to \mathbb{Z}_8$ defined by $f([a]_4) = [3a]_8$

   (c) $f : \mathbb{Z}_6 \to \mathbb{Z}_7$ by $f([a]_6) = [a+k]_7$, where k is any integer.

(18) **Multiples of $\mathbb{Z}$.** Recall that for any integer $n > 1$,

$$n\mathbb{Z} = \{nx : x \in \mathbb{Z}\}.$$

Let m and n be positive integers. Find and prove a necessary and sufficient condition for $m\mathbb{Z}$ and $n\mathbb{Z}$ to be isomorphic.

(19) **Direct sums of $\mathbb{Z}_m$ and $\mathbb{Z}_n$.** Under what circumstances is $\mathbb{Z}_n \oplus \mathbb{Z}_m$ isomorphic to $\mathbb{Z}_{mn}$? State (and prove) your answer in the form of a biconditional (if and only if) statement.

(20) Recall that, for every natural number n, $\mathcal{P}_n$ denotes the power set of $\{1, 2, \ldots, n\}$, with addition defined by symmetric difference and multiplication defined by set intersection. (See Investigation 3.)

   (a) Prove that $\mathcal{P}_2$ is isomorphic to $\mathbb{Z}_2 \oplus \mathbb{Z}_2$.

   (b) Prove that $\mathcal{P}_3$ is isomorphic to $\mathbb{Z}_2 \oplus \mathbb{Z}_2 \oplus \mathbb{Z}_2$.

   (c) Based on parts (a) and (b), make and prove a general conjecture about $\mathcal{P}_n$.

Connections

In this investigation, we studied isomorphisms of rings. If you studied group theory before ring theory, you should notice connections between isomorphisms of rings in this investigation and isomorphisms of groups in Investigation 26. The idea is the same in both contexts: isomorphic rings (or groups) are essentially the same, and an isomorphism is a bijection that preserves the structure of the algebraic set. Since there are two operations defined on a ring and only one operation in a group, the major difference between isomorphisms of rings and isomorphisms of groups is that an isomorphism of rings needs to preserve two operations, but an isomorphism of groups needs to preserve only one. The process of verifying an isomorphism is the same in both contexts, but there is an extra step required for isomorphisms of rings.

Part III

Polynomial Rings

Investigation 8

Polynomial Rings

Focus Questions

By the end of this investigation, you should be able to give precise and thorough answers to the questions listed below. You may want to keep these questions in mind to focus your thoughts as you complete the investigation.

- What is a polynomial ring over a commutative ring? How are addition and multiplication of polynomials defined?

- What are some important definitions and terminology associated with polynomials?

- When are two polynomials over a commutative ring considered to be equal?

- Under what conditions is a polynomial ring an integral domain?

- What is a polynomial function, and how are polynomial functions different than polynomials?

A good deal of time is spent studying polynomials in algebra courses that are intended to be a preparation for the study of calculus. For the time being, we can consider polynomials to be algebraic expressions such as

$$3x^2 - 5x - 7 \quad \text{and} \quad x^4 - 3x^3 + 2x^2 - 7.$$

In algebra courses, we learned how to add, subtract, and multiply polynomials. We even studied "polynomial long division" as a way to find a quotient and a remainder when one polynomial is divided by another polynomial. This is similar to the work we have done with the integers, and so it may not be surprising that we will be able to develop our study of polynomials within the context of ring theory.

The polynomials studied in elementary mathematics courses were almost always restricted to having real number coefficients. In this and subsequent investigations, we will consider polynomials that have coefficients from other commutative rings and fields. We will see that most of the familiar results from elementary algebra courses about polynomials will be true when the coefficients for the polynomials are from a field.

Preview Activity 8.1. Complete each of the following using the rules for addition and multiplication of polynomials learned in precalculus algebra courses.

(a) Let $f(x) = 2x^2 + 3x + 2$ and $g(x) = 3x^2 + 4x + 2$. Determine $f(x) + g(x)$ and $f(x)g(x)$.

(b) Let $f(x) = [2]x^2 + [3]x + [2]$ and $g(x) = [3]x^2 + [4]x + [2]$, where the coefficients of the polynomials are elements of $\mathbb{Z}_5$. Determine $f(x) + g(x)$ and $f(x)g(x)$.

(c) Let $f(x) = [2]x^2 + [3]x + [2]$ and $g(x) = [3]x^2 + [4]x + [2]$, where the coefficients of the polynomials are elements of $\mathbb{Z}_6$. Determine $f(x) + g(x)$ and $f(x)g(x)$.

Polynomials over Commutative Rings

We will now give a formal definition of a polynomial over a ring. Although the definitions could be written for any ring, we will focus our attention on polynomials over commutative rings.

Definition 8.2. Let R be a commutative ring. A **polynomial in x over R** is an expression of the form

$$a_n x^n + a_{n-1} x^{n-1} + \cdots a_2 x^2 + a_1 x^1 + a_0 x^0,$$

where n is a nonnegative integer, and $a_n, a_{n-1}, \ldots, a_2, a_1, a_0$ are elements of R. The **set of all polynomials over the ring R** will be denoted by $R[x]$.

Conventions and Terminology

One of our first goals will be to prove that $R[x]$ is a commutative ring. Before doing that, we will need to introduce some notation and terminology. Let R be a commutative ring, and let $p(x) = a_n x^n + a_{n-1} x^{n-1} + \cdots a_2 x^2 + a_1 x^1 + a_0 x^0 \in R[x]$ with $a_n \neq 0$.

- The symbol x is called an **indeterminate**. It is to be regarded as a formal symbol and not as an element of the ring R. In effect, the symbols $x^0, x^1, x^2, \ldots, x^n$ serve as placeholders for the ring elements $a_0, a_1, \ldots, a_n$.

- The expressions $a_k x^k$ in the polynomial $p(x)$ are called the **terms of the polynomial**. The elements $a_0, a_1, \ldots, a_n$ in the ring R are called the **coefficients** of the polynomial $p(x)$. We call a_k the coefficient of x^k in the representation of $p(x)$.

- When working with a polynomial, instead of writing x^1, we simply write x. In addition, we usually do not write x^0 and will write $a_0 x^0$ simply as a_0. Using these conventions, we can write $p(x)$ in the form

$$p(x) = a_n x^n + a_{n-1} x^{n-1} + \cdots a_2 x^2 + a_1 x + a_0$$

where n is a nonnegative integer, $a_n, a_{n-1}, \ldots, a_2, a_1, a_0$ are elements of R, and $a_n \neq 0$. (Although this substitution may seem obvious, there are actually some subtle issues that must be addressed in order for it to be valid. Activity 8.14 explores these subtleties in more detail.)

- We will usually omit any term having a zero coefficient from the representation of a polynomial, and if the ring R has an identity, we will write a term of the form $1_R x^k$ simply as x^k. For example, in $\mathbb{R}[x]$, instead of writing $f(x) = 3x^3 + 1x^2 + 0x + 7$, we will write $f(x) = 3x^3 + x^2 + 7$. We will also write terms of the form $(-a_k)x^k$ as $-a_k x^k$. So instead of writing $g(x) = (-3)x^2 + (-7)$, we will write $g(x) = -3x^2 - 7$.

- The coefficient a_0 is called the **constant term** of the polynomial $p(x)$. A polynomial of the form $p(x) = a$, where $a \in R$, is called a **constant polynomial**.

- The coefficient a_n is called the **leading coefficient** of the polynomial $p(x)$. If the ring R has an identity and the leading coefficient a_n is equal to 1_R, the polynomial $p(x)$ is called a **monic polynomial**.[*]

- The nonnegative integer n is called the **degree** of the polynomial $p(x)$, and we write $\deg(p(x)) = n$.

- We will denote by 0 the polynomial in $R[x]$ having all of its coefficients equal to 0. This polynomial is called the **zero polynomial**, and since it does not have a leading coefficient, the degree of the zero polynomial is undefined.

- Two polynomials in $R[x]$

$$p(x) = a_n x^n + a_{n-1} x^{n-1} + \cdots a_2 x^2 + a_1 x^1 + a_0 x^0, \text{ and}$$
$$q(x) = b_m x^m + b_{m-1} x^{m-1} + \cdots b_2 x^2 + b_1 x^1 + b_0 x^0,$$

are considered to be **equal polynomials** if both of them are the zero polynomial or if both have the same degree and all pairs of corresponding coefficients are equal.

- When we write a polynomial in the form

$$p(x) = a_n x^n + a_{n-1} x^{n-1} + \cdots a_2 x^2 + a_1 x + a_0,$$

we say that we have written the polynomial in **descending powers of x**. A polynomial in the form

$$p(x) = a_0 + a_1 x + a_2 x^2 + \cdots + a_{n-1} x^{n-1} + a_n x^n$$

is said to be written in **ascending powers of x**.

The formal definitions of addition and multiplication of polynomials are quite technical and notationally complex, but one of the purposes of Preview Activity 8.1 was to show that when working with specific polynomials, we can simply add and multiply polynomials as we did in previous mathematics courses. For example, we might "combine like terms" as follows:

$$5x^2 + 7x^2 = 12x^2.$$

Or, stated more generally:

$$a_2 x^2 + b_2 x^2 = (a_2 + b_2) x^2.$$

The next activity builds on this intuitive process and foreshadows the formal definitions of polynomial addition and multiplication.

Activity 8.3. Let R be a commutative ring with identity and let

$$p(x) = a_3 x^3 + a_2 x^2 + a_1 x + a_0 \quad \text{and} \quad q(x) = b_2 x^2 + b_1 x + b_0$$

be polynomials in $R[x]$. Using the standard rules for adding and multiplying polynomials from previous mathematics courses, determine the sum $p(x) + q(x)$ and the product $p(x)q(x)$. Write both results in descending powers of x.

---

[*]In familiar rings, it is easy to determine whether a given polynomial is monic by simply looking at its leading term. For instance, in $\mathbb{R}[x]$, both $p(x) = x^3 + 3x - 5$ and $q(x) = x^{99} + 1$ are monic polynomials since their leading terms are of the form x^n (for $n = 3$ in the first case, and $n = 99$ in the second). It is important to note, however, that the definition of a monic polynomial depends on 1_R, and so certain monic polynomials in certain rings may not "look" monic. For instance, $p(x) = [4]x^2$ is monic in $2\mathbb{Z}_6[x]$, since $[4]$ is the multiplicative identity in $2\mathbb{Z}_6$.

We will now formally define polynomial addition and multiplication. It might be helpful to use the work from Activity 8.3 as examples while studying these definitions.

Definition 8.4. Let R be a commutative ring and let $p(x), q(x) \in R[x]$ with

$$p(x) = a_n x^n + a_{n-1} x^{n-1} + \cdots + a_2 x^2 + a_1 x + a_0 \text{ with } a_n \neq 0, \text{ and}$$
$$q(x) = b_m x^m + b_{m-1} x^{m-1} + \cdots + b_2 x^2 + b_1 x + b_0 \text{ with } b_m \neq 0.$$

Since it must be true that $m \leq n$ or $n \leq m$, we can assume that $m \leq n$ without loss of generality. We will then use $b_{m+1} = b_{m+2} = \cdots b_n = 0$ and so we can write

$$p(x) = a_n x^n + a_{n-1} x^{n-1} + \cdots + a_2 x^2 + a_1 x + a_0, \text{ and}$$
$$q(x) = b_n x^n + b_{n-1} x^{n-1} + \cdots + b_2 x^2 + b_1 x + b_0.$$

The **sum of the polynomials** $p(x)$ and $q(x)$ is defined to be

$$p(x) + q(x) = (a_n + b_n) x^n + (a_{n-1} + b_{n-1}) x^{n-1} + \cdots + (a_1 + b_1) x + (a_0 + b_0).$$

The **product of the polynomials** $p(x)$ and $q(x)$ is defined to be

$$p(x)q(x) = c_{m+n} x^{m+n} + c_{m+n-1} x^{m+n-1} + \cdots + c_2 x^2 + c_1 x + c_0,$$

where for each k with $0 \leq k \leq m + n$,

$$c_k = a_k b_0 + a_{k-1} b_1 + a_{k-2} b_2 + \cdots + a_2 b_{k-2} + a_1 b_{k-1} + a_0 b_k.$$

In the definition of multiplication, notice that in each of the terms for c_k, the sum of the two subscripts is equal to k. So we can say that in the product $p(x)q(x)$, the coefficient of the x^k term is the sum of all products of the form $a_i b_j$, where $i + j = k$. We can also write

$$c_k = \sum_{i+j=k} a_i b_j = \sum_{i=0}^{k} a_{k-i} b_i.$$

For example,

$$c_0 = a_0 b_0$$
$$c_1 = a_1 b_0 + a_0 b_1$$
$$c_2 = a_2 b_0 + a_1 b_1 + a_0 b_2$$
$$\vdots \quad \vdots$$
$$c_{m+n} = a_n b_m$$

The work in Activity 8.3 illustrated these definitions with specific polynomials. Using

$$p(x) = a_3 x^3 + a_2 x^2 + a_1 x + a_0 \quad \text{and} \quad q(x) = b_2 x^2 + b_1 x + b_0,$$

we see that

$$p(x) + q(x) = a_3 x^3 + (a_2 + b_2) x^2 + (a_1 + b_1) x + (a_0 + b_0),$$

and

$$p(x)q(x) = a_3 b_2 x^5 + (a_3 b_1 + a_2 b_2) x^4 + (a_3 b_0 + a_2 b_1 + a_1 b_2) x^3$$
$$+ (a_2 b_0 + a_1 b_1 + a_0 b_2) x^2 + (a_1 b_0 + a_0 b_1) x + a_0 b_0.$$

If R is a commutative ring, then in order for $R[x]$ to be a ring, $R[x]$ must be closed under polynomial addition and polynomial multiplication. So if $p(x), q(x) \in R[x]$, we see by the very definitions of addition and multiplication that $p(x) + q(x) \in R[x]$ and $p(x)q(x) \in R[x]$. Hence, $R[x]$ is closed under addition and multiplication. In addition, using the definitions for addition and multiplication of polynomials, it is possible to prove that if R is a commutative ring, then $R[x]$ is a commutative ring.

Activity 8.5. In this activity, we will explore the proofs of the properties of addition in $R[x]$. Formal proofs of these properties would be required in a proof that $R[x]$ is a commutative ring when R is a commutative ring. To understand these proofs, it is sometimes wise to work with the properties in "simple" special cases. For polynomials, this often means trying to do the proofs for polynomials with small degrees, such as degree 1, 2, or 3. That is what we will do in this activity.

So let R be a commutative ring and let

$$p(x) = a_2x^2 + a_1x + a_0, \quad q(x) = b_2x^2 + b_1x + b_0, \quad \text{and} \quad r(x) = c_1x + c_0$$

be polynomials in $R[x]$. For the following activities (and the formal proofs), since the coefficients of a polynomial are elements of the ring R, we can use the ring properties of R to prove the corresponding properties for $R[x]$.

(a) Illustrate the commutative property of addition in $R[x]$ by using the polynomials $p(x)$ and $q(x)$.

(b) Illustrate the associative property of addition in $R[x]$ by using the polynomials $p(x)$, $q(x)$, and $r(x)$.

(c) Let $z(x) = 0$ be the zero polynomial in $R[x]$. Verify that $p(x) + z(x) = p(x)$. What does this illustrate about addition in $R[x]$?

(d) Determine an additive inverse for $p(x)$ in $R[x]$.

(e) Notice that

$$p(x)q(x) = a_2b_2x^4 + (a_2b_1 + a_1b_2)\,x^3 + (a_2b_0 + a_1b_1 + a_0b_2)\,x^2$$
$$+ (a_1b_0 + a_0b_1)\,x + a_0b_0.$$

Verify that $p(x)q(x) = q(x)p(x)$.

(f) Assume R has a multiplicative identity and let $u(x) = 1_R$. Verify that $p(x)u(x) = p(x)$. What does this illustrate about multiplication in $R[x]$?

We still have to explore the associative property for multiplication and the distributive property in $R[x]$. This will be done in Activity 8.16 and Exercise (8).

Even though we have not given a formal proof, our work in Activity 8.5, Activity 8.16, and Exercise (8) should suggest that the following theorem is true.

Theorem 8.6. *If R is a commutative ring, then $R[x]$ is a commutative ring. In addition, if the ring R has an identity, then the ring $R[x]$ has an identity.*

A formal proof of Theorem 8.6 is given in the online supplemental materials.

Polynomials over an Integral Domain

Let R be a commutative ring with identity. Since we now know that $R[x]$ is a commutative ring with identity, it is natural to consider whether $R[x]$ is also an integral domain, or perhaps even a field. The following examples might be helpful in exploring the conditions under which $R[x]$ is an integral domain. Notice that if $p(x) = 2x^2 + 4x + 2$ and $q(x) = 3x + 3$ are considered to be polynomials in $\mathbb{Z}[x]$, we can then verify that

$$p(x)q(x) = \left(2x^2 + 4x + 2\right)(3x + 3) = 6x^3 + 18x^2 + 18x + 6.$$

However, if $f(x) = [2]x^2 + [4]x + [2]$ and $g(x) = [3]x + [3]$ are polynomials in $\mathbb{Z}_6[x]$, then we see that

$$\begin{aligned} f(x)g(x) &= [6]x^3 + [18]x^2 + [18]x + [6] \\ &= [0] \end{aligned}$$

and so $f(x)$ and $g(x)$ are zero divisors in $\mathbb{Z}_6[x]$. This shows that $\mathbb{Z}_6[x]$ is not an integral domain. Basically, what we did was to use zero divisors in $\mathbb{Z}_6$ to construct zero divisors in $\mathbb{Z}_6[x]$. In Exercise (4), we will generalize this argument and prove the following theorem.

Theorem 8.7. *If R is a commutative ring that contains zero divisors, then the polynomial ring $R[x]$ also contains zero divisors.*

A question that still remains is whether or not $\mathbb{Z}[x]$ is an integral domain. The work in Activity 8.9 will help determine the answer to this question, but in a somewhat more general context. The key will be to focus on the relationship between the degrees of the two polynomials and the degree of the product of these two polynomials. In previous algebra classes, when we worked with polynomials with real number coefficients, we learned that if $\deg(p(x)) = m$ and $\deg(q(x)) = n$, then $\deg(p(x)q(x)) = m + n$. The preceding example, however, shows that this may not be the case in polynomial rings such as $\mathbb{Z}_6[x]$, as multiplying two polynomials with positive degree can yield a polynomial with undefined degree—namely, the zero polynomial. The next activity provides another example of this type of behavior.

Activity 8.8. In $\mathbb{Z}_6[x]$, let

$$f(x) = [2]x^2 + x + [5] \quad \text{and} \quad g(x) = [3]x + [2].$$

(a) Calculate (and simplify) the product $f(x)g(x)$.

(b) What is the degree of the product $f(x)g(x)$? How does this compare to $\deg(f(x)) + \deg(g(x))$?

Activity 8.8 demonstrates that polynomials over rings other than $\mathbb{R}$ may not behave exactly the way we might expect them to with regard to multiplication. The next activity explores what conditions must be placed on R in order to ensure the expected relationship between the degrees of two polynomials in $R[x]$ and the degree of their product.

Activity 8.9. Let D be an integral domain. We then know that the product of any two nonzero elements of D is also nonzero.

(a) Let a and u be nonzero elements in D and let $b, v \in D$. Write the product

$$(ax + b)(ux + v)$$

in $D[x]$ in descending powers of x. What is the degree of this product? Justify your conclusion. (Hint: Since $a \neq 0$, $u \neq 0$, and D is an integral domain, what can be concluded about the product au?)

(b) With the same notation as in part (a), let $c \in D$ and write the product

$$\left(ax^2 + bx + c\right)(ux + v)$$

in $D[x]$ in descending powers of x. What is the degree of this product? Justify your conclusion.

(c) Use the ideas from the previous two examples to help complete the proof of the following very important theorem.

Theorem 8.10. *Let m and n be nonnegative integers. If D is an integral domain, then the product of polynomials of degree m and n in $D[x]$ is a polynomial in $D[x]$ of degree $(m+n)$.*

Proof. Let m and n be nonnegative integers, and let D be an integral domain and let $p(x), q(x) \in D[x]$ with $\deg(p(x)) = n$ and $\deg(q(x)) = m$. We can then write

$$p(x) = a_n x^n + a_{n-1} x^{n-1} + \cdots a_2 x^2 + a_1 x^1 + a_0 x^0, \text{ and}$$
$$q(x) = b_m x^m + b_{m-1} x^{m-1} + \cdots b_2 x^2 + b_1 x^1 + b_0 x^0,$$

with $a_n \neq 0$ and $b_m \neq 0$. Using the definition of the product of two polynomials in Definition 8.4, we first note that the largest possible sum of the subscripts of a product of the form $a_i b_j$ is $m + n$ and hence, $\deg(p(x)q(x)) \leq m + n$. ⑨

(Now complete the proof by focusing on the coefficient of the x^{m+n} term.) ∎

(d) Explain how to use Theorem 8.10 to prove the following corollary.

Corollary 8.11. *If D is an integral domain, then $D[x]$ is an integral domain.*

(e) Determine $([2]x + [1])^2$ in $\mathbb{Z}_4[x]$. Write the product in descending powers of x. Explain why this shows that it is necessary to assume that D is an integral domain in Theorem 8.10.

(f) How would Theorem 8.10 need to be modified if D was not assumed to be an integral domain? (Hint: Look at the proof from part (c). Only part of it relies on the assumption that D is an integral domain.)

Polynomial Functions

The definitions of polynomials and addition and multiplication of polynomials were meant to formalize and generalize the way we added and multiplied polynomials in previous mathematics courses. However, there has been one big difference between our formal study of polynomials and the idea of a polynomial function used in precalculus and calculus courses. In those courses, we often think of a polynomial as function $\overline{p} : \mathbb{R} \to \mathbb{R}$ for which $\overline{p}(x) = a_n x^n + a_{n-1} x^{n-1} + \cdots a_1 x + a_0$, for some real numbers $a_0, a_1, \ldots a_{n-1}, a_n$.

The notations are similar, but there is a fundamental difference between the polynomial function $\overline{p}$ and the polynomial $p(x) = a_n x^n + a_{n-1} x^{n-1} + \cdots a_1 x + a_0$ in $\mathbb{R}[x]$. The polynomial $p(x)$ is a formal expression in which x is an indeterminate, essentially serving as a placeholder for the real numbers $a_0, a_1, \ldots a_{n-1}, a_n$. In this way, polynomials are really just sequences of numbers (or ring elements in general). The notation we use to represent polynomials, which uses powers of the indeterminate x to keep track of the coefficients, allows us to add and multiply polynomials in a more intuitive way. (Can you imagine how much harder it would be to add and multiply polynomials if we only were able to use the formal definitions of addition and multiplication, without the algebraic intuition behind them?)

On the other hand, when working with a polynomial function $\overline{p}$, the symbol x is a variable, which means that real numbers can be substituted for x, making $\overline{p}(x)$ a real number. So there is a fundamental difference—albeit a subtle one—between a polynomial and a polynomial function. In spite of this fundamental difference, however, we can still use a polynomial over a commutative ring to define a polynomial function. The next definition describes how we can do this.

Definition 8.12. Let R be a commutative ring and let

$$p(x) = a_n x^n + a_{n-1} x^{n-1} + \cdots a_1 x + a_0$$

be a polynomial in $R[x]$. The **polynomial function induced by $p(x)$** is the function $\overline{p} : R \to R$, where for each r in R,

$$\overline{p}(r) = a_n r^n + a_{n-1} r^{n-1} + \cdots a_1 r + a_0.$$

For simplicity, we often just say that $\overline{p}$ is a **polynomial function**.

The next activity will help illustrate the difference between polynomials and polynomial functions.

Activity 8.13.

(a) Let $p(x) = x^4$ and $q(x) = x^2$ be polynomials in $\mathbb{Z}_5[x]$.

    (i) Is $p(x) = q(x)$ in $\mathbb{Z}_5[x]$?

    (ii) For the polynomial function $\overline{p} : \mathbb{Z}_5 \to \mathbb{Z}_5$, determine $\overline{p}([0]), \overline{p}([1]), \overline{p}([2]), \overline{p}([3])$, and $\overline{p}([4])$.

    (iii) For the polynomial function $\overline{q} : \mathbb{Z}_5 \to \mathbb{Z}_5$, determine $\overline{q}([0]), \overline{q}([1]), \overline{q}([2]), \overline{q}([3])$, and $\overline{q}([4])$.

    (iv) Is the function $\overline{p}$ equal to the function $\overline{q}$? Explain.

(b) Let $p(x) = x^4$ and $q(x) = x^2$ be polynomials in $\mathbb{Z}_4[x]$.

    (i) Is $p(x) = q(x)$ in $\mathbb{Z}_4[x]$?

    (ii) For the polynomial function $\overline{p} : \mathbb{Z}_4 \to \mathbb{Z}_4$, determine $\overline{p}([0]), \overline{p}([1]), \overline{p}([2])$, and $\overline{p}([3])$.

    (iii) For the polynomial function $\overline{q} : \mathbb{Z}_4 \to \mathbb{Z}_4$, determine $\overline{q}([0]), \overline{q}([1]), \overline{q}([2])$, and $\overline{q}([3])$.

    (iv) Is the function $\overline{p}$ equal to the function $\overline{q}$? Explain.

Concluding Activities

Activity 8.14. The formal definition of a polynomial over a commutative ring R includes a term of the form $a_0 x^0$, where a_0 is an element of R. In order to simplify notation a bit, we decided to write a_0 in place of $a_0 x^0$. Although this replacement may seem natural, there are a few subtle points that must be considered in order for it to be valid. First, although it may be tempting to simply say that $x^0 = 1$ and so $a_0 x^0$ is in fact *equal* to a_0, we must remember that x does not denote a ring element, but rather an indeterminate. In addition, even if we did view x as a ring element, the ring R may not have an identity, and so even in this case it might not make sense to say that $x^0 = 1$. What we *can* say is that the set of polynomials of the form $a_0 x^0$ is a subring of $R[x]$ and is always isomorphic to R. Thus, the substitution of a_0 for $a_0 x^0$ is legitimate, and this substitution behaves exactly as we would expect it to with respect to polynomial addition and multiplication. In parts (a) and (b) below, you are asked to prove that this is in fact the case.

(a) Let R be a commutative ring and let $S = \{a_0 x^0 : a_0 \in R\}$. Prove that S is a subring of $R[x]$.

(b) Define a function $h : R \to S$ by $h(r) = rx^0$, for each r in R. Prove that h is a ring isomorphism.

Activity 8.15. We know that if R is a commutative ring, then $R[x]$ is a commutative ring, and that if R is a commutative ring with identity, then $R[x]$ is a commutative ring with identity. (See Theorem 8.6.) We also know that if D is an integral domain, then $D[x]$ is an integral domain. (See Corollary 8.11.) So it seems reasonable to ask, "If F is a field, then is $F[x]$ a field?"

(a) Is $\mathbb{R}[x]$ a field? (Hint: Consider the nonzero polynomial $p(x) = x$ in $\mathbb{R}[x]$.)

(b) Is $\mathbb{Z}_3[x]$ a field?

(c) If F is a field, is $F[x]$ always, sometimes, or never a field? Give a proof or a pair of examples (whichever is most appropriate) to justify your answer.

Activity 8.16. In Activity 8.5, we used special examples to illustrate the properties of addition for the ring $R[x]$ when R is a commutative ring. We will now explore the associative property of multiplication in $R[x]$. The formal proofs of this property can be notationally complex and somewhat difficult to follow. As such, here we will look at cases that are a bit simpler than the general forms. (A formal proof of the associative property in $R[x]$ is provided in the online supplemental materials, and Exercise (8) explores the distributive property for $R[x]$ from a similar perspective as this activity.)

To begin, let R be a commutative ring, and let

$$p(x) = a_2 x^2 + a_1 x + a_0, \quad q(x) = b_2 x^2 + b_1 x + b_0, \quad \text{and} \quad r(x) = c_1 x + c_0$$

be polynomials in $R[x]$.

(a) Show that

$$p(x)q(x) = a_2 b_2 x^4 + (a_2 b_1 + a_1 b_2) x^3 + (a_2 b_0 + a_1 b_1 + a_0 b_2) x^2$$
$$+ (a_1 b_0 + a_0 b_1) x + a_0 b_0$$

(b) Use your answer to part (a) to calculate $[p(x)q(x)]r(x)$. Then rewrite your result to show that

$$[p(x)q(x)]r(x) = (a_2b_2)\,c_1x^5 + [(a_2b_2)\,c_0 + (a_2b_1)\,c_1 + (a_1b_2)\,c_1]\,x^4$$
$$+ [(a_2b_1)\,c_0 + (a_1b_2)\,c_0 + (a_2b_0)\,c_1 + (a_1b_1)\,c_1 + (a_0b_2)\,c_1]\,x^3$$
$$+ [(a_2b_0)\,c_0 + (a_1b_1)\,c_0 + (a_0b_2)\,c_0 + (a_1b_0)\,c_1 + (a_0b_1)\,c_1]\,x^2$$
$$+ [(a_1b_0)\,c_0 + (a_0b_1)\,c_0 + (a_0b_0)\,c_1]\,x + (a_0b_0)\,c_0$$

(Notice that the sum of the subscripts in each term of the coefficient of x^j is equal to j.)

(c) Now use a process similar to that in parts (a) and (b) to write the product $p(x)[q(x)r(x)]$ in descending powers of x.

(d) Use the results of parts (a) – (c) to prove that

$$[p(x)q(x)]r(x) = p(x)[q(x)r(x)].$$

Exercises

(1) For each of the following polynomials, state its degree, its leading coefficient, and its constant term, and write the polynomial in ascending powers of x.

    (a) $5x^3 + 2x^2 - x + 7$ in $\mathbb{Z}[x]$.

    (b) $7x - \sqrt{2}$ in $\mathbb{R}[x]$.

    (c) $(1+i)x^4 - (2i)x^2 + x^3 + (5+2i)$ in $\mathbb{C}[x]$.

    (d) $[7]x^8 + [3]x^4 + [9]$ in $\mathbb{Z}_{10}[x]$.

(2) In each of the following, perform the indicated operations and write the result in descending powers of x.

    (a) $([3]x^2 + x + [2]) + (x^3 + x^2 + [3]x + [2])$ in $\mathbb{Z}_4[x]$.

    (b) $([2]x + [1])^2$ in $\mathbb{Z}_4[x]$.

    (c) $([3]x + [1])^2$ in $\mathbb{Z}_6[x]$.

    (d) $([4]x + [1])^2$ in $\mathbb{Z}_8[x]$.

    (e) $([2]x + [5])\,([5]x + [8])$ in $\mathbb{Z}_{10}[x]$.

    (f) $([2]x + [5])\,([5]x + [8])$ in $\mathbb{Z}_{11}[x]$.

(3) Give an example of each of the following or explain why no such example exists.

    (a) Two polynomials of degree 1 in $\mathbb{Z}_4[x]$ whose product is a polynomial of degree 1.

    (b) Two polynomials of degree 3 in $\mathbb{R}[x]$ whose sum is a polynomial of degree 1.

    (c) Two polynomials of degree 2 in $\mathbb{Z}_6[x]$ whose product is a polynomial of degree 5.

⋆ (4) Let R be a commutative ring, and let a and b be nonzero elements in R with $ab = 0$. (So R contains zero divisors.) Prove that $R[x]$ is not an integral domain by constructing two polynomials of degree 0 or 1 in $R[x]$ whose product is equal to the zero polynomial.

⋆ (5) Let R be a commutative ring, and let $p(x), q(x) \in R[x]$ with $\deg(p(x)) = m$ and $\deg(q(x)) = n$, where m and n are nonnegative integers.

    (a) Prove that the degree of $p(x) + q(x)$ is less than or equal to the maximum value of m and n.

    (b) Prove that $\deg(p(x)q(x)) \le m + n$.

(6) (a) List all polynomials of degree 2 in $\mathbb{Z}_2[x]$.

    (b) List all polynomials of degree 3 in $\mathbb{Z}_2[x]$.

    (c) List all polynomials of degree 2 in $\mathbb{Z}_3[x]$.

    (d) List all polynomials of degree 3 in $\mathbb{Z}_3[x]$.

(7) Let m be an integer with $m \ge 1$.

    (a) How many polynomials of degree m are there in $\mathbb{Z}_2[x]$?

    (b) How many polynomials of degree m are there in $\mathbb{Z}_3[x]$?

    (c) Let n be an integer with $n \ge 2$. How many polynomials of degree m are there in $\mathbb{Z}_n[x]$?

⋆ (8) Let R be a commutative ring, and let

$$p(x) = a_2 x^2 + a_1 x + a_0, \quad q(x) = b_2 x^2 + b_1 x + b_0, \quad \text{and} \quad r(x) = c_1 x + c_0$$

be polynomials in $R[x]$. Verify that $p(x)\,(q(x) + r(x)) = p(x)q(x) + p(x)r(x)$, thereby illustrating the distributive property in $R[x]$.

(9) Let R be a commutative ring, and let

$$S = \left\{ a_n x^n + a_{n-1} x^{n-1} + \cdots a_1 x + a_0 : a_k = 0 \text{ if } k \text{ is odd} \right\}.$$

Is S a subring of $R[x]$? Justify your conclusion.

(10) Let R be a commutative ring, and let

$$S = \left\{ a_n x^n + a_{n-1} x^{n-1} + \cdots a_1 x + a_0 : a_k = 0 \text{ if } k \text{ is even} \right\}.$$

Is S a subring of $R[x]$? Justify your conclusion.

(11) Let R be a commutative ring, and let

$$S = \left\{ a_n x^n + a_{n-1} x^{n-1} + \cdots a_1 x + a_0 : a_0 = 0_R \right\}.$$

Is S a subring of $R[x]$? Justify your conclusion.

(12) Let R be a commutative ring, and let

$$S = \left\{ p(x) \in R[x] : \deg(p(x)) = 2 \right\}.$$

Is S a subring of $R[x]$? Justify your conclusion.

(13) Let R and S be commutative rings, and let $h : R \to S$ be an isomorphism. Define the function $H : R[x] \to S[x]$ by

$$H\left(a_n x^n + \cdots a_1 x + a_0\right) = h\left(a_n\right) x^n + \cdots h\left(a_1\right) x + h\left(a_0\right),$$

for each $a_n x^n + a_{n-1} x^{n-1} + \cdots a_1 x + a_0$ in $R[x]$.

   (a) Prove that the function H is an isomorphism.

   (b) Deduce from part (a) that if R is isomorphic to S, then $R[x]$ is isomorphic to $S[x]$.

(14) In each of the following, a ring R is given and two polynomials $p(x)$ and $q(x)$ in $R[x]$ are given. In each case, determine if the polynomial functions $\overline{p} : R \to R$ and $\overline{q} : R \to R$ are equal.

   (a) $\mathbb{Z}_4$ and $p(x) = [2]x^2 + [2]x$ and $q(x) = [0]$.

   (b) $\mathbb{Z}_3$ and $p(x) = x^3 + [2]x$ and $q(x) = [2]x + [2]$.

   (c) $\mathbb{Z}_4$ and $p(x) = x^3 + [2]x^2 + [2]x$ and $q(x) = [3]x + [2]$.

   (d) $\mathbb{Z}_3$ and $p(x) = x^3 + x$ and $q(x) = [2]x$.

Connections

This investigation introduced polynomial rings. An important idea to remember from this investigation is that a polynomial ring with coefficients in an integral domain behaves much like the ring of integers from Investigation 1. We will see more of this behavior in subsequent investigations where we discuss the division algorithm, greatest common divisors, and unique factorization in polynomial rings.

Investigation 9

Divisibility in Polynomial Rings

Focus Questions

By the end of this investigation, you should be able to give precise and thorough answers to the questions listed below. You may want to keep these questions in mind to focus your thoughts as you complete the investigation.

- What does it mean to say that one polynomial divides another polynomial? What are some of the important properties of divisibility in $F[x]$, where F is a field?

- What is the Division Algorithm in $F[x]$, where F is a field? How is this similar to the Division Algorithm for integers, and how is it different?

- What is the greatest common divisor of two polynomials in $F[x]$, where F is a field? What results about greatest common divisors of integers also hold in polynomial rings?

- How can the Euclidean Algorithm be used to find the greatest common divisor of two polynomials in $F[x]$, where F is a field?

Preview Activity 9.1. In this investigation, we will explore the Division Algorithm and greatest common divisors within the context of polynomial rings. Let's begin with a review of the analogous ideas in the ring of integers.

(a) Write the definition of what it means to say that an integer a divides an integer b.

(b) Give a precise statement of the Division Algorithm in the integers.

(c) Write the definition for the greatest common divisor of two integers a and b, at least one of which is nonzero.

(d) What is a prime number? Give a precise definition.

(e) State the Fundamental Theorem of Arithmetic.

Introduction

In the next two investigations, we will follow a process similar to what we did for the integers in order to prove several results about divisibility within the integral domain $F[x]$, where F is a field. Our work in the integers used the Division Algorithm, greatest common divisors, and prime numbers to eventually prove the Fundamental Theorem of Arithmetic. Our work here will also lead to a unique factorization theorem in $F[x]$ and will set the stage for us to investigate roots of polynomials. We will begin by defining divisibility within $R[x]$, where R is a commutative ring. Notice the similarity between the next definition and the definition of divides in $\mathbb{Z}$.

Definition 9.2. Let R be a commutative ring and let $u(x)$ and $v(x)$ be polynomials in $R[x]$. The polynomial $u(x)$ **divides** the polynomial $v(x)$ provided that there exists a polynomial $q(x) \in R[x]$ such that $v(x) = u(x)q(x)$. In this case, we also say that $u(x)$ is a **factor** of $v(x)$ and sometimes write $u(x) \mid v(x)$.

For example, in $\mathbb{R}[x]$, let $f(x) = 3x+2$, $g(x) = x^2+4x-7$, and $h(x) = 3x^3+14x^2-13x-14$. it is easy to verify by multiplication that

$$f(x)g(x) = (3x+2)\left(x^2 + 4x - 7\right) = 3x^3 + 14x^2 - 13x - 14 = h(x),$$

and so we can say that in $\mathbb{R}[x]$, $f(x)$ divides $h(x)$ and $g(x)$ divides $h(x)$.

The use of Definition 9.2 in proofs is very similar to the use of the definition of divides in the integers. This will be illustrated with proofs of the various parts of Theorem 9.3. Notice that this theorem deals with polynomials over a field F. We will see later in the investigation why this restriction is necessary.

Theorem 9.3. *Let F be a field and let $f(x), g(x) \in F[x]$.*

(i) *If $f(x)$ divides $g(x)$ and $c \in F$ and $c \neq 0$, then $cf(x)$ divides $g(x)$.*

(ii) *If $f(x) \neq 0$, $g(x) \neq 0$, and $f(x)$ divides $g(x)$, then $\deg(f(x)) \leq \deg(g(x))$.*

(iii) *If $f(x) \neq 0$ and a_n is the leading coefficient of $f(x)$, then $a_n^{-1}f(x)$ is a monic polynomial.*

(iv) *If $f(x)$ divides $g(x)$ and $g(x)$ divides $f(x)$, then there exists $c \in F$ with $c \neq 0$ such that $f(x) = cg(x)$.*

(v) *Let $f(x)$ and $g(x)$ be monic polynomials in $F[x]$. If $f(x)$ divides $g(x)$ and $g(x)$ divides $f(x)$, then $f(x) = g(x)$.*

Proof. We will prove parts (i) and (iii). The other parts will be proved in Activity 9.4. For part (i), we assume that $f(x)$ divides $g(x)$ and $c \in F$ with $c \neq 0$. So there exists a polynomial $q(x) \in F[x]$ such that

$$g(x) = f(x)q(x).$$

We multiply both sides of this equation by c and obtain $cg(x) = cf(x)q(x)$. Since $c \neq 0$, c has a multiplicative inverse in F and so if we now multiply both sides of the equation by c^{-1}, we obtain

$$c^{-1}[cg(x)] = c^{-1}\left[cf(x)q(x)\right].$$

We know that $c^{-1}c = 1$ on the left side, and on the right side, we can write $c^{-1}\left[cf(x)q(x)\right] = [cf(x)]\left[c^{-1}q(x)\right]$. So we have

$$g(x) = [cf(x)]\left[c^{-1}q(x)\right].$$

This proves that $cf(x)$ divides $g(x)$.

For part (iii), we write $f(x) = a_n x^n + a_{n-1} x^{n-1} + \cdots + a_1 x + a_0$, where $a_0, a_1, \ldots, a_n$ are in the field F and $a_n \neq 0$. (Recall that a monic polynomial is a polynomial whose leading coefficient is equal to 1.) So if we multiply both sides of the equation by a_n^{-1}, we obtain

$$
\begin{aligned}
a_n^{-1} f(x) &= a_n^{-1} \left(a_n x^n + a_{n-1} x^{n-1} + \cdots + a_1 x + a_0 \right) \\
&= a_n^{-1} a_n x^n + a_n^{-1} a_{n-1} x^{n-1} + \cdots + a_n^{-1} a_1 x + a_n^{-1} a_0 \\
&= x^n + a_n^{-1} a_{n-1} x^{n-1} + \cdots + a_n^{-1} a_1 x + a_n^{-1} a_0
\end{aligned}
$$

and this proves that $a_n^{-1} f(x)$ is a monic polynomial.   ∎

Activity 9.4. Prove parts (ii), (iv), and (v) of Theorem 9.3. Here are some hints that might help you:

- For part (ii), use the definition of divides and Theorem 8.10. (See page 123.)

- For part (iv), use the definition of divides two times and then make a substitution in one of the resulting equations.

- Part (v) is a corollary of part (iv).

The Division Algorithm in $F[x]$

Now that we have a notion of divisibility within polynomial rings, our goal is to develop a Division Algorithm for polynomials much like the Division Algorithm for integers. However, in order to do this, we will restrict our attention to polynomial rings of the form $F[x]$, where F is a field (such as $\mathbb{Q}$, $\mathbb{R}$, or $\mathbb{Z}_p$, where p is a prime). Recall that when F is a field, we know that $F[x]$ is an integral domain. (See Corollary 8.11 on page 123.)

The Division Algorithm for $F[x]$ is stated below. You may notice the similarities between it and the Division Algorithm for the integers. (See Preview Activity 9.1.)

The Division Algorithm. *Let F be a field and let $f(x)$ and $g(x)$ be polynomials in $F[x]$ with $g(x) \neq 0$. Then there exist unique polynomials $q(x)$ and $r(x)$ in $F[x]$ such that*

$$
f(x) = g(x)q(x) + r(x) \quad \text{and}
$$

$r(x) = 0 \text{ or } \deg(r(x)) < \deg(g(x))$.

As in the integers, the Division Algorithm for polynomials guarantees the existence and uniqueness of a quotient and remainder, but says nothing about how to find these polynomials. Thus, before proving the Division Algorithm, we will illustrate a long division process for polynomials that can be used to find the quotient $q(x)$ and the remainder $r(x)$. This process is similar to long division in the integers.

In $\mathbb{R}[x]$, let

$$
f(x) = 4x^4 + 10x^3 + 7x^2 - 4x + 7 \quad \text{and} \quad g(x) = 2x^2 + 2x - 1.
$$

The long division process for $f(x)$ divided by $g(x)$ is shown below, and this process produces a quotient and a remainder. When using long division, the polynomial $f(x)$ is called the *dividend* and the polynomial $g(x)$ is called the *divisor*. In the example below, the quotient, $q(x) = 2x^2 + 3x + \dfrac{3}{2}$, is on the top line; the remainder, $r(x) = -4x + \dfrac{17}{2}$, is the last line.

Long division provides a way to obtain the quotient $q(x)$ in a step-by-step manner. The first term of the quotient, $2x^2$, is determined by finding the term that when multiplied by the first term of the divisor, $2x^2$, is equal to the first term of the dividend, $4x^4$. This first term can be found by noting that $\dfrac{4x^4}{2x^2} = 2x^2$. Notice that when we complete the subtraction in the next step and obtain $f(x) - 2x^2 g(x)$, the result is a polynomial whose degree is less than that of the dividend. We then repeat this process to find the remaining terms of the quotient. (Note that, in order for long division to work correctly, both $f(x)$ and $g(x)$ must be written in descending powers of x.)

$$
\begin{array}{r}
2x^2 + \quad 3x + \dfrac{3}{2} \qquad \leftarrow \text{quotient } q(x) \\[4pt]
2x^2 + 2x - 1 \overline{\big)\; 4x^4 + 10x^3 + 7x^2 - 4x + 7} \\
\end{array}
$$

$4x^4 + \;4x^3 - 2x^2$	$\leftarrow 2x^2 g(x)$
$6x^3 + 9x^2 - 4x + \;7$	$\leftarrow f(x) - 2x^2 g(x)$
$6x^3 + 6x^2 - 3x$	$\leftarrow 3xg(x)$
$3x^2 - \;x + \;7$	$\leftarrow \left[f(x) - 2x^2 g(x) \right] - 3xg(x)$
$3x^2 + 3x - \dfrac{3}{2}$	$\leftarrow \frac{3}{2} g(x)$
$-4x + \dfrac{17}{2}$	$\leftarrow$ remainder $r(x)$

Activity 9.5.

(a) In the division process just illustrated, use polynomial multiplication and addition to verify that $f(x) = g(x)q(x) + r(x)$ in $\mathbb{R}[x]$, where

$$f(x) = 4x^4 + 10x^3 + 7x^2 - 4x + 7 \qquad\qquad q(x) = 2x^2 + 3x + \frac{3}{2}$$

$$g(x) = 2x^2 + 2x - 1 \qquad\qquad\qquad\qquad r(x) = -4x + \frac{17}{2}$$

(b) In the Division Algorithm, if $r(x) = 0$, then we can conclude that $g(x)$ divides $f(x)$. In $\mathbb{R}[x]$, let $f(x) = 2x^4 + 9x^3 - x^2 + x - 3$ and $g(x) = 2x^2 + x + 1$. Use long division to show that $g(x)$ divides $f(x)$. What is the quotient $q(x)$ for which $f(x) = g(x)q(x)$?

The proof of the Division Algorithm will show us why we need to assume that F is a field; in particular, we will need to use multiplicative inverses of some nonzero elements. The long division process illustrated prior to Activity 9.5 gives a clue as to how to prove the Division Algorithm. Notice that after the first subtraction, we obtained a polynomial

$$f(x) - 2x^2 g(x) = 6x^3 + 9x^2 - 4x + 7,$$

which is a polynomial whose degree is less than the degree of $f(x)$. The next step in the process was a similar step using $f(x) - 2x^2 g(x)$ instead of $f(x)$. Notice that after the next subtraction, we obtained a polynomial of degree 2, which is less than the degree of $f(x) - 2x^2 g(x)$. This process of reducing the degree at each step suggests a proof by mathematical induction. Note that this technique is different than our approach in the integers, which relied on the Well-Ordering Principle. In fact, we could use the Well-Ordering Principle to prove the Division Algorithm for $F[x]$, and we could have used induction back in Investigation 1 to prove the Division Algorithm for $\mathbb{Z}$. We are using a different technique here to illustrate the various strategies for proving results of this type.

Proof of the Division Algorithm. Let F be a field and let $f(x)$ and $g(x)$ be polynomials in $F[x]$ with $g(x) \neq 0$. We will prove that there exist unique polynomials $q(x)$ and $r(x)$ in $F[x]$ such that

$$f(x) = g(x)q(x) + r(x) \quad \text{and} \quad r(x) = 0 \text{ or } \deg(r(x)) < \deg(g(x)).$$

We will first prove the existence of the polynomials $q(x)$ and $r(x)$. We will consider three cases for $f(x)$: (1) $f(x) = 0$; (2) $f(x) \neq 0$ and $\deg(f(x)) < \deg(g(x))$; or (3) $f(x) \neq 0$ and $\deg(f(x)) \geq \deg(g(x))$.

Case 1: If $f(x) = 0$, then $0 = f(x) = g(x) \cdot 0 + f(x)$. That is, we can use $q(x) = 0$ and $r(x) = f(x) = 0$.

Case 2: If $f(x) \neq 0$ and $\deg(f(x)) < \deg(g(x))$, then

$$f(x) = g(x) \cdot 0 + f(x).$$

So once again we use $q(x) = 0$ and $r(x) = f(x)$. Notice that since $r(x) = f(x)$, $\deg(r(x)) < \deg(g(x))$.

Case 3: If $f(x) \neq 0$ and $\deg(f(x)) \geq \deg(g(x))$, we will use induction on the degree of $f(x)$ to prove the existence of $q(x)$ and $r(x)$. For the basis step, we assume $\deg(f(x)) = 0$. Since $\deg(f(x)) \geq \deg(g(x))$ and $g(x) \neq 0$, we then also know that $\deg(g(x)) = 0$. This means that there exist nonzero elements a and b in F such that $f(x) = b$ and $g(x) = a$, and we see that

$$f(x) = b = a\left(a^{-1}b\right) + 0 = g(x)\left(a^{-1}b\right) + 0.$$

So we can use $q(x) = a^{-1}b$ and $r(x) = 0$. This proves the basis step.

For the inductive step, we let $n \in \mathbb{N}$ and assume that a quotient and remainder exist whenever the dividend is a polynomial of degree less than n. We will now assume $\deg(f(x)) = n$ and use this inductive assumption to prove that there exist polynomials $q(x)$ and $r(x)$ in $F[x]$ such that $f(x) = g(x)q(x) + r(x)$, and $r(x) = 0$ or $\deg(r(x)) < \deg(g(x))$. We will write

$$f(x) = a_n x^n + a_{n-1} x^{n-1} + \cdots + a_1 x + a_0$$
$$g(x) = b_m x^m + b_{m-1} x^{m-1} + \cdots + b_1 x + b_0$$

where $a_n \neq 0$, $b_m \neq 0$, and $n \geq m$. We will then use long division (as illustrated in Activity 9.5) to divide $f(x)$ by $g(x)$. Although it would be very formidable to write the entire process, we can complete the first step as follows:

$$b_m x^m + b_{m-1} x^{m-1} + \cdots + b_0 \overline{\smash{\big)}\ \begin{array}{l} b_m^{-1} a_n x^{n-m} \\ \hline a_n x^n \quad +a_{n-1}x^{n-1} \quad +\cdots \\ a_n x^n \quad +b_{m-1}b_m^{-1}a_n x^{n-1}+\cdots \end{array}} \tag{9.1}$$

Notice that we chose the first term in the quotient so that when we multiplied $g(x)$ by this term, the leading term would be $a_n x^n$. Subtracting, we obtain

$$f(x) - b_m^{-1} a_n x^{n-m} g(x) = \left(a_{n-1} - b_{m-1}b_m^{-1}a_n\right) x^{n-1} + \cdots,$$

which is a polynomial in $F[x]$ that is either 0 or has degree less than n. We can now apply the inductive assumption with this polynomial $(f(x) - b_m^{-1}a_n x^{n-m} g(x))$ as the dividend and $g(x)$ as the divisor. By induction, there exist polynomials $q_1(x)$ and $r_1(x)$ in $F[x]$ such that

$$f(x) - b_m^{-1}a_n x^{n-m} g(x) = g(x)q_1(x) + r_1(x),$$

and $r_1(x) = 0$ or $\deg(r_1(x)) < \deg(g(x))$. We can now rewrite this equation by adding $b_m^{-1}a_n x^{n-m} g(x)$ to both sides of the equation. This gives

$$f(x) = b_m^{-1}a_n x^{n-m} g(x) + g(x)q_1(x) + r_1(x)$$
$$f(x) = \left[b_m^{-1}a_n x^{n-m} + q_1(x) \right] g(x) + r_1(x)$$

The last equation shows the existence of a quotient (with $q(x) = b_m^{-1}a_n x^{n-m} + q_1(x)$) and a remainder (with $r(x) = r_1(x)$) since we already know that $r_1(x) = 0$ or $\deg(r_1(x)) < \deg(g(x))$. This completes the inductive step for the induction proof, and we can conclude that there exist polynomials $q(x)$ and $r(x)$ in $F[x]$ such that

$$f(x) = g(x)q(x) + r(x) \quad \text{and} \quad r(x) = 0 \text{ or } \deg(r(x)) < \deg(g(x)). \qquad (9.2)$$

We must now prove the uniqueness of $q(x)$ and $r(x)$. To do so, we assume that there exist polynomials $q_2(x)$ and $r_2(x)$ in $F[x]$ such that

$$f(x) = g(x)q_2(x) + r_2(x) \quad \text{and} \quad r_2(x) = 0 \text{ or } \deg(r_2(x)) < \deg(g(x)). \qquad (9.3)$$

Using Equations (9.2) and (9.3), we can conclude that

$$g(x)q(x) + r(x) = g(x)q_2(x) + r_2(x).$$

We can then rewrite this equation as

$$g(x)\left[q(x) - q_2(x)\right] = r_2(x) - r(x). \qquad (9.4)$$

Since both $r(x)$ and $r_2(x)$ are either 0 or have a degree less than the degree of $g(x)$, we can conclude that the right side of Equation (9.4) is 0 or a polynomial whose degree is less than $\deg(g(x))$. (See Exercise (5) in Investigation 8.) On the other hand, if $q(x) - q_2(x) \neq 0$, then by Theorem 8.10, the left side of Equation (9.4) is a polynomial whose degree is greater than or equal to $\deg(g(x))$. This is a contradiction and so we conclude that $q(x) - q_2(x) = 0$ or $q(x) = q_2(x)$. This means that the left side of Equation (9.4) is equal to 0 and so we conclude that $r(x) - r_2(x)$ is also equal to 0. Hence, $r(x) = r_2(x)$. This completes the proof of the uniqueness of the polynomials $q(x)$ and $r(x)$ and thus completes the proof of the Division Algorithm. ∎

Activity 9.6. Up to this point, we have only used long division for polynomials over the real numbers $\mathbb{R}$. However, we can also use the same process for dividing polynomials over any field F. In this case, all computations with the coefficients must be done in the field F. For example, in $\mathbb{Z}_3[x]$, we could ask the question, "Does $g(x) = [2]x^2 + x + [1]$ divide $f(x) = x^4 + x^3 + [2]x^2 + x + [2]$?" We can start the long division process as follows:

$$
\begin{array}{r}
[2]x^2 \\
[2]x^2 + x + [1] \overline{\big)\ x^4 + \ \ x^3 + [2]x^2 + x + [2]} \\
x^4 + [2]x^3 + [2]x^2 \\
\hline
[2]x^3 + x + [2]
\end{array}
$$

Remember that all the above calculations are being performed in $\mathbb{Z}_3[x]$, and so

$$[2]x^2\left([2]x^2 + x + [1]\right) = [4]x^4 + [2]x^3 + [2]x^2 = x^4 + [2]x^3 + [2]x^2,$$

and

$$x^4 + x^3 + [2]x^2 + x + [2] - \left(x^4 + [2]x^3 + [2]x^2\right)$$
$$= ([1] - [1])\, x^4 + ([1] - [2])\, x^3 + ([2] - [2])\, x^2 + x + [2]$$
$$= [2]x^3 + x + [2].$$

(a) Complete this long division process to find polynomials $q(x)$ and $r(x)$ in $\mathbb{Z}_3[x]$ such that $f(x) = g(x)q(x) + r(x)$ and $0 \leq \deg(r(x)) < 2$. Does $g(x)$ divide $f(x)$ in $\mathbb{Z}_3[x]$?

(b) In $\mathbb{Z}_5[x]$, let $g(x) = [2]x^2 + x + [1]$ and $f(x) = x^4 + x^3 + [2]x^2 + x + [2]$. Use long division to find polynomials $q(x)$ and $r(x)$ in $\mathbb{Z}_5[x]$ such that $f(x) = g(x)q(x) + r(x)$ and $0 \leq \deg(r(x)) < 2$. Does $g(x)$ divide $f(x)$ in $\mathbb{Z}_5[x]$?

Greatest Common Divisors of Polynomials

When we were working in $\mathbb{Z}$, the next step (after studying the Division Algorithm) was to define the greatest common divisor of two integers. To continue the analogy between $\mathbb{Z}$ and $F[x]$, where F is a field, we will soon define the greatest common divisor of two polynomials in $F[x]$. The following preview activity is intended to explore some of the subtleties that will be involved with this definition.

Preview Activity 9.7. In $\mathbb{R}[x]$, let $f(x) = 2x^2 + x - 6$ and let $g(x) = 2x^2 - 7x + 6$.

(a) Verify that $f(x) = (x+2)(2x-3)$ and $g(x) = (x-2)(2x-3)$. Using these factorizations as guides, what do you think is the greatest common divisor of $f(x)$ and $g(x)$?

(b) Verify that $f(x) = (2x+4)\left(x - \dfrac{3}{2}\right)$ and $g(x) = (2x-4)\left(x - \dfrac{3}{2}\right)$. Does this alter your opinion as to what you think is the greatest common divisor of $f(x)$ and $g(x)$?

(c) Verify that $f(x) = \left(\dfrac{1}{2}x + 1\right)(4x - 6)$ and $g(x) = \left(\dfrac{1}{2}x - 1\right)(4x - 6)$. Does this alter your opinion as to what you think is the greatest common divisor of $f(x)$ and $g(x)$?

The point of Preview Activity 9.7 is that care must be taken in defining the greatest common divisor of two polynomials. In effect, the preview activity shows that we could define the greatest common divisor of $f(x)$ and $g(x)$ to be any one of the following (as well as others):

$$\gcd(f(x), g(x)) = 2x - 3 \qquad \text{or} \qquad \gcd(f(x), g(x)) = x - \frac{3}{2} \quad \text{or}$$

$$\gcd(f(x), g(x)) = 4x - 6.$$

Because we want greatest common divisors of polynomials to be unique, we will adopt the following definition. Note the similarities between this definition and the one introduced by Theorem 1.28 in Investigation 1. (See page 19.)

Definition 9.8. Let F be a field and let $f(x), g(x) \in F[x]$ that are not both zero. The **greatest common divisor** of $f(x)$ and $g(x)$ is the polynomial $d(x) \in F[x]$ that satisfes the following three conditions:

(i) $d(x)$ divides $f(x)$ and $d(x)$ divides $g(x)$.

(ii) If $h(x) \in F[x]$ and $h(x)$ divides both $f(x)$ and $g(x)$, then $h(x)$ divides $d(x)$.

(iii) $d(x)$ is a monic polynomial. (Recall that this means that the leading coefficient of $d(x)$ is equal to 1_F.)

Although we have stated a reasonable definition of the greatest common divisor of two polynomials, there is still a question as to whether a polynomial $d(x)$ satisfying all three conditions from Definition 9.8 actually exists. We will first prove that if $d(x)$ exists, then it must be unique. We will then show that $d(x)$ exists whenever $f(x)$ and $g(x)$ are not both zero.

Activity 9.9. Let F be a field and let $f(x)$ and $g(x)$ be elements of $F[x]$ that are not both zero. Let $d(x)$ be the greatest divisor of $f(x)$ and $g(x)$, as in Definition 9.8. Assume that $c(x)$ is another polynomial in $F[x]$ that satisfies the three conditions of the greatest common divisor. That is: (i) $c(x)$ divides $f(x)$ and $c(x)$ divides $g(x)$; (ii) if $h(x) \in F[x]$ and $h(x)$ divides both $f(x)$ and $g(x)$, then $h(x)$ divides $c(x)$; and (iii) $c(x)$ is a monic polynomial.

We will now prove that $c(x) = d(x)$, which will prove that if $d(x)$ in Definition 9.8 exists, then it must be unique.

(a) Explain why $d(x)$ divides $c(x)$.

(b) Explain why $c(x)$ divides $d(x)$.

(c) Use one of the parts of Theorem 9.3 to conclude that $c(x) = d(x)$.

Because the greatest common divisor as defined in Definition 9.8 is unique (if it exists), we can use the notation $\gcd(f(x), g(x))$ for the greatest common divisor of $f(x)$ and $g(x)$ in $F[x]$. The next theorem will establish the fact that, indeed, $\gcd(f(x), g(x))$ exists whenever $f(x)$ and $g(x)$ are not both zero. This theorem is similar to the results in Investigation 1 about the greatest common divisor of two integers. (See Proposition 1.17 and Theorem 1.19 on page 13.)

Theorem 9.10. *Let F be a field and let $f(x)$ and $g(x)$ be polynomials in $F[x]$ that are not both zero. There exists a unique monic polynomial $d(x)$ in $F[x]$ of smallest degree that can be written in the form*

$$d(x) = f(x)u(x) + g(x)v(x)$$

for some polynomials $u(x)$ and $v(x)$ in $F[x]$. In addition, $d(x) = \gcd(f(x), g(x))$.

Proof. Using the notation in the theorem, let S be the set of all monic polynomials $h(x)$ in $F[x]$ such that there exist polynomials $m(x), n(x) \in F[x]$ with $h(x) = f(x)m(x) + g(x)n(x)$.

Since $f(x)$ and $g(x)$ are not both zero, we may assume without loss of generality that $f(x) \neq 0$. Let c be the leading coefficient of $f(x)$. Then $c \neq 0$ and by Theorem 9.3, $c^{-1}f(x)$ is a monic polynomial in $F[x]$. Since

$$c^{-1}f(x) = 1 \cdot c^{-1}f(x) + 0 \cdot g(x),$$

we see that $c^{-1}f(x) \in S$, and so $S \neq \emptyset$.

This conclusion implies that the degrees of the polynomials in S form a nonempty subset of the whole numbers, and hence, by the Well-Ordering Principle, there must be a polynomial in S of smallest degree. Let $d(x)$ be a polynomial of smallest degree in S. Then $d(x)$ is a monic polynomial, and there exist $u(x), v(x) \in F[x]$ such that

$$d(x) = f(x)u(x) + g(x)v(x). \tag{9.5}$$

We will first use the Division Algorithm for $F[x]$ to prove that $d(x)$ divides $f(x)$. We know that

$$f(x) = d(x)q(x) + r(x), \tag{9.6}$$

for some $q(x), r(x) \in F[x]$ with $r(x) = 0$ or $\deg(r(x)) < \deg(d(x))$. We will show that $r(x) = 0$ by showing that we get a contradiction if $r(x) \neq 0$.

If $r(x) \neq 0$, then we can use Equations (9.5) and (9.6) to write

$$
\begin{aligned}
r(x) &= f(x) - d(x)q(x) \\
&= f(x) - [f(x)u(x) + g(x)v(x)]q(x) \\
&= f(x) - f(x)u(x)q(x) - g(x)v(x)q(x) \\
&= f(x)[1 - u(x)q(x)] + g(x)[-v(x)q(x)].
\end{aligned}
$$

So if a is the leading coefficient of $r(x)$, then

$$a^{-1}r(x) = f(x)\left[a^{-1} - a^{-1}u(x)q(x)\right] + g(x)\left[-a^{-1}v(x)q(x)\right]$$

is a monic polynomial and therefore must be in S. However, $\deg(a^{-1}r(x)) < \deg(d(x))$, which is a contradiction to the assumption that $d(x)$ is the smallest degree polynomial in S. Therefore, $r(x) = 0$ and then Equation (9.6) implies that $d(x)$ divides $f(x)$.

We can use a similar argument to prove that $d(x)$ divides $g(x)$, and this establishes that $d(x)$ is a common divisor of $f(x)$ and $g(x)$.

We now assume that $s(x) \in F[x]$ and $s(x)$ divides both $f(x)$ and $g(x)$. From Equation (9.5), we see that $d(x) = f(x)u(x) + g(x)v(x)$. We can then use the result in Exercise (2) to conclude that $s(x)$ divides $f(x)u(x) + g(x)v(x)$. Thus, $s(x)$ divides $d(x)$. Because $d(x)$ is a monic polynomial, it then follows that $d(x)$ is the greatest common divisor of $f(x)$ and $g(x)$. ∎

Because we now know that if F is a field, then the greatest common divisor of two polynomials $f(x)$ and $g(x)$ in $F[x]$ actually exists and is unique (whenever $f(x)$ and $g(x)$ are not both zero), we can use the notation $\gcd(f(x), g(x))$ to represent the greatest common divisor of $f(x)$ and $g(x)$.

Relatively Prime Polynomials

Theorem 1.19 on page 13 tells us that the greatest common divisor of two integers (not both zero), is the smallest positive linear combination of these two integers. Theorem 9.10 is a similar result for $F[x]$, where F is a field. The following definitions are similar to the corresponding definitions in Investigation 1.

Definition 9.11. Let F be a field and let $f(x)$ and $g(x)$ be polynomials in $F[x]$. A **linear combination** of $f(x)$ and $g(x)$ is a polynomial in $F[x]$ that can be written as $f(x)u(x) + g(x)v(x)$ for some polynomials $u(x)$ and $v(x)$ in $F[x]$.

So using Theorem 9.10, we can say that the greatest common divisor of $f(x)$ and $g(x)$ is the monic polynomial of least degree that is a linear combination of $f(x)$ and $g(x)$. As in the integers, a particularly important case is when the greatest common divisor of two polynomials is 1.

Definition 9.12. Let F be a field. Two polynomials $f(x)$ and $g(x)$ in $F[x]$ are **relatively prime** provided that $\gcd(f(x), g(x)) = 1$.

Activity 9.13. The following results can be proved in much the same manner as their corresponding results in the integers. For this activity, let F be a field, and let $p(x)$ and $q(x)$ be polynomials in $F[x]$ that are not both zero.

(a) Prove that $p(x)$ and $q(x)$ are relatively prime if and only if

$$p(x)u(x) + q(x)v(x) = 1$$

for some polynomials $u(x), v(x) \in F[x]$. (Hint: Use Theorem 9.10.)

(b) Prove that if $a, b \in F$ with $a \neq b$, then $\gcd(x + a, x + b) = 1$. (Hint: Notice that $(x + a) - (x + b) = a - b$. Use this observation to write 1 as a linear combination of $(x + a)$ and $(x + b)$.)

(c) Let $h(x) \in F[x]$. Prove that if $p(x)$ divides $h(x)$, $q(x)$ divides $h(x)$, and $\gcd(p(x), q(x)) = 1$, then $p(x)q(x)$ divides $h(x)$. (Hint: Use Theorem 9.10 to write $p(x)u(x) + q(x)v(x) = 1$ for some $u(x), v(x) \in F[x]$. Then multiply both sides of this equation by $h(x)$ and make appropriate substitutions.)

The Euclidean Algorithm for Polynomials

Given two integers a and b (not both zero), we learned how to use the Euclidean Algorithm to find the greatest common divisor of a and b. (See Investigation 1.) The basic idea was to repeatedly use the Division Algorithm until a remainder of zero was obtained. If F is a field, then we have a Division Algorithm in $F[x]$, and so it is not surprising that we also have a Euclidean Algorithm to find the greatest common divisor of two polynomials in $F[x]$. The Euclidean Algorithm in $F[x]$ can be described as follows:

Let $f(x)$ and $g(x)$ be polynomials over the field F, with $g(x) \neq 0$. Successively apply the Division Algorithm until the first zero remainder is obtained:

$$f(x) = g(x)q_1(x) + r_1(x) \qquad \text{and} \qquad \deg(r_1(x)) < \deg(g(x));$$
$$g(x) = r_1(x)q_2(x) + r_2(x) \qquad \text{and} \qquad \deg(r_2(x)) < \deg(r_1(x));$$
$$r_1(x) = r_2(x)q_3(x) + r_3(x) \qquad \text{and} \qquad \deg(r_3(x)) < \deg(r_2(x));$$
$$\vdots \qquad\qquad\qquad \vdots \qquad\qquad \vdots$$
$$r_{n-2}(x) = r_{n-1}(x)q_n(x) + r_n(x) \qquad \text{and} \qquad \deg(r_n(x)) < \deg(r_{n-1}(x));$$
$$r_{n-1}(x) = r_n(x)q_{n+1}(x)$$

If c is the leading coefficient of the last nonzero remainder $r_n(x)$, then $c^{-1}r_n(x)$ is the greatest common divisor of $f(x)$ and $g(x)$—that is, $\gcd(f(x), g(x)) = c^{-1}r_n(x)$. For example, using the polynomials from Activity 9.5, in $\mathbb{R}[x]$, let

$$f(x) = 4x^4 + 10x^3 + 7x^2 - 4x + 7 \qquad \text{and} \qquad g(x) = 2x^2 + 2x - 1.$$

We have already used long division to obtain

$$f(x) = g(x)\left(2x^2 + 3x + \frac{3}{2}\right) + \left(-4x + \frac{17}{2}\right). \tag{9.7}$$

So we have $q_1(x) = 2x^2 + 3x + \dfrac{3}{2}$ and $r_1(x) = -4x + \dfrac{17}{2}$. We can now divide $g(x)$ by the remainder $r_1(x) = -4x + \dfrac{17}{2}$ and obtain

$$g(x) = \left(-4x + \frac{17}{2}\right)\left(-\frac{1}{2}x - \frac{25}{16}\right) + \frac{393}{32}. \tag{9.8}$$

So in the Euclidean Algorithm, $q_2(x) = -\dfrac{1}{2}x - \dfrac{25}{16}$ and $r_2(x) = \dfrac{393}{32}$. If we now divide $r_1(x) = -4x + \dfrac{17}{2}$ by $r_2(x) = \dfrac{393}{32}$, we will obtain a remainder of 0. This means that the greatest common divisor of $f(x)$ and $g(x)$ is

$$d(x) = \frac{32}{393}r_2(x) = 1.$$

As with the integers, we can also use the Euclidean Algorithm in reverse to write the greatest common divisor as a linear combination of $f(x)$ and $g(x)$. It will make the computations easier if we first write $r_2(x) = \dfrac{393}{32}$ as a linear combination of $f(x)$ and $g(x)$. However, the algebra and computations can still be quite complicated.

We first use Equation (9.8) to write

$$r_2(x) = g(x) - r_1(x)q_2(x).$$

We can now use Equation (9.7) and substitute for $r_1(x) = -4x + \dfrac{17}{2}$. This gives

$$\begin{aligned}
r_2(x) &= g(x) - [f(x) - g(x)q_1(x)]\,q_2(x) \\
&= -f(x)q_2(x) + g(x)\left[1 + q_1(x)q_2(x)\right]
\end{aligned}$$

Since $r_2(x) = \dfrac{393}{32}$, we see that

$$\begin{aligned}
1 &= \gcd(f(x), g(x)) \\
&= \frac{32}{393}\left[-f(x)q_2(x)\right] + g(x)\left[1 + q_1(x)q_2(x)\right] \\
&= f(x)\left[-\frac{32}{393}q_2(x)\right] + g(x)\left[\frac{32}{393}\left(1 + q_1(x)q_2(x)\right)\right] \\
&= f(x)\left[\frac{16}{393}x + \frac{50}{393}\right] + g(x)\left[-x^3 - \frac{37}{8}x^2 - \frac{87}{16}x - \frac{43}{32}\right].
\end{aligned}$$

Activity 9.14.

(a) In $\mathbb{R}[x]$, let $f(x) = 3x^5 + 7x^4 + 11x^3 + 15x^2 + 10x + 2$ and $g(x) = 3x^2 + 7x + 2$. Use the Euclidean Algorithm to determine $\gcd(f(x), g(x))$ and write $\gcd(f(x), g(x))$ as a linear combination of $f(x)$ and $g(x)$.

(b) In $\mathbb{Z}_3[x]$, let $f(x) = x^4 + [2]x^3 + x + [2]$ and $g(x) = [2]x^2 + [1]$. Determine $\gcd(f(x), g(x))$.

Concluding Activities

Activity 9.15. Although we stated the Division Algorithm for polynomials over a field, there is also a modified Division Algorithm for $D[x]$, where D is an integral domain. To explore this modification, first note that, in the proof of the Division Algorithm, we used the fact that F was a field to conclude that the leading coefficient of the divisor $g(x)$ has a multiplicative inverse. In particular, we let

$$f(x) = a_n x^n + a_{n-1} x^{n-1} + \cdots + a_1 x + a_0 \text{ and}$$
$$g(x) = b_m x^m + b_{m-1} x^{m-1} + \cdots + b_1 x + b_0,$$

where $a_n \neq 0$ and $b_m \neq 0$. In the proof, we made use of b_m^{-1} in the division process in Equation (9.1). This is really the only place where we used the assumption that we were working with polynomials over a field. So if we assume that the polynomials are in $D[x]$, where D is an integral domain, we can use the same proof as long as we also assume that the leading coefficient of $g(x)$, namely b_m, is a unit in D. This means that b_m^{-1} exists in D and we can proceed with the division process. The following theorem formalizes these observations, and the problems after it provide an example of polynomial division in $\mathbb{Z}[x]$.

Theorem 9.16. *Let D be an integral domain and let $f(x)$ and $g(x)$ be polynomials in $D[x]$ with $g(x) \neq 0$. If the leading coefficient of $g(x)$ is a unit in D, then there exist unique polynomials $q(x)$ and $r(x)$ in $D[x]$ such that*

$$f(x) = g(x)q(x) + r(x) \quad and$$

$r(x) = 0$ *or* $\deg(r(x)) < \deg(g(x))$.

For the problems below, assume that all polynomials are in $\mathbb{Z}[x]$.

(a) Let $f(x) = 2x^3 + 2x^2 - 5x + 3$ and $g(x) = x + 4$. Find the quotient and remainder when $f(x)$ is divided by $g(x)$.

(b) Let $f(x) = 2x^3 + 2x^2 - 5x + 3$ and $g(x) = 2x + 3$. Show that the division process does not give a unique quotient $q(x)$ and remainder $r(x)$ with $r(x) = 0$ or $0 \le \deg(r(x)) < \deg(g(x))$.

Exercises

(1) Let R be a commutative ring, and let $f(x), g(x), h(x) \in R[x]$. Prove each of the following:

    (a) If $f(x)$ divides $g(x)$ and $g(x)$ divides $h(x)$, then $f(x)$ divides $h(x)$.

    (b) If $f(x)$ divides $g(x)$ and $f(x)$ divides $h(x)$, then $f(x)$ divides $g(x) + h(x)$.

    (c) If $f(x)$ divides $g(x)$, then $f(x)$ divides $g(x)h(x)$.

* (2) Let R be a commutative ring, and let $p(x)$, $f(x)$, and $g(x)$ be polynomials in $R[x]$. Prove that if $p(x)$ divides both $f(x)$ and $g(x)$ in $R[x]$, then for any polynomials $u(x)$ and $v(x)$ in $R[x]$, $p(x)$ divides $f(x)u(x) + g(x)v(x)$.

(3) In each of the following, a field F is specified and two polynomials $f(x)$ and $g(x)$ in $F[x]$ are given. Find $q(x), r(x) \in F[x]$ so that $f(x) = g(x)q(x) + r(x)$ with $r(x) = 0$ or $\deg(r(x)) < \deg(g(x))$.

   (a) In $\mathbb{Q}[x]$, $f(x) = x^3 + 2x^2 + 2x + 1$ and $g(x) = x + 2$.

   (b) In $\mathbb{Z}_3[x]$, $f(x) = x^3 + [2]x^2 + [2]x + [1]$ and $g(x) = x + [2]$.

   (c) In $\mathbb{R}[x]$, $f(x) = 2x^4 + 4x^3 + 2x^2 + 2x + 1$ and $g(x) = x^2 + 4$.

   (d) In $\mathbb{Z}_5[x]$, $f(x) = [2]x^4 + [4]x^3 + [2]x^2 + [2]x + [1]$ and $g(x) = x^2 + [4]$.

(4) Let F be a field, and let $p(x), f(x), g(x) \in F[x]$. Prove that if $p(x)$ and $f(x)$ are relatively prime and $p(x)$ and $g(x)$ are relatively prime, then $p(x)$ and $f(x)g(x)$ are relatively prime.

(5) Let F be a field, and let $p(x), f(x), g(x) \in F[x]$. Prove that if $p(x)$ divides $f(x)g(x)$ and $\gcd(p(x), f(x)) = 1$, then $p(x)$ divides $g(x)$.

(6) Let F be a field, and let $p(x), q(x), f(x) \in F[x]$. Assume that $\gcd(p(x), q(x)) = 1$ and both $p(x)$ and $q(x)$ divide $f(x)$. Prove or disprove: $p(x)q(x)$ divides $f(x)$.

(7) Let F and K be fields such that F is a subfield of K.

   (a) Assume that $p(x), q(x) \in F[x]$ and that $p(x)$ and $q(x)$ are relatively prime in $F[x]$. Are $p(x)$ and $q(x)$ relatively prime in $K[x]$? Justify your conclusion.

   (b) Assume that $p(x), q(x) \in F[x]$ and that $p(x)$ and $q(x)$ are relatively prime in $K[x]$. Are $p(x)$ and $q(x)$ relatively prime in $F[x]$? Justify your conclusion.

(8) Use the Euclidean Algorithm to find $\gcd(f(x), g(x))$ in the indicated polynomial ring. Then write $\gcd(f(x), g(x))$ as a linear combination of $f(x)$ and $g(x)$.

   (a) $f(x) = x^4 + 2x^3 + x^2 + 1$ and $g(x) = x^2 + 2x + 4$ in $\mathbb{Q}[x]$.

   (b) $f(x) = x^6 + x^5 + x + [1]$ and $g(x) = x^3 + [2]x^2 + [2]x + [1]$ in $\mathbb{Z}_3[x]$.

   (c) $f(x) = x^4 + [4]x^3 + [2]x^2 + [4]x + [4]$ and $g(x) = x^3 + [4]x^2 + x + [4]$ in $\mathbb{Z}_5[x]$.

Connections

In Investigation 8, we saw that a polynomial ring with coefficients from an integral domain has much in common with the ring of integers. This investigation explored that connection in more detail for polynomials whose coefficients come from a field. In particular, the ideas of divisibility of polynomials and a division algorithm for such polynomials are essentially the same as divisibility and the division algorithm for integers from Investigation 1. In addition, we can define what it means for two polynomials to be relatively prime just as we did for integers in Investigation 1. We will take this connection between polynomial rings over fields and the integers a step farther in Investigation 10 when we discuss unique factorization in polynomial rings.

Investigation 10

Roots, Factors, and Irreducible Polynomials

Focus Questions

By the end of this investigation, you should be able to give precise and thorough answers to the questions listed below. You may want to keep these questions in mind to focus your thoughts as you complete the investigation.

- What does the Remainder Theorem say when a polynomial over a field F is divided by a polynomial of the form $(x - c)$, for some $c \in F$?

- What does it mean to say that an element of a field F is a root of a polynomial in $F[x]$?

- What does the Factor Theorem say about the roots of a polynomial over a field F and certain factors of that polynomial?

- How many roots can a polynomial of degree n over a field have?

- If F is a field, what does it mean to say that a polynomial in $F[x]$ is irreducible? What does it mean to say a polynomial in $F[x]$ is reducible?

- How can the Factor Theorem be used to determine if a polynomial of degree 2 or 3 over a field is irreducible or reducible?

- If F is a field, what does it mean to say that a polynomial in $F[x]$ can be factored into irreducible polynomials in a unique way?

Preview Activity 10.1. In Definition 8.12, we defined the concept of a polynomial function. If R is a commutative ring and $p(x) = a_n x^n + a_{n-1} x^{n-1} + \cdots a_1 x + a_0$ is a polynomial in $R[x]$, then the **polynomial function** induced by $p(x)$ is the function $\overline{p} : R \to R$, where for each r in R,

$$\overline{p}(r) = a_n r^n + a_{n-1} r^{n-1} + \cdots a_1 r + a_0.$$

To simplify notation, we will write $p(r)$ for $\overline{p}(r)$ and write $p : R \to R$ for the polynomial function induced by the polynomial $p(x)$.

In each of the following, the term "remainder" means the remainder according to the Division Algorithm, which can be determined by using long division.

(a) Let $f(x) = x^2 - 3x - 2$ be in $\mathbb{R}[x]$. Notice that $f(5) = 8$. Determine the remainder when $f(x)$ is divided by $(x - 5)$.

(b) Let $g(x) = x^2 - 3x - 10$ be in $\mathbb{R}[x]$. Determine $g(5)$ and determine the remainder when $g(x)$ is divided by $(x - 5)$.

(c) Let $h(x) = x^2 - 3x - 10$ be in $\mathbb{Q}[x]$. Determine $h(-2)$ and determine the remainder when $h(x)$ is divided by $(x - (-2)) = (x + 2)$.

(d) Let $p(x) = x^2 + [2]x + [2]$ be in $\mathbb{Z}_3[x]$. Determine $p([1])$ and determine the remainder when $p(x)$ is divided by $(x - [1])$.

Polynomial Functions and Remainders

In our comparison of the development of $\mathbb{Z}$ and $F[x]$ (where F is a field), the next natural step is to develop what is meant by unique factorization in $F[x]$. In this investigation, we will consider whether we can obtain something like the Fundamental Theorem of Arithmetic, but for polynomials rather than integers. The language for polynomials will be somewhat different. As we shall see, instead of using the term *prime*, we will use the term *irreducible* to describe polynomials that, loosely speaking, cannot be factored. We will give a precise meaning to this term later in this investigation.

Questions about whether or not a polynomial can be factored can sometimes be answered by using the corresponding induced polynomial function. Some care must be used when taking this approach. For example, in $\mathbb{R}[x]$, if we let $p(x) = x^2 + 4x - 7$, then:

- Since x is an indeterminate in $\mathbb{R}[x]$, the statement $p(x) = x^2 + 4x - 7 = 0$ is false because this is stating that $p(x)$ is equal to the zero polynomial, which it is not.

- However, if we work with the induced polynomial function $p : \mathbb{R} \to \mathbb{R}$, then it is reasonable to ask for which values $x \in \mathbb{R}$ is $p(x) = 0$. In this case, we are treating x as a variable over $\mathbb{R}$.

If R is a commutative ring, then it may be helpful to realize that statements about the indeterminate x occur in the polynomial ring $R[x]$, whereas questions about the variable x occur in the ring R.

The work in Preview Activity 10.1 was meant to illustrate an important relation between the value $p(r)$ of a polynomial function and the remainder when the polynomial is divided by $(x - r)$. This relationship is explored further in the next activity.

Activity 10.2. Let F be a field, let $p(x) \in F[x]$, and let $c \in F$.

(a) Suppose $f(x) \in F[x]$ and $f(x) = 0$ or $\deg(f(x)) = 0$. Explain why $f(x)$ must be of the form $f(x) = s$, for some $s \in F$.

(b) Write down the result of the Division Algorithm when $p(x)$ is divided by $(x - c)$. Pay particular attention to the remainder, and explain why the remainder must be a constant polynomial.

(c) Evaluate both sides of the equation from part (b) using $x = c$, and use your work to complete the following statement:

The remainder when $p(x)$ is divided by $(x - c)$ is equal to _____.

The work in Activity 10.2 outlines the proof of the Remainder Theorem, which can be stated formally as follows:

Theorem 10.3 (The Remainder Theorem). *Let F be a field, let $p(x) \in F[x]$, and let $c \in F$. The remainder when $p(x)$ is divided by $(x - c)$ is equal to $p(c)$. That is, there exists a unique polynomial $q(x) \in F[x]$ such that*

$$p(x) = (x - c)q(x) + p(c).$$

Proof. Let F be a field, let $p(x) \in F[x]$, and let $c \in F$. By the Division Algorithm, there exist unique $q(x)$ and $r(x)$ in $F[x]$ such that

$$p(x) = (x - c)q(x) + r(x) \quad \text{with} \quad r(x) = 0 \text{ or } \deg(r(x)) < \deg(x - c).$$

Since the degree of $(x - c)$ is 1, it follows that $r(x) = 0$ or $\deg(r(x)) = 0$. In either case, $r(x) = a$ for some $a \in F$. This then means that

$$p(x) = (x - c)q(x) + a.$$

We can now treat both sides of the last equation as polynomial functions and evaluate each at $x = c$. Doing so, we obtain:

$$p(c) = (c - c)q(c) + a = 0_F + a = a.$$

This proves that the remainder when $p(x)$ is divided by $(x - c)$ is equal to $p(c)$. ∎

Roots of Polynomials and the Factor Theorem

The study of the roots of a polynomial has been an important part of the history of algebra. As it turns out, the Remainder Theorem can be used to establish an important relationship between the roots and the factors of a polynomial. To explore this relationship more, we must first define what a root is.

Definition 10.4. Let R be a commutative ring and let $p(x) \in R[x]$. An element c of R is called a **root of the polynomial $p(x)$** provided that $p(c) = 0_R$.

The Factor Theorem, stated formally below, describes precisely the relationship between the roots of a polynomial and its monic, degree 1 factors.

Theorem 10.5 (The Factor Theorem). *Let F be a field, let $p(x) \in F[x]$, and let $c \in F$. Then c is a root of the polynomial $p(x)$ if and only if $(x - c)$ is a factor of $p(x)$.*

The following activity will help guide us through a proof of the Factor Theorem.

Activity 10.6. Let F be a field, let $p(x) \in F[x]$, and let $c \in F$. Using the Remainder Theorem, there exists a unique polynomial $q(x)$ in $F[x]$ such that

$$p(x) = (x - c)q(x) + p(c). \tag{10.1}$$

(a) Assume that c is root of $p(x)$. Use Equation (10.1) to prove that $(x - c)$ is a factor of $p(x)$.

(b) Now assume that $(x - c)$ is a factor of $p(x)$. Use the definition of divides to write $p(x)$ in the form of a product, and then use this product to show that $p(c) = 0$.

(c) Use your work from parts (a) and (b) to write a formal proof of the Factor Theorem.

The next activity shows how the Factor Theorem can be used to help factor polynomials.

Activity 10.7.

(a) Let $p(x) = x^4 - x^3 - 5x^2 + 3x + 6$ be a polynomial in $\mathbb{R}[x]$. Verify that $p(-1) = 0$ and $p(2) = 0$. What does the Factor Theorem allow you to conclude from these observations?

(b) Explain how Activity 9.13 (see page 138) allows us to conclude that $\gcd(x + 1, x - 2) = 1$ and that
$$(x + 1)(x - 2) = x^2 - x - 2$$
divides $p(x)$. Use this information and long division to factor $p(x)$.

(c) Let $p(x) = x^4 + [3]x^3 + [2]x^2 + [4]x + [4]$ be a polynomial in $\mathbb{Z}_5[x]$. Evaluate $p([x])$ for each $[x] \in \mathbb{Z}_5$. Then identify the roots of $p(x)$, and use the Factor Theorem and long division to factor $p(x)$ into a product of monic, degree 1 factors. (Note that some of these factors may be repeated.)

As illustrated in Activity 10.7, one way to find the roots of a polynomial $p(x)$ in $R[x]$, where R is a commutative ring, is to simply evaluate $p(c)$ for each $c \in R$ and find the values of c for which $p(c) = 0_R$. For example, let $p(x) = [3]x^2 + [3]$ be a polynomial in $\mathbb{Z}_6[x]$. We can easily observe the following:

$p([0]) = [3]$	$p([1]) = [0]$	$p([2]) = [3]$
$p([3]) = [0]$	$p([4]) = [3]$	$p([5]) = [0]$

So in this case, we have a polynomial $p(x)$ of degree 2 that has three roots ($[1], [3]$, and $[5]$). This may seem contrary to a well-known result from your previous algebra courses—namely, that if a polynomial has degree n, then that polynomial has at most n roots. To resolve this apparent discrepancy, it may help to remember that your prior studies of polynomials most likely dealt only with polynomials over the *field* of real numbers, $\mathbb{R}$. In this example, however, notice that we are working with $\mathbb{Z}_6$, which is not a field. In fact, $\mathbb{Z}_6$ is not even an integral domain. This distinction turns out to be very important, and we will explore it more in the next activity. Eventually, we will prove Theorem 10.9, which states that a polynomial of degree n over a field has at most n roots.

Activity 10.8. Let F be a field.

(a) Let $p(x)$ be a polynomial in $F[x]$ with $\deg(p(x)) = 0$. (So $p(x) = k$ for some $k \in F$ with $k \neq 0$.) Explain why $p(x)$ has no roots in F.

(b) Let $p(x)$ be a polynomial in $F[x]$ with $\deg(p(x)) = 1$. (So $p(x) = ax + b$ with $a, b \in F$ and $a \neq 0$.) Verify that $-a^{-1}b$ is a root of $p(x)$. Then let $r \in F$ be any root of $p(x)$. Prove that $r = -a^{-1}b$, and explain why this fact establishes that $-a^{-1}b$ is the only root of $p(x)$ in F.

(c) Let $f(x) = x^2 - 3x - 10 \in \mathbb{R}[x]$, and let $g(x) = x^2 + x + 1 \in \mathbb{R}[x]$. Notice that both of these polynomials have degree 2. How many roots do $f(x)$ and $g(x)$ have in $\mathbb{R}$? (For this part, you may use the quadratic formula.)

(d) Let $f(x) = [3]x^2 + [3] \in \mathbb{Z}_5[x]$. Notice that $\deg(f(x)) = 2$. How many roots does $f(x)$ have in $\mathbb{Z}_5$?

Theorem 10.9, which we will state shortly, formalizes the ideas from Activity 10.8. To prove it, we will use induction on the degree of the polynomial. Notice that in part (a) of Activity 10.8, we proved that a polynomial of degree 0 in $F[x]$ has no roots. In part (b), we proved that a polynomial of degree 1 has 1 root. These results establish the basis step for the induction proof. The basic idea of the induction step is that if we have a root c of a polynomial of degree $k + 1$, then we can use the Factor Theorem (Theorem 10.5) to write the polynomial as a product of $(x - c)$ and a polynomial whose degree is k. This step also makes use of Theorem 8.10. (See page 123.)

Theorem 10.9. *If F is a field and $n \in \mathbb{N}$, then a polynomial of degree n in $F[x]$ has at most n roots in F. That is, a nonzero polynomial of degree n over a field has at most n roots in the field.*

Proof. Let F be a field. We will proceed by induction on the degree of $p(x)$. In Activity 10.8, we proved that a polynomial of degree 0 in $F[x]$ has no roots and a polynomial of degree 1 in $F[x]$ has exactly one root. So the basis for the induction has been established.

Now let $k \in \mathbb{Z}$ with $k \geq 0$, and suppose that any polynomial of degree k in $F[x]$ has at most k roots. Now let $p(x) \in F[x]$ with $\deg(p(x)) = k + 1$. Then either $p(x)$ has no roots in F, or $p(x)$ has at least one root in F. We will consider each of these cases.

In the case where $p(x)$ has no roots, it is clear that $p(x)$ has at most $k + 1$ roots. So assume that $p(x)$ has at least one root in F, and let c be such a root. By the Factor Theorem, there exists a polynomial $q(x)$ in $F[x]$ such that

$$p(x) = (x - c)q(x).$$

By Theorem 8.10, $\deg(p(x)) = \deg(x - c) + \deg(q(x))$. Since $\deg(x - c) = 1$ and $\deg(p(x)) = k + 1$, it follows that $\deg(q(x)) = k$. Hence, by the induction hypothesis, we can conclude that $q(x)$ has at most k roots in F. Now notice that if r is a root of $p(x)$, then

$$0 = p(r) = (r - c)q(r).$$

Since F is a field (and an integral domain), we know that F has no zero divisors. Therefore, we know that $r - c = 0$ or $q(r) = 0$. That is, the only roots of $p(x)$ are c and the roots of $q(x)$. Since $q(x)$ has at most k roots, we conclude that $p(x)$ has at most $k + 1$ roots. This completes the proof of the inductive step, and by mathematical induction, we have proved that a polynomial in $F[x]$ of degree n has at most n roots. ∎

Irreducible Polynomials

In our study of the integers, the prime numbers were very important. In fact, we proved that every natural number (except 1) is either prime or a product of prime numbers. We will obtain a similar result for $F[x]$, where F is a field. However, when working with polynomials, we use the term *irreducible* rather than *prime*. In the following definition, the term *nonconstant polynomial* refers to a polynomial with positive degree.

Definition 10.10. Let F be a field. A nonconstant polynomial $p(x)$ in $F[x]$ is **irreducible in $F[x]$** provided that $p(x)$ cannot be factored as a product of two polynomials in $F[x]$, both of which have a positive degree. Otherwise, the polynomial $p(x)$ is called **reducible in $F[x]$**.

Basically, Definition 10.10 means that a nonconstant polynomial is irreducible if and only if the only way it can be factored is by having at least one of the factors have degree 0. That is, $p(x)$ is irreducible if and only if any factorization of $p(x)$ must be of the form $cp(x)$, for some $c \in F$.

This also means that if F is a field, then a nonconstant polynomial $p(x)$ in $F[x]$ is reducible if and only if there exist polynomials $f(x)$ and $g(x)$ such that

- $p(x) = f(x)g(x)$, and

- $\deg(f(x)) \geq 1$ and $\deg(g(x)) \geq 1$.

For example, in $\mathbb{R}[x]$:

- Even though we can write $\dfrac{3}{2}x + 4 = \dfrac{3}{2}(x + 6)$, the polynomial $\left(\dfrac{3}{2}x + 4\right)$ is irreducible in $\mathbb{R}[x]$ since it cannot be written as a product of two polynomials with positive degree.

- Since $x^4 + 5x^2 + 6 = (x^2 + 2)(x^2 + 3)$, the polynomial $(x^2 + 5x + 6)$ is reducible in $\mathbb{R}[x]$.

The next theorem states a result that is important but not terribly surprising in light of the definition of irreducibility. It can be proved easily using Theorem 8.10, and we leave this proof as an exercise.

Theorem 10.11. *Let F be a field. Any polynomial in $F[x]$ that has degree 1 is irreducible in $F[x]$.*

As you might suspect after studying the Factor Theorem, there is a relationship between the reducibility (or irreducibility) of a polynomial over a field and the roots of the polynomial. However, there are some subtleties to this relationship and some care must be taken when using it. To start, let F be a field, and let $p(x) \in F[x]$ with $\deg(p(x)) \geq 2$. If $p(x)$ has a root c in F, then by the Factor Theorem, $(x - c)$ is a factor of $p(x)$. So we know that there exists a polynomial $q(x) \in F[x]$ such that

$$p(x) = (x - c)q(x). \tag{10.2}$$

By Theorem 8.10, we can conclude that

$$\deg(p(x)) = \deg(x - c) + \deg(q(x)).$$

Since $\deg(p(x)) \geq 2$ and $\deg(x - c) = 1$, it follows that $\deg(q(x)) \geq 1$. This fact and Equation (10.2) allow us to conclude that $p(x)$ is reducible. Thus, we have proved the following theorem.

Theorem 10.12. *Let F be a field and let $p(x) \in F[x]$ with $\deg(p(x)) \geq 2$. If $p(x)$ has a root in F, then $p(x)$ is reducible in $F[x]$.*

If we write the contrapositive of the conditional statement in Theorem 10.12, we obtain the following corollary:

Corollary 10.13. *Let F be a field and let $p(x) \in F[x]$ with $\deg(p(x)) \geq 2$. If $p(x)$ is irreducible in $F[x]$, then $p(x)$ has no roots in F.*

Activity 10.14. In this activity, we will explore some of the subtleties of the relationship between roots and reducibility.

(a) Let $p(x) = x^4 + 3x^2 + 2$ be a polynomial in $\mathbb{R}[x]$. Show that $p(x)$ can be factored in $\mathbb{R}[x]$ and hence, $p(x)$ is reducible in $\mathbb{R}[x]$. Also explain why $p(x)$ has no roots in $\mathbb{R}$.

(b) Use part (a) to explain why the following converse of Corollary 10.13 is false.

> *Let F be a field and let $p(x) \in F[x]$ with $\deg(p(x)) \geq 2$. If $p(x)$ has no roots in F, then $p(x)$ is irreducible in $F[x]$.*

Although we have shown that the converse of Corollary 10.13 is false, if we restrict ourselves to polynomials of degree 2 or 3, we can show that if the polynomial has no roots, then the polynomial is irreducible. Notice that the counterexample in part (b) of Activity 10.8 used a polynomial of degree 4.

Theorem 10.15. *Let F be a field and let $p(x) \in F[x]$. If $\deg(p(x)) = 2$ or $\deg(p(x)) = 3$ and if $p(x)$ has no roots in F, then $p(x)$ is irreducible in $F[x]$.*

Proof. We will use a proof by contradiction. So we assume that $\deg(p(x)) = 2$ or $\deg(p(x)) = 3$, $p(x)$ has no roots in F, and $p(x)$ is reducible in $F[x]$. This means that there exist polynomials $f(x), g(x) \in F[x]$ such that

$$p(x) = f(x)g(x),$$
$$\deg(f(x)) \geq 1, \text{ and}$$
$$\deg(g(x)) \geq 1.$$

In addition, by Theorem 8.10, $\deg(p(x)) = \deg(f(x)) + \deg(g(x))$. Since $\deg(p(x))$ is 2 or 3, we must have $\deg(f(x)) = 1$ or $\deg(g(x)) = 1$. Without loss of generality, we may assume that $\deg(f(x)) = 1$. This means that there exist $a, b \in F$ with $a \neq 0$ and $f(x) = ax + b$. We then see that

$$f\left(-a^{-1}b\right) = a\left(-a^{-1}b\right) + b$$
$$= -\left(a^{-1}a\right)b + b$$
$$= -b + b$$
$$= 0$$

Hence, $p\left(-a^{-1}b\right) = 0$ and $p(x)$ has a root in F. This contradicts the assumption that $p(x)$ has no roots in F. Therefore, $p(x)$ is irreducible in $F[x]$. ∎

Unique Factorization in $F[x]$

In Investigation 1, we proved the Fundamental Theorem of Arithmetic, which states that every integer greater than 1 is either prime or a product of primes, and this factorization is unique up to the order of the factors. An important result that was needed to prove the Fundamental Theorem was Euclid's Lemma. (See page 10.) Here we will prove an analog of Euclid's Lemma for polynomials, which will then be used to prove a unique factorization theorem for polynomials over a field F. In this unique factorization theorem, irreducible polynomials will play a role similar to that of prime numbers in the integers.

Theorem 10.16. *Let F be a field and let $p(x), f(x), g(x) \in F[x]$. If $p(x)$ divides $f(x)g(x)$ and $p(x)$ is irreducible, then $p(x)$ divides $f(x)$ or $p(x)$ divides $g(x)$.*

The next activity will guide us through one way to prove Theorem 10.16. The method of proof is similar to the one outlined in Activity 1.26 (see page 16) for the integers.

Activity 10.17. Let F be a field, and let $p(x), f(x), g(x) \in F[x]$. Assume that $p(x)$ divides $f(x)g(x)$ and that $p(x)$ is irreducible. If $p(x)$ divides $f(x)$, then we are done. So assume that $p(x)$ does not divide $f(x)$.

(a) Explain why it must be the case that $\gcd(p(x), f(x)) = 1$.

(b) Use Theorem 9.10 (see page 136) to translate the observation from part (a) into an equation involving a linear combination of $p(x)$ and $f(x)$.

(c) Multiply both sides of your equation from part (b) by an appropriate polynomial in order to obtain an equation of the form

$$g(x) = \underline{\hspace{1cm}} + \underline{\hspace{1cm}}.$$

(d) Explain why each of the terms on the right-hand side of the equation from part (c) are divisible by $p(x)$. Then conclude that $p(x)$ divides $g(x)$. (This proves that $p(x)$ divides $f(x)$ or $p(x)$ divides $g(x)$.)

As with Euclid's Lemma for the integers, Theorem 10.16 can be generalized as follows:

Corollary 10.18. *Let F be a field and let $p(x), f_1(x), f_2(x), \ldots, f_n(x) \in F[x]$. If $p(x)$ divides $f_1(x)f_2(x) \cdots f_n(x)$ and $p(x)$ is irreducible, then $p(x)$ divides $f_i(x)$, for some $i \in \mathbb{N}$ with $1 \leq i \leq n$.*

One way to prove this generalization is to use mathematical induction. This proof is left as an exercise. (See Exercise (13).)

Before we prove the unique factorization theorem for polynomials over a field, there is one more idea that we need to introduce.

Definition 10.19. Let F be a field. A polynomial $g(x)$ in $F[x]$ is said to be an **associate** of the polynomial $f(x)$ in $F[x]$ provided that there exists a nonzero $c \in F$ such that $g(x) = cf(x)$.

For example:

- In $\mathbb{R}[x]$, $g(x) = -2x + 6$ is an associate of $f(x) = x - 3$ since $g(x) = -2f(x)$. Notice that both $f(x)$ and $g(x)$ are irreducible in $\mathbb{R}[x]$ since both have degree 1.

- In $\mathbb{Q}[x]$, $g(x) = 3x^2 + x + 4$ is an associate of $f(x) = x^2 + \frac{1}{3}x + \frac{4}{3}$ since $g(x) = 3f(x)$.

- In $\mathbb{Z}_7[x]$, $q(x) = [4]x^2 + [5]x$ is an associate of $p(x) = x^2 + [3]x$ since $g(x) = [4]p(x)$

In the next activity, we will explore some results about associates that will be helpful in proving the unique factorization theorem.

Activity 10.20. Let F be a field, and let $p(x)$ and $q(x)$ be polynomials in $F[x]$ with $q(x)$ irreducible. We know that if $\deg(q(x)) \geq 1$, then $q(x)$ cannot be factored as a product of two polynomials of positive degree. Use this fact to explain why if $p(x)$ divides $q(x)$, then either $p(x) = c$ for some $c \in F$, or $p(x)$ is an associate of $q(x)$.

We are now ready to prove the unique factorization theorem for polynomials.

Theorem 10.21 (Unique Factorization in $F[x]$). *Let F be a field. If $f(x) \in F[x]$ is a polynomial with positive degree, then $f(x)$ is either irreducible or a product of irreducible polynomials in $F[x]$. This factorization is unique in the following sense: If*

$$f(x) = p_1(x)p_2(x)\cdots p_m(x) \qquad and \qquad f(x) = q_1(x)q_2(x)\cdots q_k(x) \qquad (10.3)$$

with $p_1(x), p_2(x), \ldots p_m(x)$ and $q_1(x), q_2(x), \ldots q_k(x)$ irreducible in $F[x]$, then $m = k$ and after the $q_j(x)$ are reordered and relabeled, if necessary, $p_j(x)$ is an associate of $q_j(x)$, for each j with $1 \le j \le m$.

Proof. We will first prove the existence of a factorization into irreducibles using induction on the degree of $f(x)$. If $\deg(f(x)) = 1$, then we know that $f(x)$ is irreducible (by Theorem 10.11), and this establishes the basis step.

For the induction step, we assume that $n \in \mathbb{N}$, that $\deg(f(x)) = n$, and that all polynomials in $F[x]$ of degree $1, 2, \ldots, n-1$ are either irreducible or the product of irreducible polynomials. We will consider two cases: either $f(x)$ is irreducible or $f(x)$ is reducible.

There is nothing to prove in the case that $f(x)$ is irreducible. In the case that $f(x)$ is reducible, we have

$$f(x) = g(x)h(x) \qquad (10.4)$$

for some polynomials $g(x)$ and $h(x)$ of positive degree in $F[x]$. We can use Theorem 8.10 to conclude that $\deg(f(x)) = \deg(g(x)) + \deg(h(x))$, and hence that the degrees of both $g(x)$ and $h(x)$ are less than n. The inductive hypothesis then implies that both $g(x)$ and $h(x)$ are either irreducible or a product of irreducible polynomials. We can then use Equation (10.4) to conclude that $f(x)$ is irreducible or a product of irreducible polynomials. This completes the induction proof that any polynomial of positive degree in $F[x]$ is either irreducible or a product of irreducible polynomials.

To prove that the factorization is unique in the sense stated in the theorem, we will again use induction on the degree of $f(x)$. If $\deg(f(x)) = 1$, then $f(x)$ is irreducible, and the only way that Equation (10.3) can be true is if $f(x) = p_1(x) = q_1(x)$. This establishes the basis step.

For the inductive step, assume that $n \in \mathbb{N}$, that $\deg(f(x)) = n$, and that the factorization into irreducible polynomials is unique (as stated in the theorem) for all polynomials in $F[x]$ of degree 1, $2, \ldots, n-1$. Furthermore, assume that

$$f(x) = p_1(x)p_2(x)\cdots p_m(x) \qquad and \qquad f(x) = q_1(x)q_2(x)\cdots q_k(x) \qquad (10.5)$$

for some irreducible $p_1(x), p_2(x), \ldots p_m(x)$ and $q_1(x), q_2(x), \ldots q_k(x)$ in $F[x]$. Since

$$p_1(x)p_2(x)\ldots p_m(x) = q_1(x)q_2(x)\ldots q_k(x),$$

it follows that $p_1(x)$ divides $q_1(x)q_2(x)\cdots q_k(x)$. We can then use Corollary 10.18 to conclude that $p_1(x)$ divides $q_j(x)$ for some j. After rearranging and relabeling, if necessary, we may assume that $p_1(x)$ divides $q_1(x)$. Since $p_1(x)$ and $q_1(x)$ are irreducible, $p_1(x)$ must be an associate of $q_1(x)$. So $p_1(x) = cq_1(x)$ for some $c \in F$. Using this and the equations in (10.5), we see that

$$f(x) = cq_1(x)p_2(x)\cdots p_m(x) = q_1(x)q_2(x)\cdots q_k(x).$$

Since $F[x]$ is an integral domain, we can cancel the $q_1(x)$ on both sides of this equation and obtain

$$cp_2(x)\cdots p_m(x) = q_2(x)\cdots q_k(x).$$

Since $\deg(f(x)) = n$ and $\deg q_1(x) \ge 1$, we can conclude that the polynomials on each side of the last equation have degree less than n. After rearranging and relabeling the factors, if necessary, we can now use the inductive hypothesis to conclude that $m = k$ and that each $p_j(x)$ is an associate of $q_j(x)$. This completes the induction proof that the factorization of $f(x)$ into irreducibles is unique. ∎

Concluding Activities

Activity 10.22. In this activity, we will explore some different ways to write the factorization of a polynomial into irreducibles. Let F be a field, and let $f(x) \in F[x]$.

(a) Use Theorem 10.21 to explain why there exist distinct irreducible polynomials $p_1(x), p_2(x),$ $\ldots, p_n(x)$ and positive integers $r_1, r_2, \ldots, r_n$ such that

$$f(x) = p_1(x)^{r_1} p_2(x)^{r_2} \cdots p_n(x)^{r_n}.$$

(b) Let $c \in F$ be the leading coefficient of $f(x)$. Explain why there exist unique monic irreducible polynomials $\hat{p}_1(x), \hat{p}_2(x), \ldots, \hat{p}_n(x)$ and positive integers $r_1, r_2, \ldots, r_n$ such that

$$f(x) = c\hat{p}_1(x)^{r_1} \hat{p}_2(x)^{r_2} \cdots \hat{p}_n(x)^{r_n}.$$

Activity 10.23. In geometry, we learned that two points determine a straight line. We can view this fact in terms of linear functions with real coefficients. A linear function can be considered a polynomial of degree 1. Let $f(x) = ax + b$ and $g(x) = cx + d$ be degree 1 polynomials in $\mathbb{R}[x]$. Assume that $f(1) = g(1)$ and that $f(3) = g(3)$. (This means that the straight lines that are the graphs of these two functions both pass through the points $(1, f(1))$ and $(3, f(3))$, so the graphs are the same straight line.)

(a) Can we conclude that $f(x) = g(x)$ in $\mathbb{R}[x]$? To answer this question, first notice that:

   - $f(1) = g(1)$ implies that $a + b = c + d$; and
   - $f(3) = g(3)$ implies that $3a + b = 3c + d$.

   Use these equations to show that $a = c$ and $b = d$.

(b) Although the method from part (a) could be used to prove a similar result for degree 2 polynomials, another approach involves using Theorem 10.9. Let $f(x) = ax^2 + bx + c$ and $g(x) = px^2 + qx + r$ be degree 2 polynomials in $\mathbb{R}[x]$. Assume further that u, v, and w are three distinct real numbers, and that $f(u) = g(u)$, $f(v) = g(v)$, and $f(w) = g(w)$. Construct the polynomial $h(x) = f(x) - g(x)$ in $\mathbb{R}[x]$. Explain why $h(x)$ has three roots, and then use Theorem 10.9 to explain why $h(x)$ has to be the zero polynomial, and thus $f(x) = g(x)$.

(c) Use an argument similar to the argument in part (b) to prove the following theorem. (A proof by contradiction could be helpful.)

   Theorem 10.24. *Let F be a field, let n be a natural number, and let $u(x)$ and $v(x)$ be polynomials in $F[x]$.*

   - *If $\deg(u(x)) < n$ and $\deg(v(x)) < n$, and*
   - *if there exist n distinct elements $a_1, a_2, \ldots, a_n$ in F such that $u(a_1) = v(a_1)$, $u(a_2) = v(a_2)$, ..., $u(a_n) = v(a_n)$,*

   then $u(x) = v(x)$ in $F[x]$.

Exercises

(1) Determine all roots of $f(x) = x^2 + [3]x + [2]$ in $\mathbb{Z}_6[x]$.

(2) Let $f(x) = x^2 + [7]x$ be a polynomial in $\mathbb{Z}_{10}[x]$. Find two different factorizations of $f(x)$ into a product of two monic polynomials of degree 1. Does this contradict Theorem 10.21? Explain.

(3) Let $p(x) = x^4 - 3x^2 - 10$. Completely factor $p(x)$ in:

    (a) $\mathbb{C}[x]$ (b) $\mathbb{R}[x]$ (c) $\mathbb{Q}[x]$

(4)   (a) Factor the polynomial $p(x) = 4x^4 - 7x^3 - 25x^2 + 33x - 9$ in $\mathbb{Q}[x]$.

    (b) Factor the polynomial $p(x) = 4x^4 - 7x^3 - 25x^2 + 33x - 9$ in $\mathbb{R}[x]$.

    (c) Factor the polynomial $p(x) = x^3 + 4x^2 + 4x + 3$ in $\mathbb{Z}_{11}[x]$.

(5)   (a) Find all real numbers c so that $(x - c)$ is a factor of the polynomial $p(x) = 3x^3 + 4x^2 - 5x + c$ in $\mathbb{R}[x]$.

    (b) Find all real numbers c so that $(x - c)$ is a factor of the polynomial $p(x) = 3x^3 + 4x^2 + cx - 8$ in $\mathbb{R}[x]$.

(6) Prove Theorem 10.11. (Hint: Let $p(x)$ be a polynomial of degree 1 in $F[x]$, and write $p(x) = f(x)g(x)$. Use Theorem 8.10 to conclude that $\deg(p(x)) = \deg(f(x)) + \deg(g(x))$.)

(7) Theorem 10.15 can be useful in determining if polynomials of degree 2 or degree 3 are irreducible. In addition, the quadratic formula can often be used to find the roots of a degree 2 polynomial.

    (a) Let $f(x) = 2x^2 - 5x - 7$ be a polynomial in $\mathbb{R}[x]$. Is $f(x)$ irreducible or reducible in $\mathbb{R}[x]$? Justify your conclusion.

    (b) Let $f(x) = x^2 + x + 1$ be a polynomial in $\mathbb{R}[x]$. Is $f(x)$ irreducible or reducible in $\mathbb{R}[x]$? Justify your conclusion.

    (c) Let $f(x) = x^2 + x + 1$ be a polynomial in $\mathbb{C}[x]$. Is $f(x)$ irreducible or reducible in $\mathbb{C}[x]$? Justify your conclusion.

(8) Let $p(x) = ax^2 + bx + c$ be a polynomial in $\mathbb{C}[x]$ with $a, b, c \in \mathbb{C}$ and $a \neq 0$. Prove that $p(x)$ is reducible in $\mathbb{C}[x]$.

(9) Let $p(x) = ax^2 + bx + c$ be a polynomial in $\mathbb{R}[x]$ with $a, b, c \in \mathbb{R}$ and $a \neq 0$. Use the quadratic formula to determine a necessary and sufficient condition on the coefficients for $p(x)$ to be irreducible in $\mathbb{R}[x]$.

(10) Let F be a field. In Activity 10.8, we proved that every polynomial of degree 1 in $F[x]$ is irreducible and has exactly one root in F. Now let $p(x) \in F[x]$ with $\deg(p(x)) = 2$. Prove that $p(x)$ either has no roots in F or has two roots in F. In the latter case, prove that $p(x) = c(x - a)(x - b)$, for some $a, b, c \in F$ with $c \neq 0$.

(11) Let F be a field and let $q(x) = a_n x^{2n} + a_{n-1} x^{2(n-1)} + \cdots + a_1 x^2 + a_0$ be a polynomial in $F[x]$. Prove that if c is a root of $q(x)$ in F, then c^2 is a root of the polynomial $p(x) = a_n x^n + a_{n-1} x^{n-1} + \cdots + a_1 x + a_0$ in $F[x]$.

Suggestion: If you are having a difficult time with the notation, you might first try to prove the result for a polynomial of degree 2 or 3. For example, you could prove the result for $q(x) = a_2x^4 + a_1x^2 + a_0$, and then try the general case.

(12) Let F be a field and let $p(x) = a_nx^n + a_{n-1}x^{n-1} + \cdots + a_1x + a_0$ be a polynomial in $F[x]$. Prove that if c is a nonzero root of $p(x)$ in F, then c^{-1} is a root of the polynomial $q(x) = a_0x^n + a_1x^{n-1} + \cdots + a_{n-1}x + a_n$ in $F[x]$.

Suggestion: If you are having a difficult time with the notation, you might first try to prove the result for a polynomial of degree 2 or 3. For example, you could prove the result for $p(x) = a_2x^2 + a_1x + a_0$, and then try the general case.

$\star$ (13) Use induction to prove Corollary 10.18, which is a generalization of Theorem 10.16.

Connections

In Investigations 8 and 9, we saw that a polynomial ring with coefficients from a field has much in common with the ring of integers, including: a divides relation, a division algorithm, and the idea of relatively prime polynomials. This investigation made an additional connection regarding unique factorization. The Fundamental Theorem of Arithmetic from Investigation 1 showed that the prime numbers form the building blocks of the ring of integers. In the same way, Theorem 10.21 establishes that the irreducible polynomials in $F[x]$ play the same role as the primes do in the integers, forming the building blocks for all nonconstant polynomials. Furthermore, just like in the integers, factorization into irreducibles is unique (up to order and associates).

Investigation 11

Irreducible Polynomials

Focus Questions

By the end of this investigation, you should be able to give precise and thorough answers to the questions listed below. You may want to keep these questions in mind to focus your thoughts as you complete the investigation.

- What is an irreducible polynomial?

- What does the Fundamental Theorem of Algebra say about irreducible polynomials in $\mathbb{C}[x]$?

- How can we determine if a polynomial in $\mathbb{R}[x]$ is irreducible? What are some important results related to irreducibility and factorization of polynomials in $\mathbb{R}[x]$?

- How can do we determine if a polynomial in $\mathbb{Q}[x]$ is irreducible? What are some important results related to irreducibility and factorization of polynomials in $\mathbb{Q}[x]$?

- What are the main differences between factorization and irreducibility in $\mathbb{C}[x]$, $\mathbb{R}[x]$, and $\mathbb{Q}[x]$?

Preview Activity 11.1. In previous investigations, we defined irreducible polynomials and showed that irreducible polynomials in polynomial rings over fields play the same role as primes play in $\mathbb{Z}$. In this investigation we will explore some methods to determine when a polynomial is irreducible, with a special emphasis on polynomials with coefficients in $\mathbb{C}$, $\mathbb{R}$, and $\mathbb{Q}$. To begin, we will review the definition and a simple case. Let F be a field.

(a) Give a formal definition of what it means for a polynomial $f(x) \in F[x]$ to be irreducible in $F[x]$.

(b) Is the polynomial $x^2 - 2$ irreducible in $\mathbb{C}[x]$? Explain.

(c) Is the polynomial $x^2 - 2$ irreducible in $\mathbb{R}[x]$? Explain.

(d) Is the polynomial $x^2 - 2$ irreducible in $\mathbb{Q}[x]$? Explain.

(e) The simplest nonconstant polynomials are the linear (degree 1) polynomials. Show that any degree 1 polynomial in $F[x]$ is irreducible.

Introduction

We have seen that every nonconstant polynomial in a polynomial ring over a field can be factored into a product of irreducible polynomials. While there is no general technique to find the factorization of an arbitrary polynomial into a product of irreducible polynomials, there are some tools that are helpful in this endeavor. As we will see, the factorization of a polynomial into a product of irreducible polynomials, and exactly which polynomials are irreducible, depends heavily on the coefficient field. In this investigation, we will discuss which polynomials are irreducible over $\mathbb{C}, \mathbb{R}$, and $\mathbb{Q}$, and we will discover some techniques for determining whether polynomials in $\mathbb{C}[x], \mathbb{R}[x]$, and $\mathbb{Q}[x]$ are irreducible.

Factorization in $\mathbb{C}[x]$

We will begin our work in this investigation by considering polynomials in $\mathbb{C}[x]$. To determine which polynomials in $\mathbb{C}[x]$ are irreducible, we will invoke a very important result known as The Fundamental Theorem of Algebra.

Theorem 11.2 (The Fundamental Theorem of Algebra). *Every polynomial of degree 1 or greater in $\mathbb{C}[x]$ has a root in $\mathbb{C}$.*

There are many different proofs of the Fundamental Theorem of Algebra, and most of them require some knowledge of complex analysis that is beyond the scope of this text.

As its name would suggest, the importance of the Fundamental Theorem of Algebra cannot to be overstated. Because every nonconstant polynomial in $\mathbb{C}[x]$ has a root in $\mathbb{C}$, we say that $\mathbb{C}$ is an *algebraically closed* field. This is a fundamental difference between $\mathbb{C}$ and $\mathbb{R}$. Furthermore, since every root of a polynomial corresponds to a linear factor of the polynomial, the Fundamental Theorem of Algebra tells us the following:

Corollary 11.3. *Every polynomial of degree 1 or greater in $\mathbb{C}[x]$ can be factored as a product of linear polynomials in $\mathbb{C}[x]$.*

In other words, the only irreducible polynomials in $\mathbb{C}[x]$ are the linear polynomials. Unfortunately, although in theory we can factor any polynomial in $\mathbb{C}[x]$ as a product of linear polynomials, there are no general techniques that will perform the factorization. Even so, if a polynomial in $\mathbb{C}[x]$ is of small enough degree or has a particularly convenient form, we may be able to find its factors quite easily.

As an example, consider the polynomial $x^3 - x^2 + x \in \mathbb{C}[x]$. We can factor out a common factor of x to obtain

$$x^3 - x^2 + x = x(x^2 - x + 1).$$

The quadratic formula then tells us that the roots of $x^2 - x + 1$ are $\frac{1 \pm \sqrt{3}i}{2}$. So the Factor Theorem shows that

$$x^2 - x + 1 = \left[x - \left(\frac{1 + \sqrt{3}i}{2} \right) \right] \left[x - \left(\frac{1 - \sqrt{3}i}{2} \right) \right].$$

Therefore,

$$x^3 - x^2 + x = x \left[x - \left(\frac{1 + \sqrt{3}i}{2} \right) \right] \left[x - \left(\frac{1 - \sqrt{3}i}{2} \right) \right]. \tag{11.1}$$

Since every degree 1 polynomial is irreducible, the factorization from Equation (11.1) is the unique factorization of $x^3 - x^2 + x$ into a product of irreducible polynomials in $\mathbb{C}[x]$.

Activity 11.4. Factor $f(x) = x^4 - 1$ in $\mathbb{C}[x]$ into a product of irreducible polynomials in $\mathbb{C}[x]$.

In addition to what Corollary 11.3 tells us about irreducible polynomials in $\mathbb{C}[x]$, it also tells us something about the number of roots that a polynomial of degree n in $\mathbb{C}[x]$ must have. You may recall that Theorem 10.9 (see page 147) states that a polynomial of degree n over a field F can have at most n roots in F. Corollary 11.3 allows us to take this result one step further for polynomials in $\mathbb{C}[x]$. In particular, since every polynomial in $\mathbb{C}[x]$ can be factored as a product of linear polynomials in $\mathbb{C}[x]$, and since the Factor Theorem tells us that every linear factor corresponds to a root, it follows that every polynomial in $\mathbb{C}[x]$ has exactly n roots in $\mathbb{C}$. There is, however, one small caveat to this conclusion. Since the factorization of a polynomial in $\mathbb{C}[x]$ may include repeated linear factors, the n roots guaranteed by the Factor Theorem and Corollary 11.3 may not be distinct. So, for instance, the polynomial $p(x) = x^4 - 6x^3 + 13x^2 - 24x + 36$ can be factored as follows:

$$p(x) = (x - 2i)(x + 2i)(x - 3)^2.$$

In this case, 3 is a repeated root—that is, a root of *multiplicity* 2. In general, the multiplicity of a root c of a polynomial $f(x)$ is the largest integer k for which $(x - c)^k$ divides $f(x)$. Using this language, we obtain the following result:

Corollary 11.5. *Every polynomial of degree $n \geq 1$ in $\mathbb{C}[x]$ has exactly n roots in $\mathbb{C}$ (with repeated roots counted according to their multiplicities).*

Factorization in $\mathbb{R}[x]$

The irreducible polynomials in $\mathbb{R}[x]$ are closely related to the irreducible polynomials in $\mathbb{C}[x]$. As an example, consider again the polynomial $f(x) = x^3 - x^2 + x \in \mathbb{R}[x]$. In Equation (11.1), we factored $f(x)$ into a product of irreducible linear factors in $\mathbb{C}[x]$. Each linear factor corresponds to a root of $f(x)$ (in $\mathbb{C}$), and two of the roots of $f(x)$ are *complex conjugates* of each other.[*] Furthermore, the product of the factors corresponding to these two roots, $\left[x - \left(\frac{1+\sqrt{3}i}{2} \right) \right] \left[x - \left(\frac{1-\sqrt{3}i}{2} \right) \right]$, is a polynomial with real coefficients. This observation is the key idea in understanding which polynomials are irreducible in $\mathbb{R}[x]$.

Recall that the roots of a quadratic $ax^2 + bx + c \in \mathbb{R}[x]$ are $-\frac{b}{2a} \pm \frac{\sqrt{b^2 - 4ac}}{2a}$. If either root is complex, then its complex conjugate is also a root. That this always happens is the subject of the next theorem.

Theorem 11.6. *Let $f(x) \in \mathbb{R}[x]$. If $z \in \mathbb{C}$ is a root of $f(x)$, then so is $\bar{z}$.*

---

[*]The *complex conjugate* of a complex number $z = a + bi$ is the complex number $\bar{z} = a - bi$. Note that $\bar{\bar{z}} = z$, $z + \bar{z} = 2a$, and $z\bar{z} = a^2 + b^2$.

Proof. Let $f(x) = a_n x^n + a_{n-1} x^{n-1} + \cdots + a_1 x + a_0$ with $a_n, a_{n-1}, \ldots, a_1, a_0 \in \mathbb{R}$. Let $z \in \mathbb{C}$ be a root of $f(x)$. Then, using the facts that $\overline{u + w} = \overline{u} + \overline{w}$ and $\overline{uw} = \overline{u}\,\overline{w}$ for any complex numbers u and w, and the fact that $\overline{y} = y$ for any real number y, we see that

$$\begin{aligned}
f(\overline{z}) &= a_n \overline{z}^n + a_{n-1} \overline{z}^{n-1} + \cdots + a_1 \overline{z} + a_0 \\
&= \overline{a_n}\,\overline{z^n} + \overline{a_{n-1}}\,\overline{z^{n-1}} + \cdots + \overline{a_1 z} + \overline{a_0} \\
&= \overline{a_n z^n} + \overline{a_{n-1} z^{n-1}} + \cdots + \overline{a_1 z} + \overline{a_0} \\
&= \overline{a_n z^n + a_{n-1} z^{n-1} + \cdots + a_1 z + a_0} \\
&= \overline{f(z)} \\
&= \overline{0} \\
&= 0.
\end{aligned}$$

Thus, $\overline{z}$ is a root of $f(x)$.   ∎

We can use Theorem 11.6 in the following way. By the Fundamental Theorem of Algebra, we can factor any polynomial $f(x) \in \mathbb{R}[x] \subset \mathbb{C}[x]$ into a product of linear factors in $\mathbb{C}[x]$. Suppose $x - z$ is one such factor. Theorem 11.6 and the Factor Theorem tell us that $x - \overline{z}$ is also a factor of $f(x)$. Let $z = a + bi$. Then

$$(x - z)(x - \overline{z}) = x^2 - (z + \overline{z})x + z\overline{z} = x^2 - 2ax + (a^2 + b^2) \in \mathbb{R}[x].$$

So if we factor a polynomial $f(x) \in \mathbb{R}[x]$ into a product of irreducibles in $\mathbb{C}[x]$ and collect the linear factors corresponding to complex conjugate roots of $f(x)$, then we can write $f(x)$ as a product of irreducible linear and quadratic factors. Therefore, any polynomial in $\mathbb{R}[x]$ can be factored into a product of linear and quadratic polynomials in $\mathbb{R}[x]$. This proves the following theorem.

Theorem 11.7. *If $f(x) \in \mathbb{R}[x]$ is an irreducible polynomial, then $\deg(f(x))$ is either 1 or 2.*

We can determine which quadratic polynomials in $\mathbb{R}[x]$ are irreducible by using the quadratic formula and checking for real roots. There are similar, but more complicated, formulas for finding complex roots of cubic and quartic polynomials. More information about the cubic formula can be found in the online supplemental material.

Activity 11.8. Factor $f(x) = x^5 - 4x$ in $\mathbb{R}[x]$ into a product of irreducible polynomials in $\mathbb{R}[x]$.

Factorization in $\mathbb{Q}[x]$

Unlike in $\mathbb{R}[x]$ and $\mathbb{C}[x]$, we will see that there are irreducible polynomials of every degree in $\mathbb{Q}[x]$, and there is no general theory to tell us which polynomials are irreducible in $\mathbb{Q}[x]$. However, there are some tools we can use in addition to the Factor Theorem in order to determine if a given polynomial in $\mathbb{Q}[x]$ has roots in $\mathbb{Q}$.

Let

$$f(x) = \frac{a_n}{b_n} x^n + \frac{a_{n-1}}{b_{n-1}} x^{n-1} + \cdots + \frac{a_1}{b_1} x + \frac{a_0}{b_0} \in \mathbb{Q}[x],$$

with $a_n, a_{n-1}, \ldots, a_1, a_0, b_n, b_{n-1}, \ldots, b_1, b_0 \in \mathbb{Z}$. Let $B = b_n b_{n-1} \cdots b_1 b_0$ and $B_i = \frac{B}{b_i}$ for each i between 0 and n. We can then express $f(x)$ in the form

$$f(x) = \frac{1}{B} \left((a_n B_n)x^n + (a_{n-1}B_{n-1})x^{n-1} + \cdots + (a_1 B_1)x + (a_0 B_0) \right),$$

which is a nonzero rational number times a polynomial in $\mathbb{Z}[x]$. Hence, when looking for irreducible polynomials in $\mathbb{Q}[x]$, it suffices to study polynomials with integer coefficients.

Our first result in factoring polynomials in $\mathbb{Q}[x]$ helps us determine all possible rational roots.

Activity 11.9. Consider the polynomial $f(x) = 2x^3 + 7x^2 + 2x - 3$ in $\mathbb{Q}[x]$. Assume $\frac{p}{q} \in \mathbb{Q}$ is a root of $f(x)$, with $\frac{p}{q}$ in reduced form (that is, $\gcd(p, q) = 1$).

(a) Substitute $\frac{p}{q}$ for x in $f(x)$ and show that the equation

$$2p^3 + 7p^2q + 2pq^2 - 3q^3 = 0 \tag{11.2}$$

results.

(b) Add $3q^3$ to both sides of Equation (11.2), and factor all common factors from $2p^3 + 7p^2q + 2pq^2$. What does this tell us about p? (Hint: $\gcd(p, q) = 1$. What are the possible values for p?)

(c) Now subtract $2p^3$ from both sides of Equation (11.2), and factor all common factors from $7p^2q + 2pq^2 - 3q^3$. What does this tell us about q? (Hint: $\gcd(p, q) = 1$. What are the possible values for q?)

(d) What are the possible roots of $f(x)$ in $\mathbb{Q}$?

(e) Find all rational roots of $f(x)$.

Activity 11.9 is a specific case of the next theorem, which tells us how to find all rational roots of a polynomial in $\mathbb{Z}[x]$.

Theorem 11.10 (Rational Root Theorem). *Let*

$$f(x) = a_n x^n + a_{n-1}x^{n-1} + \cdots + a_1 x + a_0 \in \mathbb{Z}[x]$$

with $a_n \neq 0$. If $r = \frac{p}{q} \in \mathbb{Q}$ is a root of $f(x)$ with $\gcd(p, q) = 1$, then $p \mid a_0$ and $q \mid a_n$.

Proof. Let $f(x) = a_n x^n + a_{n-1}x^{n-1} + \cdots + a_1 x + a_0 \in \mathbb{Z}[x]$ and suppose $r = \frac{p}{q} \in \mathbb{Q}$ is a root of $f(x)$ with $\gcd(p, q) = 1$. Then

$$0 = f\left(\frac{p}{q}\right)$$

$$= a_n \left(\frac{p}{q}\right)^n + a_{n-1}\left(\frac{p}{q}\right)^{n-1} + \cdots + a_1 \left(\frac{p}{q}\right) + a_0$$

$$= \frac{1}{q^n} \left(a_n p^n + a_{n-1}p^{n-1}q + a_{n-2}p^{n-2}q^2 + \cdots + a_2 p^2 q^{n-2} + a_1 pq^{n-1} + a_0 q^n \right).$$

Therefore, we must have

$$0 = a_n p^n + a_{n-1}p^{n-1}q + a_{n-2}p^{n-2}q^2 + \cdots + a_2 p^2 q^{n-2} + a_1 pq^{n-1} + a_0 q^n. \tag{11.3}$$

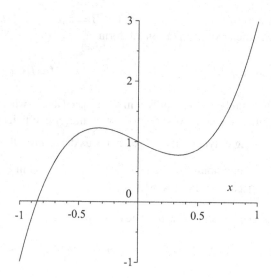

Figure 11.1
The graph of $x^3 - x + 1$ over $\mathbb{R}$.

Subtracting $a_n p^n$ from both sides of Equation (11.3) gives us

$$-a_n p^n = a_{n-1} p^{n-1} q + a_{n-2} p^{n-2} q^2 + \cdots + a_2 p^2 q^{n-2} + a_1 p q^{n-1} + a_0 q^n$$
$$= q \left(a_{n-1} p^{n-1} + a_{n-2} p^{n-2} q + \cdots + a_2 p^2 q^{n-3} + a_1 p q^{n-2} + a_0 q^{n-1} \right).$$

Thus, q divides $a_n p^n$. Since $\gcd(p, q) = 1$, we must have $q \mid a_n$.

Similarly, subtracting $a_0 q^n$ from both sides of Equation (11.3) gives us

$$-a_0 q^n = a_n p^n + a_{n-1} p^{n-1} q + a_{n-2} p^{n-2} q^2 + \cdots + a_2 p^2 q^{n-2} + a_1 p q^{n-1}$$
$$= p \left(a_n p^{n-1} + a_{n-1} p^{n-2} + a_{n-2} p^{n-3} q + \cdots + a_2 p q^{n-3} + a_1 q^{n-2} \right).$$

Thus p divides $a_0 q^n$. Since $\gcd(p, q) = 1$, we must have $p \mid a_0$. ∎

We can use Theorem 11.10 to help us determine if certain polynomials in $\mathbb{Q}[x]$ are irreducible. For example, Theorem 11.10 tells us that the possible rational roots of the polynomial $f(x) = 3x^3 - x + 1 \in \mathbb{Z}[x]$ are of the form $\frac{p}{q}$, where p divides 1 and q divides 3. So the possible rational roots of $f(x)$ are ± 1 or $\pm \frac{1}{3}$. The graph of $f(x)$, shown in Figure 11.1 indicates that these potential rational roots are not roots of $f(x)$ at all. (We could also verify this by simply evaluating $f(x)$ at $x = \pm 1$ and $x = \pm \frac{1}{3}$.) Since $\deg(f(x)) = 3$ and $f(x)$ has no roots in $\mathbb{Q}$, we can conclude that $f(x)$ is irreducible in $\mathbb{Q}[x]$.

Activity 11.11.

(a) Factor $f(x) = 18x^3 + 9x^2 - 5x - 2$ in $\mathbb{Q}[x]$.

(b) Is $f(x) = 6x^3 - 7x^2 + x - 1$ irreducible in $\mathbb{Q}[x]$?

Polynomials with No Linear Factors in $\mathbb{Q}[x]$

The Rational Root Theorem helps us find all the linear factors of a polynomial with integer coefficients. But what happens if a polynomial $f(x)$ in $\mathbb{Q}[x]$ has no linear factors? Can we still determine if $f(x)$ is irreducible in $\mathbb{Q}[x]$? The answer is yes, but the problem becomes more difficult.

Consider, for example, the polynomial $f(x) = x^4 + 2x^3 + 5x^2 + 4x + 3$ in $\mathbb{Q}[x]$. A quick use of the Rational Root Theorem shows us that $f(x)$ has no linear factors in $\mathbb{Q}[x]$. It is still possible, however, that $f(x)$ could factor into a product of two quadratic polynomials in $\mathbb{Q}[x]$. The problem would then be to determine if there are any quadratic polynomials $ax^2 + bx + c$ and $rx^2 + sx + t$ in $\mathbb{Q}[x]$ such that $f(x) = (ax^2 + bx + c)(rx^2 + sx + t)$. This would be a very hard task, since there are infinitely many combinations of coefficients that might yield such a factorization, and we can't possibly check them all. Fortunately, there is a result that makes the problem much easier.

Activity 11.12. Let $f(x) = 2x^2 + 3x + 1 \in \mathbb{Z}[x]$.

(a) Show that $f(x) = g(x)h(x)$ with $g(x) = x + \frac{1}{2}$ and $h(x) = 2x + 2$ in $\mathbb{Q}[x]$.

(b) From this factorization of $f(x)$ as $g(x)h(x)$, find associates $g_0(x)$ and $h_0(x)$ of $g(x)$ and $h(x)$ so that $g_0(x), h_0(x) \in \mathbb{Z}[x]$ and $f(x) = g_0(x)h_0(x)$.

Activity 11.12 showed how a factorization of $f(x)$ in $\mathbb{Q}[x]$ gave rise to a corresponding factorization of $f(x)$ in $\mathbb{Z}[x]$. This powerful result is the subject of the next lemma.

Lemma 11.13 (Gauss' Lemma). *Let $f(x) \in \mathbb{Z}[x]$. If there are polynomials $g(x)$ and $h(x)$ in $\mathbb{Q}[x]$ such that $f(x) = g(x)h(x)$, then there are also polynomials $g_0(x), h_0(x) \in \mathbb{Z}[x]$ so that $f(x) = g_0(x)h_0(x)$.*

Proof. Let

$$f(x) = a_n x^n + a_{n-1} x^{n-1} + \cdots + a_1 x + a_0 \in \mathbb{Z}[x].$$

Let

$$g(x) = \frac{b_m}{c_m} x^m + \frac{b_{m-1}}{c_{m-1}} x^{m-1} + \cdots + \frac{b_1}{c_1} x + \frac{b_0}{c_0}$$

and

$$h(x) = \frac{r_k}{s_k} x^k + \frac{r_{k-1}}{s_{k-1}} x^{k-1} + \cdots + \frac{r_1}{s_1} x + \frac{r_0}{s_0}$$

be polynomials in $\mathbb{Q}[x]$ so that $f(x) = g(x)h(x)$. We can factor out $c = c_m c_{m-1} \cdots c_1 c_0$ from $g(x)$ and $s = s_k s_{k-1} \cdots s_1 s_0$ from $h(x)$ to obtain polynomials

$$\hat{g}(x) = u_m x^m + u_{m-1} x^{m-1} + \cdots + u_1 x + u_0$$

and

$$\hat{h}(x) = v_k x^k + v_{k-1} x^{k-1} + \cdots + v_1 x + v_0$$

in $\mathbb{Z}[x]$ so that

$$g(x) = \frac{1}{c}\hat{g}(x) \text{ and } h(x) = \frac{1}{s}\hat{h}(x).$$

Then

$$f(x) = g(x)h(x) = \frac{1}{cs}\hat{g}(x)\hat{h}(x)$$

and

$$csf(x) = \hat{g}(x)\hat{h}(x). \tag{11.4}$$

If $cs = 1$, then we can let $g_0(x) = \hat{g}(x)$ and $h_0(x) = \hat{h}(x)$, and we are done. If $cs \neq 1$, then let $|cs| = p_1 p_2 \cdots p_l$ for prime integers $p_1, p_2, \ldots p_l$. Let t be between 1 and l, inclusive. Then p_t divides every coefficient of the polynomial $csf(x)$. Thus p_t divides every coefficient of the product $\hat{g}(x)\hat{h}(x)$.

To continue the proof, we will first need to prove the following claim:

Claim. Either p_t divides every coefficient of $\hat{g}(x)$ or p_t divides every coefficient of $\hat{h}(x)$.

Proof of Claim:. We will prove the claim by contradiction. Let $p = p_t$. Assume that there exist coefficients of both $\hat{g}(x)$ and $\hat{h}(x)$ that are not divisible by p. Let r and s be the smallest integers such that u_r and v_s are not divisible by p. Then p divides u_i for each $i < r$ and p divides v_j for each $j < s$. The coefficient of x^{r+s} in $\hat{g}(x)\hat{h}(x)$ is

$$u_{r+s}v_0 + u_{r+s-1}v_1 + u_{r+s-2}v_2$$
$$+ \cdots + u_{r+1}v_{s-1} + u_r v_s + u_{r-1}v_{s+1}$$
$$+ \cdots + u_1 v_{r+s-1} + u_0 v_{r+s}. \tag{11.5}$$

Now p divides every coefficient in $\hat{g}(x)\hat{h}(x)$, so p divides the expression in (11.5) above. Then

$$u_r v_s \equiv -(u_{r+s}v_0 + u_{r+s-1}v_1 + u_{r+s-2}v_2 + \cdots + u_{r+1}v_{s-1})$$
$$- (u_{r-1}v_{s+1} + \cdots + u_1 v_{r+s-1} + u_0 v_{r+s}) \pmod{p}.$$

Since p divides v_j for each $j < s$, we know p divides $u_{r+s}v_0 + u_{r+s-1}v_1 + u_{r+s-2}v_2 + \cdots + u_{r+1}v_{s-1}$. Similarly, since p divides u_i for $i < r$, we know p divides $u_{r-1}v_{s+1} + \cdots + u_1 v_{r+s-1} + u_0 v_{r+s}$. Therefore, p divides $u_r v_s$. Since p is a prime, p must divide u_r or p must divide v_s. Either conclusion is a contradiction to our assumption that both u_r and v_s are not divisible by p. Therefore, it must be the case that either p divides every coefficient of $\hat{g}(x)$ or p divides every coefficient of $\hat{h}(x)$. ∎

Having established that p_t divides every coefficient of $\hat{g}(x)$ or p divides every coefficient of $\hat{h}(x)$, we may assume, without loss of generality, that p_t divides every coefficient of $\hat{g}(x)$. Then we can factor out p_t from each coefficient and write $\hat{g}(x) = p_t g'(x)$ for some $g'(x) \in \mathbb{Z}[x]$. Then we can cancel the factor of p_t from both sides of Equation (11.4) to obtain

$$p_1 p_2 \cdots p_{t-1} p_{t+1} \cdots p_l f(x) = g'(x)\hat{h}(x).$$

We can continue to do this for each value of t to ultimately obtain

$$f(x) = g_0(x)h_0(x)$$

for some $g_0(x), h_0(x) \in \mathbb{Z}[x]$. ∎

The beauty of Gauss' Lemma is that it allows us to consider the problem of factoring with integer coefficients rather than rational ones.

As an example, let's return to the polynomial $f(x) = x^4 + 2x^3 + 5x^2 + 4x + 3$ in $\mathbb{Q}[x]$. To determine if $f(x)$ is irreducible, it suffices to show that there are no polynomials $g(x) = ax^2 + bx + c$

and $h(x) = rx^2 + sx + t$ in $\mathbb{Z}[x]$ with $a, r \neq 0$ so that $f(x) = g(x)h(x)$. By working with integer coefficients, we reduce the possible values of these coefficients to a manageable amount.

To see this in action, suppose there exist $g(x) = ax^2 + bx + c$ and $h(x) = rx^2 + sx + t \in \mathbb{Z}[x]$ such that

$$x^4 + 2x^3 + 5x^2 + 4x + 3$$
$$= g(x)h(x)$$
$$= (ar)x^4 + (as + br)x^3 + (at + bs + cr)x^2 + (bt + cs)x + ct.$$

By equating the coefficients, we obtain the following system of equations:

$$ar = 1$$
$$as + br = 2$$
$$at + bs + cr = 5$$
$$bt + cs = 4$$
$$ct = 3.$$

Note that we only need to find integer solutions to these equations. This leaves us with only two possibilities for a and r: either $a = r = 1$ or $a = r = -1$. Since we can always factor out a factor of -1 from $g(x)$ and $h(x)$, we can assume $a = r = 1$. The fact that $ct = 3$ gives us four possibilities: $c = 1, t = 3$; $c = -1, t = -3$; $c = 3, t = 1$; or $c = -3, t = -1$. The first case ($c = 1$, $t = 3$, and $a = r = 1$) leaves us with the following equations:

$$s + b = 2$$
$$3 + bs + 1 = 5$$
$$3b + s = 4.$$

Note that $b = s = 1$ gives a complete solution. Therefore,

$$f(x) = x^4 + 2x^3 + 5x^2 + 4x + 3 = (x^2 + x + 1)(x^2 + x + 3)$$

is reducible in $\mathbb{Q}[x]$.

Activity 11.14. Factor $x^4 + 2x^3 + 5x^2 + 6x + 6$ into a product of two quadratic polynomials in $\mathbb{Z}[x]$.

Of course, the problem of determining whether a polynomial in $\mathbb{Q}[x]$ is irreducible becomes much more involved if the degree of the polynomial is larger than 4. For example, to show that a degree 8 polynomial $f(x)$ is irreducible using the previous method, we would have to consider all the ways that $f(x)$ could be written as a product $g(x)h(x)$, with $\deg(g(x)) + \deg((h(x)) = 8$. So we would need to consider the cases where $\deg(g(x)) = 1$ and $\deg(h(x)) = 7$; $\deg(g(x)) = 2$ and $\deg(h(x)) = 6$; $\deg(g(x)) = 3$ and $\deg(h(x)) = 5$; and $\deg(g(x)) = 4$ and $\deg(h(x)) = 4$. This is a lot of work!

Reducing Polynomials in $\mathbb{Z}[x]$ Modulo Primes

Another tool in determining reducibility of polynomials with integer coefficients in $\mathbb{Q}[x]$ is reducing modulo a prime p. For example, consider the polynomial $f(x) = x^3 + x + 1 \in \mathbb{Z}[x]$. We can reduce this polynomial to the polynomial $\overline{f(x)} = [1]x^3 + [1]x + [1]$ in $\mathbb{Z}_2[x]$. Note that $\overline{f(x)}$ is irreducible in $\mathbb{Z}_2[x]$. (This follows from the Factor Theorem, since $f(x)$ is a degree 3 polynomial and has no roots in $\mathbb{Z}_2$.) So if $f(x) = g(x)h(x) \in \mathbb{Z}[x]$, then we should have $\overline{f(x)} = \overline{g(x)}\ \overline{h(x)} \in \mathbb{Z}_2[x]$. So it would appear that the irreducibility of $\overline{f(x)}$ in $\mathbb{Z}_2[x]$ implies the irreducibility of $f(x) \in \mathbb{Z}[x]$. To generalize this argument to $\mathbb{Z}_p$, we need only to require that the leading coefficient of $f(x)$ is not divisible by p.

Theorem 11.15. *Let $f(x) = a_n x^n + a_{n-1}x^{n-1} + \cdots + a_1 x + a_0$ be a polynomial with integer coefficients and $a_n \neq 0$. If there is a prime p so that $[a_n] \neq [0]$ in $\mathbb{Z}_p$ and the polynomial $\overline{f(x)} = [a_n]x^n + [a_{n-1}]x^{n-1} + \cdots + [a_1]x + [a_0]$ is irreducible in $\mathbb{Z}_p[x]$, then $f(x)$ is irreducible in $\mathbb{Q}[x]$.*

Proof. Let $f(x) = a_n x^n + a_{n-1}x^{n-1} + \cdots + a_1 x + a_0$ be a polynomial with integer coefficients. We will prove the contrapositive. That is, we assume $f(x)$ is not irreducible in $\mathbb{Q}[x]$. Then Gauss' Lemma tells us there are polynomials $g(x), h(x) \in \mathbb{Z}[x]$ such that $f(x) = g(x)h(x)$ with $1 \leq \deg(g(x)), \deg(h(x)) < \deg(f(x))$. Let b_m be the leading coefficient of $g(x)$ and c_k the leading coefficient of $h(x)$. Then $b_m c_k = a_n$. Suppose p is a prime so that $[a_n] \neq [0]$ in $\mathbb{Z}_p$. Let $\overline{f(x)}$ be the polynomial obtained by reducing the coefficients of $f(x)$ modulo p. Then $\deg(\overline{f(x)}) = \deg(f(x))$. Since $[a_n] = [b_m][c_k]$ and $\mathbb{Z}_p$ is an integral domain, $[b_m] \neq [0]$ and $[c_k] \neq [0]$ in $\mathbb{Z}_p$. Thus, $\deg(\overline{g(x)}) = \deg(g(x))$ and $\deg(\overline{h(x)}) = \deg(h(x))$. Also, $\overline{f(x)} = \overline{g(x)}\ \overline{h(x)}$, with $1 \leq \deg(\overline{g(x)}), \deg(\overline{h(x)}) < \deg(\overline{f(x)})$. So $\overline{f(x)}$ is reducible in $\mathbb{Z}_p[x]$. ∎

Note that we only need to find one prime p so that $\overline{f(x)}$ is irreducible in $\mathbb{Z}_p[x]$ to conclude that $f(x)$ is irreducible in $\mathbb{Q}[x]$. However, the relationship does not work in the other direction. For example, the polynomial $x^2 + [1]$ is reducible in $\mathbb{Z}_2[x]$, but is irreducible in $\mathbb{Q}[x]$.

Activity 11.16. Is $x^3 + 2x + 2$ irreducible in $\mathbb{Q}[x]$? Check by reducing modulo a prime.

Eisenstein's Criterion

There is one other well-known tool that is often used to determine the irreducibility of polynomials in $\mathbb{Q}[x]$.

Theorem 11.17 (Eisenstein's Criterion). *Let $f(x) = a_n x^n + a_{n-1}x^{n-1} + \cdots + a_1 x + a_0 \in \mathbb{Z}[x]$ be of positive degree. If there is a prime p so that $p \mid a_0$, $p \mid a_1$, $p \mid a_2$, ..., $p \mid a_{n-1}$, but p does not divide a_n and p^2 does not divide a_0, then $f(x)$ is irreducible in $\mathbb{Q}[x]$.*

Proof. Let $f(x) = a_n x^n + a_{n-1}x^{n-1} + \cdots + a_1 x + a_0 \in \mathbb{Z}[x]$ be of positive degree. Let p be a prime so that $p \mid a_0$, $p \mid a_1$, $p \mid a_2$, ..., $p \mid a_{n-1}$, p does not divide a_n and p^2 does not divide a_0. To prove that $f(x)$ is irreducible in $\mathbb{Q}[x]$, we proceed by contradiction. Assume there are nonconstant polynomials $g(x), h(x) \in \mathbb{Z}[x]$ (by Gauss's Lemma) so that $f(x) = g(x)h(x)$. Let

$$g(x) = b_m x^m + b_{m-1}x^{m-1} + \cdots + b_1 x + b_0$$

and

$$h(x) = c_k x^k + c_{k-1}x^{k-1} + \cdots + c_1 x + c_0,$$

with $b_m, c_k \neq 0$. Since $a_0 = b_0 c_0$, and p divides a_0, we know $p \mid b_0 c_0$. So $p \mid b_0$ or $p \mid c_0$. However, since p^2 does not divide a_0, we know that p cannot divide both b_0 and c_0. Assume without loss of generality that p does not divide b_0. We also know $a_n = b_m c_k$. Since p does not divide a_n, we cannot have p as a factor of b_m or c_k. Let l be the smallest integer so that c_l is not divisible by p. In other words, p divides c_i for each $i < l$. Note that we have $l \leq k < n$. Therefore, we know p divides a_l by hypothesis. Now

$$a_l = b_l c_0 + b_{l-1} c_1 + b_{l-2} c_2 + \cdots b_1 c_{l-1} + b_0 c_l,$$

and so

$$b_0 c_l = a_l - (b_l c_0 + b_{l-1} c_1 + b_{l-2} c_2 + \cdots b_1 c_{l-1}). \tag{11.6}$$

Notice that since p divides a_l and c_i for each $i < l$, we have p as a factor of each term on the right side of (11.6). Therefore, p must divide $b_0 c_l$. But neither b_0 nor c_l is divisible by p, so this is impossible. We conclude that no such polynomials $g(x), h(x)$ exist, and $f(x)$ is irreducible in $\mathbb{Q}[x]$. ∎

Activity 11.18.

(a) Show that $2x^5 + 27x^2 + 9x + 3$ is irreducible in $\mathbb{Q}[x]$.

(b) Show that for any prime p and any positive integer n, the polynomial $f(x) = x^n + p$ is irreducible in $\mathbb{Q}[x]$. Conclude that there are irreducible polynomials in $\mathbb{Q}[x]$ of any degree.

Factorization in $F[x]$ for Other Fields F

To factor polynomials over fields other than $\mathbb{Q}, \mathbb{R}$, or $\mathbb{C}$, trial and error is often the best option. We can first look for roots, which (by the Factor Theorem) give us linear factors. If we find roots, we can then divide by the corresponding linear factor to reduce the degree and look for roots again.

Activity 11.19. Factor $f(x) = [3]x^3 + [4]x^2 + [3]$ in $\mathbb{Z}_5[x]$.

Summary

Let $f(x) \in F[x]$. To factor $f(x)$ into a product of irreducible polynomials, we can always begin by testing for roots.

- If $f(x)$ can be written with integer coefficients, then we can use the Rational Root Theorem to check for roots in $\mathbb{Q}$.

- If $f(x) \in \mathbb{Z}_p[x]$ for some prime p, then we can just test each element in $\mathbb{Z}_p$ to see if $f(x)$ has any roots.

- If $r \in F$ is a root of $f(x)$, then we can divide $f(x)$ by $x - r$ and obtain a quotient $q(x)$ of degree $\deg(f(x)) - 1$. We can then repeat this process with $q(x)$.

- If $f(x)$ is of degree 2 or 3 and has no roots in F, then we know that $f(x)$ is irreducible in $F[x]$.

- If we can reduce $f(x)$ down to a polynomial of degree 2 in $\mathbb{Q}[x]$, $\mathbb{R}[x]$ or $\mathbb{C}[x]$, then we can use the quadratic formula to find the roots of $f(x)$ in $\mathbb{C}$.

Remember that the polynomial rings $\mathbb{Q}[x]$, $\mathbb{R}[x]$, and $\mathbb{C}[x]$ are fundamentally different with regard to the types of irreducible polynomials they contain. In particular:

- The Fundamental Theorem of Algebra tells us that the only irreducible polynomials in $\mathbb{C}[x]$ are the linear polynomials.

- The Fundamental Theorem of Algebra and Theorem 11.6 show us that there are no irreducible polynomials in $\mathbb{R}[x]$ of degree higher than 2.

- There are irreducible polynomials of any degree in $\mathbb{Q}[x]$.

Concluding Activities

Activity 11.20. Let $f(x) = ax^2 + bx + c \in \mathbb{C}[x]$ with $a \neq 0$. The discriminant of $f(x)$ is the number $D = b^2 - 4ac$.

(a) Show that the roots of $f(x)$ in $\mathbb{C}$ are $\frac{-b+\sqrt{D}}{2a}$ and $\frac{-b-\sqrt{D}}{2a}$.

(b) We can use the discriminant to completely characterize the irreducible quadratic polynomials in $\mathbb{R}[x]$. If $f(x) = ax^2 + bx + c \in \mathbb{R}[x]$ with $a \neq 0$, show that $f(x)$ is irreducible in $\mathbb{R}[x]$ if and only if $D < 0$.

Activity 11.21. We have used the quadratic formula to find roots of polynomials with complex coefficients, and we have seen that there is a cubic formula as well. As this activity will show, there is a quadratic formula for fields other than $\mathbb{C}$. Let F be any field, and let $a, b, c \in F$ with $a \neq 0$. Suppose also, that for each $t \neq 0$ in F, we have $t + t = 2t \neq 0$ as well. (Note that this assumption implies that F does not have characteristic 2.)

(a) Show that if there is an element $q \in F$ so that $q^2 = b^2 - 4ac$, then the polynomial $f(x) = ax^2 + bx + c$ has the two roots $r_1 = (2a)^{-1}(-b + q)$ and $r_2 = (2a)^{-1}(-b - q)$. Explain why these are all the roots of $f(x)$.

(b) Prove the converse of the previous problem. That is, show that if there is no element $q \in F$ so that $q^2 = b^2 - 4ac$, then the polynomial $f(x) = ax^2 + bx + c$ has no roots in F. (Hint: Complete the square and perform a little algebra.)

(c) Apply the previous results to find the roots of the polynomial $x^2 + [2]x + [2]$ in $\mathbb{Z}_5$. (Hint: Compute q^2 for each element $q \in \mathbb{Z}_5$. Use the method above instead of computing the roots directly.)

(d) Apply the previous results to show that the polynomial $[3]x^2 + x + [2]$ is irreducible in $\mathbb{Z}_5[x]$.

(e) Compare the results of the first two parts of this problem to the well-known quadratic formula.

(f) Why did we insist that $t + t = 2t \neq 0$ for each nonzero $t \in F$?

Exercises

(1) Factor $f(x) = 18x^3 + 9x^2 - 5x - 2$ into a product of irreducible polynomials in $\mathbb{Q}[x]$.

(2) Factor $f(x) = 2x^4 + x^3 + 4x^2 + x + 2$ into a product of irreducible polynomials in $\mathbb{Q}[x]$.

(3) Factor $f(x) = x^3 - 2x^2 - 3x + 6$ into a product of irreducible polynomials in $\mathbb{R}[x]$.

(4) Factor $f(x) = x^3 - 4x^2 + 6x - 4$ into a product of irreducible polynomials in $\mathbb{C}[x]$.

(5) Factor $f(x) = [3]x^3 + [4]x^2 + [3]$ into a product of irreducible polynomials in $\mathbb{Z}_5[x]$.

(6) Is $f(x) = 235x^3 + 110x + 59$ irreducible in $\mathbb{Q}[x]$?

(7) Is $f(x) = 6x^3 - 7x^2 + x - 6$ irreducible in $\mathbb{Q}[x]$?

(8) Is $f(x) = x^4 + 3x^2 + 2$ irreducible in $\mathbb{Q}[x]$?

(9) Is $f(x) = 2x^8 + 15x^5 - 21x^2 + 9x + 3$ irreducible in $\mathbb{Q}[x]$?

(10) Is $f(x) = [6]x^3 + x + [2]$ irreducible in $\mathbb{Z}_7[x]$?

(11) Find all the roots in $\mathbb{C}$ of the given polynomial. (You might find the cubic formula in the online supplemental material to be helpful.)

    (a) $x^3 + 3x^2 + 3x + 12$

    (b) $x^3 + 6x^2 + 18x + 22$

(12) Let $f(x) \in \mathbb{R}[x]$.

    (a) If $\deg(f(x)) = 2$, then how many real roots can $f(x)$ have? Explain.

    (b) If $\deg(f(x)) = 3$, then how many real roots can $f(x)$ have? Explain.

    (c) If $\deg(f(x)) = 4$, then how many real roots can $f(x)$ have? Explain.

    (d) If $\deg(f(x)) = 5$, then how many real roots can $f(x)$ have? Explain.

(13) Let F be a field. We know that

- Any monic polynomial of degree 2 in $F[x]$ can be written in the form $x^2 + bx + c$ for some elements b and c in F.

- Any monic polynomial of degree 2 that is reducible can be written in the form $(x - r)(x - s)$ for some elements r and s in F.

    (a) How many monic polynomials of degree 2 in $\mathbb{Z}_2[x]$ are irreducible in $\mathbb{Z}_2[x]$. Justify your conclusion. Note: It might be easier to count the total number of monic polynomials of degree 2 in $\mathbb{Z}_2[x]$ and the monic polynomials of degree 2 in $\mathbb{Z}_2[x]$ that are reducible.

    (b) How many monic polynomials of degree 2 in $\mathbb{Z}_3[x]$ are irreducible in $\mathbb{Z}_3[x]$. Justify your conclusion. (Hint: It might be easier to count the total number of monic polynomials of degree 2 in $\mathbb{Z}_3[x]$ and the monic polynomials of degree 2 in $\mathbb{Z}_3[x]$ that are reducible.)

(c) If p is a prime number, then how many monic polynomials of degree 2 in $\mathbb{Z}_p[x]$ are irreducible in $\mathbb{Z}_p[x]$? (Your answer should be given in terms of p.) Justify your conclusion.

(14) Can an odd degree polynomial of degree greater than one in $\mathbb{R}[x]$ be irreducible? If yes, find one. If no, explain why.

(15) Let u and w be complex numbers and y a real number. Prove each of the following.

  (a) $\overline{u + w} = \overline{u} + \overline{w}$

  (b) Extend part (a) to show that if n is a positive integer and $w_1, w_2, \ldots, w_n$ are complex numbers, then
  $$\overline{w_1 + w_2 + \cdots + w_n} = \overline{w_1} + \overline{w_2} + \cdots \overline{w_n}.$$

  (c) $\overline{uw} = \overline{u}\,\overline{w}$

  (d) Extend part (c) to show that if n is a positive integer and $w_1, w_2, \ldots, w_n$ are complex numbers, then
  $$\overline{w_1 w_2 \cdots w_n} = \overline{w_1}\,\overline{w_2}\,\cdots\,\overline{w_n}.$$

  (e) $\overline{uw} = \overline{u}\,\overline{w}$

  (f) $\overline{y} = y$

(16) Let F be a field and $c \in F$. Show that a polynomial $f(x) \in F[x]$ is irreducible in $F[x]$ if and only if the polynomial $f(x + c)$ is irreducible in $F[x]$.

(17) Prove that for any prime number p, the polynomial $f(x) = x^{p-1} + x^{p-2} + \cdots + x + 1$ is irreducible in $\mathbb{Q}[x]$. (Hint: Consider the polynomial $g(x) = f(x + 1)$, and use the fact that $\sum_{i=0}^{n} x^i = \frac{1 - x^{n+1}}{1 - x}$.)

(18) We showed that the factorization of a polynomial into a product of irreducible polynomials is unique when the quotient ring is a field. This need not be true if the quotient ring is not a field. For example, find a polynomial $f(x) \in \mathbb{Z}_8[x]$ that can be factored in at least two different ways as a product of linear polynomials.

(19) Show that for any prime integer p and any integer $n \geq 2$, the number $\sqrt[n]{p}$ is not a rational number. This will show that there are infinitely many irrational numbers. (Hint: Consider an appropriate polynomial equation with integer coefficients.)

(20) You might wonder if the converse of Theorem 11.15 is true. That is, if $f(x) \in \mathbb{Z}[x]$ and the polynomial $\overline{f(x)} \in \mathbb{Z} - p[x]$ is reducible for each prime p, must it be the case that $f(x)$ is reducible in $\mathbb{Z}[x]$? We will answer that question in this exercise. Let $f(x) = x^4 + 1 \in \mathbb{Z}[x]$.

  (a) Show that $\overline{f(x)}$ is reducible in $\mathbb{Z}_2[x]$.

  (b) Now we will see how to show that $\overline{f(x)}$ is reducible in $\mathbb{Z}_p[x]$ for any odd prime p. Let p be an odd prime and let U_p be the set of units in $\mathbb{Z}_p$. Let $U_p^2 = \{[a]^2 : [a] \in U_p\}$ be the set of squares in U_p.

    (i) Why is $p = 2k + 1$ for some integer k?

    (ii) Explain why $U_p^2 = \{[1]^2, [2]^2, \ldots, [k-1]^2, [k]^2\}$.

    (iii) Show that if $a^2 \equiv b^2 \pmod{p}$, then $[a] = [b]$ or $[a] = -[b]$ in $\mathbb{Z}_p$. Use this result to conclude that U_p^2 contains exactly k elements.

(iv) Let $[a]$ be in $U_p \setminus U_p^2$ and define $[a]U_p^2$ as

$$[a]u_p^2 = \{[a][b] : [b] \in U_p^2\}.$$

Prove that $[a]U_p^2 \cap U_p^2 = \emptyset$ and $[a]U_p^2 \cup U_p^2 = U_p$.

(v) Show that the product of two nonsquares in U_p is a square. (Hint: Based on the previous part of this exercise, what form do all nonsquares have?)

(vi) Explain why at least one of $[-1]$, $[2]$, or $[2]$ is in U_p^2. Show that in each case, $\overline{f(x)}$ is reducible in $\mathbb{Z}_p[x]$.

(vii) Show that $f(x)$ is irreducible in $\mathbb{Q}[x]$. Conclude that the converse of Theorem 11.15 is false.

(c) Do some research and determine if there are other polynomials in $\mathbb{Z}[x]$ that are irreducible in $\mathbb{Q}[x]$ but reducible in $\mathbb{Z}_p[x]$ for every prime p. If there are other such polynomials, list at least three, if possible.

Connections

In Investigation 10 we defined irreducible polynomials and learned that any polynomial with coefficients in a field can be factored into a product of irreducible polynomials. We also discovered that polynomials of degree 2 or 3 over a field are irreducible if and only if they have no roots in the field. In this investigation we explored more general methods for determining when a polynomial is irreducible over certain types of fields. These methods made connections to the Fundamental Theorem of Algebra for polynomials over $\mathbb{C}$, $\mathbb{R}$, and $\mathbb{Q}$, and to modular arithmetic from Investigation 2 for polynomials with coefficients in $\mathbb{Z}$.

Investigation 12

Quotients of Polynomial Rings

Focus Questions

By the end of this investigation, you should be able to give precise and thorough answers to the questions listed below. You may want to keep these questions in mind to focus your thoughts as you complete the investigation.

- How is congruence modulo a polynomial defined? How is this definition similar to congruence modulo a positive integer n in the integers?

- What is a quotient of a polynomial ring, and are such quotients constructed? How are quotient rings of polynomials similar to $\mathbb{Z}_n$?

- When is a quotient of a polynomial ring a field?

- How can quotients of polynomial rings be used to find roots of irreducible polynomials?

- What is an algebraic number, and what is the structure of the set of all algebraic numbers?

Preview Activity 12.1. Recall that we defined the ring $\mathbb{Z}_n$ to be the set of congruence classes of integers modulo the positive integer n, where congruence on $\mathbb{Z}$ is defined by $a \equiv b \pmod{n}$ if and only if $n \mid (b - a)$. We then defined addition and multiplication on the set of congruence classes to make $\mathbb{Z}_n$ into a ring. This construction only depended on having a notion of divisibility, so we can attempt the same construction in other sets where we have a divides relation—for example, in polynomial rings.

Consider the polynomial $x^2 + 1$ in $\mathbb{Q}[x]$. Define a relation on $\mathbb{Q}[x]$ as follows:

The polynomial $f(x)$ is congruent to the polynomial $g(x)$ modulo $x^2 + 1$ if $x^2 + 1$ divides $f(x) - g(x)$ in $\mathbb{Q}[x]$.

(a) Find three distinct polynomials congruent to $x - 1$ modulo $x^2 + 1$ in $\mathbb{Q}[x]$.

(b) Characterize (that is, describe in a precise way) all polynomials congruent to $x - 1$ modulo $x^2 + 1$ in $\mathbb{Q}[x]$.

Introduction

In previous investigations, we have seen different ways in which polynomial rings over fields are similar to the ring of integers. The similarities have included a notion of divisibility and a Division Algorithm, and a decomposition of elements into products of irreducible elements. We will now take another step and use the divides relation to define congruence in polynomial rings just like we did in $\mathbb{Z}$. We will then study the resulting equivalence class structures, which will allow us to better understand Kronecker's Theorem (see page 91) and roots of polynomials in general.

Congruence Modulo a Polynomial

We have a definition of divides in a polynomial ring over a field, and so we can use this idea (just as we did in $\mathbb{Z}$) to define congruence modulo a polynomial.

Definition 12.2. Let F be a field and $f(x) \in F[x]$ a nonconstant polynomial. The polynomial $g(x) \in F[x]$ is **congruent to the polynomial** $h(x) \in F[x]$ **modulo** $f(x)$ if and only if $f(x) \mid (g(x) - h(x))$.

If $g(x)$ is congruent to $h(x)$ modulo $f(x)$, we will denote this relationship by writing

$$g(x) \equiv h(x) \ (\mathrm{mod} f(x)).$$

The polynomial $f(x)$ is called the *modulus*. Note that congruence modulo a nonconstant polynomial is a relation on the set $F[x]$, just as congruence modulo n (where n is a positive integer) is a relation on $\mathbb{Z}$. The next activity asks some natural questions pertaining to the congruence relation on $F[x]$.

Activity 12.3. Let F be a field and $f(x)$ a nonconstant polynomial in $F[x]$.

(a) Is congruence modulo $f(x)$ a reflexive relation on $F[x]$? Prove your answer.

(b) Is congruence modulo $f(x)$ a symmetric relation on $F[x]$? Prove your answer.

(c) Is congruence modulo $f(x)$ a transitive relation on $F[x]$? Prove your answer.

(d) Why do we assume $f(x)$ is a nonconstant polynomial in Definition 12.2?

Congruence Classes of Polynomials

In Activity 12.3, we saw that congruence of polynomials is an equivalence relation on $F[x]$. The equivalence classes corresponding to the congruence relation on $F[x]$ are called congruence classes, defined formally as follows:

Definition 12.4. Let F be a field and $f(x)$ a nonconstant polynomial in $F[x]$. The **congruence class of the polynomial** $g(x)$ **modulo** $f(x)$ is the set

$$\{h(x) \in F[x] : h(x) \equiv g(x) \,(\mathrm{mod}\, f(x))\}.$$

We will use the notation $\overline{g(x)}_{\langle f(x) \rangle}$ to represent the congruence class of the polynomial $g(x)$ modulo the polynomial $f(x)$. Note the change in notation from $\mathbb{Z}_n$, where we used brackets (for instance, $[1]$) to denote congruence classes. The difference in notation is intended to avoid confusion. For example, if we choose $\mathbb{Z}_5$ as our coefficient field, we want to be able to distinguish between the coefficients of our polynomials in $\mathbb{Z}_5[x]$ and the congruence class of a polynomial in $\mathbb{Z}_5[x]$. When the modulus $f(x)$ is clear from the context, we usually drop the subscript and denote the congruence class of $g(x)$ as $\overline{g(x)}$. As with any equivalence relation, Theorem 2.6 (see page 27) tells us that two congruence classes are either equal or disjoint and that the union of all congruence classes is the entire polynomial ring $F[x]$. Or, stated differently, each element of $F[x]$ belongs to exactly one congruence class modulo $f(x)$.

Activity 12.5.

(a) Find all distinct congruence classes modulo $f(x) = x^2 - [1]$ in $\mathbb{Z}_3[x]$. (Hint: The Division Algorithm guarantees that the remainder when a polynomial is divided by $f(x)$ will either be zero or have a degree less than $\deg(f(x))$.)

(b) Find all distinct congruence classes modulo $p(x) = x^3 + [1]$ in $\mathbb{Z}_2[x]$.

(c) Based on your work in parts (a) and (b), make a conjecture to answer the following question: How many distinct congruence classes are there in $\mathbb{Z}_n[x]$ modulo a polynomial of degree m? Explain your conjecture and how you might prove it.

The Set $F[x]/\langle f(x) \rangle$

After we defined the set of congruence classes in $\mathbb{Z}$ modulo a positive integer n, we then defined addition and multiplication on those congruence classes in order to ultimately define the ring $\mathbb{Z}_n$ of integers modulo n. Now that we have a congruence relation on polynomials in a polynomial ring over a field, it seems reasonable to attempt the same constructions here. If F is a field and $f(x) \in F[x]$ is a nonconstant polynomial, we let $F[x]/\langle f(x) \rangle$ represent the set of distinct congruence classes modulo $f(x)$. That is,

$$F[x]/\langle f(x) \rangle = \{\overline{g(x)} : g(x) \in F[x]\}.$$

We read $F[x]/\langle f(x) \rangle$ as "F bracket x mod $f(x)$."

Analogous to our work with congruence in $\mathbb{Z}$, we can define addition and multiplication on $F[x]/\langle f(x) \rangle$ using the addition and multiplication from $F[x]$. In particular, if $\overline{g(x)}, \overline{h(x)} \in F[x]/\langle f(x) \rangle$, then

$$\overline{g(x)} + \overline{h(x)} = \overline{g(x) + h(x)} \text{ and } \overline{g(x)}\,\overline{h(x)} = \overline{g(x)h(x)}. \tag{12.1}$$

Note that these definitions involve calculating $g(x) + h(x)$ and $g(x)h(x)$, using the standard addition and multiplication operations in $F[x]$. The next activity provides a few examples and poses an important question.

Activity 12.6. For parts (a) and (b), write the result of each operation in the form $\overline{h(x)}$, where $h(x)$ has as small a degree as possible.

(a) $\overline{x^2 + [2]x + [1]} + \overline{[2]x^3 + x + [3]}$ in $\mathbb{Z}_5[x]/\langle x^2 + x \rangle$

(b) $\left(\overline{x^2 + [2]x + [1]} \right) \left(\overline{[2]x^3 + x + [3]} \right)$ in $\mathbb{Z}_5[x]/\langle x^2 + x \rangle$

(c) In general, is addition as we have defined it in $F[x]/\langle f(x) \rangle$ well-defined? Prove your answer. (Hint: You may want to refer back to Definition 2.16 on page 32 and the subsequent discussion.)

(d) In general, is multiplication as we have defined it in $F[x]/\langle f(x) \rangle$ well-defined? Prove your answer.

(e) Why are these last two questions important? Why do we need to consider whether addition and multiplication in $F[x]/\langle f(x) \rangle$ are well-defined?

We will now look at the structure of $F[x]/\langle f(x) \rangle$ with a specific example.

Activity 12.7. In this activity, we will work in the set $\mathbb{Z}_3[x]/\langle x^2 - x \rangle$.

(a) Why is $\overline{x^2} = \overline{x}$ in $\mathbb{Z}_3[x]/\langle x^2 - x \rangle$?

(b) Show that the element $\overline{x^3 + [3]x^2 + x + [2]}$ can be written as $\overline{h(x)}$ in $\mathbb{Z}_3[x]/\langle x^2 - x \rangle$ for some polynomial $h(x) \in \mathbb{Z}_3[x]$ of degree 1.

(c) Explain why $\mathbb{Z}_3[x]/\langle x^2 - x \rangle$ contains exactly nine elements. Find all of these elements.

(d) Write the addition table for $\mathbb{Z}_3[x]/\langle x^2 - x \rangle$.

(e) Write the multiplication table for $\mathbb{Z}_3[x]/\langle x^2 - x \rangle$.

(f) What properties of addition and multiplication properties appear to hold in $\mathbb{Z}_3[x]/\langle x^2 - x \rangle$? What algebraic structure do you believe $\mathbb{Z}_3[x]/\langle x^2 - x \rangle$ has?

Now that we have some experience working with elements of $\mathbb{Z}_3[x]/\langle x^2 - x \rangle$, we will move on to the general context and examine the algebraic structure of the set $F[x]/\langle f(x) \rangle$ for an arbitrary field F and an arbitrary nonconstant polynomial $f(x) \in F[x]$.

Activity 12.8. Let F be a field and $f(x)$ a nonconstant polynomial in $F[x]$.

(a) Why is $F[x]/\langle f(x) \rangle$ closed under the addition and multiplication defined in (12.1)?

(b) Rewrite all the properties of integer arithmetic listed on page 5 in the context of $F[x]/\langle f(x) \rangle$.

(c) Verifying the properties of integer arithmetic in the context of $F[x]/\langle f(x) \rangle$ is really no different than verifying them in $\mathbb{Z}_n$. For example, to prove that left multiplication distributes over addition in $F[x]/\langle f(x) \rangle$, we can use the distributive laws from $F[x]$. To illustrate, let $g(x), h(x)$, and $k(x)$ be elements in $F[x]/\langle f(x) \rangle$. Then

$$
\begin{aligned}
\overline{g(x)} \left(\overline{h(x)} + \overline{k(x)} \right) &= \overline{g(x)} \left(\overline{h(x) + k(x)} \right) \\
&= \overline{g(x)(h(x) + k(x))} \\
&= \overline{g(x)h(x) + g(x)k(x)} \\
&= \overline{g(x)h(x)} + \overline{g(x)k(x)} \\
&= \overline{g(x)}\ \overline{h(x)} + \overline{g(x)}\ \overline{k(x)}.
\end{aligned}
$$

Prove the remaining properties of integer arithmetic are valid in the context of $F[x]/\langle f(x)\rangle$. (Note the similarity between Activity 2.12 and your work here.)

The next theorem is the consequence of Activity 12.8.

Theorem 12.9. *Let F be a field and $f(x)$ a nonconstant polynomial in $F[x]$. Then $F[x]/\langle f(x)\rangle$ is a commutative ring with identity under the operations defined in (12.1).*

The ring $F[x]/\langle f(x)\rangle$ is called a quotient of $F[x]$ by $\langle f(x)\rangle$, or simply a quotient ring. Quotient rings can be defined in general (that is, for rings other than polynomial rings), and we will do so in Investigation 13.

Special Quotients of Polynomial Rings

Recall that if p is prime, then $\mathbb{Z}_p$ is a field, but if n is composite, then $\mathbb{Z}_n$ is not even an integral domain. So $\mathbb{Z}_n$ has a nicer structure for some values of n than for others. Given the similarities between $\mathbb{Z}_n$ and $F[x]/\langle f(x)\rangle$, it is natural to ask if the structure of $F[x]/\langle f(x)\rangle$ depends on certain properties of $f(x)$.

Activity 12.10. In Activity 12.7 we investigated the quotient ring $\mathbb{Z}_3[x]/\langle x^2 - x\rangle$. In this activity, we will do the same with $\mathbb{Z}_3[x]/\langle x^2 - [1]\rangle$ and $\mathbb{Z}_3[x]/\langle x^2 + x + [2]\rangle$. We will then compare the two rings.

(a) Find all the elements in $\mathbb{Z}_3[x]/\langle x^2 - [1]\rangle$.

(b) Find all the elements in $\mathbb{Z}_3[x]/\langle x^2 + x + [2]\rangle$.

(c) Explain why the addition tables for the rings $\mathbb{Z}_3[x]/\langle x^2 - x\rangle$, $\mathbb{Z}_3[x]/\langle x^2 - [1]\rangle$, and $\mathbb{Z}_3[x]/\langle x^2 + x + [2]\rangle$ have the same structure.

(d) The multiplication tables for the rings $\mathbb{Z}_3[x]/\langle x^2 - [1]\rangle$ and $\mathbb{Z}_3[x]/\langle x^2 + x + [2]\rangle$ are given in Tables 12.1 and 12.2. Explain in detail how the entry for $\left(\overline{[2]x + [1]}\right)\left(\overline{[2]x + [2]}\right)$ is obtained in each.

(e) There is a significant difference between the rings $\mathbb{Z}_3[x]/\langle x^2 - [1]\rangle$ and $\mathbb{Z}_3[x]/\langle x^2 + x + [2]\rangle$. What is the difference, and why do you think it happens?

The result of Activity 12.10 shows that $F[x]/\langle f(x)\rangle$ has a nicer structure for some polynomials $f(x)$ than for others. You might notice the similarity between this observation and Activity 4.28 on page 65, where we saw that $\mathbb{Z}_p$ is a field if and only if p is prime. The next theorem is an analogous result for polynomials over a field.

Theorem 12.11. *Let F be a field and $f(x)$ a nonconstant polynomial in $F[x]$. The ring $F[x]/\langle f(x)\rangle$ is a field if and only if $f(x)$ is irreducible in $F[x]$.*

Proof. Let F be a field and $f(x)$ a nonconstant polynomial in $F[x]$. We will prove the forward implication first. Suppose $F[x]/\langle f(x)\rangle$ is a field. To show that $f(x)$ is irreducible in $F[x]$, we will proceed by contradiction and assume $f(x)$ is reducible in $F[x]$. So we can find polynomials $g(x), h(x)$ with $1 \leq \deg(g(x), \deg(h(x)) < \deg(f(x))$, so that $f(x) = g(x)h(x)$. Since $1 \leq$

$\deg(g(x)) < \deg(f(x))$ and $1 \leq \deg(h(x)) < \deg(f(x))$, we know $\overline{g(x)}$ and $\overline{h(x)}$ are nonzero in $F[x]/\langle f(x)\rangle$. But $\overline{g(x)}\ \overline{h(x)} = \overline{g(x)h(x)} = \overline{f(x)} = \overline{0}$, and so $\overline{g(x)}$ is a zero divisor in $F[x]/\langle f(x)\rangle$. However, this cannot happen in a field, so we conclude $f(x)$ is irreducible in $F[x]$.

To prove the reverse implication, we assume $f(x)$ is irreducible in $F[x]$. Let $\overline{g(x)} \in F[x]$ with $\overline{g(x)} \neq \overline{0}$. So $f(x)$ does not divide $g(x)$. Since $f(x)$ is irreducible in $F[x]$, we know $\gcd(f(x), g(x)) = 1$. Thus, there exist polynomials $s(x), t(x)$ in $F[x]$ so that

$$1 = s(x)f(x) + t(x)g(x).$$

Then

$$\overline{1} = \overline{s(x)f(x) + t(x)g(x)} = \overline{s(x)}\ \overline{f(x)} + \overline{t(x)}\ \overline{g(x)} = \overline{t(x)}\ \overline{g(x)}.$$

Therefore, $\overline{g(x)}$ is a unit in $F[x]/\langle f(x)\rangle$, and $F[x]/\langle f(x)\rangle$ is a field. ∎

Theorem 12.11 gives us many ways to construct new types of fields. Also, note that any field of the form $\mathbb{Z}_p[x]/\langle q(x)\rangle$, where $q(x)$ is an irreducible polynomial of degree n in $\mathbb{Z}_p[x]$, has p^n elements. It turns out that if F is a finite field, then $|F| = p^n$ for some prime p and positive integer n, although we won't prove that fact here.

There is a critically important application of quotient rings to the problem of finding roots of polynomials. To illustrate, let's look at an example. The polynomial $f(x) = x^2 + 1$ is irreducible in $\mathbb{Q}[x]$. Thus, the quotient $E = \mathbb{Q}[x]/\langle x^2 + 1\rangle$ is a field. We can consider $\mathbb{Q}$ as a subset of E in a natural way, so $f(x) = x^2 + 1$ is also a polynomial in $E[x]$. In E, we have $f(\overline{x}) = \overline{x}^2 + 1 = \overline{x^2 + 1}$. But in E, we know that $\overline{x^2 + 1} = \overline{0}$, and so $f(\overline{x}) = \overline{0}$, and $\overline{x}$ is a root of $f(x)$ in E. As a matter of convention, we typically refer to $\overline{x}$ as i. Note that since $\overline{x}^2 + 1 = \overline{0}$, the familiar property that $i^2 = -1$ is satisfied.

In general, suppose we have a polynomial $f(x)$ with coefficients in a field F. The argument in the previous paragraph shows how we can always find a field in which $f(x)$ has a root. The process works by first factoring $f(x)$ into a product of irreducible polynomials. Let $p(x)$ be one of the irreducible factors of $f(x)$. Theorem 12.11 shows that $E = F[x]/\langle p(x)\rangle$ is a field. Consider the element $\overline{x} \in E$. In E we have

$$p(\overline{x}) = \overline{p(x)} = \overline{0}.$$

So $\overline{x}$ is a root of $p(x)$ in E and, consequently, a root of $f(x)$ in E. This proves the following important theorem referred to in Investigation 6.

Theorem 12.12 (Kronecker's Theorem). *Let F be a field and $p(x)$ an irreducible polynomial in $F[x]$. There exists a field E containing F such that $p(x)$ has a root in E.*

We can continue to apply this idea repeatedly to ultimately factor any polynomial into a product of linear factors in some large field.

Activity 12.13. Let $p(x) = x^2 + x + [2]$ in $\mathbb{Z}_3[x]$

(a) Show that $p(x)$ is an irreducible polynomial in $\mathbb{Z}_3[x]$.

(b) Let $E = \mathbb{Z}_3[x]/\langle x^2 + x + [2]\rangle$. Show explicitly that $\overline{x}^2 + \overline{x} + \overline{[2]} = \overline{[0]}$ in E. Then use this observation to determine exactly which element of E is guaranteed to be a root of $p(x)$.

Algebraic Numbers

Kronecker's Theorem shows that every polynomial with coefficients in a field has a root in some larger field. If we restrict ourselves to polynomials with rational coefficients, then the Fundamental Theorem of Algebra shows that all such polynomials can be completely factored into linear factors with coefficients from the field $\mathbb{C}$ of complex numbers. We will now use this idea to define algebraic numbers and find another field that is an extension of $\mathbb{Q}$.

First, we need to review a relationship between polynomials in $\mathbb{Q}[x]$ and polynomials in $\mathbb{Z}[x]$. (This relationship is closely related to Gauss' Lemma (see page 161), and you may notice similarities between the following argument and parts of the proof of Gauss' Lemma.) Suppose

$$p(x) = \left(\frac{r_n}{s_n}\right) x^n + \left(\frac{r_{n-1}}{s_{n-1}}\right) x^{n-1} + \cdots + \left(\frac{r_1}{s_1}\right) x + \frac{r_0}{s_0}$$

is a polynomial in $\mathbb{Q}[x]$. Then, with a little bit of algebra, we can rewrite $p(x)$ as a rational number times a polynomial $q(x) \in \mathbb{Z}[x]$. In particular, let $S = s_n s_{n-1} \cdots s_1 s_0$ and $S_i = \frac{S}{s_i} = s_n s_{n-1} \cdots s_{i+1} s_{i-1} s_{i-2} \cdots s_1 s_0$ for each i between 0 and n, inclusive. Then

$$p(x) = \left(\frac{1}{S}\right) \left(S_n r_n x^n + S_{n-1} r_{n-1} x^{n-1} + \cdots + S_1 r_1 x + S_0 r_0\right),$$

with $q(x) = S_n r_n x^n + S_{n-1} r_{n-1} x^{n-1} + \cdots + S_1 r_1 x + S_0 r_0$. Now if $p(r) = 0$—that is, if r is a root of $p(x)$—then $q(r) = 0$ as well. Also, if $q(r) = 0$, then $p(r) = 0$. So roots of polynomials with rational coefficients correspond to roots of polynomials with integer coefficients. In other words, when considering roots of polynomials in $\mathbb{Q}[x]$, it suffices to consider roots of polynomials in $\mathbb{Z}[x]$. The advantage of this perspective is that polynomials with integer coefficients are easier to work with than polynomials with rational coefficients.

As we saw in Investigation 6, we can attach roots of polynomials to $\mathbb{Q}$ to build extension fields of $\mathbb{Q}$. We can extend this idea by considering all the roots of polynomials with integer coefficients, which leads us to the definition of an algebraic number.

Definition 12.14. A real number a is an **algebraic** number if $p(a) = 0$ for some polynomial $p(x) \in \mathbb{Z}[x]$.

A real number that is not algebraic is said to be transcendental. Examples of algebraic numbers include any rational number $\frac{r}{s}$ (using $p(x) = sx - r$), $\sqrt{7}$ (using $p(x) = x^2 - 7$), and $\sqrt[3]{2}$ (using $p(x) = x^3 - 2$). Examples of transcendental numbers include π, e, and $2^{\sqrt{2}}$.

Let $\mathcal{A}$ denote the set of algebraic numbers. Once we have defined a set of numbers like this, it is natural to ask what kind of structure this set possesses. For example, is the sum of two algebraic numbers algebraic? What about the product of two algebraic numbers?

One way to answer these questions about combinations of algebraic numbers is with linear algebra. Recall that a set of vectors spans a space if every vector in the space can be written as a linear combination of vectors in the set. Also, the dimension of a space is the number of vectors in a minimal spanning set. With that in mind, consider the question of whether the number $z = \sqrt{2} + \sqrt{7}$ is algebraic. Since $\sqrt{2}$ is a zero of the polynomial $x^2 - 2$ and $\sqrt{7}$ is a zero of $x^2 - 7$, we see that both $\sqrt{2}$ and $\sqrt{7}$ are algebraic. To determine if z is algebraic, we need to look for a polynomial with z as a zero. The next activity demonstrates how we might do so.

Activity 12.15.

(a) Explain why $z = \sqrt{2} + \sqrt{7}$ cannot be the root of any linear polynomial with integer coefficients.

(b) Calculate z^2, and simplify your answer as much as possible.

(c) Use your answer to part (b) to find an integer k such that $(z^2 - k)^2 \in \mathbb{Z}$.

(d) Use your answer to part (c) to show that z is a root of a degree 4 polynomial with integer coefficients.

The argument from Activity 12.15 will be difficult to repeat for every sum of algebraic numbers, but the general idea is not. Since z^4 could be written in terms of linear combinations of lower powers of z, we were able to combine those powers to make a polynomial that had z as a zero. In other words, since z^4 can be written as a linear combination of $1, z, z^2, z^3$, the elements $1, z, z^2, z^3$ form a spanning set for a four-dimensional vector space V over $\mathbb{Q}$. We will exploit this idea in the following theorem.

Theorem 12.16. *Let a and b be algebraic numbers. Then $a + b$, ab, and $-a$ are also algebraic. In addition, if $a \neq 0$, then $\frac{1}{a}$ is algebraic.*

Proof. Let a and b be algebraic numbers. If either a or b is zero, then it is clear that $a + b$ and ab are algebraic. So suppose that a and b are both nonzero. Then there are polynomials

$$p(x) = a_k x^k + a_{k-1} x^{k-1} + \cdots a_1 x + a_0$$

and

$$q(x) = b_m x^m + b_{m-1} x^{m-1} + \cdots b_1 x + b_0$$

in $\mathbb{Z}[x]$ so that $a_k, b_m \neq 0$ and $p(a) = q(b) = 0$. So

$$a^k = \frac{-1}{a_k} \left(a_{k-1} x^{k-1} + \cdots a_1 x + a_0 \right) \quad \text{and} \tag{12.2}$$

$$b^m = \frac{-1}{b_m} \left(b_{m-1} x^{m-1} + \cdots b_1 x + b_0 \right). \tag{12.3}$$

First, we will show that $-a \in \mathcal{A}$. Notice that

$$a_0 - a_1(-a) + a_2(a^2) - a_3(-a)^3 + \cdots + (-1)^k a_k(-a)^k$$
$$= a_0 + a_1(a) + a_2(a^2) + \cdots + a_k(a^k)$$
$$= p(a)$$
$$= 0,$$

so $-a$ is a root of the polynomial $a_0 - a_1 x + a_2 x^2 - a_3 x^3 + \cdots + (-1)^k a_k x^k$. Thus, $-a$ is an algebraic number, and the set $\mathcal{A}$ contains the additive inverse of each of its elements.

Now we will show that $a^{-1} \in \mathcal{A}$. Note that

$$a_0 \left(\frac{1}{a} \right)^k + a_1 \left(\frac{1}{a} \right)^{k-1} + a_2 \left(\frac{1}{a} \right)^{k-2} + \cdots + a_k$$

$$= \left(\frac{1}{a} \right)^k \left(a_0 + a_1(a) + a_2(a^2) + a_3(a^3) + \cdots + a_k(a^k) \right)$$

$$= \left(\frac{1}{a} \right)^k p(a) = 0,$$

so a^{-1} is a root of the polynomial $a_0x^k + a_1x^{k-1} + a_2x^{k-2} + \cdots + a_{k-1}x + a_k$. Thus, a^{-1} is an algebraic number and $\mathcal{A}$ contains the inverse of each of its nonzero elements.

Next we will show that $a + b$ is algebraic. Let $n = \max\{k, m\}$.

Let $z = a + b$. By the Binomial Theorem (see Exercise (10) on page 81 of Investigation 5), we know that

$$z^n = (a+b)^n$$
$$= a^n + na^{n-1}b + \binom{n}{2}a^{n-2}b^2 + \cdots + \binom{n}{n-2}a^2b^{n-2} + nab^{n-1} + b^n$$
$$= \sum_{i=0}^{n}\binom{n}{i}a^{n-i}b^i.$$

By (12.2), all the powers a^r for $r \geq k$ can be written as linear combinations of the powers $1, a, a^2, a^3, \ldots, a^{k-1}$. Similarly, (12.3) shows that all the powers b^s for $s \geq m$ can be written as linear combinations of the powers $1, b, b^2, b^3, \ldots, b^{m-1}$. Therefore, each monomial of the form a^ib^j in the expansion of z^n can be written as a product of linear combinations of $1, a, a^2, a^3, \ldots, a^{k-1}$ and $1, b, b^2, b^3, \ldots, b^{m-1}$. We can conclude that the elements of the form a^ib^j for $0 \leq i \leq k-1$ and $0 \leq j \leq m-1$ form a spanning set of a vector space V over $\mathbb{Q}$ of which z is a member. So the elements $1, z, z^2, z^3, \ldots, z^{km}$ have to be linearly dependent and satisfy some polynomial relationship over $\mathbb{Z}$. This shows that $a + b$ is algebraic.

The proof that ab is algebraic is similar and left as an exercise. (See Exercise (6).) ∎

Activity 12.17. What does Theorem 12.16 tell us about the structure of the set $\mathcal{A}$ of algebraic numbers? Be specific and explain why.

In a sense, the proof given for Theorem 12.16 that $a + b$ is algebraic is constructive. Let's illustrate with an example. Let $a = \sqrt{2}$, $b = \sqrt[3]{3}$, and $z = a + b$. We know a is a zero of $x^2 - 2$ and b is a zero of $x^3 - 3$. In this case, $k = 2$ and $m = 3$, so z must satisfy some polynomial equation of degree no more than $km = 6$. Now we know that

$$z^0 = 1$$
$$z^1 = a + b$$
$$z^2 = (a+b)^2 = 2 + b^2 + 2ab$$
$$z^3 = (a+b)^3 = 3 + 2a + 6b + 3ab^2$$
$$z^4 = (a+b)^4 = 4 + 12a + 3b + 12b^2 + 8ab$$
$$z^5 = (a+b)^5 = 60 + 4a + 20b + 3b^2 + 15ab + 20ab^2$$
$$z^6 = (a+b)^6 = 17 + 120a + 90b + 60b^2 + 24ab + 18ab^2.$$

If z satisfies a polynomial equation of the form

$$c_6x^6 + c_5x^5 + c_4x^4 + c_3x^3 + c_2x^2 + c_1x + c_0 = 0,$$

then we have

$$0 = c_6 \left(17 + 24ab + 60b^2 + 120a + 90b + 18ab^2\right)$$
$$+ c_5 \left(60 + 4a + 20b + 20ab^2 + 15ab + 3b^2\right)$$
$$+ c_4 \left(4 + 8ab + 12b^2 + 12a + 3b\right)$$
$$+ c_3 \left(3 + 2a + 6b + 3ab^2\right)$$
$$+ c_2 \left(2 + 2ab + b^2\right)$$
$$+ c_1 (a + b)$$
$$+ c_0.$$

Equating like terms gives us the system of equations

$$17c_6 + 60c_5 + 4c_4 + 3c_3 + 2c_2 + c_0 = 0$$
$$120c_6 + 4c_5 + 12c_4 + 2c_3 + c_1 = 0$$
$$90c_6 + 20c_5 + 3c_4 + 6c_3 + c_1 = 0$$
$$60c_6 + 3c_5 + 12c_4 + c_2 = 0$$
$$24c_6 + 15c_5 + 8c_4 + 2c_2 = 0$$
$$18c_6 + 20c_5 + 3c_3 = 0.$$

We can solve this system using row reduction to obtain the solution $c_0 = 1, c_1 = -36, c_2 = 12, c_3 = -6, c_4 = -6, c_5 = 0, c_6 = 1$. So $\sqrt{2} + \sqrt[3]{3}$ is a zero of the polynomial $p(x) = x^6 - 6x^4 - 6x^3 + 12x^2 - 36x + 1$. Check it if you dare!

Concluding Activities

Activity 12.18. Let F be a field and $f(x)$ a nonconstant polynomial in $F[x]$.

(a) Show that the congruence class $\overline{f(x)}_{\langle f(x) \rangle}$ is a subring of $F[x]$.

(b) Show that if $f_0(x) \in \overline{f(x)}_{\langle f(x) \rangle}$ and $h(x) \in F[x]$, then both $h(x)f_0(x)$ and $f_0(x)h(x)$ are in $\overline{f(x)}_{\langle f(x) \rangle}$. (Any subset of a ring that satisfies the properties in this activity is called an *ideal*.)

Activity 12.19. Find a field E over which the polynomial $f(x) = x^3 + x + [1] \in \mathbb{Z}_3[x]$ factors completely. Give the complete factorization of $f(x)$ into a product of linear factors in $E[x]$.

Activity 12.20. Use the methods of this investigation to find a polynomial in $\mathbb{Z}[x]$ that has $\sqrt{3} + \sqrt{6}$ as a root.

Exercises

(1) In this problem, we will work within the field $\mathbb{Z}_2$.

(a) Find an irreducible polynomial in $\mathbb{Z}_2[x]$ of degree 3.

(b) Use the polynomial from part (a) to construct a field E with 8 elements. List the elements in E and then create the operations tables for E.

(c) Find a field with 16 elements and list its elements.

(2) Find a field with 25 elements and list its elements.

(3) Let F be a field and $p(x)$ an irreducible polynomial in $F[x]$. In this investigation we showed that $E = F[x]/\langle p(x) \rangle$ is a field, and we implied that F is a subfield of E. Now we will examine what we mean by that statement.

(a) There is a natural mapping ι from F to E. Identify this mapping (ι is called the *inclusion* mapping). Show that ι preserves the structure of F. Is ι an isomorphism? Explain.

(b) Explain how E contains an isomorphic copy of F. (It is in this sense that we say F is a subfield of E. This subfield of E that is isomorphic to F is called an *embedding* of F in E.)

(4) In a previous investigation, we saw that the field $\mathbb{Q}(\sqrt{2}) = \{a + b\sqrt{2} : a, b \in \mathbb{Q}\}$ is an extension of $\mathbb{Q}$ that contains a root of the irreducible polynomial $x^2 - 2 \in \mathbb{Q}[x]$. In this investigation, we saw that $\mathbb{Q}[x]/\langle x^2 - 2 \rangle$ is also an extension of $\mathbb{Q}$ that contains a root of the irreducible polynomial $x^2 - 2$. In this problem, we will explore the connection between the fields $\mathbb{Q}[x]/\langle x^2 - 2 \rangle$ and $\mathbb{Q}(\sqrt{2})$.

(a) Since $\mathbb{Q}[x]/\langle x^2 - 2 \rangle$ contains a root of $x^2 - 2$, there is an element in $\mathbb{Q}[x]/\langle x^2 - 2 \rangle$ that behaves like $\sqrt{2}$. What element is this? Why?

(b) Explain why every element in $\mathbb{Q}[x]/\langle x^2 - 2 \rangle$ has the form $\overline{a + bx}$ for some $a, b \in \mathbb{Q}$.

(c) Use the previous parts of this exercise to define a function φ from $\mathbb{Q}[x]/\langle x^2 - 2 \rangle$ to $\mathbb{Q}(\sqrt{2})$. Show that φ is an isomorphism. What is the relationship between $\mathbb{Q}[x]/\langle x^2 - 2 \rangle$ and $\mathbb{Q}(\sqrt{2})$?

(5) Use the methods of this investigation to find a polynomial in $\mathbb{Z}[x]$ that has $\sqrt[3]{2} + \sqrt{6}$ as a root.

★ (6) **Closure of algebraic numbers under multiplication.** Prove that if a and b are algebraic numbers, then ab is an algebraic number.

(7) **Constructible numbers.** We are familiar with the classification of real numbers as rational and irrational numbers, and in this investigation we learned about algebraic numbers. There are other important classifications that are also of interest. Much of the mathematics of ancient times, even pre-dating the Greek mathematicians of Euclid's age, was very geometric in nature. The number theory of the time dealt with geometric constructions of numbers. The constructions were performed with straightedge (unmarked) and compass. In fact, the compass of the time was a collapsible one; it would not hold its position when lifted from the page.

With the tools of an unmarked ruler and collapsible compass, every integer can be constructed. It is also possible to build a rich variety of construction techniques, which allow one to see that many other objects are constructible. A sample of such techniques includes:

- Given points P and Q, we can construct a ray emanating from P that passes through Q.
- Given points P and Q, we can construct a circle with center P and radius PQ.
- Given a point P and a constructible length r, we can construct a circle with center P and radius r.

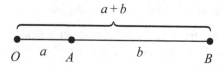

Figure 12.1
Constructing the sum of constructible a and b.

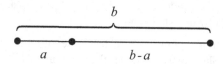

Figure 12.2
Constructing the difference of constructible a and b.

- Given a point P and a line l, we can construct a line through P perpendicular to l.
- Given a point P and a line l, we can construct a line through P parallel to l.

We now call numbers that can be constructed with these tools *constructible numbers*. It is not difficult to show that every rational number is constructible. In fact, as we will see in this exercise, the quotient of any two constructible numbers is a constructible number.

Suppose a and b are constructible numbers. We can then construct a from an origin point O. Let A be the other endpoint of this constructed segment. We can extend the constructible segment to a line by constructing the line through A parallel to the segment. At A, construct a circle with radius b. Let B be the point of intersection of this circle with the ray $\overrightarrow{OA}$. Then $\overline{OB}$ has length $a + b$, and $a + b$ is constructible. (See Figure 12.1.) So the sum of two constructible numbers is constructible.

(a) Figure 12.2 illustrates how the difference of two constructible numbers is constructed. Explain how this construction works.

(b) Figure 12.3 illustrates how the product of two constructible numbers is constructed. Explain how this construction works.

(c) Figure 12.4 illustrates how the quotient of two constructible numbers is constructed. Explain how this construction works.

(d) Let S be the set of constructible numbers. Explain why S is a subfield of $\mathbb{R}$ that contains $\mathbb{Q}$.

(e) The Greeks were very interested in geometric constructions. There are three problems from this time that were extremely influential in the subsequent development of geometry and algebra. These problems are labeled as: squaring the circle, trisecting an angle, and doubling the cube. To square the circle means to construct a square of the same area as that of a given circle. To trisect an angle means to subdivide a given angle into three congruent angles. To double the cube means to construct a cube whose volume is twice that of a given cube. All of these constructions are to be performed using only a straightedge and compass. Whether these constructions could actually be performed with the given tools went unanswered for a very long time. The answers to

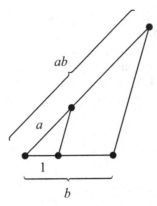

Figure 12.3
Constructing the product of constructible a and b.

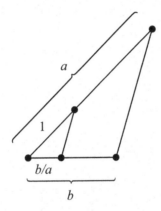

Figure 12.4
Constructing the quotient of constructible a and b.

these problems (squaring the circle, trisecting an angle, and doubling the cube) depend on the constructibility of the numbers π, $\cos(20°)$, and $\sqrt[3]{2}$. Research this problem to find sources that prove that π, $\cos(20°)$, and $\sqrt[3]{2}$ are not constructible. Explain then why it is impossible to square the circle, trisect an angle, or double the cube using only an unmarked straightedge and collapsible compass.

Connections

Given a polynomial in a polynomial ring $F[x]$, we can define congruence modulo a polynomial and form the quotient ring $F[x]/\langle f(x) \rangle$. Quotient structures are useful in mathematics in that they often have a simpler structure than the original set and can therefore provide important information about the original set. We are familiar with congruence and quotient structures in the integers, namely $\mathbb{Z}_n$ (from Investigation 2). If you studied group theory before ring theory, you should notice connections between the topics in this investigation and those in Investigation 24. In particular, the set $F[x]/\langle f(x) \rangle$ represents the set of distinct congruence classes modulo $f(x)$ in the same way that the set G/N represents the set of left cosets of N in G. Although G/N is a group only if N is normal in G, we don't have that problem in $F[x]$. Since $F[x]$ is an Abelian group under addition, any subgroup is normal.

Table 12.1
Multiplication table for $\mathbb{Z}_3[x]/\langle x^2 - [1]\rangle$.

$\times$	$\overline{[0]}$	$\overline{[1]}$	$\overline{[2]}$	$\overline{x}$	$\overline{x+[1]}$	$\overline{x+[2]}$	$\overline{[2]x}$	$\overline{[2]x+[1]}$	$\overline{[2]x+[2]}$
$\overline{[0]}$	$\overline{[0]}$	$\overline{[0]}$	$\overline{[0]}$	$\overline{[0]}$	$\overline{[0]}$	$\overline{[0]}$	$\overline{[0]}$	$\overline{[0]}$	$\overline{[0]}$
$\overline{[1]}$	$\overline{[0]}$	$\overline{[1]}$	$\overline{[2]}$	$\overline{x}$	$\overline{x+[1]}$	$\overline{x+[2]}$	$\overline{[2]x}$	$\overline{[2]x+[1]}$	$\overline{[2]x+[2]}$
$\overline{[2]}$	$\overline{[0]}$	$\overline{[2]}$	$\overline{[1]}$	$\overline{[2]x}$	$\overline{[2]x+[2]}$	$\overline{[2]x+[1]}$	$\overline{x}$	$\overline{x+[2]}$	$\overline{x+[1]}$
$\overline{x}$	$\overline{[0]}$	$\overline{x}$	$\overline{[2]x}$	$\overline{[1]}$	$\overline{x+[1]}$	$\overline{[2]x+[1]}$	$\overline{[2]}$	$\overline{x+[2]}$	$\overline{[2]x+[2]}$
$\overline{x+[1]}$	$\overline{[0]}$	$\overline{x+[1]}$	$\overline{[2]x+[2]}$	$\overline{x+[1]}$	$\overline{[2]x+[2]}$	$\overline{[0]}$	$\overline{[2]x+[2]}$	$\overline{[0]}$	$\overline{x+[1]}$
$\overline{x+[2]}$	$\overline{[0]}$	$\overline{x+[2]}$	$\overline{[2]x+[1]}$	$\overline{[2]x+[1]}$	$\overline{[0]}$	$\overline{x+[2]}$	$\overline{x+[2]}$	$\overline{[2]x+[1]}$	$\overline{[0]}$
$\overline{[2]x}$	$\overline{[0]}$	$\overline{[2]x}$	$\overline{x}$	$\overline{[2]}$	$\overline{[2]x+[2]}$	$\overline{x+[2]}$	$\overline{[1]}$	$\overline{[2]x+[1]}$	$\overline{x+[1]}$
$\overline{[2]x+[1]}$	$\overline{[0]}$	$\overline{[2]x+[1]}$	$\overline{x+[2]}$	$\overline{x+[2]}$	$\overline{[0]}$	$\overline{[2]x+[1]}$	$\overline{[2]x+[1]}$	$\overline{x+[2]}$	$\overline{[0]}$
$\overline{[2]x+[2]}$	$\overline{[0]}$	$\overline{[2]x+[2]}$	$\overline{x+[1]}$	$\overline{[2]x+[2]}$	$\overline{x+[1]}$	$\overline{[0]}$	$\overline{x+[1]}$	$\overline{[0]}$	$\overline{[2]x+[2]}$

Table 12.2
Multiplication table for $\mathbb{Z}_3[x]/\langle x^2 + x + [2]\rangle$.

$\times$	$\overline{[0]}$	$\overline{[1]}$	$\overline{[2]}$	$\overline{x}$	$\overline{x+[1]}$	$\overline{x+[2]}$	$\overline{[2]x}$	$\overline{[2]x+[1]}$	$\overline{[2]x+[2]}$
$\overline{[0]}$	$\overline{[0]}$	$\overline{[0]}$	$\overline{[0]}$	$\overline{[0]}$	$\overline{[0]}$	$\overline{[0]}$	$\overline{[0]}$	$\overline{[0]}$	$\overline{[0]}$
$\overline{[1]}$	$\overline{[0]}$	$\overline{[1]}$	$\overline{[2]}$	$\overline{x}$	$\overline{x+[1]}$	$\overline{x+[2]}$	$\overline{[2]x}$	$\overline{[2]x+[1]}$	$\overline{[2]x+[2]}$
$\overline{[2]}$	$\overline{[0]}$	$\overline{[2]}$	$\overline{[1]}$	$\overline{[2]x}$	$\overline{[2]x+[2]}$	$\overline{[2]x+[1]}$	$\overline{x}$	$\overline{x+[2]}$	$\overline{x+[1]}$
$\overline{x}$	$\overline{[0]}$	$\overline{x}$	$\overline{[2]x}$	$\overline{[2]x+[1]}$	$\overline{[1]}$	$\overline{x+[1]}$	$\overline{x+[2]}$	$\overline{[2]x+[2]}$	$\overline{[2]}$
$\overline{x+[1]}$	$\overline{[0]}$	$\overline{x+[1]}$	$\overline{[2]x+[2]}$	$\overline{[1]}$	$\overline{x+[2]}$	$\overline{[2]x}$	$\overline{[2]}$	$\overline{x}$	$\overline{[2]x+[1]}$
$\overline{x+[2]}$	$\overline{[0]}$	$\overline{x+[2]}$	$\overline{[2]x+[1]}$	$\overline{x+[1]}$	$\overline{[2]x}$	$\overline{[2]}$	$\overline{[2]x+[2]}$	$\overline{[1]}$	$\overline{x}$
$\overline{[2]x}$	$\overline{[0]}$	$\overline{[2]x}$	$\overline{x}$	$\overline{x+[2]}$	$\overline{[2]}$	$\overline{[2]x+[2]}$	$\overline{[2]x+[1]}$	$\overline{x+[1]}$	$\overline{[1]}$
$\overline{[2]x+[1]}$	$\overline{[0]}$	$\overline{[2]x+[1]}$	$\overline{x+[2]}$	$\overline{[2]x+[2]}$	$\overline{x}$	$\overline{[1]}$	$\overline{x+[1]}$	$\overline{[2]}$	$\overline{[2]x}$
$\overline{[2]x+[2]}$	$\overline{[0]}$	$\overline{[2]x+[2]}$	$\overline{x+[1]}$	$\overline{[2]}$	$\overline{[2]x+[1]}$	$\overline{x}$	$\overline{[1]}$	$\overline{[2]x}$	$\overline{x+[2]}$

Part IV

More Ring Theory

Investigation 13

Ideals and Homomorphisms

Focus Questions

By the end of this investigation, you should be able to give precise and thorough answers to the questions listed below. You may want to keep these questions in mind to focus your thoughts as you complete the investigation.

- What is an ideal of a ring?

- What is a principal ideal domain? What is a Euclidean domain? What is the relationship between principal ideal domains and Euclidean domains?

- How is congruence modulo an ideal defined? How can congruence modulo an ideal be used to construct quotient rings?

- What are prime and maximal ideals, and how are they related to certain properties of quotient rings?

- What is a homomorphism of rings? What are the similarities and differences between homomorphisms and isomorphisms?

- What are the kernel and image of a homomorphism? Why are the kernel and image important?

- What important relationship is established by the First Isomorphism Theorem for rings?

In Investigation 12, we considered quotients of polynomial rings. In this investigation, we will expand on these ideas to construct quotients of other rings. To begin, let's consider an example.

Preview Activity 13.1. Let I be the subset of $\mathbb{Z}_{12}$ defined by $I = \{[0], [3], [6], [9]\} = \{k[3] : k \in \mathbb{Z}\}$.

(a) Show that I is a subring of $\mathbb{Z}_{12}$.

(b) Show that if $[a] \in \mathbb{Z}_{12}$ and $[b] \in I$, then $[a][b] \in I$ and $[b][a] \in I$.

(c) Define a relation $\equiv_I$ on $\mathbb{Z}_{12}$ by

$$[a] \equiv_I [b] \text{ if and only if } [b] - [a] \in I.$$

    (i) Describe all the elements in $\mathbb{Z}_{12}$ that are related to $[0]$.

    (ii) Describe all the elements in $\mathbb{Z}_{12}$ that are related to $[1]$.

(iii) Describe all the elements in $\mathbb{Z}_{12}$ that are related to $[2]$.

(iv) Describe all the elements in $\mathbb{Z}_{12}$ that are related to $[7]$.

(v) It can be shown that $\equiv_I$ is an equivalence relation. Find all disjoint equivalence classes corresponding to $\equiv_I$.

Introduction

In Investigation 12, we saw that for a polynomial ring $F[x]$ over a field F, there is a quotient ring $F[x]/\langle f(x)\rangle$ for any nonconstant polynomial $f(x) \in F[x]$. We showed in Activity 12.18 (see page 180) that the congruence class $\overline{f(x)}_{\langle f(x)\rangle}$ is a subring of $F[x]$ with the property that both $h(x)f_0(x)$ and $f_0(x)h(x)$ are in $\overline{f(x)}_{\langle f(x)\rangle}$ for any $f_0(x) \in \overline{f(x)}_{\langle f(x)\rangle}$ and any $h(x) \in F[x]$. Any subset of a ring that satisfies these properties is called an *ideal*. For any ideal of R, we can construct a quotient structure analogous to $F[x]/\langle f(x)\rangle$. In this investigation, we will study ideals and the corresponding quotient structures. We will also introduce a special type of function called a homomorphism of rings, and we will study the relationship between homomorphisms and ideals.

Ideals

In an arbitrary ring R, the subsets that allow us to create a quotient structure are called *ideals*.

Definition 13.2. An **ideal** I in a ring R is a subring of R such that $rx \in I$ and $xr \in I$ for all $r \in R$ and $x \in I$.

As an example, we saw in Preview Activity 13.1 that $I = \{[0], [3], [6], [9]\}$ is an ideal of $\mathbb{Z}_{12}$. Notice that an ideal I is not only closed under multiplication by elements of I, but also is closed under multiplication by elements of the larger ring R. This property (that rx and xr are in I for any ring $r \in R$ and any $x \in I$) is called the *absorbing property*, or *closure under outside multiplication*.

Recall that the Subring Test (Theorem 6.4 on page 86) shows that we only need to establish three conditions to show that a subset of a ring is a subring. We can use the Subring Test, along with the absorbing property, to develop a relatively simple test to determine if a subset I of a ring R is an ideal.

Activity 13.3. Let R be a ring and I a subset of R. Assume that I is nonempty, closed under subtraction, and that $ra \in I$ and $ar \in I$ for every $r \in R$ and $a \in I$.

(a) Explain why I must be closed under multiplication.

(b) A formal statement of the Ideal Test is the following:

> **Theorem 13.4** (The Ideal Test). *Let R be a ring. A subset I of R is an ideal of R if and only if:*
>
> (i) *I is nonempty;*

(ii) $a - b \in I$ *for every* $a, b \in I$*; and*

(iii) $ra \in I$ *and* $ar \in I$ *for every* $r \in R$ *and* $a \in I$.

Explain how we have proved this theorem.

Now some applications of this test are in order.

Activity 13.5. For each of the following parts, determine if the set I is an ideal of the ring R. Use the Ideal Test to justify your answer.

(a) $R = \mathbb{Z}$ and $I = \{0\}$.

(b) $R = \mathbb{Z}$ and $I = \{2n : n \in \mathbb{Z}\}$.

(c) $R = \mathbb{R}$ and $I = \mathbb{Q}$.

(d) R is a commutative ring, $a \in R$, and $I = \{ra : r \in R\}$.

If R is any ring with additive identity 0_R, then the set $\{0_R\}$ is always an ideal of R. This ideal is called the *trivial ideal* and is the simplest of all ideals. An ideal I in R is a *proper ideal* if $I \neq R$. If an ideal I in a commutative ring R contains a nonidentity element a, then the absorbing property of I shows that ra and ar are in I for any $r \in R$. So the ideal $\{ra : r \in R\}$ is the smallest ideal of a commutative ring R containing the element a. These ideals are the next simplest ideals (compared to the trivial ideal) and are given a special name.

Definition 13.6. An ideal I in a commutative ring R is a **principal ideal** if $I = \{ra : r \in R\}$ for some $a \in R$.

We say that the ideal $I = \{ra : r \in R\}$ is *generated by* a and denote this ideal as $\langle a \rangle$. For example, the ideal $\langle 3 \rangle$ in $\mathbb{Z}$ is a principal ideal. This ideal consists of all the integer multiples of 3, which we have denoted by $3\mathbb{Z}$ in previous investigations. In general, the principal ideal $\langle k \rangle$ in $\mathbb{Z}$ consisting of all integer multiples of k is also denoted by $k\mathbb{Z}$. (See Exercise (7) on page 66 of Investigation 4.)

For some familiar rings, the only ideals are the principal ones.

Activity 13.7. Let I be an ideal of $\mathbb{Z}$.

(a) If I is the trivial ideal, explain why I is a principal ideal.

(b) Now assume I is a nontrivial ideal. Use the fact that I contains a nonzero element b to show that I contains a positive integer. (Hint: The integer b is either positive or negative.)

(c) Let $S = \{x \in I : x > 0\}$. Explain why S contains a smallest element a.

(d) Explain why it must be the case that $\langle a \rangle \subseteq I$.

(e) Let $y \in I$. Use the Division Algorithm to divide y by a, and then use the fact that I is an ideal to show that the remainder r after this division must be an element of I. What can we conclude about r, and why does this imply that $I \subseteq \langle a \rangle$?

(f) Explain how we have proved the following theorem:

Theorem 13.8. *Every ideal of* $\mathbb{Z}$ *is a principal ideal.*

Integral domains like $\mathbb{Z}$ in which every ideal is a principal ideal are given a special name.

Definition 13.9. An integral domain R is a **principal ideal domain** (PID) if every ideal of R is a principal ideal.

The ring of integers is not the only principal ideal domain. We have seen that the polynomial ring $F[x]$ is like $\mathbb{Z}$ in many ways, and so it is natural to ask if $F[x]$ is also a PID. The next theorem provides the answer.

Theorem 13.10. *Let F be a field. Then $F[x]$ is a principal ideal domain.*

Proof. Let F be a field and I an ideal of $F[x]$. Let 0 denote the additive identity in F and 1 the multiplicative identity. If $I = \{0\}$, then I is generated by 0. Suppose $I \neq \{0\}$. Then I contains a nonzero element $b(x)$. Let $S = \{\deg(p(x)) : p(x) \in I\}$. Since I contains a nonzero element, we know that S is not empty. The Well-Ordering Principle tells us that S contains a smallest element n. Let $a(x)$ be a polynomial in I of degree n. Note that if the leading coefficient a_n of $a(x)$ is not equal to 1, then the polynomial $\frac{1}{a_n}a(x)$ is in I. So we can assume $a(x)$ is a monic polynomial. We will show that I is generated by $a(x)$—that is, I is equal to the ideal $J = \{f(x)a(x) : f(x) \in F[x]\}$. First, we will demonstrate that $J \subseteq I$. Let $p(x) \in J$. Then $p(x) = h(x)a(x)$ for some $h(x) \in F[x]$. Since $a(x) \in I$ we know $h(x)a(x) \in I$ and $p(x) \in I$. Thus, $J \subseteq I$. For the reverse containment, let $g(x) \in I$. By the Division Algorithm, there are polynomials $q(x), r(x)$ such that $g(x) = q(x)a(x) + r(x)$ with $0 \leq \deg(r(x)) < \deg(a(x))$ or $r(x) = 0$. Now $g(x) \in I$ and $q(x)a(x) \in I$, so $r(x) = g(x) - q(x)a(x) \in I$. But $a(x)$ is a polynomial in S of smallest degree, so $\deg(r(x)) < \deg(a(x))$ would be a contradiction. Therefore, we must have $r(x) = 0$ and $g(x) = q(x)a(x) \in J$. So $I = J$, and I is generated by a single element. ∎

One more example of a principal ideal domain is the ring of Gaussian integers $\mathbb{Z}[i]$. (See Exercise (2) on page 52 of Investigation 3.) Recall that $\mathbb{Z}[i]$ is the subring of the field of complex numbers consisting of all elements of the form $a + bi$, where $a, b \in \mathbb{Z}$ and $i^2 = -1$.

Lemma 13.11. *The ring $\mathbb{Z}[i]$ is a principal ideal domain.*

Proof. Let $I \subset \mathbb{Z}[i]$ be an ideal. We need to show that there is an element $a + bi \in \mathbb{Z}[i]$ such that $I = \langle a + bi \rangle$. If $I = \{0\}$, then $I = \langle 0 \rangle$. So assume $I \neq \langle 0 \rangle$. As in the proofs of Theorems 13.8 and 13.10, we will look for an element in I that is smallest in some sense. In this case, we will use the complex norm of elements in $\mathbb{Z}[i]$ as our measure of size. In other words, we will define a function $\delta : \mathbb{Z}[i] \to \mathbb{Z}^+ \cup \{0\}$ by $\delta(a + bi) = a^2 + b^2$. It is left to the reader to show that $\delta((a + bi)(c + di)) = \delta(a + bi)\delta(c + di)$. Now let $S = \{\delta(u + vi) > 0 : u + vi \in I\}$. By definition, S is bounded below. We know S is not empty because I contains a nonzero element. Thus, by the Well-Ordering Principle, we know that S contains a smallest positive integer of the form $\delta(a + bi) = a^2 + b^2$ for some $a + bi \in I$.

To show that $I = \langle a + bi \rangle$, we need to show that every element in I is a multiple of $a + bi$. Let $c + di \in I$. We will show that when $c + di$ is divided by $a + bi$, the remainder is 0. But how do we divide $a + bi$ into $c + di$ in $\mathbb{Z}[i]$? First, we will do the division in $\mathbb{C}$:

$$\frac{c + di}{a + bi} = \frac{(c + di)(a - bi)}{a^2 + b^2} = \left(\frac{ca + db}{a^2 + b^2}\right) + \left(\frac{da - cb}{a^2 + b^2}\right)i.$$

Let $x = \frac{ca + db}{a^2 + b^2}$ and $y = \frac{da - cb}{a^2 + b^2}$, so that $\frac{c + di}{a + bi} = x + yi$. Now x and y are rational numbers, so there exist integers m and n such that

$$|m - x| \leq \frac{1}{2} \text{ and } |n - y| \leq \frac{1}{2}.$$

Now $\frac{c+di}{a+bi} = x + yi$ implies

$$
\begin{aligned}
c + di &= (a + bi)(x + yi) \\
&= (a + bi)[(x - m + m) + (y - n + n)i] \\
&= (a + bi)[(m + ni) + ((x - m) + (y - n)i)] \\
&= (a + bi)(m + ni) + (a + bi)((x - m) + (y - n)i).
\end{aligned}
$$

Let $q = m + ni$ and $r = (a + bi)((x - m) + (y - n)i)$. We know $q \in \mathbb{Z}[i]$ and $r = (c + di) - q(a + bi)$, so $r \in \mathbb{Z}[i]$. Also, we know $a + bi, c + di \in I$, so $r \in I$. However, we have

$$
\begin{aligned}
\delta(r) &= \delta(a + bi)\delta((x - m) + (y - n)i) \\
&= (a^2 + b^2)((x - m)^2 + (y - n)^2) \\
&\leq (a^2 + b^2)\left(\frac{1}{4} + \frac{1}{4}\right) \\
&< a^2 + b^2,
\end{aligned}
$$

which means that r is in I but has a smaller δ value than $a + bi$. This is a contradiction unless $r = 0$. Thus, $(a + bi)q = c + di$, and so $c + di \in \langle a + bi \rangle$. It follows that $I = \langle a + bi \rangle$ is a principal ideal. ∎

The three examples of principal ideal domains ($\mathbb{Z}$, $F[x]$, and $\mathbb{Z}[i]$) all have something in common. In the proof of Lemma 13.11, we essentially showed that there is a division algorithm in $\mathbb{Z}[i]$ just like in $\mathbb{Z}$ and $F[x]$. Integral domains in which there is a division algorithm are called *Euclidean domains*. In a Euclidean domain, we can define greatest common divisors, and the function δ helps establish a Euclidean algorithm to find greatest common divisors. The next definition formalizes these ideas.

Definition 13.12. Let D be an integral domain with additive identity 0_D. Then D is a **Euclidean domain** if there is a function $\delta : D - \{0_D\} \to \mathbb{Z}^+ \cup \{0\}$ such that the following conditions hold:

(i) If $a, b \in D$ are nonzero, then $\delta(a) \leq \delta(ab)$.

(ii) If $a, b \in D$ with a nonzero, then there exist elements $q, r \in D$ such that

$$
b = aq + r \text{ with either } r = 0_D \text{ or } \delta(r) < \delta(a).
$$

The function δ in Definition 13.12 is called a **norm** for D.

Some of our previous examples can be used to illustrate Definition 13.12. In particular, note the following:

- In $\mathbb{Z}$, $\delta(x) = |x|$.

- In $F[x]$, where F is a field, $\delta(f(x)) = \deg(f(x))$.

- In $\mathbb{Z}[i]$, $\delta(a + bi) = a^2 + b^2$.

As we have seen, all of these rings are principal ideal domains. The arguments we used to prove this fact can be generalized to establish the following theorem:

Theorem 13.13. *Every Euclidean domain is a principal ideal domain.*

There are principal ideal domains that are not Euclidean domains, but none that we have encountered in our investigations up to this point.* It is also important to note that not every integral domain is a principal ideal domain. Consider the ring $\mathbb{Z}[x]$ and the ideal I generated by 2 and x—that is,

$$I = \langle 2, x \rangle = \{2f(x) + xg(x) : f(x), g(x) \in \mathbb{Z}[x].\}$$

(As an exercise, you may want to verify that I is in fact an ideal.) Suppose that I was a principal ideal with generator $a(x)$. Since $2 \in I$, we must have $2 = c(x)a(x)$ for some $c(x) \in \mathbb{Z}[x]$. It follows that $\deg(a(x)) = \deg(c(x)) = 1$. So $a(x) = a \in \mathbb{Z}$. We also have $x \in I$, so $x = b(x)a$ for some $b(x) \in \mathbb{Z}[x]$. Comparing degrees gives us $\deg(b(x)) = 1$, so $b(x) = b_1 x + b_0$ for some $b_1, b_0 \in \mathbb{Z}$. Then $x = (b_1 x + b_0)a$ implies $b_1 a = 1$. Therefore, we can conclude that $a = 1$ or $a = -1$. In each of these cases, a is a unit in $\mathbb{Z}[x]$, which implies that $I = \langle a \rangle = \mathbb{Z}[x]$, and so $1 \in I$. By the definition of I, there must then exist polynomials $p(x), q(x) \in \mathbb{Z}[x]$ such that

$$1 = 2p(x) + xq(x).$$

If p_0 is the constant term of $p(x)$, it follows that the constant term of $2p(x) + xq(x)$ is $2p_0$. Thus, we have $2p_0 = 1$ for some $p_0 \in \mathbb{Z}$, a contradiction. Therefore, I is not a principal ideal of $\mathbb{Z}[x]$.

Given two principal ideals, one might ask if they can be equal without having their generators be equal.

Activity 13.14.

(a) How are the two principal ideals $\langle 2 \rangle$ and $\langle -2 \rangle$ in $\mathbb{Z}$ related?

(b) There is an integer k so that $2 = k(-2)$. What is this integer and what special property does it have in $\mathbb{Z}$?

(c) Let R be any commutative ring with identity, and let $a, b \in R$ with $a = ub$ for some unit $u \in R$. What conclusion can you draw about $\langle a \rangle$ and $\langle b \rangle$? Explain.

Elements a and b that are related as in part (c) of Activity 13.14 are given a special name.

Definition 13.15. Let R be a commutative ring with identity. Elements $a, b \in R$ are said to be **associates** if $a = ub$ for some unit $u \in R$.

The result of the previous activity suggests the following lemma, which is almost—but not quite—a biconditional statement.

Lemma 13.16. *Let R be a commutative ring with identity, and let $a, b \in R$.*

(i) *If a and b are associates, then $\langle a \rangle = \langle b \rangle$.*

(ii) *If R is an integral domain and $\langle a \rangle = \langle b \rangle$, then a and b are associates.*

Proof. Let R be a commutative ring with additive identity 0_R and multiplicative identity 1_R. Let $a, b \in R$.

To prove (i), assume a and b are associates. Then $a = ub$ for some unit $u \in R$. To show $\langle a \rangle = \langle b \rangle$, we will prove that $\langle a \rangle \subseteq \langle b \rangle$ and $\langle b \rangle \subseteq \langle a \rangle$. Let $x \in \langle a \rangle$. Then $x = ra$ for some $r \in R$. So $x = r(ub) = (ru)b \in \langle b \rangle$, and $\langle a \rangle \subseteq \langle b \rangle$. Now let $y \in \langle b \rangle$. Then $y = sb$ for some $s \in R$. So $y = s\left(u^{-1}a\right) = \left(su^{-1}\right)a \in \langle a \rangle$, and $\langle b \rangle \subseteq \langle a \rangle$. The two containments show that $\langle a \rangle = \langle b \rangle$.

---

*For an example, see the paper "The Euclidean Algorithm" by T. Motzkin, *Bulletin of the American Mathematical Society*, **55**(12), 1949, pp. 1142–1146.

To prove the forward implication, assume $\langle a \rangle = \langle b \rangle$. If $\langle a \rangle = \langle b \rangle = \{0_R\}$, then we must have $a = b = 0_R$, and so $a = 1_R b$. Thus, a and b are associates. Now assume a and b are nonzero elements in R. Since $a \in \langle a \rangle$, it follows that $a \in \langle b \rangle$. Similarly, since $b \in \langle b \rangle$, we know $b \in \langle a \rangle$. Thus, $a = ub$ and $b = va$ for some $u, v \in R$. Then $a = u(va) = (uv)a$. Therefore, $a(1_R - uv) = 0_R$. The fact that R is an integral domain and $a \neq 0_R$ implies that $1_R - uv = 0_R$ or $uv = 1_R$. We can therefore conclude that u and v are units and that a and b are associates. ∎

That Lemma 13.16 is not a biconditional is not so easy to see. However, Exercise (15) provides an example.

Congruence Modulo an Ideal

In Investigation 12, we defined congruence modulo a polynomial and constructed quotients of polynomial rings based on this relation. The congruence class of a polynomial is an ideal, so now we extend that construction to congruence modulo an arbitrary ideal of a ring R. The definition is analogous to Definition 12.2.

Definition 13.17. Let R be a ring and I an ideal of R. The element $a \in R$ is **congruent modulo** I to $b \in R$ if $b - a \in I$.

If a is congruent to b modulo I, we denote this relation by writing

$$a \equiv b \pmod{I}.$$

The next activity provides an example.

Activity 13.18. Let $R = \mathbb{Z}_{12}$ and $I = \langle [3] \rangle$ as in Preview Activity 13.1.

(a) Find all the elements of R that are congruent to $[0]$ modulo I.

(b) Find all the elements of R that are congruent to $[1]$ modulo I.

(c) Find all the elements of R that are congruent to $[2]$ modulo I.

(d) What can you say about the three sets of elements you have just found? What might we expect to be true about the relation of congruence modulo an ideal?

As with any relation, there are three natural questions to ask.

Activity 13.19. Let R be a ring and I an ideal of R.

(a) Is congruence modulo I a reflexive relation? Prove your answer.

(b) Is congruence modulo I a symmetric relation? Prove your answer.

(c) Is congruence modulo I a transitive relation? Prove your answer.

The collection of distinct equivalence classes of R under the congruence modulo I relation is denoted R/I (read as "R mod I"). As we did with congruence classes of polynomials, we can make R/I into a ring. To do so, we first need to understand some properties of the congruence relation.

Activity 13.20. Let R be a ring and I an ideal. Let $a, b, c, d \in R$ such that $a \equiv c \pmod{I}$ and $b \equiv d \pmod{I}$. (Note the similarities between this activity, Activity 2.25 on page 35, and Activity 12.6 on page 174.)

(a) Show that $(a + b) \equiv (c + d) \pmod{I}$.

(b) Show that $(ab) \equiv (cd) \pmod{I}$.

We can make the set R/I into a ring by defining addition and multiplication on the set of equivalence classes. We denote by $r + I$ the equivalence class of r modulo I, so that $R/I = \{a + I : a \in R\}$. We can then define addition and multiplication on R/I in a natural way; in particular, for $r + I, s + I \in R/I$, we define

$$(r + I) + (s + I) = (r + s) + I \tag{13.1}$$
$$(r + I)(s + I) = rs + I. \tag{13.2}$$

We will now investigate the structure that addition and multiplication impose on the set R/I.

Activity 13.21. Let $R = \mathbb{Z}_{12}$ and $I = \langle [3] \rangle$, as in Preview Activity 13.1.

(a) Find all the elements in R/I.

(b) Construct the addition table for R/I.

(c) Construct the multiplication table for R/I.

(d) What kind of structure does R/I appear to have? Explain.

Of course, one question we must ask about R/I is whether the operations defined in (13.1) and (13.2) are well-defined. This is a question we have asked many times by now, so answering it should entail a familiar process.

Activity 13.22. Let R be a ring and I an ideal of R.

(a) Formally state what it means for the operation defined in (13.1) to be well-defined. Then prove that this operation is well-defined.

(b) Formally state what it means for the operation defined in (13.2) to be well-defined. Then prove that this operation is well-defined.

As we might expect from our work in this section, the set R/I of congruence classes is a ring under the operations defined in (13.1) and (13.2).

Activity 13.23. Let R be a ring and I an ideal of R.

(a) Why is R/I closed under the operation in (13.1)?

(b) Why is R/I closed under the operation in (13.2)?

(c) What is the additive identity in R/I? Prove your conjecture.

(d) What is the additive inverse of the element $a + I$ in R/I? Prove your conjecture.

(e) Complete the proof that R/I is a ring. (Note the similarity between this proof and that from Activity 12.8 on page 174.)

(f) If R is a commutative ring, must R/I be a commutative ring? Prove your conjecture.

(g) If R contains an identity, must R/I also be a ring with identity? Prove your conjecture.

The result of Activity 13.23 is the following theorem.

Theorem 13.24. *Let R be a ring and I an ideal of R. The set R/I of congruence classes modulo I is a ring under the operations (13.1) and (13.2). Moreover, If R is a commutative ring, then R/I is also a commutative ring. If R is a ring with identity, then R/I is also a ring with identity.*

In Investigation 6, we saw how field extensions and direct sums can be used to construct new rings from old ones. Theorem 13.24 shows that congruence modulo an ideal can be used for the same purpose. As in Investigation 12, the resulting rings are called *quotient rings*.

Maximal and Prime Ideals

Recall that the ring $\mathbb{Z}_n$ is a field for prime values of n, but is not even an integral domain when n is composite. A similar type of behavior happens with quotient rings.

Activity 13.25. In Activity 13.21, we created the operation tables for the quotient ring $\mathbb{Z}_{12}/I$, where $I = \langle [3] \rangle$. We will now compare that quotient ring to another.

(a) Let $J = \langle [6] \rangle \subset \mathbb{Z}_{12}$. Construct the addition and multiplication tables for $\mathbb{Z}_{12}/J$.

(b) The ring $\mathbb{Z}_{12}/I$ is isomorphic to a familiar ring R_1. Identify this familiar ring and exhibit an isomorphism between $\mathbb{Z}_{12}/I$ and R_1.

(c) The ring $\mathbb{Z}_{12}/J$ is isomorphic to a familiar ring R_2. Identify this familiar ring and exhibit an isomorphism between $\mathbb{Z}_{12}/J$ and R_2.

(d) The rings R_1 and R_2 are fundamentally different. Explain the differences and why you think they might occur. (Note that $J \subset I$.)

Activity 13.25 shows that some quotient rings have more algebraic structure than others. Of the various types of rings we have considered, fields have the most structure. This being the case, it would be useful to know when a quotient ring is a field. In Activity 13.25, we saw that because J was contained (properly) in another proper ideal I, the ring $\mathbb{Z}_{12}/J$ was not even an integral domain. This observation suggests that proper ideals that are not contained in any larger ideals may play an important role in determining when a quotient ring is a field.

Definition 13.26. An ideal I in a commutative ring R is a **maximal ideal** if $I \neq R$ and there is no proper ideal J in R such that $I \subset J$.

In fact, maximal ideals are important because of the following theorem.

Theorem 13.27. *Let R be a commutative ring with identity and I an ideal of R. Then R/I is a field if and only if I is a maximal ideal.*

Proof. Let R be a commutative ring with identity and I a proper ideal of R. To prove this biconditional statement, first assume that R/I is a field. We will proceed by contradiction and assume that there is a proper ideal J in R with $I \subset J$. Let $b \in J$ such that $b \notin I$. Then $b + I \neq I$ and $b + I$ is a unit in R/I. So there exists $c \in R$ such that $c + I = (b+I)^{-1}$. Thus, $1 + I = (b+I)(c+I) = bc + I$ and $1 - bc \in I$. Now $I \subset J$, so $1 - bc \in J$. Let $1 - bc = j$. Recall that $b \in J$, so we also have $bc \in J$. This gives us $1 = bc + j \in J$. But if $1 \in J$, then $J = R$, a contradiction. Therefore, we cannot have a proper ideal of R that contains I as a proper subset.

To prove the reverse implication, assume I is a maximal ideal of R. We know that R/I is a commutative ring with identity, so it remains to show that every nonzero element in R/I is a unit. Let $a + I$ be a nonzero element in R/I. Let $\overline{J} = \langle a + I \rangle$. Since $a + I \neq I$, we know $\overline{J}$ is a nontrivial ideal of R/I. Let $J = \{r \in R : r + I \in \overline{J}\}$. Note that $a \in J$, so J is nonempty. Also, if $r, s \in J$, then

$$(r + I) - (s + I) = (r - s) + I \in \overline{J}$$

because $\overline{J}$ is an ideal. Thus, we have $r - s \in J$. Finally, if $x \in R$, then

$$(x + I)(r + I) = xr + I \in \overline{J}$$

and $xr \in J$. Therefore, J is an ideal of R. If $i \in I$, then $i + I = I \in \overline{J}$, so $i \in J$. Thus, we have $I \subseteq J$. However, we know that $a \in J$ and $a \notin I$, so $I \subset J$. Since I is a maximal ideal, we conclude that $J = R$. Thus, $1 \in J$ and $1 + I \in \overline{J}$. This implies that there is an element $c \in R$ such that $1 + I = (c + I)(a + I)$, which makes $a + I$ a unit in R/I. Therefore, we conclude that R/I is a field if I is a maximal ideal of R. ■

The next activity shows that maximal ideals of $\mathbb{Z}$ and $F[x]$ (for a field F) are easy to recognize.

Activity 13.28.

(a) Let $m \in \mathbb{Z}^+$ be a composite integer. Show that there is a proper ideal I in $\mathbb{Z}$ that contains $\langle m \rangle$. Use this fact to classify the maximal ideals of $\mathbb{Z}$.

(b) Let F be a field. We have seen many instances where $F[x]$ shares the same properties as $\mathbb{Z}$. With that in mind, write a problem analogous to that in part (a), but for $F[x]$ instead of $\mathbb{Z}$. Provide a solution that includes a classification of the maximal ideals of $F[x]$.

Theorem 13.27 provides us with a method of constructing new fields, similar to what we saw in Investigation 12.

Activity 13.29. Let $I = \langle 3 \rangle$ in $\mathbb{Z}[i]$.

(a) Find all the elements in $\mathbb{Z}[i]/I$. (Hint: Remember that $i^2 = -1$.)

(b) We know that $\mathbb{Z}[i]/I$ is a commutative ring with identity. Is it true that every nonzero element in $\mathbb{Z}[i]/I$ is a unit? If so, explain why. If not, verify your statement.

(c) Is I a maximal ideal of $\mathbb{Z}[i]$? Why or why not?

(d) Do you know of any fields with nine elements? Explain.

At this point, it might be natural to ask if it is necessary that I be a maximal ideal in order for R/I to be an integral domain. The next activity provides an answer to this question.

Activity 13.30. Let $R = \mathbb{Z}[x]$ and let $I = \langle 3 \rangle$.

(a) Describe the elements in R/I.

(b) The ring R/I is isomorphic to a familiar ring S. What ring is S? Why is S an integral domain?

(c) Explain why I is not a maximal ideal of R.

Activity 13.30 shows that we can have R/I be an integral domain without I being a maximal ideal. The next activity explores conditions on the ideal I under which this happens in $\mathbb{Z}[x]$.

Activity 13.31. Let $R = \mathbb{Z}[x]$ and let $I = \langle 3 \rangle$.

(a) Suppose $f(x), g(x) \in R$ and $f(x)g(x) \in I$. How does this relate 3 to the product $f(x)g(x)$?

(b) What conclusion can you draw from the previous question?

The property illustrated in Activity 13.31 suggests the following definition:

Definition 13.32. A proper ideal I in a commutative ring R is a **prime ideal** if for any $a, b \in R$, whenever $ab \in I$ then $a \in I$ or $b \in I$.

You may notice a similarity between the definition of a prime ideal and the statement of Euclid's Lemma from Investigation 1. (See page 10.) In fact, Euclid's Lemma can be restated in terms of prime ideals as follows:

Euclid's Lemma. *Let p be any prime number. Then $\langle p \rangle$ is a prime ideal of $\mathbb{Z}$.*

As discussed above, we defined prime ideals with the goal of classifying the ideals I for which R/I is an integral domain. The next activity accomplishes this goal.

Activity 13.33. Let R be a commutative ring with identity and I an ideal of R.

(a) Give a formal statement of what a zero divisor looks like in R/I.

(b) Assume I is a prime ideal of R. Use the fact that I is a prime ideal to show that a nonzero element $a + I \in R/I$ cannot be a zero divisor in R/I.

(c) Assume R/I is an integral domain and let $a, b \in R$. Use the fact that R/I is an integral domain to show that if $ab \in I$, then $a \in I$ or $b \in I$.

(d) Explain how we have proven the following theorem:

> **Theorem 13.34.** *Let R be a commutative ring with identity and I a proper ideal of R. Then R/I is an integral domain if and only if I is a prime ideal.*

Note that a prime ideal need not be maximal, as our example of the ideal $\langle 3 \rangle$ in $\mathbb{Z}[x]$ (from Activity 13.30) shows. However, Theorems 13.27 and 13.34 together establish the following result:

Theorem 13.35. *Let R be a commutative ring with identity. Then every maximal ideal of R is a prime ideal.*

Homomorphisms

Investigation 7 introduced us to isomorphisms of rings. Recall that a ring isomorphism is a bijective function that preserves the additive and multiplicative structure of a ring, thereby identifying isomorphic rings as being essentially the same. It is also worth considering functions that preserve the structure of a ring but are not necessarily bijections. Such functions are called *homomorphisms* and will be the focus of the remainder of this investigation.

The formal definition of a homomorphism is as follows:

Definition 13.36. Let R and S be rings. A function $\varphi : R \to S$ is a **homomorphism** of rings if

$$\varphi(a + b) = \varphi(a) + \varphi(b) \text{ and } \varphi(ab) = \varphi(a)\varphi(b)$$

for all $a, b \in R$.

Certain types of homomorphisms are given special names. In particular, suppose R and S are rings and let $\varphi : R \to S$ be a function.

- If φ is a homomorphism and an injection, then φ is called a *monomorphism.*

- If φ is a homomorphism and a surjection, then φ is called an *epimorphism.*

- If φ is a homomorphism and a bijection, then φ is called an *isomorphism.* In this case, we say that the rings R and S are isomorphic rings.

If φ maps the ring R onto the ring S, then S inherits the additive and multiplicative structure of R. We call S a *homomorphic image* of R. The next activity provides some examples.

Activity 13.37. Determine if the given function is a homomorphism. If the function is a homomorphism, is it a monomorphism, an epimorphism, an isomorphism, or none of the above?

(a) Let R and S be any rings, and define $\varphi : R \to S$ by $\varphi(r) = 0_S$ (where 0_S is the identity in S).

(b) Let R be any ring, and define $id_R : R \to R$ by $id_R(r) = r$.

(c) Let $n \in \mathbb{Z}^+$, $n > 1$, and let $\varphi : \mathbb{Z} \to \mathbb{Z}_n$ be defined by $\varphi(k) = [k]$.

(d) Let R be a field and $r \in R$. Let $ev_r : R[x] \to R$ be defined by $ev_r(f(x)) = f(r)$. (This map is called the *evaluation map.*)

(e) Let $\varphi : \mathbb{Z}_{12} \to \mathbb{Z}_6$ be defined by $\varphi([k]_{12}) = [4k]_6$, where $[k]_n$ denotes the congruence class of k modulo n.

In Investigation 7, we saw that isomorphisms preserve certain properties of rings called *invariants.* Homomorphisms also preserve some properties of rings, as the next activity shows.

Activity 13.38. Let $\varphi : R \to S$ be a ring homomorphism, and let 0_R and 0_S be the additive identities in R and S, respectively.

(a) Isomorphisms preserve the additive identity. Use the fact that $0_R = 0_R + 0_R$ to show that $\varphi(0_R) = 0_S$. Conclude that homomorphisms also preserve additive identities.

(b) Isomorphisms preserve additive inverses. Let $a \in R$. Use the fact that φ is a homomorphism to show that $\varphi(-a) = -\varphi(a)$. Conclude that homomorphisms also preserve additive inverses.

(c) Isomorphisms preserve differences. Let $a, b \in R$. Use the result from part (b) and the fact that φ is a homomorphism to show that $\varphi(a - b) = \varphi(a) - \varphi(b)$. Conclude that homomorphisms also preserve differences.

(d) If A is a ring with identity 1_A and $\psi : A \to B$ is an isomorphism, then B is a ring with identity $1_B = \psi(1_A)$. Is it true that if R is a ring with identity 1_R, then S is also a ring with identity $1_S = \varphi(1_R)$? Explain. (Hint: Consider $R = \mathbb{Z}_3$, $S = \mathbb{Z}_6$, and $\varphi([k]_3) = [3k]_6$.)

(e) If A is a ring with identity 1_A, $\psi : A \to B$ is an isomorphism, and $u \in A$ is a unit, then $\psi(u)$ is a unit in B and $\psi(u)^{-1} = \psi\left(u^{-1}\right)$. Is it true that if R is a ring with identity 1_R and u is a unit in R, then $\varphi(u)$ is a unit in S and $\psi(u)^{-1} = \psi\left(u^{-1}\right)$? Explain. (Hint: Again consider $R = \mathbb{Z}_3$, $S = \mathbb{Z}_6$, and $\varphi([k]_3) = [3k]_6$.)

Activity 13.38 shows that homomorphisms don't preserve all the invariants that isomorphisms do. However, under certain conditions, homomorphisms do preserve these invariants, as demonstrated in the next lemma.

Theorem 13.39. *Let $\varphi : R \to S$ be a ring homomorphism. Let 0_R and 0_S be the additive identities in R and S, respectively.*

(i) *$\varphi(0_R) = 0_S$.*

(ii) *If $a \in R$, then $-\varphi(a) = \varphi(-a)$.*

(iii) *If $a, b \in R$, then $\varphi(a - b) = \varphi(a) - \varphi(b)$.*

(iv) *If R has identity 1_R, φ is an epimorphism, and S is not the trivial ring (that is, $S \neq \{0_S\}$), then S has identity 1_S and $\varphi(1_R) = 1_S$.*

(v) *If R has identity 1_R, φ is an epimorphism, and S is not the trivial ring (that is, $S \neq \{0_S\}$), then $\varphi(u)$ is a unit in S for any unit u in R and $(\varphi(u))^{-1} = \varphi\left(u^{-1}\right)$.*

Proof. Let $\varphi : R \to S$ be a ring homomorphism and let 0_R and 0_S be the additive identities in R and S, respectively. The proofs of the first three properties were part of Activity 13.38. Thus, we will focus on the latter properties.

For property (iv), let $s \in S$. Since φ is an epimorphism, there is an element $r \in R$ such that $\varphi(r) = s$. Then

$$\varphi(1_R)s = \varphi(1_R)\varphi(r) = \varphi(1_R r) = \varphi(r) = \varphi(r 1_R) = \varphi(r)\varphi(1_R) = s\varphi(1_R),$$

and so $\varphi(1_R)$ is an identity in S.

Finally, to verify property (v), choose u to be a unit in R. So there exists $u^{-1} \in R$ such that $uu^{-1} = u^{-1}u = 1_R$. Then

$$\varphi(u)\varphi\left(u^{-1}\right) = \varphi\left(uu^{-1}\right) = \varphi(1_R) = \varphi\left(u^{-1}u\right) = \varphi\left(u^{-1}\right)\varphi(u).$$

Therefore, we have $(\varphi(u))^{-1} = \varphi\left(u^{-1}\right)$. ∎

The Kernel and Image of a Homomorphism

There are two important sets that are related to a ring homomorphism: the kernel and image. To understand why the kernel is important, consider the homomorphism $\varphi : \mathbb{Z}_{12} \to \mathbb{Z}_6$ defined by

$\varphi([k]_{12}) = [4k]_6$ (from Activity 13.37). As with all homomorphisms, we have $\varphi([0]_{12}) = [0]_6$. If φ was a monomorphism, then $[0]_{12}$ would be the only element in $\mathbb{Z}_{12}$ that φ sends to $[0]_6$. However, if we look at the preimage $\varphi^{-1}([0]_6)$ of $[0]_6$ (that is, all the elements in $\mathbb{Z}_{12}$ that map to $[0]_6$), we see that $\varphi^{-1}([0]_6) = \{[0]_{12}, [3]_{12}, [6]_{12}, [9]_{12}\}$. As we saw in our preview activity, this preimage is an ideal of $\mathbb{Z}_{12}$. For convenience, let $K = \varphi^{-1}([0]_6)$. Recall that whenever I is an ideal of a ring R and $a \in R$, $a + I = \{a + r : r \in I\}$. Now notice that

- $\varphi^{-1}([0]_6) = \{[0]_{12}, [3]_{12}, [6]_{12}, [9]_{12}\} = K$;

- $\varphi^{-1}([1]_6) = \emptyset$;

- $\varphi^{-1}([2]_6) = \{[2]_{12}, [5]_{12}, [8]_{12}, [11]_{12}\} = [2]_{12} + K$;

- $\varphi^{-1}([3]_6) = \emptyset$;

- $\varphi^{-1}([4]_6) = \{[1]_{12}, [4]_{12}, [7]_{12}, [10]_{12}\} = [1]_{12} + K$; and

- $\varphi^{-1}([5]_6) = \emptyset$.

Thus, every nonempty preimage is just a translation of the ideal K. So if $K = \{[0]_{12}\}$, then each preimage will be a one element set, and φ will be a monomorphism. In this sense, the set $\varphi^{-1}([0]_6)$ tells us how close φ is to being a monomorphism. This particular preimage is called the *kernel* of φ, denoted $\mathrm{Ker}(\varphi)$.

Definition 13.40. Let $\varphi : R \to S$ be a homomorphism of rings. The **kernel** of φ is the set

$$\mathrm{Ker}(\varphi) = \{r \in R : \varphi(r) = 0_S\}.$$

In our example of $\varphi : \mathbb{Z}_{12} \to \mathbb{Z}_6$ defined by $\varphi([k]_{12}) = [4k]_6$, we have $\mathrm{Ker}(\varphi) = \{[0]_{12}, [3]_{12}, [6]_{12}, [9]_{12}\}$. As we saw in this example, the kernel was an ideal of R. It is natural to ask if this is always the case.

Activity 13.41. Let $\varphi : R \to S$ be a ring homomorphism. Let 0_R be the additive identity in R and 0_S the additive identity in S.

(a) Theorem 13.39 shows that $0_R \in \mathrm{Ker}(\varphi)$. Use Theorem 13.39 and the fact that φ is a homomorphism to show that $\mathrm{Ker}(\varphi)$ is closed under subtraction.

(b) Let $a \in R$ and $k \in \mathrm{Ker}(\varphi)$. Use the fact that φ is a homomorphism to explain why $ak \in \mathrm{Ker}(\varphi)$ and $ka \in \mathrm{Ker}(\varphi)$.

(c) Explain how we have proved the following result:

Theorem 13.42. *If $\varphi : R \to S$ is a ring homomorphism, then $\mathrm{Ker}(\varphi)$ is an ideal of R.*

As mentioned earlier, the kernel of a homomorphism φ tells us how close φ is to being a monomorphism. The next theorem formalizes this statement.

Theorem 13.43. *Let $\varphi : R \to S$ be a ring homomorphism. Then φ is a monomorphism if and only if $\mathrm{Ker}(\varphi) = \{0_R\}$, where 0_R is the additive identity in R.*

Proof. Let $\varphi : R \to S$ be a ring homomorphism. Let 0_R and 0_S be the additive identities in R and S, respectively. We will prove the forward implication first. Assume φ is a monomorphism. Theorem 13.39 shows that $0_R \in \mathrm{Ker}(\varphi)$. Suppose $k \in \mathrm{Ker}(\varphi)$. Then $\varphi(k) = 0_S = \varphi(0_R)$. Since φ is a monomorphism, we must have $k = 0_R$. Thus, $\mathrm{Ker}(\varphi) = \{0_R\}$.

For the reverse implication, assume $\text{Ker}(\varphi) = \{0_R\}$. To show that φ is a monomorphism, let $a, b \in R$ such that $\varphi(a) = \varphi(b)$. Then $\varphi(a) - \varphi(b) = 0_S$, which implies $\varphi(a - b) = 0_S$. Thus, we have $a - b \in \text{Ker}(\varphi)$. Since $\text{Ker}(\varphi) = \{0_R\}$, we can conclude that $a - b = 0_R$, or $a = b$. This proves that φ is a monomorphism. ∎

The second important set related to a homomorphism is its image. In our example of $\varphi : \mathbb{Z}_{12} \to \mathbb{Z}_6$ defined by $\varphi([k]_{12}) = [4k]_6$, note that

- $\varphi([0]_{12}) = \varphi([3]_{12}) = \varphi([6]_{12}) = \varphi([9]_{12}) = [0]_6$,

- $\varphi([1]_{12}) = \varphi([4]_{12}) = \varphi([7]_{12}) = \varphi([10]_{12}) = [4]_6$, and

- $\varphi([2]_{12}) = \varphi([5]_{12}) = \varphi([8]_{12}) = \varphi([11]_{12}) = [2]_6$.

In this case, φ maps onto the set $\{[0]_6, [4]_6, [2]_6\}$. It is not difficult to see that this set is in fact a subring of $\mathbb{Z}_6$. The set of all images of elements under the homomorphism φ is called the *image* (or range) of φ and is denoted as $\text{Im}(\varphi)$.

Definition 13.44. Let $\varphi : R \to S$ be a ring homomorphism. The **image** of φ is the set

$$\text{Im}(\varphi) = \{\varphi(r) : r \in R\}.$$

In our example, we saw that the image of φ was a subring of $\mathbb{Z}_6$. Is it always the case that the image of a homomorphism is a ring?

Activity 13.45. Let $\varphi : R \to S$ be a ring homomorphism. Let 0_R be the additive identity in R and 0_S the additive identity in S.

(a) Theorem 13.39 shows that $0_S \in \text{Im}(\varphi)$. Use Theorem 13.39 and the fact that φ is a homomorphism to show that $\text{Im}(\varphi)$ is closed under subtraction.

(b) Show that $\text{Im}(\varphi)$ is closed under multiplication.

(c) Explain how we have proved the following result:

Theorem 13.46. *If $\varphi : R \to S$ is a ring homomorphism, then $\text{Im}(\varphi)$ is a subring of S.*

The First Isomorphism Theorem for Rings

If $\varphi : R \to S$ is a ring homomorphism, then there is an important relationship between R, $\text{Ker}(\varphi)$, and $\text{Im}(\varphi)$. We will investigate this relationship in the next activity.

Activity 13.47. Let $R = \mathbb{Z}_{12}$, $S = \mathbb{Z}_6$, and $\varphi : R \to S$ defined by $\varphi([k]_{12}) = [4k]_6$. Recall that $\text{Ker}(\varphi) = \{[0]_{12}, [3]_{12}, [6]_{12}, [9]_{12}\}$ and $\text{Im}(\varphi) = \{[0]_6, [4]_6, [2]_6\}$.

(a) Construct the addition and multiplication tables for $\text{Im}(\varphi)$.

(b) In Activity 13.21, we constructed the addition and multiplication tables for R/I, where $I = \langle[3]\rangle$. Show in detail that R/I is isomorphic to $\text{Im}(\varphi)$.

The important general relationship uncovered in Activity 13.47 between R, $\text{Ker}(\varphi)$, and $\text{Im}(\varphi)$ is given in the next theorem.

Theorem 13.48 (First Isomorphism Theorem for Rings). *Let $\varphi : R \to S$ be a ring homomorphism. Then $R/Ker(\varphi) \cong Im(\varphi)$.*

Proof. Let $\varphi : R \to S$ be a ring homomorphism. For ease of notation, let $K = \text{Ker}(\varphi)$. Define a function $\Phi : R/K \to \text{Im}(\varphi)$ by $\Phi(a + K) = \varphi(a)$. We will show that Φ is a well-defined isomorphism. First, we will show that Φ is well-defined. Suppose $a + K = a' + K$ for some $a, a' \in R$. Thus, $a = a' + k$ for some $k \in K$. Then

$$\Phi(a' + K) = \varphi(a') = \varphi(a') + 0_s = \varphi(a') + \varphi(k) = \varphi(a' + k) = \varphi(a) = \Phi(a + K).$$

Therefore, Φ is well-defined.

Next we will show that Φ is a ring homomorphism. Let $a + K, b + K \in R/K$. Then

$$\begin{aligned}
\Phi((a + K) + (b + K)) &= \Phi((a + b) + K) \\
&= \varphi(a + b) \\
&= \varphi(a) + \varphi(b) \\
&= \Phi(a + K) + \Phi(b + K).
\end{aligned}$$

Furthermore,

$$\begin{aligned}
\Phi((a + K)(b + K)) &= \Phi((ab) + K) \\
&= \varphi(ab) \\
&= \varphi(a)\varphi(b) \\
&= \Phi(a + K)\Phi(b + K).
\end{aligned}$$

Thus, Φ is a homomorphism.

It remains to show that Φ is an isomorphism. To show that Φ is an injection, suppose $\Phi(a+K) = \Phi(b + K)$. Then $\varphi(a) = \varphi(b)$. So $\varphi(a) + (-\varphi(b)) = 0_S$. Recall that $-\varphi(b) = \varphi(-b)$, so we have $\varphi(a - b) = 0_S$. Therefore, $a - b \in K$, and so $a + K = b + K$. Thus, Φ is an injection. To conclude, we will show that Φ is a surjection. Let $s \in \text{Im}(\varphi)$. So $s = \varphi(r)$ for some $r \in R$. Then $\Phi(r + K) = \varphi(r) = s$, which proves that Φ is a surjection.

Since Φ is a bijective ring homomorphism, it follows that Φ is an isomorphism and $R/\text{Ker}(\varphi) \cong \text{Im}(\varphi)$. ∎

In addition to Theorem 13.48, there are two other ring isomorphism theorems that are introduced in Exercises (22) and (23).

Concluding Activities

Activity 13.49. Let $n \in \mathbb{Z}^+$, and define $\varphi : \mathbb{Z} \to \mathbb{Z}_n$ by $\varphi(k) = [k]$.

(a) Show that $\text{Ker}(\varphi) = \langle n \rangle$. We also denote this ideal as $n\mathbb{Z}$.

(b) Find $\text{Im}(\varphi)$.

(c) Explain why $\mathbb{Z}_n \cong \mathbb{Z}/n\mathbb{Z}$. (Note that this gives us another way of constructing the ring of integers modulo n.)

Activity 13.50. Let R be a commutative ring with identity, and let 0 denote the additive identity in R. Let $ev_0 : R[x] \to R$ be the evaluation map (see Activity 13.37) defined by $ev_0(f(x)) = f(0)$.

(a) Explain why $\text{Ker}(ev_0) = \langle x \rangle$.

(b) What is $\text{Im}(ev_0)$? Explain.

(c) What is $R[x]/\langle x \rangle$? Explain.

(d) Under what conditions is $\langle x \rangle$ a prime ideal of $R[x]$? Give a precise answer with justification.

(e) Under what conditions is $\langle x \rangle$ a maximal ideal of $R[x]$? Give a precise answer with justification.

(f) Can $\langle x \rangle$ ever be a prime ideal of $R[x]$ but not a maximal ideal? Explain.

Exercises

(1) If an ideal I in a ring R with identity contains a unit, what must I be? Prove your answer.

(2) Find all ideals of the following rings.

    (a) $\mathbb{Z}_3$

    (b) $\mathbb{Z}_6$

    (c) $\mathbb{Z}_{12}$

    (d) $\mathbb{Z}$

    (e) $\mathbb{R}$

(3) Determine which of the following sets I is an ideal of the indicated ring.

    (a) Let R be any ring, and let I be the collection of all polynomials in $R[x]$ with a constant term equal to 0_R, the additive identity in R.

    (b) Let $R = M_{2\times 2}(\mathbb{Z})$ and $I = \left\{ \begin{pmatrix} a & 0 \\ b & 0 \end{pmatrix} : a, b \in \mathbb{Z} \right\}$.

    (c) Let R be a commutative ring and $a_1, a_2, \ldots, a_k \in R$ for some $k \in \mathbb{Z}^+$. Let $I = \{r_1 a_1 + r_2 a_2 + \cdots + r_k a_k : r_1, r_2, \ldots, r_k \in R\}$.

    (d) Let $R = \mathbb{Z}[x]$, and let I be the collection of all polynomials in $R[x]$ with even coefficients.

(4) Determine if the given function is a homomorphism. If the function is a homomorphism, is it a monomorphism, epimorphism, isomorphism, or none of the above?

    (a) Let $n \in \mathbb{Z}^+$, $n > 1$, and let $\varphi : \mathbb{Z} \to \mathbb{Z}$ be defined by $\varphi(k) = nk$.

    (b) Let $\varphi : \mathbb{Z}_3 \to \mathbb{Z}_{12}$ be defined by $\varphi([k]_3) = [4k]_{12}$.

(c) Let $\varphi : \mathbb{Z}_{12} \to \mathbb{Z}_3$ be defined by $\varphi([k]_{12}) = [k]_3$.

(d) Let $\varphi : \mathbb{R} \to M_{2\times 2}(\mathbb{R})$ be defined by $\varphi(x) = \begin{bmatrix} x & 0 \\ -x & 0 \end{bmatrix}$.

(e) Let $\varphi : \mathbb{Z}_3 \to \mathbb{Z}_3$ be defined by $\varphi([n]) = [n^3]$.

(5) Let $S = \left\{ \begin{bmatrix} a & b \\ 0 & c \end{bmatrix} \mid a, b, c \in \mathbb{R} \right\}$ and define $f : S \to \mathbb{R}$ by $f \left(\begin{bmatrix} a & b \\ 0 & c \end{bmatrix} \right) = a$ for all $\begin{bmatrix} a & b \\ 0 & c \end{bmatrix} \in$ S.

(a) Prove that S is a subring of $\mathcal{M}_{2\times 2}(\mathbb{R})$.

(b) Prove that f is a surjective homomorphism that is not an isomorphism.

(c) Does the result in part (b) prove that S is not isomorphic to $\mathbb{R}$? Explain.

(d) Determine all matrices $A \in S$ such that $A^2 = I = \begin{bmatrix} 1 & 0 \\ 0 & 1 \end{bmatrix}$. (Any matrix A that satisfies $A^2 = I$ is called an *idempotent matrix*.)

(e) Can the result in part (d) be used to prove that the ring S is not isomorphic to $\mathbb{R}$? If it is possible, write a complete proof. If it is not possible, explain why it is not possible.

(6) Let R be a ring.

(a) Is every ideal of R a subring of R? Explain.

(b) Is every subring of R an ideal of R? Explain.

(7) Let $R = \left\{ \begin{pmatrix} a & b \\ 0 & a \end{pmatrix} : a, b \in \mathbb{R} \right\}$ and let $I = \left\{ \begin{pmatrix} 0 & c \\ 0 & 0 \end{pmatrix} : c \in \mathbb{R} \right\}$.

(a) Prove that I is an ideal of R.

(b) Show that $R/I \cong \mathbb{R}$ by exhibiting a specific homomorphism from R onto $\mathbb{R}$ with kernel I. Be sure to verify that your map is a homomorphism.

(8) Let $\varphi : R \to S$ be a ring homomorphism.

(a) Is it the case that $\text{Im}(\varphi)$ is always an ideal of S? Prove your answer.

(b) Show that if φ is a surjection, then $\text{Im}(\varphi)$ is an ideal of S. Is it possible for $\text{Im}(\varphi)$ to be an ideal of S even if φ is not a surjection?

(9) Let $R = \mathbb{Z}_5[x]$.

(a) Show that the polynomial $f(x) = x^2 + [2]x + [3]$ is irreducible in R.

(b) Let I be the ideal generated by $f(x)$. What kind of structure must R/I have? Explain.

(c) Is the element $([2]x + [3]) + I$ in R/I a unit? If yes, find its multiplicative inverse. If no, explain why.

(10) Let $R = \mathbb{Z}_{24}$. For parts (a) – (d), construct the addition and multiplication tables for R/I for the given ideal I.

(a) $I = \langle [3] \rangle$

(b) $I = \langle [4] \rangle$

(c) $I = \langle [6] \rangle$

(d) $I = \langle [8] \rangle$

(e) Which, if any, of the ideals from parts (a) – (d) is a prime/maximal ideal? Explain.

(f) Show that the only principal ideals of $\mathbb{Z}_{24}$ are the ideals generated by $[0]$, $[1]$, $[2]$, $[3]$, $[4]$, $[6]$, $[8]$, and $[12]$. (Hint: You can do so by proving the more general result that if $n \in \mathbb{Z}^+$, then the only nontrivial principal ideals of $\mathbb{Z}_n$ are those of the form $\langle [a] \rangle$, where a is a divisor of n.)

(11) Construct the addition and multiplication tables for $\mathbb{Z}_2[x]/\langle x^3 \rangle$.

(12) Let $R = \mathbb{Z}_3[x]$ and $I = \langle x^2 \rangle$.

(a) Construct the addition and multiplication tables for R/I.

(b) We know of two rings with nine elements: $\mathbb{Z}_9$ and $\mathbb{Z}_3 \oplus \mathbb{Z}_3$. Is R/I isomorphic to either of these rings? Justify your answer.

(c) Is I a prime ideal of R? Why or why not?

(d) Explain why I is not a maximal ideal of R and find an ideal J of R that is between I and R.

(13) Show that the quotient ring $\mathbb{Z}[x]/\langle 2x - 1 \rangle$ is isomorphic to the ring $\mathbb{Z}\left[\frac{1}{2}\right] = \left\{\frac{p}{q} : p \in \mathbb{Z} \text{ and } q = 2^m \text{ for some nonnegative integer } m\right\}$. (Hint: Consider the evaluation homomorphism $ev_{1/2}$.)

(14) Is every ideal of $\mathbb{Z}_n$ a principal ideal? Verify your conjecture.

$\star$ (15) Prove that Lemma 13.16 is not a biconditional with the following example.[†] Let R be the ring of all continuous real-valued functions on the interval $[0, 3]$. Let

$$a(t) = \begin{cases} 1 - t, & \text{if } 0 \le t \le 1, \\ 0, & \text{if } 1 < t \le 2, \\ t - 2, & \text{if } 2 < t \le 3. \end{cases}$$

and let

$$b(t) = a(t) \text{ on } [0, 2] \text{ and } b(t) = -a(t) \text{ on } (2, 3].$$

(a) Find a continuous function $k(t)$ defined on $[0, 3]$ so that $a(t) = k(t)b(t)$. Explain how this shows that $\langle a(t) \rangle \subseteq \langle b(t) \rangle$.

(b) Find a continuous function $m(t)$ defined on $[0, 3]$ so that $b(t) = m(t)a(t)$. Explain how this shows that $\langle a(t) \rangle = \langle b(t) \rangle$.

(c) Prove that $a(t)$ and $b(t)$ are not associates in R. (Hint: Show that no unit in R can ever attain the value 0, and then use the Intermediate Value Theorem from Calculus.)

(16) **Ideals of fields.** Prove that if R is a field, then the only ideals of R are $\{0_R\}$ and R. Is the converse true if R is a commutative ring with identity? Prove your answer.

[†]Based on the paper "Elementary divisors and modules" by I. Kaplansky, *Transactions of the American Mathematical Society*, **66**(2), 1949, pp. 464–491.

(17) **Simple rings.** As Exercise (16) shows, a commutative ring R with identity is field if and only if the only ideals of R are $\{0_R\}$ and R. As this exercise will demonstrate, this result is not true for noncommutative rings. In general, we say that a ring R is **simple** if R is nonzero and the only ideals of R are $\{0_R\}$ and R. So every field is a simple ring. In this exercise, we illustrate a noncommutative simple ring.

Recall that H (the ring of quaternions from Exercise (7) on page 95 of Investigation 6) is a division ring—that is, H is a noncommutative ring in which every nonzero element is a unit. There are other examples of division rings, but H is the only one we have encountered so far. Let D be any division ring. In this exercise we will prove that $R = M_{n \times n}(D)$ is a noncommutative simple ring.

(a) Let E_{ij} be the elementary matrix in R whose entries are all 0 except the (i,j) entry, which is 1. Note that if $A = [a_{ij}] \in R$, then

$$A = \sum_{ij} a_{ij} E_{ij}$$

and every element in R can be written as a linear combination of these elementary matrices. Show that

$$E_{ij}E_{rs} = \begin{cases} 0 & \text{if } j \neq r \\ E_{is} & \text{if } j = r. \end{cases}$$

(b) Now we will show that if I is a nonzero ideal of R, then $I = R$. So let I be a nonzero ideal of R and let $A = [a_{ij}]$ be a nonzero matrix in I with nonzero entry a_{rs}.

(i) Let u and v be integers between 1 and n. Explain why $E_{ur}AE_{sv} \in I$.

(ii) Use the fact that $A = \sum_{ij} a_{ij}E_{ij}$ and the product of elementary matrices to show that

$$E_{ur}AE_{sv} = a_{rs}E_{uv}.$$

Explain how this shows that $E_{uv} \in I$. Why does this complete our proof that R is a noncommutative simple ring?

(18) Prove that if R is a ring and I is an ideal of R, then $I[x]$ is an ideal of $R[x]$.

(19) **Intersections and unions of ideals.** Ideals are subsets of the rings that contain them, so we can form the intersection and union of two ideals of the same ring.

(a) Let $I = \{k[4] : k \in \mathbb{Z}\}$ and $J = \{m[6] : m \in \mathbb{Z}\}$ be ideals of $\mathbb{Z}_{48}$. Find the elements in $I \cap J$ and $I \cup J$. Is either an ideal of $\mathbb{Z}_{48}$?

(b) In general, if I and J are ideals of a ring R, must it be true that $I \cap J$ is an ideal of R? Prove your answer.

(c) In general, if I and J are ideals of a ring R, must it be true that $I \cup J$ is an ideal of R? Is $I \cup J$ ever an ideal of R? If yes, find necessary and sufficient conditions so that $I \cup J$ is an ideal of R. Prove your answer.

(20) Ideals are subsets of the rings that contain them, so we can add and multiply elements in ideals of the same ring. If I and J are ideals of a ring R, define the sum $I + J$ and product IJ of I and J as follows:

$$I + J = \{i + j : i \in I, j \in J\}$$
$$IJ = \{i_1 j_1 + i_2 j_2 + \cdots + i_k j_k : i_k \in I, j_k \in J, k \in \mathbb{Z}^+\}.$$

(a) Let $I = \{k[4] : k \in \mathbb{Z}\}$ and $J = \{m[6] : m \in \mathbb{Z}\}$ be ideals of $\mathbb{Z}_{48}$. Find the elements in $I + J$ and IJ. Is either an ideal of $\mathbb{Z}_{48}$?

(b) In general, if I and J are ideals of a commutative ring R, must it be true that $I + J$ is an ideal of R? Prove your answer.

(c) In general, if I and J are ideals of a commutative ring R, must it be true that IJ is an ideal of R? Prove your answer.

(21) Characterize all ring homomorphisms $\varphi : \mathbb{Z} \to \mathbb{Z}$.

$\star$ (22) There is a **Second Isomorphism Theorem** that we will investigate in this problem. Let R be a ring, I a subring of R, and J an ideal of R. For each of the results you are asked to prove, also illustrate the result with the example $R = \mathbb{Z}_{24}$, $I = \{4[a] : [a] \in \mathbb{Z}_{24}\}$, and $J = \{[0], [6], [12], [18]\}$.

(a) Prove that $I \cap J$ is an ideal of I.

(b) Define $I + J$ to be the set $\{i + j : i \in I, j \in J\}$. Prove that $I + J$ is a ring and that J is an ideal of $I + J$.

(c) Define $\varphi : I \to (I + J)/J$ by $\varphi(a) = a + J$. Prove that φ is a homomorphism.

(d) Find $\text{Ker}(\varphi)$.

(e) Prove the following theorem:

Theorem 13.51 (The Second Isomorphism Theorem). *Let R be a ring, I a subring of R, and J an ideal of R. Then*

$$I/(I \cap J) \cong (I + J)/J.$$

$\star$ (23) There is a **Third Isomorphism Theorem** that we will investigate in this problem. Let R be a ring, and let I and J be ideals of R with $J \subset I$. For each of the results you are asked to prove, also illustrate the result with the example $R = \mathbb{Z}_{24}$, $I = \langle [3] \rangle$, and $J = \langle [6] \rangle$.

(a) Show that J is an ideal of I.

(b) Explain how I/J is a subset of R/J. Prove that I/J is an ideal of R/J.

(c) Define $\varphi : R \to \frac{R/J}{I/J}$ by $\varphi(a) = (a + J) + I/J$. Prove that φ is a homomorphism.

(d) Find $\text{Ker}(\varphi)$.

(e) Prove the following theorem:

Theorem 13.52 (The Third Isomorphism Theorem). *Let R be a ring, and let I and J be ideals of R with $J \subset I$. Then*

$$R/I \cong \frac{R/J}{I/J}.$$

Connections

Given an ideal I in a ring R, we can define congruence modulo I and form the quotient ring R/I. Quotient structures are useful in mathematics in that they often have a simpler structure than the original set and can therefore provide important information about the original set. We are familiar with congruence and quotient structures in the integers—namely $\mathbb{Z}_n$ (from Investigation 2)—and quotient rings of the form $F[x]/\langle f(x) \rangle$ in a polynomial ring (from Investigation 12). If you studied group theory before ring theory, you should notice connections between the topics in this investigation and those in Investigation 24. In particular, the set R/I represents the set of distinct congruence classes modulo the ideal I in the same way that the set G/N represents the set of left cosets of N in G. Although G/N is a group only if N is normal in G, we don't have that problem in a ring R. Since R is an Abelian group under addition, and any ideal I is a subgroup under addition, I is also a normal subgroup of R under addition.

A homomorphism from a ring R to a ring S is a structure preserving map. The isomorphism theorems for rings show us different ways that homomorphisms can determine isomorphisms. If you studied group theory before ring theory, you should notice the connections between ring homomorphisms in this investigation and group homomorphisms in Investigation 27. A group homomorphism preserves the structure of a group, but since groups have only one operation, there is a bit less structure to preserve. In addition, a careful perusal of the ring isomorphism theorems shows that they are essentially the same as the corresponding group isomorphism theorems.

Investigation 14

Divisibility and Factorization in Integral Domains

Focus Questions

By the end of this investigation, you should be able to give precise and thorough answers to the questions listed below. You may want to keep these questions in mind to focus your thoughts as you complete the investigation.

- What is a Euclidean domain, and what are some examples of Euclidean domains?

- What is the difference between prime and irreducible elements in an integral domain?

- What is a unique factorization domain?

- What is the relationship between Euclidean domains, principal ideal domains, and unique factorization domains?

- What are two different ways of proving that every Euclidean domain is a unique factorization domain?

Preview Activity 14.1. Look back to the proof of the Fundamental Theorem of Arithmetic on page 3 of Investigation 1. List all the results from Investigation 1 that were needed for the proof. Your list should include one theorem that is central to all the others. Identify this theorem, and explain why it is so important.

Introduction

We began our investigations of abstract algebra by studying arithmetic, divisibility, and factorization in the integers. Later, we generalized these ideas to polynomial rings. In this investigation, we will take our work one step further by showing that a result analogous to the Fundamental Theorem of Arithmetic holds in any integral domain that admits a division algorithm. This shouldn't come as too much of a surprise; after all, it was the Division Algorithm in $\mathbb{Z}$ that gave us the Euclidean Algorithm. We used the Euclidean Algorithm to prove Bezout's Identity, which in turn allowed us to prove Euclid's Lemma—the key ingredient in our proof of the uniqueness of prime factorizations. What we will see in this activity is that each of these steps can be generalized. In particular, we will argue that every *Euclidean domain*—that is, every integral domain that admits a division algorithm—is a *unique factorization domain*. We will consider two different proofs of this

211

important result: one that mirrors our work in the integers and does not require the machinery of ideals, and one that uses ideals to prove a more general result. The former approach relies on material from Investigations 1 – 4 and references a few results from Investigations 9 and 10. The latter additionally relies on concepts from Investigation 13. For those who have not yet studied ideals, some of the relevant definitions from Investigation 13 are repeated here; these definitions are sufficient to complete most of the investigation, with the exception of the second proof of our main result.

Divisibility and Euclidean Domains

To begin our study of divisibility in integral domains, we must first specify more generally what it means for one element of an integral domain to divide another. The next definition does just that and is analogous to those we used in the integers and in polynomial rings.

Definition 14.2. Let D be an integral domain, and let $a, b \in D$. Then a **divides** b, denoted $a \mid b$, if there is an element $q \in D$ such that $b = aq$.

Next, we must define precisely what it means for an integral domain to "admit a division algorithm."

Definition 14.3. Let D be an integral domain with additive identity 0_D. Then D is a **Euclidean domain** if there is a function $\delta : D - \{0_D\} \to \mathbb{Z}^+ \cup \{0\}$ such that the following conditions hold:

(i) If $a, b \in D$ are nonzero, then $\delta(a) \leq \delta(ab)$.

(ii) If $a, b \in D$ with a nonzero, then there exist elements $q, r \in D$ such that

$$b = aq + r \text{ with either } r = 0_D \text{ or } \delta(r) < \delta(a).$$

The function δ in Definition 14.3 is called a **norm** for D.

Activity 14.4.

(a) How is the second condition in Definition 14.3 related to the Division Algorithm for $\mathbb{Z}$? What are the similarities and differences?

(b) Is $\mathbb{Z}$ a Euclidean domain? If so, what function serves as a norm for $\mathbb{Z}$, and why does this function satisfy the conditions listed in Definition 14.3?

(c) Let F be a field. Is $F[x]$ a Euclidean domain? If so, what function serves as a norm for $F[x]$, and why does this function satisfy the conditions listed in Definition 14.3?

Primes and Irreducibles

In order to generalize prime factorization, we will need to first revisit the way we defined prime numbers in our study of the integers. Recall that we defined a prime number in $\mathbb{Z}$ to be an integer

$p > 1$ whose only positive divisors are 1 and p. In general, the word *prime* is defined slightly differently, and what we've called prime in the past is actually closer to what is called *irreducible* in other contexts. It turns out that in the integers, the notions of prime and irreducible are equivalent. They are not, however, equivalent in all settings, as we will see shortly.

Definition 14.5. Let D be an integral domain.

- A nonzero element $a \in D$ is said to be **irreducible** provided that a is not a unit, and if $a = bc$ for some $b, c \in D$, then b or c is a unit.

- A nonzero element $a \in D$ is said to be **prime** provided that a is not a unit, and for all b, $c \in D$, if $a \mid bc$, then $a \mid b$ or $a \mid c$.

- Two elements $a, b \in D$ are said to be **associates** if $a = ub$ for some unit $u \in D$.

Activity 14.6.

(a) Explain why all the integers that we called prime in Investigation 1 are in fact irreducible according to Definition 14.5.

(b) Explain why all the integers that we called prime in previous activities are also prime according to Definition 14.5.

(c) List at least three irreducible integers that we did not previously consider to be prime. Would each of these integers be considered prime according to Definition 14.5?

(d) The definition of prime should look familiar to you. How is this definition related to a result we used in Investigation 1?

(e) Find two integers that are associates and two that are not.

(f) Find two real numbers that are associates and two that are not.

It is important to note that the definition of prime from Definition 14.5 is more general than the definition we considered for $\mathbb{Z}$ in Investigation 1. In fact, as Activity 14.6 demonstrates, there are many integers that we did not consider prime in Investigation 1 that are considered prime according to Definition 14.5. Also, as noted earlier, the notions of prime and irreducible are equivalent in $\mathbb{Z}$. The next theorem formalizes this result.

Theorem 14.7. *An integer a is prime if and only if a is irreducible.*

To prove Theorem 14.7, we will first argue that if an integer a is prime, then a is irreducible. In fact, we will actually prove the more general result described in the next activity.

Activity 14.8. Let D be an integral domain, and let a be a prime element of D. Use the definition of prime to show that if $a = bc$ for some elements $b, c \in D$, then either b or c is a unit. Explain why this proves the following theorem:

Theorem 14.9. *Let D be an integral domain, and let $a \in D$. If a is prime, then a is irreducible.*

To prove the reverse implication of Theorem 14.7, first note that the statement we need to prove is nearly identical to Euclid's Lemma. (See page 10.) Thus, our proof of Euclid's Lemma should generalize nicely. The main difference is that we will need to replace our prior assumption of primality (in the sense that we defined prime in $\mathbb{Z}$) with the new notion of irreducibility.

Activity 14.10. Modify the proof of Euclid's Lemma from Activity 1.26 on page 16 to complete the proof of Theorem 14.7.

So now we have shown that the notions of prime and irreducible are equivalent in the integers, and that Euclid's Lemma works even with the more general assumption of irreducibility. The next natural question for us to consider is whether there are integral domains in which the notions of prime and irreducible are not equivalent. Since we have already shown that in *any* integral domain, every prime element is irreducible, we will be looking for an element of an integral domain that is irreducible, but not prime.

Activity 14.11. Let D be the set of all polynomials in $\mathbb{Z}[x]$ for which the coefficient on the linear term is zero. That is,

$$D = \{a_n x^n + a_{n-1} x^{n-1} + \cdots + a_2 x^2 + a_0\} \subseteq \mathbb{Z}[x].$$

Prove that D is an integral domain, and find a polynomial of the form $p(x) = x^n$ (for some $n \in \mathbb{Z}^+$) that is irreducible, but not prime.*

Activity 14.11 demonstrates that there are integral domains in which not all irreducible elements are prime. However, we will see later in the investigation that the notions of prime and irreducible are equivalent in certain types of integral domains—specifically, principal ideal domains.

Unique Factorization Domains

We are now ready to generalize our prior results about unique factorization—namely, the Fundamental Theorem of Arithmetic (see page 10) and the unique factorization theorem for polynomials over a field (Theorem 10.21 on page 151). We'll begin with a definition.

Definition 14.12. An integral domain D is said to be a **unique factorization domain** if the following conditions hold:

(i) Every nonzero, nonunit element of D is either irreducible or can be written as a finite product of irreducible elements of D.

(ii) Factorization into irreducibles is unique up to associates and the order of the factors. In particular, if $x \in D$ can be factored as a product of irreducibles in two different ways, say

$$x = p_1 p_2 \cdots p_m \quad \text{and} \quad x = q_1 q_2 \cdots q_k$$

for some irreducible elements $p_1, p_2, \ldots, p_m, q_1, q_2, \ldots, q_k$ of D, then $m = k$ and the factors of x can be reordered so that p_i and q_i are associates for each i.

Condition (ii) of Definition 14.12 warrants some clarification. The idea behind it can be illustrated by a simple example.

Note that the integer 30 can be factored in several different ways. For instance,

$$30 = 2 \cdot 3 \cdot 5 \quad \text{and} \quad 30 = -3 \cdot 5 \cdot -2.$$

---

*Gerald Wildenberg, "An integral domain lacking unique factorization into irreducibles," *Mathematics Magazine*, **80**, 2007, pp. 75–76.

Note also that each factor in the first product can be paired with an associate factor in the second product, and vice versa. In particular, 2 can be paired with -2, 3 with -3, and 5 with 5. Since each pair consists of two associate elements of $\mathbb{Z}$, the two factorizations are considered to be the same up to associates and the order of the factors.

Activity 14.13. Look back at the proof of the Fundamental Theorem of Arithmetic. (See page 15.) Can this proof be modified to show that $\mathbb{Z}$ is a unique factorization domain? If so, what modifications would need to be made, and why is the modified proof still valid? If not, why does the proof not generalize? Is $\mathbb{Z}$ a unique factorization domain?

And now, for the punchline—the theorem that generalizes our prior investigations of divisibility and factorization:

Theorem 14.14. *Every Euclidean domain is a unique factorization domain.*

Proof 1: Generalizing Greatest Common Divisors

One way to prove Theorem 14.14 is to recognize the observation from Preview Activity 14.1 that all the work leading up to the Fundamental Theorem of Arithmetic (in Investigation 1) relied on the Division Algorithm. The same could be said for factorization of polynomials: Theorem 10.21 (see page 151) ultimately rests on the existence of a division algorithm for polynomials (page 131). In order to generalize our work from these two settings, we will need one more definition.

Definition 14.15. Let D be a Euclidean domain with norm function δ, and let a and b be elements of D, not both zero. Then a **greatest common divisor** of a and b is an element $d \in D$ such that:

(i) $d \mid a$ and $d \mid b$; and

(ii) for all $d' \in D$, if $d' \mid a$ and $d' \mid b$, then $\delta(d') \leq \delta(d)$.

The elements a and b are said to be **relatively prime** if 1 is a greatest common divisor of a and b.

Note that condition (i) of Definition 14.15 is what makes d a common divisor of a and b, whereas condition (ii) makes d a *greatest* common divisor in the sense that d has a norm greater than or equal to any other common divisor of a and b. Note also that we refer to *a* greatest common divisor of a and b instead of *the* greatest common divisor of a and b. This suggests that greatest common divisors, at least according to Definition 14.15, are not unique. The next activity confirms this suspicion.

Activity 14.16. Find two distinct greatest common divisors of 6 and 8 in $\mathbb{Z}$. What is the relationship between these two greatest common divisors? (Recall that the absolute value function is a norm for $\mathbb{Z}$.)

As Activity 14.16 demonstrates, it is possible for two elements a and b of a Euclidean domain to have more than one greatest common divisor. However, any two greatest common divisors of a and b must be associates. (See Exercise (4).) It can also be shown, as the next activity suggests, that greatest common divisors always exist, as long as at least one of the elements involved is nonzero.

Activity 14.17. Prove that if a and b are elements of a Euclidean domain D, not both zero, then there is at least one greatest common divisor of a and b. (Hint: 1 necessarily divides both a and b, and so there is at least one common divisor. Use the definition of the norm function to argue that at least one of these common divisors must be maximal.)

Now that we have formally defined greatest common divisors for any Euclidean domain, we are ready to proceed with the proof of Theorem 14.14. Rather than providing a complete proof, we will state a sequence of theorems that can be used to build up the proof one step at a time. These theorems should look familiar to you. In fact, each one is a generalization of an analogous result that we proved for the integers. Thus, you should be able to prove each of the theorems by making small modifications to the proofs of the analogous results from Investigation 1. Together, they establish Theorem 14.14—that every Euclidean domain is a unique factorization domain.

Theorem 14.18. *Let D be a Euclidean domain with norm function δ, and let a and b be elements of D, not both zero. Furthermore, let q and r be elements of D such that $b = aq + r$ and $\delta(r) < \delta(a)$. Then d is a greatest common divisor of a and b if and only if d is a greatest common divisor of a and r.*

Theorem 14.19. *Let D be a Euclidean domain, and let a and b be elements of D, not both zero. Then the last nonzero remainder produced by the Euclidean algorithm (starting with a and b) is a greatest common divisor of a and b.*

Theorem 14.20. *Let D be a Euclidean domain, let a and b be elements of D, not both zero, and let d be the greatest common divisor of a and b produced by the Euclidean algorithm. Then there exist $x, y \in D$ such that $ax + by = d$.*

Theorem 14.21. *Let D be a Euclidean domain, and let p and a be relatively prime elements of D. For all $b \in D$, if $p \mid ab$, then $p \mid b$.*

Theorem 14.22. *Let D be a Euclidean domain, and let $p \in D$ be irreducible. For all $a, b \in D$, if $p \mid ab$, then $p \mid a$ or $p \mid b$.*

Theorem 14.23. *Let D be a Euclidean domain. Then every nonzero, nonunit element of D is either irreducible or can be written as a finite product of irreducible elements of D.*

Theorem 14.24. *Let D be a Euclidean domain, and let $x \in D$. Suppose*

$$x = p_1 p_2 \cdots p_m \quad and \quad x = q_1 q_2 \cdots q_k$$

for some irreducible elements $p_1, p_2, \ldots, p_m, q_1, q_2, \ldots, q_k$ of D. Then $m = k$, and the factors of x can be reordered so that p_i and q_i are associates for each i.

Proof 2: Principal Ideal Domains

If you have studied ideals (in Investigation 13), then a second way to approach the proof of Theorem 14.14 is to prove the following more general result:

Theorem 14.25. *Every principal ideal domain is a unique factorization domain.*

Note that, by Theorem 13.13 (see page 193), every Euclidean domain is a principal ideal domain. Therefore, if we can prove Theorem 14.25, we will have also established that every Euclidean domain is a unique factorization domain.

The first step in proving Theorem 14.25 is to show that every principal ideal domain satisfies a condition known as the *ascending chain condition*. To illustrate the ascending chain condition, recall that $\mathbb{Z}$ is a principal ideal domain. Therefore, any ideal of $\mathbb{Z}$ has the form $\langle m \rangle$ for some positive integer m. Consider, for example, the ideal $I = \langle 24 \rangle$. Note that the positive divisors of 24 are 1, 2, 3, 4, 6, 12, and 24. For each such divisor k, let $I_k = \langle k \rangle$. Then

$$I = I_{24} \subset I_{12} \subset I_6 \subset I_3 \subset I_1 = \mathbb{Z}$$

and

$$I = I_{24} \subset I_{12} \subset I_4 \subset I_2 \subset I_1 = \mathbb{Z}.$$

No matter which set of factors of 24 we use, the corresponding "chain" of ideals must always eventually terminate—that is, it must contain only a finite number of distinct ideals. Since any given integer has only a finite number of factors, the same idea applies to any ideal in $\mathbb{Z}$. Because of this, we say that $\mathbb{Z}$ satisfies the ascending chain condition on ideals, defined formally below.

Definition 14.26. A commutative ring R is said to satisfy the **ascending chain condition** on ideals[†] if there does not exist an infinite sequence of ideals of R, each of which is a proper subset of the next. In other words, R satisfies the ascending chain condition if whenever

$$I_1 \subseteq I_2 \subseteq I_3 \subseteq \cdots$$

is a chain of ideals of R, then there is some integer m for which $I_k = I_m$ for all $k \geq m$.

Are there rings that do not satisfy the ascending chain condition? The answer is yes, although we have to dig a little deeper to find them.

Activity 14.27. Let $\mathcal{C}(\mathbb{R})$ denote the set of all continuous functions from $\mathbb{R}$ to $\mathbb{R}$, with addition and multiplication defined by

$$(f + g)(x) = f(x) + g(x) \quad \text{and} \quad (fg)(x) = f(x)g(x)$$

for all $f, g \in \mathcal{C}(\mathbb{R})$. It can be shown fairly easily that $\mathcal{C}(\mathbb{R})$ is a commutative ring with identity. For every integer $n \geq 0$, let I_n denote the set of all functions f in $\mathcal{C}(\mathbb{R})$ for which $f(x) = 0$ for all $x \geq n$.

(a) Show that each I_n is an ideal of $\mathcal{C}(\mathbb{R})$.

(b) Show that I_n is a proper subset of I_{n+1} for all n.

(c) Explain how parts (a) and (b) establish that $\mathcal{C}(\mathbb{R})$ does not satisfy the ascending chain condition.

That $\mathbb{Z}$ satisfies the ascending chain conditions, while other rings, like $\mathcal{C}(\mathbb{R})$, do not is a consequence of the fact that $\mathbb{Z}$ is a principal ideal domain. The next theorem states this result formally, and the activity that follows outlines its proof.

Theorem 14.28. *Every principal ideal domain satisfies the ascending chain condition.*

Activity 14.29. Let D be a principal ideal domain, and let $I_1 \subseteq I_2 \subseteq I_3 \subseteq \cdots$ be a chain of ideals of D. Define the set I to be the union of all of these ideals—that is,

$$I = \bigcup_{i=1}^{\infty} I_i.$$

Note that $x \in I$ if and only if $x \in I_i$ for some i.

[†]Rings that satisfy the ascending chain condition on ideals are also called *Noetherian* rings, named after German mathematician Emmy Noether (1882–1935), who is known for her many contributions to abstract algebra and theoretical physics.

(a) Use the Ideal Test (see page 190) to show that I is an ideal of D.

(b) Use the fact that D is a principal ideal domain to conclude that $I = \langle a \rangle$ for some $a \in D$. Then use the definition of I to deduce that $a \in I_m$ for some m.

(c) Explain why it must be that $I_m = \langle a \rangle$, and so $I_k = I_m = I$ for all $k \geq m$. Deduce that D satisfies the ascending chain condition.

The ascending chain condition allows us to establish that every nonzero, nonunit element in a principal ideal domain can be factored into irreducibles. In particular:

Theorem 14.30. *Let D be a principal ideal domain. Then every nonzero, nonunit element of D is either irreducible or can be written as a finite product of irreducible elements of D.*

Proof. Let D be a principal ideal domain, and let a be a nonzero, nonunit element of D. If a is irreducible, then we are done. Therefore, suppose that $a = a_1 a_2$ for some nonunit elements a_1, $a_2 \in D$. If both a_1 and a_2 are irreducible, then, once again, we are done. So suppose that one of a_1 or a_2—say, a_1—is not irreducible. Then there exist $a_{1,1}, a_{1,2} \in D$, with neither $a_{1,1}$ nor $a_{1,2}$ a unit, such that $a_1 = a_{1,1} a_{1,2}$. Continue factoring in this manner for as long as possible—that is, continue to factor the factors of a, and their factors, and so on. This process will terminate if and only if a can be written as a finite product of irreducible elements. Now suppose the process does not terminate. That is, suppose there is a sequence $a_1, a_{1,1}, a_{1,1,1}, \ldots$ of nonunit elements of D such that

$$\cdots \mid a_{1,1,1} \mid a_{1,1} \mid a_1 \mid a.$$

Then we would have an infinite chain of ideals

$$\langle a \rangle \subset \langle a_1 \rangle \subset \langle a_{1,1} \rangle \subset \langle a_{1,1,1} \rangle \subset \cdots,$$

each of which is properly contained in the next. (The proof that these containments are proper is left to you in Activity 14.35.) But this is a contradiction to the ascending chain condition. Therefore, it must be that a can be written as a finite product of irreducible elements of D, as desired. ∎

Now that we have established the existence of factorizations into irreducibles in any principal ideal domain, the next step in proving Theorem 14.25 is to show that these factorizations are unique. As in the integers, we will need a result analogous to Euclid's Lemma. We will prove such a result by first showing that the notions of prime and irreducible are equivalent in every principal ideal domain.

Theorem 14.31. *Let D be a principal ideal domain, and let $a \in D$. Then a is prime if and only if a is irreducible.*

Proof. Let D be a principal ideal domain, and let $a \in D$. By Theorem 14.9, if a is prime, then a is irreducible. Thus, it remains to show that if a is irreducible, then a is prime. Let $a \in D$ be irreducible. First, note that a is prime if and only if $\langle a \rangle$ is a prime ideal. Furthermore, by Theorem 13.35 (see page 199), every maximal ideal is prime. Therefore, we can show that a is prime by showing that $\langle a \rangle$ is maximal. Let I be any ideal containing $\langle a \rangle$. Since D is a principal ideal domain, $I = \langle b \rangle$ for some $b \in D$. Furthermore, since $a \in I$, it must be that $a = bd$ for some $d \in D$. But a is irreducible, and so either b or d must be a unit. If b is a unit, then $I = \langle b \rangle = D$. If d is a unit, then $I = \langle b \rangle = \langle a \rangle$. Therefore, the only ideals containing $\langle a \rangle$ are D and $\langle a \rangle$. It follows that $\langle a \rangle$ is maximal, and so a is prime. ∎

A corollary of Theorem 14.31 provides the analog of Theorem 14.22 for principal ideal domains. In particular:

Corollary 14.32. *Let D be a principal ideal domain, and let $p \in D$ be irreducible. For all a, $b \in D$, if $p \mid ab$, then $p \mid a$ or $p \mid b$.*

Corollary 14.32 can then be used to prove the following analog of Theorem 14.24, which establishes the uniqueness of factorizations into irreducibles in any principal ideal domain. Its proof is completely analogous to the proof of the uniqueness portion of the Fundamental Theorem of Arithmetic from Investigation 4 (see page 15) and is therefore left as an exercise. (See Exercise (5).)

Theorem 14.33. *Let D be a principal ideal domain, and let $x \in D$. Suppose*

$$x = p_1 p_2 \cdots p_m \text{ and } x = q_1 q_2 \cdots q_k$$

for some irreducible elements $p_1, p_2, \ldots, p_m, q_1, q_2, \ldots, q_k$ of D. Then $m = k$, and the factors of x can be reordered so that p_i and q_i are associates for each i.

Taken together, Theorems 14.30 and 14.33 establish that every principal ideal domain is a unique factorization domain. And, as noted earlier, since every Euclidean domain is a principal ideal domain, Theorem 14.14 then follows as a corollary.

Concluding Activities

Activity 14.34. Prove Theorems 14.18 – 14.24. By doing so, you will prove Theorem 14.14.

Activity 14.35. Let D be a principal ideal domain, and let $a \in D$ such that $a \neq 0$ and $a = a_1 a_2$ for some nonunit elements a_1, $a_2 \in D$. Prove that $\langle a \rangle \subset \langle a_1 \rangle \subset D$, where each containment is proper.

Activity 14.36. We have now seen that every principal ideal domain, and hence every Euclidean domain, is a unique factorization domain. So if we want to find an integral domain that is not a unique factorization domain, we will need to consider integral domains that are neither Euclidean domains nor principal ideal domains. How can we look for such integral domains? One way is to use Theorem 14.31.

(a) Use Theorem 14.31 to explain why the integral domain D in Activity 14.11 is not a Euclidean domain.

(b) Prove that D is not a unique factorization domain.

Exercises

(1) In this exercise we will prove that the set $\mathbb{Z}[i]$ of Gaussian integers (see Exercise (2) on page 52) is a Euclidean domain. The referenced exercises show that $\mathbb{Z}[i]$ is an integral domain.

We define $\delta : \mathbb{Z}[i] - \{0\} \to \mathbb{Z}^+ \cup \{0\}$ by

$$\delta(a + bi) = a^2 + b^2.$$

That is,

$$\delta(a + bi) = |a + bi|^2,$$

where $|z|$ is the complex norm of the complex number z. Let $a = x + iy$ and $b = u + vi$ be in $\mathbb{Z}[i]$ with $a \neq 0$. We need to show that there are elements q and r in $\mathbb{Z}[i]$ such that $b = qa + r$ with $r = 0$ or $\delta(r) < \delta(a)$.

(a) Show that $\delta(ab) = \delta(a)\delta(b)$. Use this equality to show that $\delta(a) \leq \delta(ab)$.

(b) Write $\frac{b}{a}$ in the form $c + di$, with c and d in $\mathbb{Q}$. Explicitly identify c and d.

(c) Explain why there exist integers m and n such that

$$|m - c| \leq \frac{1}{2} \text{ and } |n - d| \leq \frac{1}{2}.$$

(d) Let $r = b - qa$, where $q = m + ni$. Then $b = qa + r$. Explain why q and r are in $\mathbb{Z}[i]$.

(e) If $r = 0$, we are done. So assume $r \neq 0$. To complete our exercise, show that $\delta(r) < \delta(a)$. (Hint: We can define δ as a mapping from $\mathbb{C}$ to $\mathbb{R}$. Use this δ to calculate $\delta\left(\frac{b}{a} - q\right)$. Then use the fact that $r = \left(\frac{b}{a} - q\right)a$.)

(2) Let

$$D = \left\{x \in \mathbb{C} \mid x = a + b\sqrt{-6}, \text{for some } a, b, \in \mathbb{Z}\right\}.$$

Note that D is a subset of the field $\mathbb{C}$ of complex numbers.

(a) Prove that D is a subring of $\mathbb{C}$ and that D is an integral domain.

(b) For $x \in D$ with $x = a + b\sqrt{-6}$ and $a, b \in \mathbb{Z}$, the **norm** of x is defined to be the nonnegative integer $N(x)$ where

$$N(x) = a^2 + 6b^2.$$

(i) Prove that for each $x \in D$, $N(x) = 0$ if and only if $x = 0$ and that for each $x \in D$, $N(x) = 1$ if and only if $x = \pm 1$.

(ii) Prove that for all $x, y \in D$, $N(xy) = N(x)N(y)$. [Note: In this case, we say that the norm function N is multiplicative.]

(c) A nonzero element p of D is called **prime** provided that $p \neq \pm 1$ and if, whenever $p = xy$ for some $x, y \in D$, then $x = \pm 1$ or $y = \pm 1$. Which of the following elements of D are prime? Justify all your conclusions.

2	$5 + \sqrt{-6}$	$3 + 2\sqrt{-6}$
10	$10 + 2\sqrt{-6}$	$3 - 2\sqrt{-6}$

(d) Show that in D, there are two different factorizations of 33 into primes.

(3) **Primes and irreducibles in Euclidean domains.** Generalize Theorem 14.7 by proving that in any Euclidean domain, every irreducible element is prime. Deduce that the notions of prime and irreducible are equivalent in Euclidean domains.

* (4) **Uniqueness of greatest common divisors.** Let D be a Euclidean domain, and let $a, b \in D$, not both zero. Prove that if d and d' are both greatest common divisors of a and b, then d and d' are associates.

* (5) Prove Theorem 14.33.

(6) (a) Show that the ring $\mathcal{C}(\mathbb{R})$ (from Activity 14.27) is not an integral domain.

   (b) Since $\mathcal{C}(\mathbb{R})$ is not an integral domain, it follows that $\mathcal{C}(\mathbb{R})$ is not a unique factorization domain or a principal ideal domain. Find, if possible, an ideal in $\mathcal{C}(\mathbb{R})$ that is not a principal ideal.

(7) Prove that every field is a Euclidean domain.

(8) In this investigation, we showed that every principal ideal domain is a unique factorization domain, but does the converse hold? That is, is every unique factorization domain a principal ideal domain? As it turns out, the answer is no, and $\mathbb{Z}[x]$ is a canonical example of a unique factorization domain that is not a principal ideal domain.

   (a) Construct an example to show that $\mathbb{Z}[x]$ is not a principal ideal domain. (That is, find an ideal of $\mathbb{Z}[x]$ that is not principal.)

   (b) Find a proof, either online or in a textbook, that $\mathbb{Z}[x]$ is a unique factorization domain. How do the techniques used in the proof you found compare to those in this investigation, Investigation 1, and Investigation 10?

Connections

In Investigation 1 we proved the Division Algorithm for integers and then used the Division Algorithm to establish several other important results, all leading up to the Fundamental Theorem of Arithmetic. We followed a similar progression for polynomials in Investigations 9 and 10, proving the Division Algorithm for polynomials over a field and ultimately showing that all such polynomials have unique factorizations into irreducibles. Unique factorization in each of these settings is a consequence of the more general result we explored in this investigation—namely, that every Euclidean domain is a unique factorization domain. This result can be proved by either adapting the proof of the Fundamental Theorem of Arithmetic (and the results leading up to it) or—if you studied ideals in Investigation 13—by proving the stronger result that every principal ideal domain is a unique factorization domain.

Investigation 15

From $\mathbb{Z}$ to $\mathbb{C}$

Focus Questions

By the end of this investigation, you should be able to give precise and thorough answers to the questions listed below. You may want to keep these questions in mind to focus your thoughts as you complete the investigation.

- What is the field of fractions of an integral domain, and how is this field constructed?

- What are real numbers, and how can they be constructed from the rational numbers? What special property do the real numbers have that the rational numbers do not?

- How can we formally define the set of complex numbers? What special property do the complex numbers have that the real numbers do not?

- What makes $\mathbb{Z}$ unique among all ordered integral domains?

Preview Activity 15.1. Let $Q = \{(a, b) : a, b \in \mathbb{Z} \text{ and } b \neq 0\}$. Define a relation $\sim$ on Q as follows:

$$(a, b) \sim (c, d) \text{ if } ad = bc.$$

(a) Find three elements in Q that are related to $(1, 2)$.

(b) Find all elements in Q that are related to $(0, 1)$.

(c) Is $\sim$ a reflexive relation? Explain.

(d) Is $\sim$ a symmetric relation? Explain.

(e) Is $\sim$ a transitive relation? Explain.

(f) Is there a natural way to define what it means for an element of Q to be positive, negative, or zero? Explain.

(g) Is there a natural way to define an ordering on Q? Explain.

(h) Compare Q (with the relation $\sim$) to the set $\mathbb{Q}$ of rational numbers. What similarities and differences do you see?

Introduction

Throughout previous investigations, we have studied many different number systems, including the integers, the rational numbers, the real numbers, and the complex numbers. We have defined these number systems with varying degrees of formality, but we have also relied on our intuition to some extent. In this investigation, we will consider some of the formal details that we have glossed over in our prior work. Preview Activity 15.1 gives a hint of what this process will look like. It suggests one way to *formally* define the rational numbers as equivalence classes of ordered pairs of integers. And just as the rational numbers can be constructed from the integers, the real numbers can be constructed from the rational numbers, and the complex numbers from the real numbers. But before we get to any of that, we must start with the most basic of all numbers: the counting numbers.

From $\mathbb{W}$ to $\mathbb{Z}$

Counting techniques have been important for ages.

- Ancient cultures used pebbles or notches on sticks to identify a number.

- The Mayas, Aztecs, and Celts adopted base 20 counting systems (using both fingers and toes).

- The Babylonians "invented" the number 0 and used a counting system with base 60. (To this we owe the division of the hour into 60 minutes and the division of the central angle of a circle into 360 angles of measure 1 degree.)

While many of the counting techniques used by these civilizations, and others, were quite sophisticated, none of them wrestled with the logical underpinnings of number systems. While this lack of formality does not present a big problem from a practical standpoint, from the mathematical point of view, it prohibits a formal development of the mathematical ideas that are based on these number systems. In this sense, much of the mathematics we take for granted rests on the logical foundation of number systems, which, as it turns out, was not developed until the late 1800s.

Giuseppe Peano (1858–1932) was an Italian mathematician whose mathematical works spanned the fields of analysis, differential equations, mathematical logic, and set theory.[*] In 1889, Peano published his famous axioms, now called the Peano axioms, which constructed a formal foundation upon which whole numbers could be defined in terms of sets. The Peano Axioms are as follows (where $\mathbb{W}$ is the set of whole numbers):

I. $0 \in \mathbb{W}$.

II. There is a function S from $\mathbb{W}$ to $\mathbb{W}$ (called the successor function).

III. Suppose $U \subseteq \mathbb{W}$ with $0 \in U$. If $n \in U$ implies $S(n) \in U$, then $U = \mathbb{W}$. (This property is the Principle of Mathematical Induction.)

---

[*]Peano is especially noted for his invention (or discovery) of space filling curves. These are continuous mappings from the closed interval [0,1] onto the unit square in the plane. The existence of such curves has been described as one of the most remarkable facts in set theory.

IV. If $n \in \mathbb{W}$, then $S(n) > 0$.

V. If $n, m \in \mathbb{W}$ and $S(n) = S(m)$, then $n = m$.

In these axioms, the whole number $n + 1$ is the successor of the whole number n, and the successor function S maps each whole number to its successor.

As it turns out, all the structure of the integers can be derived from the Peano axioms.[†] These axioms created a logical foundation for our integer, real, and complex number systems and at one time were considered to be the fountainhead of all mathematical knowledge. From this point forward, we will assume the Peano axioms and the subsequent construction of the ring of integers. More specifically, we will assume the familiar properties of the set of integers $\mathbb{Z}$ under the standard addition and multiplication operations, as described below.

(1) $a + b \in \mathbb{Z}$ for every $a, b \in \mathbb{Z}$. (The set $\mathbb{Z}$ is closed under addition.)

(2) $a + b = b + a$ for every $a, b \in \mathbb{Z}$. (Addition is commutative in $\mathbb{Z}$.)

(3) $(a + b) + c = a + (b + c)$ for all $a, b, c \in \mathbb{Z}$. (Addition is associative in $\mathbb{Z}$.)

(4) $\mathbb{Z}$ contains an element 0 so that $0 + a = a$ for all $a \in \mathbb{Z}$. (0 is an additive identity in $\mathbb{Z}$.)

(5) For each $a \in \mathbb{Z}$ there is an element $-a \in \mathbb{Z}$ so that $a + (-a) = 0$. (Each element in $\mathbb{Z}$ has an additive inverse in $\mathbb{Z}$.) Using the additive inverse, we can define subtraction on $\mathbb{Z}$ by $a - b = a + (-b)$.

(6) $ab \in \mathbb{Z}$ for every $a, b \in \mathbb{Z}$. ($\mathbb{Z}$ is closed under multiplication.)

(7) $(ab)c = a(bc)$ for all $a, b, c \in \mathbb{Z}$. (Multiplication is associative in $\mathbb{Z}$.)

(8) $\mathbb{Z}$ contains an element $1 \neq 0$ so that $1a = a = a1$ for all $a \in \mathbb{Z}$. (1 is a multiplicative identity in $\mathbb{Z}$.)

(9) $a(b + c) = ab + ac$ for all $a, b, c \in \mathbb{Z}$. (Multiplication in $\mathbb{Z}$ distributes over addition in $\mathbb{Z}$.)

(10) There is a subset of $\mathbb{Z}$, denoted $\mathbb{Z}^+$ and called the positive integers, so that for each $a \in \mathbb{Z}$, exactly one of the following is true:

$$a \in \mathbb{Z}^+, \quad a = 0, \quad \text{or} \quad -a \in \mathbb{Z}^+.$$

(This property is called the **trichotomy** principle.) Using the trichotomy principle, we can define a relation $<$ on $\mathbb{Z}$ by

$$a < b \quad \text{if and only if} \quad b - a \in \mathbb{Z}^+.$$

This inequality relation satisfies all the standard properties with which we are familiar. Other inequalities ($>$, $\geq$, and $\leq$) are defined in a similar way.

Note that these axioms mirror those that were introduced in Investigation 1, with the exception of trichotomy, which is stated somewhat differently. As we will see shortly, this variation of trichotomy implies the version that we stated for the less than ($<$) relation on page 6.

The integers are a natural place to begin the study of algebra, in part because the integers are the smallest ring with identity and characteristic zero. (More formally, every ring with identity and

---

[†]The axioms were published in a pamphlet *Arithmetices principia, nova methodo exposita*, which was lauded as a landmark in the history of mathematical logic and the foundations of mathematics.

characteristic zero contains a subring isomorphic to the integers.) In addition, the foundation of algebra is the study of polynomials and solutions to polynomial equations, so to study algebra we might begin by asking what kinds of polynomial equations we can solve within the set $\mathbb{Z}$. We can certainly solve some simple polynomial equations (e.g., $x - 2 = 0$) in $\mathbb{Z}$, but there are many others that we cannot solve. For example, the polynomial equation $2x - 1 = 0$ has no integer solutions. To solve this and other polynomial equations with integer coefficients, we will need to have bigger sets from which to draw solutions. In particular, to solve equations like $ax + b = 0$, where a and b are integers with $a \neq 0$, we will need to find a set that contains the multiplicative inverse of each nonzero integer. To do so, we will formally construct what is called the *field of fractions* of $\mathbb{Z}$, which turns out to be just the rational numbers. Before we do that, however, we will need to take a short digression to talk about ordered rings.

Ordered Rings

The less than ($<$) relation on the set of integers allows us to compare integers so that the entire set of integers can be organized on a number line. In particular, the trichotomy principle implies that any integer can be compared to the integer 0; the integers that are greater than 0 are placed to the right of 0 on the number line and the integers that are less than 0 are placed to the left of 0. To determine how two different integers a and b are placed relative to each other, we just need to know whether $a - b$ is positive or $b - a$ is positive. The placement of every integer is thus determined by the positive integers. As it turns out, this idea generalizes to other rings as well. Whenever a ring contains a set of elements that behaves like the positive integers, it can be made into an *ordered ring*, defined formally as follows:

Definition 15.2. An **ordered ring** is a commutative ring R containing a subset P such that:

(i) P is not empty;

(ii) if $a \in P$ and $b \in P$, then $a + b \in P$ and $ab \in P$; and

(iii) for any $a \in R$, exactly one of following conditions holds:

$$a \in P, \quad a = 0, \quad \text{or} \quad -a \in P.$$

The elements in P are called the *positive* elements of the ordered ring. If $a \in P$, then we will also write $a > 0$. The integers form the standard example of an ordered ring. But the integers are also an integral domain, and this is not a requirement in the definition of an ordered ring. An ordered ring that is also an integral domain is called an **ordered domain**. However, it turns out that an ordered ring cannot have any zero divisors, and so an ordered ring with identity must be an ordered domain. The next theorem formalizes this result.

Theorem 15.3. *An ordered ring contains no zero divisors.*

Proof. Suppose R is an ordered ring, and let $a, b \in R$ with both a and b nonzero. If $a, b > 0$, then $ab > 0$ and so $ab \neq 0$. Suppose one of a or b is positive and the other negative. Without loss of generality, assume $a > 0$ and $b < 0$. Then $-b > 0$ and so $(a)(-b) > 0$. Thus, $-(ab) > 0$, which means that $ab \neq 0$. The final case is when $-a > 0$ and $-b > 0$. Then $ab = (-a)(-b) > 0$, and so $ab \neq 0$. ∎

We know of several ordered rings: $\mathbb{Z}$, $\mathbb{Q}$, and $\mathbb{R}$, for example. As a less familiar example, let R be an ordered ring, and define the set P to be the polynomials in $R[x]$ with positive leading coefficient. Then the ring $R[x]$ with positive elements P is an ordered ring. (See Exercise (9).)

There are some standard properties of elements in ordered rings that we might expect to be true, and a few are given in the next lemma.

Lemma 15.4. *Let R be an ordered ring.*

(i) *If $a \in R$ is not the zero element, then $a^2 > 0$.*

(ii) *If R contains an identity, then $1 > 0$.*

(iii) *If R contains an identity, then $(-1) < 0$.*

The proofs of the second and third parts of Lemma 15.4 are left for you as exercises. (See Exercise (3).) The next activity provides some guidance for proving the first part of the lemma.

Activity 15.5. Let R be an ordered ring, and let $a \in R$. We will show that $a^2 > 0$ by considering cases.

(a) If $a \neq 0$ in R, what else can we say about a? Explain.

(b) If $a > 0$, why must a^2 be positive?

(c) Complete the proof of part 1 of Lemma 15.4 by assuming that $(-a) > 0$ and proving that $a^2 > 0$.

The positive elements in an ordered ring allow us to compare elements to 0, but we know in the integers that we can compare any two elements to each other. For example, we know that $4 > 2$ since $4 - 2 > 0$. We can extend this idea to any ordered ring. If R is an ordered ring and $a, b \in R$, then we know by trichotomy that exactly one of the following must be true: $a - b > 0$, $a - b = 0$, or $-(a - b) > 0$. When $a - b > 0$ we will say that a is greater than b and write $a > b$. In the case where $a - b = 0$, we have that $a = b$. When $-(a - b) > 0$, we say that b is greater than a and write $b > a$. If $b > a$, we will also say that a is less than b and write $a < b$. In this way, we can use the set of positive elements to define a less than relation (an ordering) on the entire ring. There are many familiar properties satisfied by the less than relation $<$ on $\mathbb{Z}$ (and also by the greater than relation). We might expect these properties to hold in any ordered ring, and they in fact do. The next theorem formalizes this result. You may recognize the first four parts from the ordering axioms on page 6 of Investigation 1. Part (v) is a generalization of Exercise (7) from the same investigation. (See page 17.)

Theorem 15.6. *Let R be an ordered ring. The following conditions are satisfied.*

(i) **Trichotomy.** *For all a, $b \in R$, exactly one of the conditions*

$$a < b, \ a = b, \ a > b$$

is satisfied.

(ii) **Transitivity.** *For all a, b, $c \in R$, if $a > b$ and $b > c$, then $a > c$.*

(iii) **Translation invariance.** *For all a, b, $c \in R$, if $a > b$, then $a + c > b + c$.*

(iv) **Scaling.** *For all a, b, $c \in R$, if $a > b$ and $c > 0$, then $ac > bc$.*

(v) *For all a, b, c ∈ R, if ac > bc and c > 0, then a > b.*

Proof. Let $a, b \in R$. By condition (iii) of Definition 15.2, we know that exactly one of $a - b > 0$, $a - b = 0$, or $-(a - b) > 0$ must be true. Now $a - b > 0$ if and only if $a > b$; $a - b = 0$ if and only if $a = b$; and $-(a - b) > 0$ if and only if $b - a > 0$ or, equivalently, $b > a$. This completes the proof of the trichotomy condition.

Now let $a, b, c \in R$. Assume $a > b$ and $b > c$. Then $a - b > 0$ and $b - c > 0$. So $(a-b)+(b-c) > 0$. But this means that $a - c > 0$ or, equivalently, $a > c$, proving transitivity.

For part (iii), assume $a > b$, and let c be any element in R. Then $a - b > 0$. Note that $0 < a - b = (a + c) - (b + c)$, and so $a + c > b + c$.

The proof of the last two parts is left for the exercises. (See Exercise (4).) ∎

Now that we have considered ordered rings, we can return to our problem of constructing the field of rational numbers from the ring of integers.

From $\mathbb{Z}$ to $\mathbb{Q}$

In this section, we will investigate how to formally construct the field of rational numbers from the integers. It turns out that the same construction works for any integral domain, so we will make the argument in that more general context. In particular, we will show that for any integral domain D, we can construct a field that contains an isomorphic copy of D and also contains solutions to all linear equations in $D[x]$. The details of this construction are essentially contained in Preview Activity 15.1. To translate the intuition from that activity to our more general construction, it might be helpful throughout this section to think of the integral domain D as the ring of integers.

Recall that rational numbers are fractions made from pairs of integers. So we can think of a rational number as an ordered pair (a, b), where $a, b \in \mathbb{Z}$ with $b \neq 0$. Unfortunately, these ordered pairs do not give us the rational numbers because, for example, the ordered pairs $(1, 2)$ and $(2, 4)$ are different, but the rational numbers $\frac{1}{2}$ and $\frac{2}{4}$ are equal. To rectify this problem, we define a relation on the set of ordered pairs so that (a, b) is related to (c, d) if $ad = bc$. In this way, equivalent ordered pairs represent "equal" rational numbers. This same idea works in any integral domain.

Let D be an integral domain, and let

$$D' = \{(a, b) : a, b \in D \text{ with } b \neq 0\}.$$

We will define a relation $\sim$ on D' by saying that $(a, b) \sim (c, d)$ in D' if and only if $ad = bc$ in D. As with any relation, it is natural to consider what kind of relation $\sim$ is.

Activity 15.7.

(a) Prove that $\sim$ is a reflexive relation.

(b) Prove that $\sim$ is a symmetric relation.

(c) Prove that $\sim$ is a transitive relation.

(d) What do your answers to parts (a) – (c) allow you to conclude about $\sim$?

Since $\sim$ is an equivalence relation, $\sim$ partitions D' into disjoint equivalence classes. (See Theorem 2.6 on page 27.) We will denote the equivalence class of $(a, b) \in D'$ as $[(a, b)]$. Let $Q(D)$ be the collection of equivalence classes of elements in D'—that is,

$$Q(D) = \{[(a, b)] : a, b \in D \text{ with } b \neq 0\}.$$

Note that $[(a, b)] = [(c, d)]$ in $Q(D)$ if and only if $ad = bc$ in D.

To make $Q(D)$ into a field like $\mathbb{Q}$, we will need to define addition and multiplication operations on $Q(D)$ that mimic addition and multiplication in $\mathbb{Q}$. Recall that the sum $\frac{a}{b} + \frac{c}{d}$ and product $\left(\frac{a}{b}\right)\left(\frac{c}{d}\right)$ of two rational numbers $\frac{a}{b}$ and $\frac{c}{d}$ are defined as

$$\frac{a}{b} + \frac{c}{d} = \frac{ad + bc}{bd} \quad \text{and} \quad \left(\frac{a}{b}\right)\left(\frac{c}{d}\right) = \frac{ac}{bd}.$$

We can easily translate these definitions to $Q(D)$. For $[(a, b)], [(c, d)] \in Q(D)$, define the sum and product by

$$[(a, b)] + [(c, d)] = [(ad + bc, bd)] \quad \text{and} \quad [(a, b)][(c, d)] = [(ac, bd)]. \tag{15.1}$$

Activity 15.8.

(a) Let $D = \mathbb{Z}$. Compute the sums $[(1, 3)] + [(4, 5)]$ and $[(2, 6)] + [(-4, -5)]$ in $Q(\mathbb{Z})$. What do you notice about these sums? Why should we have expected this?

(b) Let $D = \mathbb{Z}$. Compute the products $[(1, 3)][(4, 5)]$ and $[(2, 6)][(-4, -5)]$ in $Q(\mathbb{Z})$. What do you notice about these products? Why should we have expected this?

(c) Describe in symbols what it would mean for addition in $Q(D)$ to be well-defined.

(d) Describe in symbols what it would mean for multiplication in $Q(D)$ to be well-defined.

(e) Verify that addition in $Q(D)$ is well-defined, and explain why it is important to do so.

(f) Verify that multiplication in $Q(D)$ is well-defined

Now that we have well-defined addition and multiplication operations on $Q(D)$, we need to determine if $Q(D)$ is a field.

Activity 15.9. Let D be an integral domain.

(a) What element would you expect to be the additive identity in $Q(D)$? Prove that you have the right element.

(b) What element would you expect to be the additive inverse of the element $[(a, b)]$ in $Q(D)$? Prove that you have the right element.

(c) What element would you expect to be the multiplicative identity in $Q(D)$? Prove that you have the right element.

(d) Now we will determine the units in $Q(D)$.

    (i) In order for an element to be a unit in $Q(D)$, it first has to be nonzero. What can we say about $a \in D$ if the element $[(a, b)]$ is nonzero in $Q(D)$? Prove your answer.

    (ii) If $[(a, b)]$ is nonzero in $Q(D)$, must $[(a, b)]$ be a unit? If so, determine $[(a, b)]^{-1}$, and prove your answer. If not, give an example of a nonzero element of $Q(D)$ that is not a unit.

Activity 15.9 establishes some of the details that we need to prove the next theorem.

Theorem 15.10. *The set $Q(D)$, with the operations in (15.1), is a field.*

We should have learned in Activity 15.9 that $[(0,1)]$ is the additive identity in $Q(D)$, that $[(-a,b)]$ is the additive inverse of the element $[(a,b)]$ in $Q(D)$, that $[(1,1)]$ is the multiplicative identity in $Q(D)$, and that $[(b,a)]$ is the multiplicative inverse of the nonzero element $[(a,b)]$ in $Q(D)$. To complete the proof of Theorem 15.10, we need to know that $Q(D)$ is closed under its operations, addition and multiplication are commutative and associative in $Q(D)$, and that multiplication distributes over addition in $Q(D)$. We will verify that multiplication distributes over addition in $Q(D)$ and leave the proofs of the remaining properties as an exercise. (See Exercise (7).)

To show that multiplication distributes over addition in $Q(D)$, let $[(a,b)]$, $[(c,d)]$, and $[(e,f)]$ be in $Q(D)$. The commutative, associative and distributive laws from D, along with Exercise (6), show that

$$\begin{aligned}
[(a,b)]([(c,d)] + [(e,f)]) &= [(a,b)][(cf+de,df)] \\
&= [(a(cf+de),b(df))] \\
&= [(acf+ade,bdf)] \\
&= [(b(acf+dae),b(bdf))] \\
&= [((ac)(bf)+(bd)(ae),(bd)(bf))] \\
&= [(ac,bd)] + [(ae,bf)] \\
&= [(a,b)][(c,d)] + [(a,b)][(e,f)].
\end{aligned}$$

Thus, left multiplication distributes over addition in $Q(D)$. The commutativity of multiplication in $Q(D)$ shows that right multiplication also distributes over addition in $Q(D)$.

The field $Q(D)$ is called the *field of fractions* or the *quotient field* of the integral domain D. Since we defined the elements of $Q(D)$ to be equivalence classes of *pairs* of elements of D, the ring D is not a subring of $Q(D)$. However, if we let

$$\overline{D} = \left\{ \frac{a}{1} : a \in D \right\},$$

then $\overline{D}$ is a subring of $Q(D)$ that is isomorphic to D. (See Activity 15.24.) So when we say that D is contained in $Q(D)$, we mean that there is a subring of $Q(D)$ that is isomorphic to D in a natural sense. One final note about the field of fractions of an integral domain: any field that contains an integral domain D will have to contain a multiplicative inverse of each of D's elements, so $Q(D)$ is the smallest field that contains a copy of D. From this point forward, we will use the standard fraction notation $\frac{a}{b}$ to represent the element $[(a,b)]$ in $Q(D)$. It is important to remember, though, that the fraction $\frac{a}{b}$ is really an equivalence class of ordered pairs of elements from the integral domain D.

Ordering on $\mathbb{Q}$

Not every integral domain is an ordered ring. (See Exercise (14), for example.) However, the integers are an ordered ring, and so we might wonder if we can extend the ordering on $\mathbb{Z}$ to an ordering on its field of fractions $\mathbb{Q}$. We can expand this question to ask if the field of fractions of an ordered integral domain can always be made into an ordered field. As we address this question in this section, it may again be helpful to think of the ordered domain D as the integers $\mathbb{Z}$ and the field of fractions $Q(D)$ as $\mathbb{Q}$.

Activity 15.11. Let D be an ordered integral domain. For $Q(D)$ to be an ordered field, there needs to be a nonempty subset P of $Q(D)$ (the positive elements) that is closed under addition and multiplication so that for any $a \in Q(D)$, exactly one of $a \in P$, $a = 0$, or $-a \in P$ is true.

(a) What subset of $Q(D)$ should we define as our set P? Note that since we are attempting to define an ordering on $Q(D)$, your definition of P cannot use such an ordering and can depend only on the ordering in D.

(b) Explain why the set P defined in part (a) is nonempty.

(c) Show that the set P from part (a) is closed under both addition and multiplication. (If you cannot prove this, then you may need to rethink how you defined P.)

(d) To prove the last condition on P, let $\frac{a}{b} \in Q(D)$. By the trichotomy principle in D, we know that exactly one of $ab < 0$, $ab = 0$, or $ab > 0$ is satisfied.

  (i) Explain why $\frac{a}{b} > 0$ if $ab > 0$.

  (ii) Explain why $\frac{a}{b} = 0$ if $ab = 0$.

  (iii) Explain why $-\frac{a}{b} > 0$ if $ab < 0$.

The result of Activity 15.11 is the following theorem:

Theorem 15.12. *If D is an ordered integral domain, then $Q(D)$ is an ordered field.*

A specific consequence of Theorem 15.12 is that $\mathbb{Q}$ is an ordered field.

As in every ordered ring, we can use the set P to define an ordering on $Q(D)$. We say that $\frac{a}{b} > \frac{c}{d}$ in $Q(D)$ if $\left(\frac{a}{b} - \frac{c}{d}\right) \in P$. The other inequalities are defined in a similar manner, and from this point on we will treat these inequalities as the old friends they are.

One final question we should ask about the ordering we have defined on $Q(D)$ is whether it really extends the ordering on D. As noted earlier, you will be asked to show in Activity 15.24 that the set $\overline{D} = \left\{\frac{a}{1} : a \in D\right\}$ is a subring of $Q(D)$ isomorphic to D (where the natural isomorphism is given by sending $a \in D$ to $\frac{a}{1}$ in $\overline{D}$). So we want to know that if $a > 0$ in D, it must follow that $\frac{a}{1} > 0$ in $Q(D)$. But we know that $1 > 0$, so if $a > 0$, then it must be that $(a)(1) > 0$ in D. This, however, implies (by part (d) of Activity 15.11) that $\frac{a}{1} > 0$ in $Q(D)$. So the ordering we defined on $Q(D)$ really does extend the ordering on D.

Now that we have the rational numbers, we can find roots of all linear polynomials in $D[x]$. If we want to find solutions to equations like $x^2 - 2 = 0$, however, the rationals are not enough. This leads us to the field of real numbers.

From $\mathbb{Q}$ to $\mathbb{R}$

The real number system was first given a rigorous treatment by Richard Dedekind, who defined irrational numbers as certain types of sets (called *Dedekind cuts*). Dedekind was a German mathematician who lived between 1831 and 1916. His work on the real number system took place in 1858, though it was not published until 1872.

A **Dedekind cut** is a subset α of $\mathbb{Q}$ satisfying:

(1) $\alpha \neq \emptyset$ and $\alpha \neq \mathbb{Q}$;

(2) if $p \in \alpha$, $q \in \mathbb{Q}$ and $q < p$, then $q \in \alpha$; and

(3) if $p \in \alpha$, then $p < r$ for some $r \in \alpha$.

It is important to note that this construction of the real numbers depends on the fact that there is an ordering on the set of rational numbers. In this way, rational numbers can be used to define any irrational number by identifying the irrational number with the set α of all rational numbers less than that irrational number. With some formal mathematics, we can then define the entire set of real numbers and prove all the relevant field properties. The work of Dedekind essentially filled in the holes in the number line not occupied by the rational numbers, thus creating a continuous number line.[‡] After Dedekind, we could all feel much more comfortable with the mathematics of the real number system. Thank you, Richard!

It would take us quite a ways off course to rigorously construct the real numbers from the rationals in this way, and the construction is not really algebraic in nature, so we will content ourselves with a more informal discussion of the real numbers using sequences.

Activity 15.13. In the episode *Wolf in the Fold* of the original Star Trek television series, the Enterprise's computer system was taken over by an alien entity. To force the entity out of the computer system, Mr. Spock directed the computer to calculate the exact value of the number π. As the computer chugged away at the problem, it consumed more and more resources which eventually drove the alien entity from the computer.

In this activity, we will examine some of the characteristics of this number π and, as a result, understand some of the important structure of the real number system. For reference, an approximation of π to 25 decimal places is

$$\pi \approx 3.1415926535897932384626433.$$

(a) As presented in Table 15.1, the decimal representations for some approximations to π are presented along with their equivalent representations as rational numbers. For example, the decimal 3.14 is equal to the rational number $\frac{314}{100}$. Complete Table 15.1.

(b) Explain what happens to the numbers in the last two columns of the completed Table 15.1 as we allow the number, n, of digits in the decimal approximation to increase without bound. You may use whatever language from calculus you need. Be very specific.

(c) (i) Explain how we could construct a sequence $\{p_n\}$ of rational numbers whose values get as close to the value of π as we want.

(ii) Suppose our universe of numbers consisted **only** of rational numbers. Explain why, in this universe, the sequence $\{p_n\}$ would **not** have a limit.

[‡]Dedekind describes the essence of continuity in the principle: "If all points of the straight line fall into two classes such that every point of the first class lies to the left of every point of the second class, then there exists one and only one point which produces this division of all points into two classes, this severing of the straight line into two portions" (from *Stetigkeit und irrationale Zahlen (Continuity and Irrational Numbers)*, which can be found at various sources online as part of *Essays on the Theory of Numbers*).

Table 15.1

Approaching π.

Number of digits in decimal expansion of π	Decimal expansion	Fractional form of decimal expansion
2	3.14	$\frac{314}{100}$
3	3.141	$\frac{3141}{1000}$
4		
5		
6		
7		
8		
9		
10		

(iii) It is important to note that computers are only able to calculate with finite decimal expansions—that is, with rational numbers. How would you explain to a fellow student why the Enterprise computer would be unable to complete the task it was given by Mr. Spock (calculating the exact value of π)?

The property of sequences described in Activity 15.13 illustrates a very important property that the set of rational numbers lacks. There are convergent sequences of rational numbers whose entries appear to approach a fixed number that is **not rational**. For this reason, we say that the set of rational numbers is not *complete*. The objects that are the limits of convergent sequences of rational numbers that are not themselves rational numbers are called *irrational numbers*. Since the rational numbers all have terminating or repeating decimal expansions (see Exercise (18)), the irrational numbers are those numbers (like π) with nonterminating, nonrepeating decimal expansions.

Activity 15.14.

(a) Explain how any irrational number can be represented as a limit of a sequence of rational numbers. Illustrate your answer by finding the first six entries in a sequence of rational numbers whose limit is $\sqrt{2}$.

(b) Explain how any rational number can be represented as a limit of a sequence of rational numbers. Illustrate your answer by finding the first six entries in a sequence of rational numbers whose limit is $\frac{2}{5}$.

(c) Let $\mathbb{R}$ be the set of all limits of convergent sequences of rational numbers. Explain as best you can in your own words how this set $\mathbb{R}$ represents the set of all real numbers. Include in your explanation a discussion of how the set of real numbers is fundamentally different than the set of rational numbers.

So we have informally defined the set of real numbers as the collection of all limits of convergent sequences of rational numbers. This is not a rigorous definition, so we won't attempt to formally define addition and multiplication on $\mathbb{R}$. Instead, we will simply accept these operations (and their familiar properties) as we have used them in the past. Note that the set of constant sequences of rational numbers in $\mathbb{R}$ is a subset of $\mathbb{R}$ that is isomorphic to $\mathbb{Q}$, so the reals contain an isomorphic copy of the rationals, although we usually just say that $\mathbb{Q}$ itself is a subfield of $\mathbb{R}$.

We have seen that $\mathbb{Z}$ and $\mathbb{Q}$ are ordered rings, and the same is true of $\mathbb{R}$. The set $\mathbb{R}$ contains a subset of positive reals that satisfies the conditions given in Definition 15.2, which in turn defines the familiar relation $>$ that makes $\mathbb{R}$ into an ordered field. The result of Activity 15.14 is that the field of real numbers is a complete ordered field. This makes $\mathbb{R}$ a very nice field indeed, but $\mathbb{R}$ is still lacking in an important algebraic way. Recall that we began our construction of $\mathbb{Q}$ and $\mathbb{R}$ in order to solve polynomial equations. While we can solve all linear equations in $\mathbb{R}$, and we can even solve other equations like $x^3 - 2 = 0$ (which we couldn't solve in $\mathbb{Q}$), there are still some very simple equations—such as $x^2 + 1 = 0$—that do not have solutions in $\mathbb{R}$. To solve these kinds of equations, we will need the complex numbers, which we define formally in the next section.

From $\mathbb{R}$ to $\mathbb{C}$

Complex numbers are often introduced as a tool to solve the quadratic equation $x^2 + 1 = 0$. However, that is not how complex numbers first came to light. The story actually involves solutions to the general cubic equation, and there are many sources that discuss this interesting history.[§]

A complex number (which we usually write in the form $a + bi$) can be identified with a pair (a, b) of real numbers. So we can define the set of complex numbers $\mathbb{C}$ as

$$\mathbb{C} = \{(a, b) : a, b \in \mathbb{R}\}.$$

(Note that, unlike the construction of $\mathbb{Q}$ from $\mathbb{Z}$, we don't need to define an additional equivalence relation on this set since two complex numbers $a + bi$ and $c + di$ are equal exactly when $a = c$ and $b = d$.) To work with complex numbers, we will need to be able to add and multiply them. Fortunately, the familiar operations from our prior discussions of complex numbers carry over nicely to the formal definition of $\mathbb{C}$ (as a collection of ordered pairs of real numbers).

Activity 15.15.

(a) Explain how the operations

$$(a, b) + (c, d) = (a + c, b + d) \quad (a, b)(c, d) = (ac - bd, ad + bc) \qquad (15.2)$$

on $\mathbb{C}$ mimic the familiar sum and product of complex numbers as you know them.

(b) Explain why the operations defined in (15.2) are well-defined.

As we might expect, the set $\mathbb{C}$, with the operations defined above, forms the familiar field of complex numbers.

Theorem 15.16. *The set $\mathbb{C} = \{(a, b) : a, b \in \mathbb{R}\}$, with addition and multiplication defined as*

$$(a, b) + (c, d) = (a + c, b + d) \quad \text{and} \quad (a, b)(c, d) = (ac - bd, ad + bc),$$

is a field.

[§]See, for example, *Journey Through Genius, The Great Theorems of Mathematics* by William Dunham, John Wiley & Sons, Inc., 1990.

We will prove certain parts of Theorem 15.16 in the next activity and leave the remainder for the exercises. (See Exercise (8).)

Activity 15.17. Let $\mathbb{C} = \{(a,b) : a,b \in \mathbb{R}\}$ with operations defined as in (15.2).

(a) What element would you expect to be the additive identity in $\mathbb{C}$? Prove that you have the right element.

(b) What element would you expect to be the additive inverse of the element (a,b) in $\mathbb{C}$? Prove that you have the right element.

(c) What element would you expect to be the multiplicative identity in $\mathbb{C}$? Prove that you have the right element.

(d) Now we will determine the units in $\mathbb{C}$. What is the form of a nonzero element in $\mathbb{C}$? If (a,b) is nonzero in $\mathbb{C}$, must (a,b) be a unit? If so, determine $(a,b)^{-1}$, and prove your answer. If not, give an example of a nonzero element of $\mathbb{C}$ that is not a unit.

From this point on, we will use the standard notation $a + bi$ for the complex number (a,b), where $i^2 = -1$ or, equivalently, $i = \sqrt{-1}$. This notation is helpful in that if we treat i as though it has the properties of a real number, then addition and multiplication of complex numbers follow naturally from the properties of the corresponding operations in $\mathbb{R}$.

Since we defined the elements of $\mathbb{C}$ to be ordered pairs of real numbers, the field $\mathbb{R}$ is not a subfield of $\mathbb{C}$. However, if we let

$$\overline{R} = \{a + 0i : a \in \mathbb{R}\},$$

then $\overline{R}$ is a subfield of $\mathbb{C}$ that is isomorphic to $\mathbb{R}$. (See Activity 15.25.) So when we say that $\mathbb{R}$ is contained in $\mathbb{C}$, we mean that there is a subfield of $\mathbb{C}$ that is isomorphic to $\mathbb{R}$ in a natural sense.

Recall that we constructed the rational numbers in order to solve linear equations in $\mathbb{Z}[x]$. Then we needed the real numbers in order to solve more equations—for example, $x^2 - 2 = 0$ in $\mathbb{Q}[x]$— and to have numbers that represent limits of convergent sequences of rational numbers. As we have seen, the complex numbers take us one step farther. In $\mathbb{R}[x]$, there are polynomials that have no roots in $\mathbb{R}$. (The polynomial $x^2 + 1$ is the canonical example.) However, if $f(x) \in \mathbb{C}[x]$ is a nonconstant polynomial, then $f(x)$ must have a root in $\mathbb{C}$, as we saw in the Fundamental Theorem of Algebra (Theorem 11.2 on page 156). This property makes $\mathbb{C}$ what we call an *algebraically closed* field. However, we do give something up when we move from $\mathbb{R}$ to $\mathbb{C}$; in particular, the field $\mathbb{C}$ is no longer an ordered field. (See Exercise (12).) There are some other important attributes of complex numbers that we have seen in previous investigations (like the polar form of a complex number) and will study more in Exercise (2).

A Characterization of the Integers

We will conclude this investigation with a discussion of the uniqueness of the set of integers. As we will see, the Principle of Mathematical Induction (and the equivalent Well-Ordering Principle) makes $\mathbb{Z}$ unique among all ordered integral domains. To begin, we need to know what it means for an ordered ring to be well-ordered.

Definition 15.18. Let S be a nonempty subset of an ordered ring R. An element a in S is called a **smallest element of** S provided that $a \leq x$ for each $x \in S$.

Note that if S contains a smallest element, then that smallest element is unique. (See Exercise (16).) In the set of integers, we are used to the fact that 1 is the smallest positive integer. That leads us to ask if 1 is always the smallest element in any ordered integral domain, provided its set of positive elements contains a smallest element. We will investigate this question in the next activity.

Activity 15.19. Let D be an ordered integral domain with D^+ as its set of positive elements. Assume D^+ contains a smallest element s. To prove that $s = 1$, we will proceed by contradiction and assume $s \neq 1$.

(a) Recall that 1 is positive in D. Explain why it must be the case that $0 < 1$ and $s < 1$.

(b) Explain why $0 < s^2$ and $s^2 < s$.

(c) Explain why $s^2 < s$ provides us with the contradiction we need to conclude that $s = 1$.

We can now define what it means for a set to be *well-ordered*.

Definition 15.20. A subset S of an ordered ring R is said to be **well-ordered** provided that every nonempty subset of S contains a smallest element.

The Principle of Mathematical Induction is equivalent to the set of positive integers being well-ordered (proofs of which can be found in the online supplemental material). Since we are assuming the Principle of Mathematical Induction through the Peano axioms, we also take the following as an axiom.

Axiom 15.21 (The Well-Ordering Principle). *The set of natural numbers is a well-ordered subset of the integers.*

The Well-Ordering Principle is not a trivial assumption, since there are many familiar sets that are not well-ordered.

Activity 15.22.

(a) Is the set of positive rational numbers a well-ordered subset of $\mathbb{Q}$? Explain.

(b) Is the set of positive real numbers a well-ordered subset of $\mathbb{R}$? Explain.

As Activity 15.22 demonstrates, not every ordered ring has a well-ordered subset of positive elements. In fact, it is exactly this property that characterizes the integers, as the next theorem demonstrates.

Theorem 15.23. *Let D be an ordered integral domain in which the set of positive elements D^+ is well-ordered. Then D is isomorphic to $\mathbb{Z}$.*

Proof. Let D be an ordered integral domain in which the set of positive elements D^+ is well-ordered. To prove that D is isomorphic to $\mathbb{Z}$, we will exhibit an isomorphism $\varphi : \mathbb{Z} \to D$. Since every integer is a multiple of 1, we can define φ from $\mathbb{Z}$ into D by

$$\varphi(n) = n \cdot 1_D,$$

where 1_D is the identity in D.

First, we will prove that φ is a homomorphism. Let m and n be integers. Using properties from Theorem 5.5, we have the following:

$$
\begin{aligned}
\varphi(m+n) &= (m+n) \cdot 1_D \\
&= m \cdot 1_D + n \cdot 1_D \\
&= \varphi(m) + \varphi(n)
\end{aligned}
\qquad\qquad
\begin{aligned}
\varphi(mn) &= (mn) \cdot 1_D \\
&= (mn) \cdot (1_D 1_D) \\
&= (m 1_D) \cdot (n 1_D) \\
&= \varphi(m) \cdot \varphi(n)
\end{aligned}
$$

Thus, φ is a homomorphism.

To show that φ is an injection, assume that $m, n \in \mathbb{Z}$ with $\varphi(m) = \varphi(n)$. Without loss of generality, we can assume that $m \geq n$. Since $\varphi(m) = \varphi(n)$ it follows that $m \cdot 1_D = n \cdot 1_D$ and

$$(m - n) \cdot 1_D = m \cdot 1_D - n \cdot 1_D = 0_D,$$

where 0_D is the additive identity in D. We know that 1_D is positive, and so if $m-n > 0$, Exercise (5) shows that $(m-n)1_D > 0$, a contradiction to trichotomy. We can thus conclude that $(m-n) = 0$, or that $m = n$. Thus, φ is an injection.

To complete the proof, we must show that φ is a surjection. To prove this result, we need to show that every element of D is of the form $m \cdot 1_D$ for some integer m. We will do so by using cases based on trichotomy.

Case 1. We will first prove that for each $y \in D^+$, there exists an integer m such that $y = m \cdot 1_D$. We will use a proof by contradiction. Assume that there exists an element y in D^+ such that y cannot be written in the form $n \cdot 1_D$ for any integer n. Let S be the set of all such elements—that is,

$$S = \{z \in D^+ \mid z \neq n \cdot 1_D \text{ for every } n \in \mathbb{Z}\}.$$

By the definition of y, we know that $y \in S$ and so S is not empty. Since D^+ is well-ordered, the set S contains a smallest element s. Now 1_D is the smallest element of D^+ and $1_D = 1 \cdot 1_D$, so $1_D \notin S$. Therefore, 1_D is not equal to s, and so $s > 1_D$. It follows that $s - 1_D > 0$, and so $s - 1_D \in D^+$. In addition,

$$s - (s - 1_D) = 1_D,$$

and so $s > (s - 1_D)$. This means that

$$(s - 1_D) \notin S,$$

and so there exists an integer k such that

$$s - 1_D = k \cdot 1_D.$$

But then

$$
\begin{aligned}
(k + 1) \cdot 1_D &= (k \cdot 1_D) + (1 \cdot 1_D) \\
&= (s - 1_D) + 1_D \\
&= s,
\end{aligned}
$$

and so $s \notin S$, a contradiction. It follows that every element in D^+ is of the form $m \cdot 1_D$ for some $m \in \mathbb{Z}$.

Case 2. If $y = 0_D$, then $y = 0 \cdot 1_D$.

Case 3. Now assume that $-y \in D^+$. By our previous work, there exists an integer k such that

$$-y = k \cdot 1_D.$$

But then

$$
\begin{aligned}
(-k) \cdot 1_D &= -(k \cdot 1_D) \\
&= -(-y) \\
&= y,
\end{aligned}
$$

and so y has the form $(-k) \cdot 1_D$.

We have shown that for each $y \in D$, there exists an integer m such that $y = m \cdot 1_D$. It follows that if $y \in D$ and $y = m \cdot 1_D$, then $\varphi(m) = m \cdot 1_D = y$, which proves that φ is a surjection. We can therefore conclude that φ is an isomorphism, and so $\mathbb{Z} \cong D$. ∎

In conclusion, $\mathbb{Z}$ is the only ordered integral domain with a well-ordered set of positive elements. So the Principle of Mathematical Induction (or the Well-Ordering Principle) is a determining property of $\mathbb{Z}$.

Concluding Activities

Activity 15.24. Let D be an integral domain, and let $\overline{D} = \left\{ \frac{a}{1} : a \in D \right\}$ in $Q(D)$.

(a) Prove that $\overline{D}$ is a subring of $Q(D)$.

(b) Prove that D is isomorphic to $\overline{D}$.

Activity 15.25. Let $\overline{R}$ be the subset of $\mathbb{C}$ defined by $\overline{R} = \{a + 0i : a \in \mathbb{R}\}$.

(a) Prove that $\overline{R}$ is a subfield of $\mathbb{C}$.

(b) Prove that $\mathbb{R}$ is isomorphic to $\overline{R}$.

Exercises

(1) Why must D be an integral domain instead of a commutative ring with identity in order to construct the field of quotients F? Be as thorough as possible in your answer. Basically, this means that you should determine what steps in the construction of the field of quotients used the assumption that D is an integral domain.

* (2) **Properties of complex numbers.** Let $z = a + bi$ be a complex number.

    i.   We define the real part of z to be the real number a.

    ii. We define the imaginary part of z to be the real number b.

iii. The complex conjugate $\bar{z}$ of z is the complex number $a - bi$.

iv. The modulus (or norm, or absolute value) of z, denoted $|z|$, is the real number $\sqrt{a^2 + b^2}$.

(a) Let $w = 2 + 3i$ and $z = -1 + 5i$.

(i) Find $\bar{w}$ and $\bar{z}$.

(ii) Compute $|w|$ and $|z|$.

(iii) Compute $w\bar{w}$ and $z\bar{z}$.

(b) Let z be an arbitrary complex number. There is a relationship between $|z|$, z, and $\bar{z}$. Find and prove this relationship.

(c) What is $\bar{z}$ if $z \in \mathbb{R}$?

$\star$ (3) Let R be an ordered ring with identity. Prove that $1 > 0$ and $(-1) < 0$ in R.

$\star$ (4) Let R be an ordered ring.

(a) Prove that for all a, b, and c in R, if $a > b$ and $c > 0$, then $ac > bc$.

(b) Prove that for all a, b, and c in R, if $ac > bc$ and $c > 0$, then $a > b$.

$\star$ (5) Prove that if x is a positive element of an ordered ring and $n \in \mathbb{Z}$, then $nx > 0$ in R if and only if $n > 0$ in $\mathbb{Z}$. (Hint: Recall how the element nx is defined, and then use induction.)

$\star$ (6) One important property of rational numbers is that if $\frac{a}{b}$ is a rational number and m is any nonzero integer, then $\frac{a}{b} = \frac{am}{bm}$. We use this property often, and so it is natural to ask if it holds in the field of fractions of *any* arbitrary integral domain. Let D be an integral domain. The equivalent formulation of this statement about rational numbers in the context of the set $Q(D)$ is the following:

If $[(x, y)] \in Q(D)$ and m is a nonzero element in D, then $[(x, y)] = [(mx, my)]$.

Prove this statement. (Hint: Consider what we need to do to show that $[(x, y)] = [(mx, my)]$ in $Q(D)$.)

$\star$ (7) Complete the proof of Theorem 15.10 by verifying the properties below. Throughout, let D be an integral domain.

(a) Show that $Q(D)$ is closed under addition and multiplication.

(b) Prove addition is commutative and associative in $Q(D)$.

(c) Prove that multiplication is commutative and associative in $Q(D)$.

$\star$ (8) Let $\mathbb{C} = \{(a, b) : a, b \in \mathbb{R}\}$ with operations defined as in (15.2). Complete the proof of Theorem 15.16 by carrying out the following steps:

(a) Explain why $\mathbb{C}$ is closed under addition and multiplication.

(b) Prove that addition and multiplication are commutative in $\mathbb{C}$.

(c) Prove that multiplication distributes over addition in $\mathbb{C}$.

$\star$ (9) (a) Let R be an ordered ring, and define a relation $>$ on $R[x]$ where $f(x) > g(x)$ if $f(x) - g(x)$ has a positive leading coefficient. Show that $R[x]$ is an ordered ring with this relation.

(b) Is the set of positive elements (the polynomials with positive leading coefficient) a well-ordered subset of $\mathbb{Z}[x]$? Explain.

(10) **Rational functions.** If R is an integral domain, then $R[x]$ is also an integral domain. Describe as best you can the elements of $Q(R[x])$, the field of quotients of $R[x]$. When $R = F$ is a field, this field of quotients of $F[x]$ is usually denoted as $F(x)$ and is called the **field of rational functions** over F.

(11) Let $\mathbb{Q}(i)$ be the subset of $\mathbb{C}$ defined by $\mathbb{Q}(i) = \{r + si : r, s \in \mathbb{Q}\}$.

   (a) Show that $\mathbb{Q}(i)$ is a subfield of $\mathbb{C}$.

   (b) What is the specific relationship between $\mathbb{Q}(i)$ and the ring of Gaussian integers introduced in Exercise (2) of Investigation 3? (See page 52.) Explain.

* (12) Can we define an ordering on $\mathbb{C}$ to make $\mathbb{C}$ an ordered ring? Explain.

(13) We are used to having exactly one set of positive elements in $\mathbb{Z}$, $\mathbb{Q}$, and $\mathbb{R}$. But is it possible that an ordered ring could contain more than one set of positive elements?

   (a) Let r be a real number, and let

   $$P_r = \{f(x) \in \mathbb{Z}[x] : f(r) > 0\}.$$

   (Note that the inequality $f(r) > 0$ takes place in R.)

      (i) Show that P_r satisfies the conditions of a set of positive elements in $\mathbb{Z}[x]$. You may use all the properties of the standard ordering on $\mathbb{R}$.

      (ii) Show that the sets P_r are not all the same. Conclude that an ordered ring may be ordered with more than one choice of a set of positive elements.

   (b) How many subsets of positive elements can $\mathbb{Z}$ contain? Prove your answer.

* (14) **The characteristic of an ordered ring.** Let R be an ordered ring.

   (a) What are the possibilities for the characteristic of R? Prove your answer.

   (b) Is there a prime p for which $\mathbb{Z}_p$ is an ordered ring? Explain.

(15) Let R be an ordered ring. Under what conditions on a and b in R is $ab > 0$? Under what conditions is $ab < 0$? Prove your answers.

* (16) Let S be a subset of an ordered ring R such that S contains a smallest element. Show that this smallest element is unique.

(17) Let R be any ordered ring. For all $a \in R$, we can define the **absolute value** of a as

$$|a| = \begin{cases} a, & \text{if } a \geq 0 \\ -a, & \text{otherwise.} \end{cases}$$

Prove that this absolute value function satisfies the following properties for any $a, b \in R$.

   (a) $|ab| = |a|\,|b|$

   (b) $|a| \geq 0$ with $|a| = 0$ if and only if $a = 0$.

   (c) $|-a| = |a|$

   (d) $|a + b| \leq (|a| + |b|)$

* (18) **Decimal expansions of real numbers.** A real number x has a decimal expansion

$$x = N + \sum_{i \geq 1} 10^{-i} x_i,$$

where N is an integer and each x_i is an integer between 0 and 9. We will also write this decimal expansion of x as

$$x = N.x_1 x_2 x_3 \ldots .$$

The decimal expansion for x *terminates* if $x_i = 0$ for all i larger than some integer m. The decimal expansion for x is *repeating* or *periodic* if there are integers m and k so that $x_{k+(sm+t)} = x_{k+t}$ for all $s \geq 1$ and $1 \leq t \leq m$. In this case, we write

$$x = N.x_1 x_2 \ldots x_{k-1} \overline{x_k x_{k+1} \ldots x_{k+m}}.$$

In this exercise, we will show that a real number x is a rational number if and only if the decimal expansion for x terminates or is periodic.

(a) We will first consider the implication that every rational number has a terminating or repeating decimal expansion.

    (i) Find the decimal representations of $\frac{81}{500}$ and $\frac{5}{11}$. Explain your process.

    (ii) Prove the forward implication by showing that every rational number has a terminating or repeating decimal expansion.

(b) Next, we will consider the implication that every terminating or periodic decimal is a rational number.

    (i) Express each of the following decimals as rational numbers: 0.213 and $0.42\overline{123}$. Explain your process. (Hint: Use the formula for the sum of a geometric series for the latter.)

    (ii) Show that any real number with a terminating or repeating decimal expansion is a rational number.

Connections

In this investigation, we began with the ring of integers, as introduced in Investigation 1. The desire to solve polynomial equations led us to integral domains and fields (namely the field of fractions of an integral domain, as well as the fields $\mathbb{R}$ and $\mathbb{C}$), as introduced in Investigation 4. In each construction we saw that the previous structure could be identified with an isomorphic copy (as in Investigation 7) of the new structure. The constructions of new number systems culminated with an algebraically closed field—namely, the field of complex numbers $\mathbb{C}$. However, this field $\mathbb{C}$ contains much more than just solutions to polynomial equations over $\mathbb{R}$; it also contains algebraic, transcendental, and constructible numbers, as discussed in Investigation 12.

Part V

Groups

Part V

Groups

Investigation 16

Symmetry

Focus Questions

By the end of this investigation, you should be able to give precise and thorough answers to the questions listed below. You may want to keep these questions in mind to focus your thoughts as you complete the investigation.

- What is a rigid motion, and what is a symmetry? How are the two related and how are they different?

- How can permutation notation be used to describe the symmetries of a regular polygon?

- What operation can be performed on a collection of symmetries? What structure does the resulting set of symmetries have?

Preview Activity 16.1. The four-petal flower at right (you may recognize this figure as the graph of $r = \cos(2\theta)$ in polar coordinates) exhibits several kinds of symmetry. Identify all the ways in which the figure is symmetric, and describe the symmetries as best you can.

Introduction

Symmetry is a basic element of design. We also see symmetry in nature—for instance, in crystals, plants, and animals. Throughout history, different cultures have embraced the use of symmetry in their architecture and art.

The major topic in this part of the text is group theory. In a sense, group theory can be thought of as the study of symmetry, and so that is where we will begin. In subsequent investigations, we will define groups, examine a large variety of different groups, and investigate the structure of groups.

In a translation, everything is moved the same amount.	
In a rotation, there is a point (called the *rotocenter* or *center of rotation*) around which everything is spun by a fixed amount (called the *rotation angle*).	
In a reflection, there is a line (called the *axis of reflection*), and the reflection consists of the mirror images of all points across this axis.	
A glide reflection consists of a reflection, followed by a translation parallel to the axis of reflection.	

Figure 16.1
Rigid motions in the plane.

Symmetries

To introduce the idea of a group, we will first study the symmetries of certain objects. We will see that we can define an operation on the symmetries of a given object to form a set that has an interesting structure.

In order to talk about symmetries, we must first discuss rigid motions in the plane. Rigid motions preserve the shape of an object, but not necessarily its location or orientation. In other words, after performing a rigid motion on an object, the resulting object must be congruent to the original, but may be in a different position. A formal definition of a rigid motion is as follows:

Definition 16.2. A **rigid motion** in the plane is a bijective function $f : \mathbb{R}^2 \to \mathbb{R}^2$ such that, for all $x, y \in \mathbb{R}^2$, the distance between $f(x)$ and $f(y)$ is the same as the distance between x and y.

In other words, a rigid motion is a distance-preserving function, or an *isometry*. For example, if we simply move an object from one location to another, we have performed an isometry, as we have not altered the distances between points in that object. It turns out (though we won't prove it) that there are four rigid motions in the plane: translation, rotation, reflection, and glide reflection. These rigid motions are illustrated on the dancing man in Figure 16.1.

For our purposes, we will only be concerned with certain types of rigid motions—namely, symmetries. For the purposes of the next definition, a *geometric object* is simply any subset of $\mathbb{R}^2$. (For example, the octagon in Figure 16.2 is a geometric object.)

Definition 16.3. A **symmetry** of a geometric object O is a rigid motion f so that $f(O) = O$.

Note that every symmetry is either a rotation or a reflection, since translations and glide reflections change the location of the object and are therefore not symmetries. To visualize some of the symmetries of the octagon, we can label its vertices with the numbers $1, 2, \ldots, 8$, as shown in Figure 16.2. (Note that the figure gives examples, but does not include all the symmetries of the octagon.)

Symmetries of Regular Polygons

To more easily identify the symmetries of regular polygons (like the octagon), we will introduce some special notation. Notice that we can completely identify a symmetry by what it does to each vertex. For example, the reflection r_V permutes the vertices of the octagon as follows:

- vertex 1 is left alone;

- vertex 2 is moved to the original position of vertex 8;

- vertex 3 is moved to the original position of vertex 7;

- vertex 4 is moved to the original position of vertex 6;

- vertex 5 is left alone;

- vertex 6 is moved to the original position of vertex 4;

- vertex 7 is moved to the original position of vertex 3; and

- vertex 8 is moved to the original position of vertex 2.

Notice that r_V permutes the vertices of the octagon in a particular way. We can describe this permutation concisely by constructing an array in which the original vertices are listed in the top row, and their corresponding images are listed in the second row. Doing so, we can represent r_V as follows:

$$r_V = \begin{pmatrix} 1 & 2 & 3 & 4 & 5 & 6 & 7 & 8 \\ 1 & 8 & 7 & 6 & 5 & 4 & 3 & 2 \end{pmatrix}.$$

Using the same notation to describe I, r_H, and R_{45}, we obtain:

$$I = \begin{pmatrix} 1 & 2 & 3 & 4 & 5 & 6 & 7 & 8 \\ 1 & 2 & 3 & 4 & 5 & 6 & 7 & 8 \end{pmatrix}, \quad r_H = \begin{pmatrix} 1 & 2 & 3 & 4 & 5 & 6 & 7 & 8 \\ 5 & 4 & 3 & 2 & 1 & 8 & 7 & 6 \end{pmatrix}, \text{ and}$$

$$R_{45} = \begin{pmatrix} 1 & 2 & 3 & 4 & 5 & 6 & 7 & 8 \\ 2 & 3 & 4 & 5 & 6 & 7 & 8 & 1 \end{pmatrix}.$$

A permutation is really a function from a set to itself (the set of vertices of an octagon in our example), so we can combine permutations through composition. For example, $R_{45}(4) = 5$ and

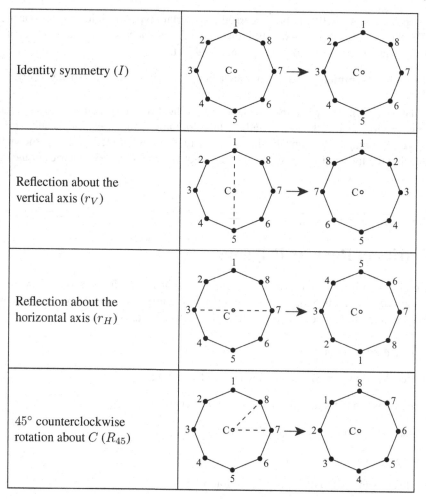

Identity symmetry (I)	
Reflection about the vertical axis (r_V)	
Reflection about the horizontal axis (r_H)	
$45°$ counterclockwise rotation about C (R_{45})	

Figure 16.2
Symmetries of an octagon.

$r_H(5) = 1$, so in the composite $r_H \circ R_{45}$ we have $(r_H \circ R_{45})(4) = r_H(R_{45}(4)) = r_H(5) = 1$. Continuing with the other vertices gives

$$r_H \circ R_{45} = \begin{pmatrix} 1 & 2 & 3 & 4 & 5 & 6 & 7 & 8 \\ 4 & 3 & 2 & 1 & 8 & 7 & 6 & 5 \end{pmatrix}.$$

Activity 16.4. Is $r_H \circ R_{45}$ a symmetry of the regular octagon? Why or why not?

For any polygon, choose one vertex of the polygon to label as vertex 1. Label the remaining vertices in order, proceeding counterclockwise from vertex 1 as shown in Figure 16.3. In addition:

- let r_i denote the reflection about the line through the origin and the vertex i;

- let $\bar{r}_i$ denote the reflection about the perpendicular to the segment joining adjacent vertices i and $i+1$ through the midpoint of that segment; and

- let R_k denote a counterclockwise rotation of the polygon about its center by an angle of $\left(\frac{360}{n}\right) k$ degrees. (Note that the identity symmetry, I, is equal to R_0.)

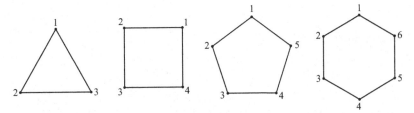

Figure 16.3
Equilateral triangle, square, regular pentagon, and regular hexagon.

As an example, the symmetries of the square (with vertices labeled as shown in Figure 16.3) can be represented as follows:

$$I = \begin{pmatrix} 1 & 2 & 3 & 4 \\ 1 & 2 & 3 & 4 \end{pmatrix} \quad r_1 = \begin{pmatrix} 1 & 2 & 3 & 4 \\ 1 & 4 & 3 & 2 \end{pmatrix}$$

$$r_2 = \begin{pmatrix} 1 & 2 & 3 & 4 \\ 3 & 2 & 1 & 4 \end{pmatrix} \quad \overline{r}_1 = \begin{pmatrix} 1 & 2 & 3 & 4 \\ 2 & 1 & 4 & 3 \end{pmatrix}$$

$$\overline{r}_2 = \begin{pmatrix} 1 & 2 & 3 & 4 \\ 4 & 3 & 2 & 1 \end{pmatrix} \quad R_1 = \begin{pmatrix} 1 & 2 & 3 & 4 \\ 2 & 3 & 4 & 1 \end{pmatrix}$$

$$R_2 = \begin{pmatrix} 1 & 2 & 3 & 4 \\ 3 & 4 & 1 & 2 \end{pmatrix} \quad R_3 = \begin{pmatrix} 1 & 2 & 3 & 4 \\ 4 & 1 & 2 & 3 \end{pmatrix}$$

The operation table (with the operation of function composition) for the set of symmetries of a square is shown in Table 16.1.

In the next activity, we will study the common properties that all sets of symmetries share. In particular, we will determine the symmetries of and create the operation tables for an equilateral triangle, a regular pentagon, and a regular hexagon.

Table 16.1
Symmetries of a square.

$\circ$	I	$\overline{r}_1$	$\overline{r}_2$	r_1	r_2	R_1	R_2	R_3
I	I	$\overline{r}_1$	$\overline{r}_2$	r_1	r_2	R_1	R_2	R_3
$\overline{r}_1$	$\overline{r}_1$	I	R_2	R_1	R_3	r_1	$\overline{r}_2$	r_2
$\overline{r}_2$	$\overline{r}_2$	R_2	I	R_3	R_1	r_2	$\overline{r}_1$	r_1
r_1	r_1	R_3	R_1	I	R_2	$\overline{r}_2$	r_2	$\overline{r}_1$
r_2	r_2	R_1	R_3	R_2	I	$\overline{r}_1$	r_1	$\overline{r}_2$
R_1	R_1	r_2	r_1	$\overline{r}_1$	$\overline{r}_2$	R_2	R_3	I
R_2	R_2	$\overline{r}_2$	$\overline{r}_1$	r_2	r_1	R_3	I	R_1
R_3	R_3	r_1	r_2	$\overline{r}_2$	$\overline{r}_1$	I	R_1	R_2

Table 16.2

Symmetries of an equilateral triangle.

∘	I	r_1	r_2	r_3	R_1	R_2
I	I	r_1	r_2	r_3	R_1	R_2
r_1	r_1		R_1	R_2		r_3
r_2	r_2	R_2	I			r_1
r_3	r_3		R_2	I		r_2
R_1	R_1	r_3			R_2	I
R_2	R_2	r_2		r_1		R_1

Activity 16.5.

(a) (i) Describe all the symmetries of an equilateral triangle (as labeled in Figure 16.3) using permutation notation and the notation from the preceding example.

   (ii) A partial operation table for the set of symmetries of an equilateral triangle is given in Table 16.2. Complete the table.

(b) (i) Describe all the symmetries of a regular pentagon (as labeled in Figure 16.3) using permutation notation and the notation from the preceding example.

   (ii) A partial operation table for the set of symmetries of a regular pentagon is given in Table 16.3. Complete the table.

(c) (i) Describe all the symmetries of a regular hexagon (as labeled in Figure 16.3) using permutation notation and the notation from the preceding example.

   (ii) A partial operation table for the set of symmetries of a regular hexagon is given in Table 16.4. Complete the table.

Table 16.3

Symmetries of a regular pentagon.

∘	I	r_1	r_2	r_3	r_4	r_5	R_1	R_2	R_3	R_4
I	I	r_1	r_2	r_3	r_4	r_5	R_1	R_2	R_3	R_4
r_1	r_1	I	R_3	R_1	R_4	R_2	r_3	r_5	r_2	r_4
r_2	r_2	R_2	I	R_3	R_1	R_4	r_4	r_1	r_3	r_5
r_3	r_3		R_2	I		R_1	r_5	r_2	r_4	r_1
r_4	r_4	R_1	R_4	R_2	I	R_3	r_1	r_3	r_5	r_2
r_5	r_5	R_3	R_1	R_4	R_2	I	r_2	r_4	r_1	r_3
R_1	R_1		r_5	r_1	r_2		R_2	R_3	R_4	I
R_2	R_2	r_2	r_3	r_4	r_5	r_1	R_3	R_4	I	R_1
R_3	R_3	r_5	r_1	r_2	r_3			I	R_1	R_2
R_4	R_4	r_3	r_4	r_5	r_1	r_2	I	R_1	R_2	R_3

Table 16.4

Symmetries of a regular hexagon.

$\circ$	I	r_1	r_2	r_3	$\bar{r}_1$	$\bar{r}_2$	$\bar{r}_3$	R_1	R_2	R_3	R_4	R_5
I	I	r_1	r_2	r_3	$\bar{r}_1$	$\bar{r}_2$	$\bar{r}_3$	R_1	R_2	R_3	R_4	R_5
r_1	r_1	I	R_4	R_2	R_5	R_3	R_1	$\bar{r}_3$	r_3		r_2	$\bar{r}_1$
r_2	r_2	R_2	I	R_4	R_1	R_5	R_3	$\bar{r}_1$	r_1	$\bar{r}_3$	r_3	$\bar{r}_2$
r_3	r_3	R_4	R_2	I	R_3	R_1		$\bar{r}_2$	r_2		r_1	$\bar{r}_3$
$\bar{r}_1$	$\bar{r}_1$	R_1	R_5	R_3	I	R_4	R_2	r_1	$\bar{r}_3$	r_3	$\bar{r}_2$	r_2
$\bar{r}_2$	$\bar{r}_2$	R_3	R_1	R_5	R_2	I	R_4	r_2	$\bar{r}_1$	r_1	$\bar{r}_3$	r_3
$\bar{r}_3$	$\bar{r}_3$	R_5	R_3	R_1	R_4	R_2	I	r_3	$\bar{r}_2$	r_2	$\bar{r}_1$	r_1
R_1	R_1	$\bar{r}_1$	$\bar{r}_2$	$\bar{r}_3$	r_2	r_3	r_1	R_2	R_3	R_4	R_5	I
R_2	R_2	r_2		r_1	$\bar{r}_2$		$\bar{r}_1$	R_3	R_4	R_5	I	R_1
R_3	R_3	$\bar{r}_2$	$\bar{r}_3$	$\bar{r}_1$	r_3	r_1	r_2	R_4	R_5	I	R_1	R_2
R_4	R_4	r_3		r_2	$\bar{r}_3$	$\bar{r}_1$	$\bar{r}_2$		I	R_1	R_2	R_3
R_5	R_5	$\bar{r}_3$	$\bar{r}_1$	$\bar{r}_2$	r_1	r_2	r_3	I	R_1	R_2	R_3	R_4

(d) What are some properties that all the operation tables from parts (a) – (c) have in common? List as many properties as you can find.

The properties that you identified in Activity 16.5 describe the group structure that we will study in the next investigations. As we proceed, we will discover many other sets that have the same properties, including the sets of symmetries of all the regular polygons.

Concluding Activities

Activity 16.6. Find, via a library or Internet search, an object (building, tiling, painting, sculpture, mosaic, fractal, rug, etc.) that has significant symmetry. Then complete the following.

(a) Identify the object and the source through which you found it. Choose something other than a simple polygon; that is, find an object that is interesting to you and that possesses at least six symmetries, including both rotational and reflective symmetry.

(b) Describe all the symmetries possessed by your object. Choose six symmetries (including at least one nontrivial rotation and one nontrivial reflection), and make a copy of the picture of your object for each symmetry. Find a convenient way to label your object so that you can use permutation notation to represent each symmetry. Then illustrate each symmetry on one of the copies of your picture.

(c) Choose three of the symmetries, and find all compositions of these three symmetries. Is each composition a symmetry of your object? Explain.

Activity 16.7. Let G be the set of symmetries of an object.

(a) Is G closed under the operation of function composition? Prove your answer.

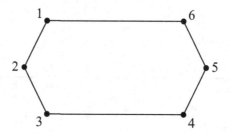

Figure 16.4
The object O.

(b) Is there an identity element for composition in G? If yes, what is it?

(c) Does each element in G have an inverse in G under composition? If yes, what is the inverse of each element?

(d) Is the operation in G associative? If yes, prove it. If no, provide an example to illustrate. (Hint: Is there a more general argument we can use here that involves functions?)

Exercises

(1) Let O be the object shown in Figure 16.4.

    (a) Write the permutation notation for the reflection α of O around the line through the points labeled 2 and 5.

    (b) Write the permutation notation for the reflection β of O around perpendicular bisector of the segment connecting the points labeled 1 and 6.

    (c) Determine $\beta\alpha$ and $\alpha\beta$.

    (d) Write the permutation notation for the 180° counterclockwise rotation γ of O around its center.

    (e) Construct the operation table for the set of symmetries of O.

(2) The permutations

$$\alpha = \begin{pmatrix} 1 & 2 & 3 & 4 & 5 & 6 & 7 & 8 & 9 & 10 \\ 3 & 4 & 5 & 6 & 7 & 8 & 9 & 10 & 1 & 2 \end{pmatrix}$$

and

$$\beta = \begin{pmatrix} 1 & 2 & 3 & 4 & 5 & 6 & 7 & 8 & 9 & 10 \\ 4 & 3 & 2 & 1 & 10 & 9 & 8 & 7 & 6 & 5 \end{pmatrix}$$

are two symmetries of a regular decagon.

    (a) Identify α and β as either rotations or reflections. Explain your choices.

    (b) Find $\alpha\beta$ and $\beta\alpha$, and identify them as reflections or rotations. Explain.

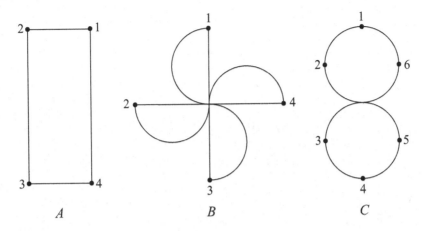

Figure 16.5
Three geometric objects.

      (c) Does α have an inverse? If so, describe the inverse of α geometrically and using permutation notation.

      (d) Does β have an inverse? If so, describe the inverse of β geometrically and using permutation notation.

(3) (a) Find all symmetries of the letter B, and create the operation table for the set of symmetries of B.

      (b) Find all symmetries of the letter T, and create the operation table for the set of symmetries of T.

      (c) Find all symmetries of the letter Z, and create the operation table for the set of symmetries of Z.

      (d) Compare the operation tables for B, T, and Z. Describe all the similarities and differences you observe.

(4) Is composition of symmetries a commutative operation? Prove your answer.

(5) Let A, B, and C be the objects shown in Figure 16.5.

      (a) For each object, find all the symmetries. Describe the symmetries in words and using the permutation notation introduced in this investigation.

      (b) Create the operation table for the set of symmetries of each object.

      (c) Describe the similarities and differences in the operation tables you made in part (b). Your description should include not only obvious attributes like the number of elements, but also how the elements interact within a given set of symmetries.

(6) **Mattress flipping.** Mattress manufacturers always recommend that users periodically flip their mattresses around in order to promote even wear. A flip of a mattress is not necessarily just a flip, but rather a symmetry of the mattress. Consequently, we should be able to describe all possible ways to rearrange a mattress using permutation notation. Label the corners of a rectangular mattress as 1, 2, 3, 4. Find all symmetries of the mattress. Then create the operation table for the set of symmetries of the mattress. What properties does this set of symmetries have?

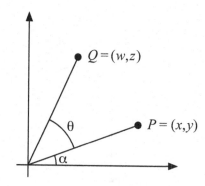

Figure 16.6
A rotation in the plane.

* (7) **Symmetries of a circle.** The figures we have considered in this investigation have all possessed only finitely many symmetries. In contrast, the unit circle (or any circle) has infinitely many rotational and reflective symmetries. We will examine those symmetries in this problem.

(a) We will first consider rotational symmetries. Let θ be an angle. We can explicitly represent the counterclockwise rotation around the origin by the angle θ using linear algebra. Let $P = (x, y) = (\cos(\alpha), \sin(\alpha))$ and $Q = (w, z) = (\cos(\alpha + \theta), \sin(\alpha + \theta))$ be points on the unit circle, as illustrated in Figure 16.6.

   (i) Use suitable trigonometric identities to show that

$$w = \cos(\theta)x - \sin(\theta)y$$
$$z = \sin(\theta)x + \cos(\theta)y.$$

   (ii) Explain why the counterclockwise rotation around the origin by an angle θ can be represented by left multiplication by the matrix $\begin{bmatrix} \cos(\theta) & -\sin(\theta) \\ \sin(\theta) & \cos(\theta) \end{bmatrix}$.

(b) We will now examine the reflective symmetries by finding a matrix that performs a reflection across the line l specified by the parametric equations $x = at$, $y = bt$. We can use the previous result about rotations to complete this problem. Assume the line l makes an angle θ with the positive x-axis and that we want to reflect the point P across l to the point Q, as shown top left in Figure 16.7.

   (i) We will first rotate everything clockwise around the origin by an angle of θ to make the x-axis the axis of reflection, as shown top right in Figure 16.7. Find the matrix M_1 that performs this rotation. Under this rotation, our original point P gets transformed to the point P_{Rot}.

   (ii) Now we will reflect the point P_{Rot} across the x-axis (our transformed line of reflection) to the point Q_{Rot}, as shown bottom left in Figure 16.7. Show that the matrix $M_2 = \begin{bmatrix} 1 & 0 \\ 0 & -1 \end{bmatrix}$ performs this reflection across the x-axis.

   (iii) Finally, we will rotate everything around the origin counterclockwise by an angle of θ to obtain the reflection Q of our original point P across the line l, as shown bottom right in Figure 16.7. Find the matrix M_3 that performs this rotation.

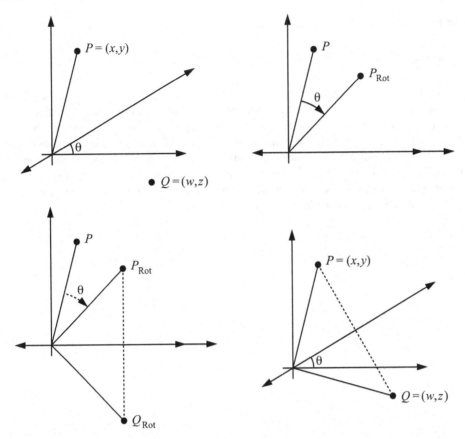

Figure 16.7
Finding a reflection matrix.

(iv) Put parts (i) – (iii) together to find the matrix $R_{(a,b)}$ that performs a reflection across the line with parameterization $x = at$, $y = bt$. Use appropriate trigonometric identities to show that

$$R_{(a,b)} = \begin{bmatrix} \cos(2\theta) & \sin(2\theta) \\ \sin(2\theta) & -\cos(2\theta) \end{bmatrix}.$$

(v) We can also write the matrix $R_{(a,b)}$ in terms of a and b. Show that

$$R_{(a,b)} = \frac{1}{a^2 + b^2} \begin{bmatrix} a^2 - b^2 & 2ab \\ 2ab & b^2 - a^2 \end{bmatrix}.$$

(c) Show that the composition of two rotations is a rotation.

(d) What is the composition of a rotation and a reflection? Explain geometrically and by multiplying appropriate matrices. Does order matter? Why or why not?

(e) What is the composition of two reflections? Explain geometrically and by multiplying appropriate matrices. Does order matter? Why or why not?

Connections

This investigation introduced the concept of symmetries of an object. We saw that the set of symmetries is closed under composition. We know that composition of functions is an associative operation. In addition, using composition as the operation, there is an identity symmetry, and each symmetry has an inverse symmetry. So the set of symmetries of an object has some of the same algebraic structure as a ring under addition, as introduced in Investigation 4. One difference is that composition of symmetries is not a commutative operation, and another is that there is only a single operation on a set of symmetries. As we will see in Investigation 17, the set of symmetries of an object is an example of an algebraic object called a group.

Investigation 17

An Introduction to Groups

Focus Questions

By the end of this investigation, you should be able to give precise and thorough answers to the questions listed below. You may want to keep these questions in mind to focus your thoughts as you complete the investigation.

- What is a group, and what are some examples of groups?

- What is an Abelian group? What are some examples of Abelian and non-Abelian groups?

- What are some basic properties that hold in all groups and that can be proved from the group axioms?

- What uniqueness properties are satisfied by identities and inverses in groups, and why do these properties hold?

- What is the order of a group? What is the difference between finite and infinite groups?

- What is the group of units of a set, and for which types of sets can a group of units be defined?

In Investigation 3, we considered a variety of different number systems, some familiar and some not. While there were significant differences between these number systems, there were also common features that seemed to be shared by all of them. In this investigation, we will focus on some of these common features and their implications.

Preview Activity 17.1. In Investigation 16, we studied the set of symmetries of an object under the operation of function composition. Compare and contrast this set to each of the following sets with the given operation. Which properties do they share, and which do they not share?

(a) $\mathbb{Z}$, the set of integers under addition

(b) $\mathbb{E}$, the set of even integers under addition

(c) $\mathbb{Q}$, the set of rational numbers under addition

(d) $\mathbb{R}^+$, the set of all positive real numbers under multiplication

(e) $\mathbb{Z}_n$, the set of all congruence classes of integers modulo n under addition

(f) $\{-1, 1\}$, the two element subset of the integers under integer multiplication

(g) $\mathcal{M}_{n \times n}(\mathbb{R})$, the set of $n \times n$ matrices with real entries under addition

(h) $GL_n(\mathbb{R})$, the set of all invertible $n \times n$ matrices with real entries under the operation of multiplication

(i) $\mathcal{F}$, the set of all functions mapping the reals to the reals under the operation of addition of functions

Groups

In Preview Activity 17.1, we identified a set of properties that seemed to be satisfied in a variety of different sets including $\mathbb{Z}$, $\mathbb{E}$, $\mathbb{Q}$, $\mathbb{R}^+$, $\mathbb{C}$, $\mathbb{Z}_n$, $\mathcal{M}_{n \times n}(\mathbb{R})$, and the set of symmetries of an object. All of these sets are examples of a special type of algebraic structure known as a *group*,[*] defined formally as follows:

Definition 17.2. A **group** is a set G on which one binary operation, denoted $\cdot$, is defined such that the following axioms hold:

The Group Axioms

- **The set G is closed under its operation**, meaning that $a \cdot b \in G$ for all $a, b \in G$.

- **The operation $\cdot$ is associative in G**, meaning that $(a \cdot b) \cdot c = a \cdot (b \cdot c)$ for all $a, b, c \in G$.

- **The set G contains an identity element**, meaning that there exists an element $e \in G$ such that $a \cdot e = a = e \cdot a$ for all $a \in G$.

- **The set G contains an inverse for each of its elements**, meaning that for each $a \in G$, there exists an element $b \in G$ such that $a \cdot b = e = b \cdot a$.

Note that the operation in G need not be commutative. However, as we have seen, the operation is commutative in some instances. This gives rise to the next definition.

Definition 17.3. A group G is an **Abelian group** if $a \cdot b = b \cdot a$ for all $a, b \in G$ (in other words, if the operation is commutative in G).

An Abelian group is also called a *commutative* group.[†] Consistent with the usual convention, we will often omit the symbol for the group operation ($\cdot$), writing ab instead of $a \cdot b$. In some cases, it is more natural to use the $+$ sign to denote the operation in a group; our choice will depend on the situation and should be clear from the context.

---

[*]The French mathematician Evariste Galois appears to be the first to use the word *group* in the expression *groupe de l'équation* in his paper *Mémoire sur les conditions de résolubilité des équations par redicaux* when referring to a subset of the collection of permutations of the roots of a polynomial. His idea of group was slightly different than the modern one.

[†]Abelian groups are named after the Norwegian mathematician Neils Henrik Abel (1802 – 1829), famous for (among other things) proving the impossibility of solving the general quintic polynomial equation by radicals.

Examples of Groups

We know of many sets on which we can define a binary operation, but not all of these sets are groups.

Activity 17.4. Which of the sets in the following list is a group under the indicated operation? If a set forms a group, is it an Abelian group? Explain.

- The set of symmetries of an object under composition

- $\mathbb{Z}$, the set of integers under addition

- $\mathbb{Z}$, the set of integers under multiplication

- any of the sets $\mathbb{Q}$, $\mathbb{R}$, $\mathbb{Z}_n$, $\mathcal{M}_{n \times n}(\mathbb{R})$ under their standard addition operation

- any of the sets $\mathbb{Q}$, $\mathbb{R}$, $\mathbb{Z}_n$, $\mathcal{M}_{n \times n}(\mathbb{R})$ under their standard multiplication operation

- the set $\mathrm{GL}_n(\mathbb{R})$ of all invertible $n \times n$ matrices with real entries under the operation of matrix multiplication (in other words, the units in $\mathcal{M}_{n \times n}(\mathbb{R})$)

- $\mathbb{Q}^*$, the set of all nonzero rational numbers under multiplication

- $\mathbb{R}^+$, the set of all positive real numbers under multiplication

- $\{-1, 1\}$, the two element subset of the integers under integer multiplication (in other words, the units in $\mathbb{Z}$)

- $\mathcal{B}$, the set of all bijections mapping the reals to the reals under the operation of composition of functions.

As the next activity illustrates, there are also unfamiliar sets that turn out to be groups.

Activity 17.5. Let $S = \{a_1, a_2, a_3, a_4, a_5, a_6\}$ be the set on which an operation is defined by Table 17.1.

(a) How can we tell if the set S is closed under its operation?

(b) What is the identity element in S? How can we tell?

Table 17.1
Operation table for the set S.

$\cdot$	a_1	a_2	a_3	a_4	a_5	a_6
a_1	a_1	a_2	a_3	a_4	a_5	a_6
a_2	a_2	a_1	a_5	a_6	a_3	a_4
a_3	a_3	a_6	a_1	a_5	a_4	a_2
a_4	a_4	a_5	a_6	a_1	a_2	a_3
a_5	a_5	a_4	a_2	a_3	a_6	a_1
a_6	a_6	a_3	a_4	a_2	a_1	a_5

Table 17.2
Properties of groups.

	$\mathbb{R}$	$\mathbb{R}^+$	$GL_2(\mathbb{R})$	$\mathcal{B}$	$\mathcal{S}$
Identity					
Inverse of x					
Commutative?					

(c) What is the inverse of element a_5? How can we tell? What is the inverse of a_3?

(d) The operation in S is associative, although it is difficult to see this from the table. Verify the associative property in one case by computing $a_2(a_3 a_4)$ and $(a_2 a_3)a_4$.

The next activity compares various group properties for a few groups.

Activity 17.6. The sets listed in Table 17.2 under their standard operations (addition for $\mathbb{R}$, multiplication for $\mathbb{R}^+$, matrix multiplication for $GL_2(\mathbb{R})$, function composition for $\mathcal{B}$, and the operation from Table 17.1 for S) can easily be seen to satisfy the four group axioms. (Verifying the associative law in each set can be a bit of work, but it is fairly straightforward in each case.) Complete Table 17.2, by determining: (i) the identity in each group; (ii) the inverse of the arbitrary element x; and (iii) whether the group's operation is commutative (answer (Y)es or (N)o). If the operation is not commutative, provide a counterexample. No proofs are required here, but you should be able to provide justification for your responses if asked.

Basic Properties of Groups

The group axioms should seem quite familiar since we have used them both implicitly and explicitly throughout previous investigations. But what about the other properties we have studied? You may have noticed that at least a few of these properties (including some that were satisfied by all the number systems from Investigation 3) were not included in Definition 17.2. Part of this is due to the fact that many of the systems we encountered in Investigation 3 came with two binary operations and a group has only one binary operation. But even if we focus on only one binary operation at a time, there are certain properties that the sets in Investigation 3 satisfy that are not part of our definition of a group. Do these properties in fact hold for all groups, and if so, why are they not included in our list of group axioms?

To answer this question, we must think back to our discussion in Investigation 1 regarding the difference between axioms and theorems. There are certain properties, such as cancellation, that are in fact satisfied in all groups. These properties, however, can be proved from the group axioms. Thus, to include them would cause our axiom system to be redundant. The primary benefit of our definition of a group is that it provides a minimal set of axioms from which numerous other algebraic properties and theorems can be proved. Moreover, any property that we can prove using only the group axioms must necessarily hold in *every* group. Thus, the theory of groups gives us a way to study algebra more abstractly, instead of just within the context of specific number systems. In fact, the entire field of abstract algebra revolves around the study of general algebraic structures, such as groups, and their applications.

With that said, let's now formally state and prove the cancellation law we just mentioned. The integer version of this property was part of Activity 1.2 on page 6). As it turns out, the argument we used there generalizes easily to the context of arbitrary groups. As you read the proof below, see if you can fill in the missing details and provide an explanation or justification wherever you see the ⑦ symbol.

Theorem 17.7 (Group Cancellation Law). *Let G be a group, and let a, b, $c \in G$. If $ac = bc$, then $a = b$. Similarly, if $ab = ac$, then $b = c$.*

Proof. Let G be a group with identity e and let a, b, $c \in G$. Since G is a group, G contains an inverse d for c. Thus,
$$(ac)d = (bc)d,$$
which implies that
$$a(cd) = b(cd).^{⑦}$$
Thus,
$$ae = be, ^{⑦}$$
and so $a = b$,⑦ as desired. The proof that $ab = ac$ implies $b = c$ is left for you to complete. (Do you see why we need to prove both?) ∎

Identities and Inverses in a Group

As we have mentioned, our list of group axioms leaves out a few important properties that we might expect to be satisfied in groups. In this section, we will verify some of these properties for arbitrary groups.

Activity 17.8.

(a) How many additive identities does $\mathbb{Z}$ have? Do you think your statement generalizes to arbitrary groups? In other words, how many identity elements does an arbitrary group have? Prove your answer. (Hint: Suppose a group G has two identities, e and e'. Evaluate ee' in two different ways, and compare.)

(b) How many additive inverses does each element in $\mathbb{Z}$ have? Does this property generalize to arbitrary groups? In other words, how many inverses does an element in an arbitrary group have? Prove your answer. (Hint: Begin, as in part (a), by assuming that an element $a \in G$ has two inverses, say b and c.)

If a is an element of a group G, then the inverse element b of a whose existence is guaranteed by the fourth group axiom is unique by Activity 17.8 so we can call this element *the* inverse of a. The notation we will use to denote the inverse of a group element a is either $-a$ (if we are using additive notation for our operation) or a^{-1} (if we are using multiplicative notation for our operation).

Activity 17.9. Let G be a group. Which of the following properties do you believe are satisfied in every group? Prove or disprove each property. (Hint: Use the definition of inverse from the group axioms.)

(a) $\left(a^{-1}\right)^{-1} = a$ for all $a \in G$,

(b) $(ab)^{-1} = a^{-1}b^{-1}$ for all $a, b \in G$,

(c) $(ab)^{-1} = b^{-1}a^{-1}$ for all $a, b \in G$.

The Order of a Group

In our examples, we have seen that some groups have an infinite number of elements ($\mathbb{Z}$, $\mathbb{Q}$, $\mathbb{R}$, $GL_n(\mathbb{R})$) and some contain a finite number of elements ($\mathbb{Z}_n$, the symmetries of a square). This leads us to define the *order* of a group.

Definition 17.10. Let G be a group. If G contains only a finite number of elements, then G has **finite order** and we say G is a **finite group**. If G contains exactly m distinct elements, then the **order** of G, denoted $|G|$, is m. If G contains infinitely many elements, then G has **infinite order** and we say G is an **infinite group**.

For example, the groups $\mathbb{Z}$, $\mathbb{Q}$, $\mathbb{R}$, $\mathbb{Q}^*$, $\mathbb{R}^+$, $GL_n(\mathbb{R})$ all have infinite order, while $\mathbb{Z}_n$ has order n. In other words, $|\mathbb{Z}_n| = n$.

Activity 17.11.

(a) What is the order of the group of symmetries of an equilateral triangle?

(b) What is the order of the group of symmetries of square?

(c) What is the order of the group of symmetries of a regular pentagon?

(d) Do you see a pattern in the previous examples? If so, what do you expect the order of the group of symmetries of a regular n-gon to be?

We will see many other examples of finite groups as we proceed through our investigations.

Groups of Units

In earlier investigations we defined the units in the sets $\mathbb{Z}$, $\mathbb{Q}$, $\mathbb{C}$, $\mathbb{Z}_n$, and $\mathcal{M}_{n \times n}(\mathbb{R})$. Recall that a unit in a set that has both addition and multiplication operations is an element that has a multiplicative inverse. To define units in general, all we need is a set with an associative multiplication operation and an identity element.

Definition 17.12. Let S be a set on which an associative binary operation of multiplication is defined such that S contains an identity element 1_S. An element $u \in S$ is a **unit** in S if there is an element $v \in S$ such that
$$uv = 1_S = vu.$$

The element v for which $uv = vu = 1$ is unique (the argument is the same as the uniqueness of inverses in a group), so we call this element the *inverse* of u in S and denote it as u^{-1}. Note also that if $uv = vu = 1_S$, then $v = u^{-1}$ and $u = v^{-1}$.

Activity 17.13.

(a) Let n be a positive integer. In Investigation 2, we classified all units in $\mathbb{Z}_n$. How can we tell if a class $[a]$ is a unit in $\mathbb{Z}_n$? How can we find the inverse of a unit in $\mathbb{Z}_n$?

(b) Find or describe all units in each of the following sets:

  (i) $\mathbb{Z}_5$

  (ii) $\mathbb{Z}_8$

  (iii) $\mathbb{Q}$

  (iv) $\mathcal{M}_2(\mathbb{R})$

(c) Let U_n be the set of units in $\mathbb{Z}_n$. Construct the multiplication tables for the sets U_3, U_4, U_5, U_6, U_7 and U_8.

The examples above indicate that the set of units in $\mathbb{Z}_n$ forms a group under multiplication. In the next activity, we will decide if this is true in general.

Activity 17.14. Let S be a set on which an associative binary operation of multiplication is defined such that S contains an identity element 1_S. Let $U(S)$ be the set of units in S.

(a) Does $U(S)$ contain an identity element? If so, what is it? Prove your answer.

(b) Is $U(S)$ closed under multiplication? Explain. (Be careful not to assume that multiplication in S is commutative.)

(c) Why is multiplication associative in $U(S)$?

(d) Does $U(S)$ contain an inverse for each of its elements? Explain.

(e) Explain how we have just proved the following theorem.

  Theorem 17.15. *Let S be a set on which an associative binary operation of multiplication is defined such that S contains an identity element 1_S. Let $U(S)$ be the set of units in S. Then $U(S)$ is a group under the operation of multiplication.*

We have seen that the number systems $\mathbb{Z}, \mathbb{Z}_n, \mathbb{Q}, \mathbb{R}, \mathbb{C}, \mathcal{M}_{n \times n}(\mathbb{R})$, and $\mathcal{P}_n$ are all groups under their additive operations. They are not, however, groups under their multiplicative operations. (Do you see why?) The result of Activity 17.14 is that the *sets of units* in the number systems $\mathbb{Z}, \mathbb{Z}_n, \mathbb{Q}, \mathbb{R}, \mathbb{C}, \mathcal{M}_{n \times n}(\mathbb{R})$, and $\mathcal{P}_n$ are groups under their multiplicative operations.

Concluding Activities

Activity 17.16. Let G be a group. Show that in the operation table for G, every element in G appears once and only once in each row and column.

Activity 17.17.

(a) In a group G with identity e, if $ab = e$ for some $a, b \in G$ must it follow that $b = a^{-1}$? Prove your answer.

(b) In a group G with identity e, if $ba = e$ for some $a, b \in G$ must it follow that $b = a^{-1}$? Prove your answer.

(c) Let f and g be functions from a set S to S. Let I be the identity function on S—that is $I(x) = x$ for all x in S. Show by example that it is possible to have $fg = I$, but $f \neq g^{-1}$. Does this violate part (a)? Explain.

Activity 17.18. Recall that a symmetry of an object O is a bijective, distance-preserving function f such that $f(O) = O$. In this activity, we will verify that the set S of symmetries of an object O forms a group under the operation of composition, called the **group of symmetries of** O.

(a) Let f and g be bijective, distance-preserving functions with $f(O) = O$ and $g(O) = O$. To show that S is closed, we need to verify that $f \circ g$ is a bijective, distance-preserving function with $(f \circ g)(O) = O$.

  (i) Prove that $|(f \circ g)(x) - (f \circ g)(y)| = |x - y|$ for all x, y in the domain of $f \circ g$.

  (ii) Use part (i) to deduce that $f \circ g$ is an injection.

  (iii) Show that $(f \circ g)(O) = O$. Deduce that $f \circ g$ is a surjection.

  (iv) Explain how parts (a) – (c) establish that S is closed under composition.

(b) Prove that composition of functions is an associative operation.

(c) What is the identity element of S? Verify your answer.

(d) If $f \in S$, what is the inverse of f in S? Verify your answer. (Please note that there is quite a bit to do to complete this problem.)

(e) Explain why S is a group. Is S an Abelian group? Explain.

Exercises

(1) Assume G is a group. Suppose that, due to a printer error, the operation table for G was printed with several entries missing, as shown below:

	a	b	c	d
a				
b			d	
c				b
d		c		

Complete the table using only the group axioms and consequent properties. Specifically explain how you determined each element.

(2) In a group, is it true that $AB = BC$ implies $A = C$? (Such a property is called a "cross-cancellation" property.) If the answer is no, find a specific counterexample. If the answer is yes, then prove this property.

(3) Let $\mathbb{Z}^\star$ be the number system consisting of the set of all integers.

    (a) Define an operation $\oplus$ on $\mathbb{Z}^\star$ by

$$a \oplus b = a + b - 1.$$

    Note that $+$ denotes the normal operation of addition in $\mathbb{Z}$. Which of the group axioms are satisfied by $\mathbb{Z}^\star$ using the operation $\oplus$, and which are not? Is $\mathbb{Z}^\star$ a group with the operation $\oplus$? If so, is $\mathbb{Z}^\star$ an Abelian group? Prove your answers.

    (b) Now define a different operation $\otimes$ on $\mathbb{Z}^\star$ by

$$a \otimes b = a + b - a \cdot b.$$

    Here $+$ and $\cdot$ denote the normal operations of addition and multiplication in $\mathbb{Z}$. Which of the group axioms are satisfied by $\mathbb{Z}^\star$ using the operation $\otimes$, and which are not? Is $\mathbb{Z}^\star$ a group with the operation $\otimes$? If so, is $\mathbb{Z}^\star$ an Abelian group? Prove your answers.(See Exercise (9) in Investigation 2.)

(4) Determine if the set G is a group under the indicated operation. If G is a group, verify that each group property is satisfied. If G is not a group, provide examples that show which of the group properties are not satisfied.

    (a) G is the set of odd integers under addition.

    (b) $G = \{[2], [4], [6], [8]\} \subset \mathbb{Z}_{10}$, with the operation of multiplication of congruence classes.

    (c) $G = \{[0], [2], [4], [6], [8]\} \subset \mathbb{Z}_{10}$, with the operation of addition of congruence classes.

    (d) $G = \{q \in \mathbb{Q} : q \neq 1\}$, with the operation $*$ defined by $a * b = a + b - ab$.

    (e) $G = \{[x] \in \mathbb{Z}_9 : x = 1, 2, 4, 5, 7, \text{ or } 8\}$, with the operation $[x] * [y] = [x][y]$.

(5) Let $\mathbb{R}^+$ denote the set of positive real numbers.

    (a) Is $\mathbb{R}^+$ a group using standard multiplication on $\mathbb{R}$? Prove your answer.

    (b) Is $\mathbb{R}^+$ a group using standard division on $\mathbb{R}$? Prove your answer.

(6) Is $\mathbb{R}^-$, the set of negative real numbers, a group using the operation $x * y = -(xy)$, where xy is the standard product in $\mathbb{R}$? Prove your answer.

(7) Let k be an integer, and let $Z(k)$ be the set of integers on which an operation $\oplus_k$ is defined as follows:

$$a \oplus_k b = a + b - k,$$

where $a + b$ denotes the standard sum of a and b in $\mathbb{Z}$. Note that the set $Z(0)$ is the group of integers under the standard addition. For which values of k is $Z(k)$ a group under the operation $\oplus_k$? Prove your answer.

(8) Prove that a group G is Abelian if and only if $(ab)^2 = a^2b^2$ for all $a, b \in G$.

(9) Consider a strip of three equally spaced I's:

I I I

Describe the group of symmetries of this strip. Is the group of symmetries of the strip Abelian?

(10) Consider an infinitely long strip of equally spaced I's:

$\cdots$ I I I I $\cdots$

Describe the group of symmetries of this strip. Is the group of symmetries of the strip Abelian?

(11) (From a GRE Practice Exam) Let p and q be distinct primes. Suppose that H is a proper subset of the integers and H is a group under addition that contains exactly three elements of the set $\{p, p+q, pq, p^q, q^p\}$. Determine which of the following are the three elements in H.

   (i) pq, p^q, q^p

   (ii) $p+q, pq, p^q$

   (iii) $p, p+q, pq$

   (iv) p, p^q, q^p

   (v) p, pq, p^q

(12) Prove that a group G is Abelian if and only if $(ab)^{-1} = a^{-1}b^{-1}$ for all $a, b \in G$.

(13) Let n be a nonnegative integer, and let $n\mathbb{Z} = \{nx : x \in \mathbb{Z}\}$, with addition defined as in $\mathbb{Z}$. Is $n\mathbb{Z}$ a group under this addition? If so, is $n\mathbb{Z}$ an Abelian group? Does your answer depend on the value of n? Prove your answers.

(14) Let n and k be natural numbers, both greater than 1. Let $M_{n \times n}(\mathbb{Z}_k)$ be the set of all $n \times n$ matrices whose entries are in $\mathbb{Z}_k$.

   (a) How many elements does $M_{n \times n}(\mathbb{Z}_k)$ have?

   (b) Is $M_{n \times n}(\mathbb{Z}_k)$ a group under standard addition of matrices?

   (c) Is $M_{n \times n}(\mathbb{Z}_k)$ a group under standard multiplication of matrices?

   (d) Find all units in $M_{2 \times 2}(\mathbb{Z}_2)$ using standard multiplication of matrices. Does this collection of units form an Abelian group? Explain.

(15) Let $\mathcal{F}(\mathbb{R})$ denote the set of all functions from $\mathbb{R}$ to $\mathbb{R}$. Define addition and multiplication on $\mathcal{F}(\mathbb{R})$ as follows:

- For all $f, g \in \mathcal{F}(\mathbb{R})$, $(f + g) : \mathbb{R} \to \mathbb{R}$ is the function defined by

$$(f + g)(x) = f(x) + g(x)$$

for all $x \in \mathbb{R}$.

- For all $f, g \in \mathcal{F}(\mathbb{R})$, $(fg) : \mathbb{R} \to \mathbb{R}$ is the function defined by

$$(fg)(x) = f(x)g(x)$$

for all $x \in \mathbb{R}$.

(a) Prove that $\mathcal{F}(\mathbb{R})$ is an Abelian group under addition.

(b) Does $\mathcal{F}(\mathbb{R})$ have an identity element for multiplication?

(c) Find an element in $\mathcal{F}(\mathbb{R})$ that does not have a multiplicative inverse in $\mathcal{F}(\mathbb{R})$. Explain how this shows $\mathcal{F}(\mathbb{R})$ is not a group under multiplication.

(d) Find necessary and sufficient conditions for an element in $\mathcal{F}(\mathbb{R})$ to be a unit in $\mathcal{F}(\mathbb{R})$. State your result in a lemma of the form "The function $f \in \mathcal{F}(\mathbb{R})$ is a unit in $\mathcal{F}(\mathbb{R})$ if and only if ...". Your lemma must say something more than just a rehash of the definition of a unit; rather, it must actually characterize the functions that are invertible under multiplication in $\mathcal{F}(\mathbb{R})$.

(16) Define a **blip** to be a pair of integers, denoted $\langle a, b \rangle$, and define two blips $\langle a, b \rangle$ and $\langle x, y \rangle$ to be equal whenever $a + b = x + y$ (so that, for instance, $\langle 3, 5 \rangle$ and $\langle 10, -2 \rangle$ would be considered equal since $3 + 5 = 8 = 10 + (-2)$). Define an operation $*$ on the set B of all blips as follows:

$$\langle a, b \rangle * \langle c, d \rangle = \langle a + c, b + d \rangle.$$

Is B a group under the operation $*$? If so, prove it. If not, determine which of the group axioms are satisfied and which are not.

(17) Let x and y be units in $\mathbb{Z}$. Prove or disprove: $x + y$ is a unit in $\mathbb{Z}$.

Connections

This investigation introduced the concept of a group. Groups are algebraic objects that share the same basic additive structure as the integers and the different number systems discussed in Investigation 3. There is a great deal of power to be found in recognizing the features these number systems have in common and then creating a larger category (groups) that encapsulates all of these features. Indeed, we can then learn about all of these number systems at one time by studying arbitrary groups.

If you studied ring theory before group theory, you should notice connections between the topics in this investigation and those in Investigation 4. In particular, groups and rings are both algebraic objects—that is, sets on which an operation or operations are defined, yielding an algebraic structure of some sort. The main difference between a group and a ring is that a group comes with one binary operation, while a ring comes with two. In fact, every ring is a group under its addition operation, but not all groups can be made into rings in a natural way.

Since there is only one operation in a group, groups are simpler objects than rings. For that reason, a good argument can be made that the study of abstract algebra should begin with groups. From a structural standpoint, however, rings may be more familiar to us than groups in that many of the sets with which we have worked in our mathematical pasts (e.g., $\mathbb{Z}$, $\mathbb{Q}$, $\mathbb{R}$, and sets of polynomials) are all rings. For this reason, starting our exploration of abstract algebra with rings is also a reasonable choice. In either case, many of the concepts we will encounter in these investigations will apply to both rings and groups.

Investigation 18

Integer Powers of Elements in a Group

> **Focus Questions**
>
> *By the end of this investigation, you should be able to give precise and thorough answers to the questions listed below. You may want to keep these questions in mind to focus your thoughts as you complete the investigation.*
>
> - How can integer exponentiation be defined in an arbitrary group?
>
> - What properties are satisfied by integer exponentiation in groups?

Our study of groups began in Investigation 16, where we learned about the set of symmetries of an object. We then saw how many familiar sets, like $\mathbb{Z}$, $\mathbb{Z}_n$, and sets of invertible square matrices, all had a structure that was similar to that of a set of symmetries—namely, the structure of a group. One of the defining axioms of a group is that it is closed under its operation. In this investigation, we will define and study a familiar shorthand notation for repeatedly applying a group's operation.

Preview Activity 18.1. Use your intuition to calculate the quantities listed below. Throughout your calculations, you will be applying the definitions that we will formally develop in this investigation. Recall that G is the group of symmetries of a square (with the operation of composition), $\mathbb{Z}_6$ is the set of integers modulo 6 (with the operation of addition of congruence classes), and $\mathrm{GL}_2(\mathbb{R})$ is the set of all invertible square matrices with real entries (with the operation of matrix multiplication).

In G:	In $\mathbb{Z}_6$:	In $\mathrm{GL}_2(\mathbb{R})$:
$\begin{pmatrix} 1 & 2 & 3 & 4 \\ 2 & 3 & 4 & 1 \end{pmatrix}^0$	$0[4]$	$\begin{bmatrix} 1 & \pi \\ -\pi & 3 \end{bmatrix}^0$
$\begin{pmatrix} 1 & 2 & 3 & 4 \\ 4 & 3 & 2 & 1 \end{pmatrix}^2$	$2[5]$	$\begin{bmatrix} 2 & 1 \\ 1 & 1 \end{bmatrix}^2$
$\begin{pmatrix} 1 & 2 & 3 & 4 \\ 2 & 1 & 4 & 3 \end{pmatrix}^{-2}$	$(-2)[3]$	$\begin{bmatrix} 1 & 1 \\ 0 & 1 \end{bmatrix}^{-2}$

Introduction

In Preview Activity 18.1, we began to intuitively develop the notions of integer exponentiation for groups. You may have performed the requested calculations by simply thinking of exponentiation as repeatedly applying the group operation. For instance, in the group of symmetries of the square, we can calculate $\begin{pmatrix} 1 & 2 & 3 & 4 \\ 2 & 3 & 4 & 1 \end{pmatrix}^3$ as follows:

$$\begin{pmatrix} 1 & 2 & 3 & 4 \\ 2 & 3 & 4 & 1 \end{pmatrix}^3 = \left[\begin{pmatrix} 1 & 2 & 3 & 4 \\ 2 & 3 & 4 & 1 \end{pmatrix} \begin{pmatrix} 1 & 2 & 3 & 4 \\ 2 & 3 & 4 & 1 \end{pmatrix} \right] \begin{pmatrix} 1 & 2 & 3 & 4 \\ 2 & 3 & 4 & 1 \end{pmatrix}$$
$$= \begin{pmatrix} 1 & 2 & 3 & 4 \\ 3 & 4 & 1 & 2 \end{pmatrix} \begin{pmatrix} 1 & 2 & 3 & 4 \\ 2 & 3 & 4 & 1 \end{pmatrix}$$
$$= \begin{pmatrix} 1 & 2 & 3 & 4 \\ 4 & 1 & 2 & 3 \end{pmatrix}.$$

This intuitive formulation of integer exponentiation makes sense as long as we are exponentiating by a positive integer. For nonpositive integers, however, we will need to be a bit more careful. Furthermore, in order to prove that integer exponentiation works the way we expect it to, we will need to make use of a more formal definition.

We will develop such a definition in the next section, and we will use this definition to prove several fundamental properties of integer exponentiation in groups. As we move forward, it is important to note that in groups, which have only one operation, we will use the term *exponentiation* to refer to the intuitive idea of repeatedly applying a group's operation, regardless of whether that operation is multiplication (as in $\mathrm{GL}_n(\mathbb{R})$), addition (as in $\mathbb{Z}_n$), or something else (for example, composition in the group of symmetries of an object). When the group operation is addition, it is more natural to write mx instead of x^m. (In this case, integer exponentiation could alternatively be called integer multiplication.)

Powers of Elements in a Group

Let G be a group (with the operation written multiplicatively), and let $a \in G$. The elements $a, aa, aaa, \ldots, a^{-1}, a^{-1}a^{-1}, a^{-1}a^{-1}a^{-1}, \ldots$ are also in G. To represent these types of elements in a more convenient and natural way, we will let a^0 be the identity element. We will then define

$$a^1 = a, \quad a^2 = a^1 a, \quad a^3 = a^2 a, \quad a^4 = a^3 a,$$

and so on. Similarly,

$$a^{-1} = a^{-1}, \quad a^{-2} = a^{-1}a^{-1} = \left(a^{-1}\right)^2, \quad a^{-3} = a^{-1}a^{-1}a^{-1} = \left(a^{-1}\right)^3,$$

and so on.

To be more formal, we can define powers of a recursively as follows:

Definition 18.2. Let G be a group with identity e, and let $a \in G$. Then for each integer m, we define a^m as follows:

- $a^0 = e$.

- $a^1 = a$.

- If m is a positive integer, then $a^m = a^{m-1}a$.

- If m is a positive integer, then $a^{-m} = (a^{-1})^m$.

When we use additive notation for the operation in our group, a^m is written ma and these definitions can be written as $0a = e$, $1a = a$, $ma = (m-1)a + a$ and $(-m)a = m(-a)$.

Throughout your studies of mathematics, you have undoubtedly used exponentiation by an integer in many settings. This familiarity should raise a number of questions about Definition 18.2 . For example, is it true that $a^m = aa^{m-1}$ for positive integers m? Is it true that $a^m = a^{m-1}a$ if m is a negative integer? Is it true in a group that $a^m a^n = a^{m+n}$? If $b \in G$, is it true that $(ab)^m = a^m b^m$? We will address these questions in the remainder of this investigation and in the exercises.

Let's begin with the important question of whether $(ab)^m = a^m b^m$.

Activity 18.3. Let G be the symmetries of the square, and let $a = \begin{pmatrix} 1 & 2 & 3 & 4 \\ 3 & 4 & 1 & 2 \end{pmatrix}$, $b = \begin{pmatrix} 1 & 2 & 3 & 4 \\ 4 & 3 & 2 & 1 \end{pmatrix}$, and $c = \begin{pmatrix} 1 & 2 & 3 & 4 \\ 2 & 3 & 4 & 1 \end{pmatrix}$.

(a) Calculate $(bc)^2$ and $b^2 c^2$. Are they the same?

(b) Calculate $(ab)^2$ and $a^2 b^2$. Are they the same?

(c) In one of the preceding parts, we have $(xy)^2 = x^2 y^2$, and in the other we don't. There is a significant difference in the relationship between x and y that accounts for this difference. How is the relationship between x and y different in the two parts, and why does this difference affect whether $(xy)^2 = x^2 y^2$?

Activity 18.3 illustrates an important point when working in groups. In general we **cannot** assume that $(ab)^m = a^m b^m$. However, if we know an additional fact about a and b, then we can use this property.

Theorem 18.4. *Let G be a group with identity e, and let $a, b \in G$ such that $ab = ba$. Then $(ab)^m = a^m b^m$ for every integer m.*

When $ab = ba$ in a group G, we say that the elements a and b *commute* (or commute with each other). An outline of the proof of Theorem 18.4 is given below. As you read the proof, try to fill in the missing details and provide additional explanation or justification where indicated.

Proof of Theorem 18.4. Let G be a group with identity e, and let a, b elements in G such that $ab = ba$. First, we will show that $(ab)^m = a^m b^m$ for every positive integer m. To do so, we will use induction on m. We know that

$$(ab)^1 \stackrel{\text{①}}{=} ab \stackrel{\text{②}}{=} a^1 b^1,$$

so our theorem is true for $m = 1$. Now assume that $(ab)^m = a^m b^m$ for some integer $m \geq 1$. We will show that $(ab)^{m+1} = a^{m+1} b^{m+1}$. Note that

$$(ab)^{m+1} \stackrel{\text{①}}{=} (ab)^m (ab) \stackrel{\text{②}}{=} (a^m b^m)(ab) \stackrel{\text{③}}{=} a^m (b^m a) b. \tag{18.1}$$

To complete this portion of the proof, we need to know that a commutes with b^m, or that $b^m a = ab^m$. This will also require an induction proof.

Claim. If n is a positive integer, then $b^n a = ab^n$.

272

Investigation 18. Integer Powers of Elements in a Group

Proof of Claim. Notice that

$$b^1 a \stackrel{\odot}{=} ba \stackrel{\odot}{=} ab \stackrel{\odot}{=} ab^1,$$

so the claim is true for $n = 1$. For the induction step, we will assume that $b^n a = ab^n$ for some integer $n \geq 1$. We will show that $b^{n+1} a = ab^{n+1}$. Now

$$b^{n+1} a \stackrel{\odot}{=} (b^n b)a \stackrel{\odot}{=} b^n (ba) \stackrel{\odot}{=} b^n (ab) \stackrel{\odot}{=} (b^n a)b \stackrel{\odot}{=} (ab^n)b \stackrel{\odot}{=} a(b^n b) \stackrel{\odot}{=} ab^{n+1},$$

and so we have verified that $b^n a = ab^n$ for every positive integer n. ∎

We will now return to our proof that $(ab)^{m+1} = a^{m+1}b^{m+1}$. Continuing with (18.1), we have

$$
\begin{aligned}
(ab)^{m+1} &= (ab)^m(ab) \\
&= (a^m b^m)(ab) \\
&= a^m (b^m a)b \\
&\stackrel{\odot}{=} a^m (ab^m)b \\
&\stackrel{\odot}{=} (a^m a)(b^m b) \\
&\stackrel{\odot}{=} a^{m+1}b^{m+1}.
\end{aligned}
$$

Thus we have that $(ab)^m = a^m b^m$ for all _____ m.

To complete the proof of Theorem 18.4, we need to verify that $(ab)^m = a^m b^m$ for $m = 0$ and $m < 0$. Since

$$(ab)^0 \stackrel{\odot}{=} e \stackrel{\odot}{=} ee \stackrel{\odot}{=} a^0 b^0,$$

the theorem is true if $m = 0$. We now need to consider the case where m is negative.

Suppose $m = -1$. Since $ab = ba$, we have

$$a^{-1}b^{-1} \stackrel{\odot}{=} (ba)^{-1} \stackrel{\odot}{=} (ab)^{-1} \stackrel{\odot}{=} b^{-1}a^{-1},$$

so we see that $(ab)^{-1} = a^{-1}b^{-1}$. Moreover, we have shown that a^{-1} and b^{-1} commute with each other. Now let m be a negative integer and let $k = -m$. Then

$$
\begin{aligned}
(ab)^m &\stackrel{\odot}{=} (ab)^{-k} \\
&\stackrel{\odot}{=} \left((ab)^{-1}\right)^k \\
&\stackrel{\odot}{=} \left((ba)^{-1}\right)^k \\
&\stackrel{\odot}{=} \left(a^{-1}b^{-1}\right)^k \\
&\stackrel{\odot}{=} \left(a^{-1}\right)^k \left(b^{-1}\right)^k \\
&\stackrel{\odot}{=} a^{-k}b^{-k} \\
&\stackrel{\odot}{=} a^m b^m.
\end{aligned}
$$

Therefore, we can conclude that _____ = _____ for all _____ m. This, along with our previous cases, completes the proof. ∎

Now that we have established that $(ab)^m = a^m b^m$ whenever a and b commute, let's consider some other familiar properties of exponentiation. For example, is it true that $(a^m)^{-1} = (a^{-1})^m$? To answer this question, note that since $aa^{-1} = e = a^{-1}a$, we see that a and a^{-1} commute. So

$$\left(a^{-1}\right)^m a^m = \left(a^{-1}a\right)^m = \left(aa^{-1}\right)^m = a^m \left(a^{-1}\right)^m.$$

It follows that $\left(a^{-1}\right)^m$ is an inverse of a^m. Since inverses are unique, we can conclude that

$$\left(a^{-1}\right)^m = (a^m)^{-1}$$

for any integer m.

Recall that, by Definition 18.2, $a^{-m} = \left(a^{-1}\right)^m$ whenever m is a positive integer. If m is nonpositive, is this also true?

We know that $\left(a^{-1}\right)^{-1} = a$ for every $a \in G$, so if m is a negative integer, then

$$\left(a^{-1}\right)^m = \left[\left(a^{-1}\right)^{-1}\right]^{-m} = a^{-m}.$$

Also,

$$a^0 = e = \left(a^{-1}\right)^0.$$

Putting all of this together, we arrive at part (i) of the next theorem.

Theorem 18.5. *Let G be a group. For every $a \in G$ and every $m, n \in \mathbb{Z}$:*

(i) $a^{-m} = \left(a^{-1}\right)^m = (a^m)^{-1}$ (or, if the group operation is written using additive notation, $(-m)a = m(-a) = -(ma)$).

(ii) $a^m a^n = a^{m+n}$ (or, if the group operation is written using additive notation, $ma + na = (m+n)a$).

(iii) $(a^m)^n = a^{mn}$ (or, if the group operation is written using additive notation, $n(ma) = (nm)a$).

The proofs of parts (ii) and (iii) of Theorem 18.5 are more complicated than part (i). Induction seems like a natural approach for these results, but there is a catch: both statements involve two arbitrary integers instead of just one. There are several different ways to proceed, and we will explore one approach to proving part (ii) ($a^m a^n = a^{m+n}$) in Activity 18.6. The proof of part (iii) is left as an exercise for you.

Activity 18.6. The statement in part (ii) of Theorem 18.5—namely, that $a^m a^n = a^{m+n}$ for all $a \in G$ and all $m, n \in \mathbb{Z}$—involves three universal quantifiers. Our approach will be to choose an arbitrary $a \in G$ and $m \in \mathbb{Z}$, and then use induction on the integer n to resolve the cases where n is nonnegative. We will then use a separate argument to deal with the case where $n < 0$. So let $a \in G$, and let $m \in \mathbb{Z}$.

(a) To perform our inductive argument, we first need to establish a base case. Apply Definition 18.2 to prove that $a^m a^n = a^{m+n}$ when $n = 0$.

(b) Now let $n > 0$, and let $P(n)$ be the predicate, "$a^m a^n = a^{m+n}$." Below is an argument showing that $P(1)$ is true. Explain the rationale for each equality marked with the ⑦ symbol.

If $n = 1$, then
$$a^m a^n = a^m a^1 \overset{⑦}{=} a^m a \overset{⑦}{=} a^{m+1} \overset{⑦}{=} a^{m+n}.$$

(c) To continue our induction proof, let n be a positive integer and assume that $P(n)$ (as defined in part (b)) is true. We then need to prove that $P(n+1)$ is true—that is,

$$a^m a^{n+1} = a^{m+(n+1)}.$$

Use the assumption that $P(n)$ is true, along with Definition 18.2 and whatever group axioms you need, to show that $P(n+1)$ is true.

(d) Deduce from parts (a) – (c) that the statement $a^m a^n = a^{m+n}$ is true for all $a \in G$, all integers m, and all nonnegative integers n.

(e) To complete our proof, we need to consider the case where $n < 0$. So assume that n is a negative integer. Give a justification for each step in the following argument:

$$\left(a^{-1}\right)^{-m} \left(a^{-1}\right)^{-n} = \left(a^{-1}\right)^{(-m)+(-n)} \ ⑦$$
$$a^m a^n = a^{-((-m)+(-n))} \ ⑦$$
$$a^m a^n = a^{m+n}. \ ⑦$$

(f) Use your work from the previous parts to write a complete, clear, and convincing proof that $a^m a^n = a^{m+n}$ for all $a \in G$ and all integers m and n.

Notice that our work in Activity 18.6 also answers (in the affirmative) the questions we asked earlier about whether $a^m = a^{m-1}a = aa^{m-1}$ for all $a \in G$ and all $m \in \mathbb{Z}$. (Do you see why?) Although these properties may seem obvious, they are not immediate consequences of Definition 18.2, which simply states that $a^m = a^{m-1}a$ for all $a \in G$ and all *positive* $m \in \mathbb{Z}$.

Concluding Activities

Activity 18.7. Let G be a group and a an element of G. Show that any two powers of a commute. That is, show that $a^m a^n = a^n a^m$ for any integers m and n.

Activity 18.8. Let G be a group and $a \in G$. Prove part (iii) of Theorem 18.5—that is, prove that $(a^m)^n = a^{mn}$ for all integers m and n.

Exercises

(1) Let G be a group.

    (a) Let $a, b, c \in G$. What element is $(abc)^{-1}$? Prove your answer.

    (b) Let m be a positive integer, and let $a_1, a_2, \ldots, a_m$ be elements in G. What element is $(a_1 a_2 \cdots a_m)^{-1}$? Prove your answer.

* (2) Prove that if G is a group with identity e in which $a^2 = e$ for every $a \in G$, then G is an Abelian group. Is the converse true? Explain.

(3) Let G be a group and $a \in G$. A **conjugate** of a in G is any element of the form bab^{-1} for some $b \in G$. Show that if $c = bab^{-1}$ is a conjugate of a in G and n is any integer, then $c^n = ba^n b^{-1}$.

(4) We can generalize Theorem 18.4 as follows: Let G be a group and $a_1, a_2, \ldots, a_m$ elements in G for some positive integer $m \geq 2$ so that $a_i a_j = a_j a_i$ for all i and j. Prove that

$$(a_1 a_2 \cdots a_m)^n = a_1^n a_2^n \cdots a_m^n$$

for any $n \in \mathbb{Z}$.

(5) **Fibonacci sequences in groups.** The Fibonacci numbers F_n are defined recursively by $F_0 = 0$, $F_1 = 1$, and $F_n = F_{n-1} + F_{n-2}$ for $n \geq 2$. The definition of this sequence only depends on a binary operation. Since every group comes with a binary operation, we can define Fibonacci-type sequences in any group. Let G be a group, and define the sequence $\{f_n\}$ in G as follows: Let a_0, a_1 be elements of G, and define

$$f_0 = a_0, \quad f_1 = a_1, \quad \text{and} \quad f_n = a_{n-1} a_{n-2} \quad \text{for } n \geq 2.$$

D.D. Wall writes the following about his introduction to these sequences: [*]

> The problem arose in connection with a method for generating random numbers, but it turned out to be unexpectedly intricate, and so quickly became of interest in its own right.

In an interesting application of Fibonacci sequences in groups, Iannis Xenakis [†] uses Fibonacci sequences in groups to create "Fibonacci motions," which are sequences of musical properties such as pitch, volume, and timbre that give the composition its framework. We will see in this problem that these Fibonacci sequences become intricate quite quickly.

(a) Explain why $f_n = F_n$ if $G = \mathbb{Z}$, $a_0 = 0$, and $a_1 = 1$.

(b) Find the elements in $\{f_n\}$ if $G = \mathbb{Z}_3$, $a_0 = [1]$ and $a_1 = [2]$.

(c) Some of the sequences $\{f_n\}$ are periodic—that is, the same list of elements repeats in the same order. The number of elements in one cycle is called the **period** of the sequence.

  (i) If a is a nonidentity element of a group with identity e and $a^2 = e$, what is the period of the sequence that begins $a\ a\ \ldots$?

  (ii) If a is a nonidentity element of a group with identity e and $a^3 = e$, what is the period of the sequence that begins $a\ a\ \ldots$?

  (iii) If a is a nonidentity element of a group with identity e and $a^4 = e$ but $a^2 \neq e$, what is the period of the sequence that begins $a\ a\ \ldots$?

  (iv) If a is a nonidentity element of a group with identity e and $a^5 = e$, what is the period of the sequence that begins $a\ a\ \ldots$?

  (v) In general, the pattern of periods of the sequences that begin $a\ a\ \ldots$ is not so easy to see. While we won't establish the pattern here (see Wall's paper for details), we can begin to see how to approach the problem. Explain how the general sequence of group elements that begins $a\ a\ \ldots$ is specifically related to the Fibonacci sequence of integers.

---

[*] "Fibonacci Series Modulo m", *American Mathematical Monthly*, Vol. 67, 1960

[†] *Formalized Music*, Indiana University Press, 1971

Connections

In this investigation, we studied integer powers of elements in groups (or integer multiples if we use additive notation). If you studied ring theory before group theory, you should notice connections between the topics in this investigation and those in Investigation 5. The major difference between groups and rings in this context is that there is only one operation in a group but two in a ring. As a result, we need to understand both integer multiples (under addition) and integer powers (under multiplication) of ring elements. With only one operation in a group, we only need one of these ideas. However, we still use integer multiples when we represent our group operation as addition and integer powers when we write our group operation multiplicatively. Consequently, we need to understand both notations, even though we will only use one of them in any given group.

Investigation 19

Subgroups

Focus Questions

By the end of this investigation, you should be able to give precise and thorough answers to the questions listed below. You may want to keep these questions in mind to focus your thoughts as you complete the investigation.

- What is a subgroup of a group? What conditions must be verified in order to show that a subset of a group is a subgroup?

- What is the center of a group, and what kind of subgroup is it?

- What is the subgroup generated by an element of a group? What is a cyclic group? How are the two related?

- How can cyclic groups be used to define the order of an element in a group?

Preview Activity 19.1. Throughout mathematics, the relationship between mathematical objects and their sub-objects is of central importance. For instance, in linear algebra, we study vector spaces and their subspaces. In discrete mathematics, many graph theory problems can be solved by finding a subgraph that is optimal in some sense. Furthermore, many other applied optimization problems involve minimizing or maximizing a certain function subject to certain constraints. These constraints define what is known as a *feasible region*, which is nothing more than a subset of the space of all possible solutions.

In light of these examples and our recent investigations of groups, it seems natural that we would be interested in defining and characterizing *subgroups*. To begin thinking along these lines, consider the set $\mathbb{E}$ of even integers as a subset of the set $\mathbb{Z}$ of integers. Since $\mathbb{E}$ is a subset of a group, it is natural to ask which (if any) of the group axioms $\mathbb{E}$ satisfies.

(a) Critique the following proof that addition is associative in $\mathbb{E}$. Is the proof correct? If so, could it be improved in any way? If not, what is the main error in the argument?

> *Proof.* Let $a, b, c \in \mathbb{E}$. Since $\mathbb{E} \subseteq \mathbb{Z}$, it follows that $a, b, c \in \mathbb{Z}$ as well. Thus,
>
> $$a + (b + c) = (a + b) + c$$
>
> since addition is associative in $\mathbb{Z}$. This proves that addition is associative in $\mathbb{E}$. ∎

(b) Critique the following proof that $\mathbb{E}$ is closed under addition. Is the proof correct? If so, could it be improved in any way? If not, what is the main error in the argument?

Proof. Let $a, b \in \mathbb{E}$. Since $\mathbb{E} \subseteq \mathbb{Z}$, it follows that $a, b \in \mathbb{Z}$ as well. But $\mathbb{Z}$ is closed under addition, so $a + b \in \mathbb{Z}$. This shows that $\mathbb{E}$ is closed under addition. ∎

(c) Which of the group axioms (using the operation of addition) does the set $\mathbb{E}$ satisfy?

Introduction

In Preview Activity 19.1, we argued that the subset $\mathbb{E}$ of the set of integers was not only a subset of $\mathbb{Z}$, but also a group itself using the same operation as in $\mathbb{Z}$. The next definition formalizes this terminology.

Definition 19.2. A subset H of a group G is a **subgroup** of G if H is a group using the same operation as in G.

Definition 19.2 is not terribly surprising. Nevertheless, there is one important caveat to note— namely, the condition that both H and G use the same operation. The operation imposes a structure on the set, and different operations impose different structures. For example, the set $\mathbb{R}^*$ of nonzero real numbers is a subset of the group $\mathbb{R}$, but we would not want to consider the group $\mathbb{R}^*$ under multiplication as a subgroup of $\mathbb{R}$ under addition because the elements behave so differently with respect to the two operations. (As an example, $3 \cdot 3 = 9$ in $\mathbb{R}^*$, but $3 + 3 = 6$ in $\mathbb{R}$.)

Activity 19.3. For each part below, decide whether the set H is a subgroup of the group G. (You may assume that all the groups listed use their standard addition operation.) Explain your answers.

(a) $H = \mathbb{Z}, G = \mathbb{R}$

(b) $H = \mathbb{Z}^+, G = \mathbb{R}$

(c) $H = \{3n : n \in \mathbb{Z}\}, G = \mathbb{Z}$

(d) $H = \{3n + 1 : n \in \mathbb{Z}\}, G = \mathbb{Z}$

(e) $H = \mathbb{Z}_5, G = \mathbb{Z}$

(f) $H = \mathbb{Z}_4, G = \mathbb{Z}_8$

The Subgroup Test

To prove that a subset H of a group G is a subgroup of G, it appears that we need to verify that H satisfies all four of the defining group properties described in Definition 17.2. (See page 258.) As it turns out, we can use the fact that H is a subset of G to simplify our work a bit.

As you may have observed in Preview Activity 19.1, the associativity of the operation in a group is a property of the operation itself and not of the underlying set. Thus, as long as we use the same operation on a subset of the group, the operation retains its associativity. A proof of this fact is essentially given in part (a) of Preview Activity 19.1.

What about the remaining three axioms: closure, the existence of an identity element, and the existence of inverses? What do they have in common, and why do they require more attention than the others?

The answer to this question lies in one small phrase: "there exists." Notice that two of the three axioms mentioned above contain this phrase, which, as you may remember from previous courses, is called an *existential quantifier*. Note also that the condition

$$ab \in G,$$

which appears in the closure axiom, is equivalent to the following:

There exists $c \in G$ such that $ab = c$.

Thus, the closure axiom can be rephrased in an equivalent form as follows:

For each $a, b \in G$, there exists $c \in G$ such that $c = ab$.

In other words, these three axioms all assert that the set itself contains certain elements. Since these are properties of sets, they are not automatically inherited by subsets.

So, to summarize, there is one group axiom—namely, associativity—that asserts a property of the group operation and will therefore automatically be satisfied in any subset of a group that uses the same operation. The remaining properties all contain an existential quantifier and assert that the set itself contains certain elements. These existential axioms (as we might call them) need to be established to show that a subset of a group G is in fact a subgroup. The following theorem states these observations more formally:

Theorem 19.4 (The Subgroup Test). *A subset H of a group G is a subgroup of G if and only if*

(i) *H is closed under the operation from G;*

(ii) *H contains the identity element e from G; and*

(iii) *H contains the inverse of each of its elements—that is, if $h \in H$ and h^{-1} is the inverse of h in G, then $h^{-1} \in H$.*

Activity 19.5. Use the Subgroup Test to prove that the indicated subset H is a subgroup of the group G.

(a) $H = \{[0], [2], [4]\}$, $G = \mathbb{Z}_6$

(b) $H = \left\{ \begin{pmatrix} a & 0 \\ b & 0 \end{pmatrix} : a, b \in \mathbb{R} \right\}$, $G = M_{2 \times 2}(\mathbb{R})$

While Theorem 19.4 may seem straightforward, there are a couple of details that need to be verified. For example, if we assume H is a subgroup of G, then H will contain an identity element. However, there is no reason we can assume that the identity element in H is the same as the identity element in G. A similar statement holds for inverses. We will deal with these issues in the proof of the Subgroup Test, which is given below. You may notice that, unlike previous proofs we have considered, we have not included any instances of the ⑦ symbol. This is because you have now gained enough experience reading proofs to be able to decide for yourself where additional details or explanations are necessary. Although we will still occasionally use the ⑦ symbol throughout the remainder of the text, we will do so less frequently than in past investigations. You should still try to fill in missing details and add clarifying information to the proofs we consider, even when you are not explicitly prompted to do so. You may even want to use the ⑦ symbol as we have in the past to remind yourself where these additional details are necessary.

Table 19.1

Symmetries of a square.

$\circ$	I	$\bar{r}_1$	$\bar{r}_2$	r_1	r_2	R_1	R_2	R_3
I	I	$\bar{r}_1$	$\bar{r}_2$	r_1	r_2	R_1	R_2	R_3
$\bar{r}_1$	$\bar{r}_1$	I	R_2	R_1	R_3	r_1	$\bar{r}_2$	r_2
$\bar{r}_2$	$\bar{r}_2$	R_2	I	R_3	R_1	r_2	$\bar{r}_1$	r_1
r_1	r_1	R_3	R_1	I	R_2	$\bar{r}_2$	r_2	$\bar{r}_1$
r_2	r_2	R_1	R_3	R_2	I	$\bar{r}_1$	r_1	$\bar{r}_2$
R_1	R_1	r_2	r_1	$\bar{r}_1$	$\bar{r}_2$	R_2	R_3	I
R_2	R_2	$\bar{r}_2$	$\bar{r}_1$	r_2	r_1	R_3	I	R_1
R_3	R_3	r_1	r_2	$\bar{r}_2$	$\bar{r}_1$	I	R_1	R_2

Proof of the Subgroup Test. Let G be a group, and let H be a subset of G. For the forward implication, we assume that H is a subgroup of G. So, by definition, H is closed. Now H must also contain an identity element, but we cannot assume that the identity element in H is the same as the identity element in G. Let e_H be the identity element in H and e_G the identity element in G. Then $e_H e_H = e_H e_G$ in G and so the cancellation law in G (Theorem 17.7 on page 261) shows that $e_H = e_G$. Thus, the identity element in H is the same as the identity element in G and H contains the identity element from G. So condition (ii) is satisfied. To verify condition (iii), let $h \in H$. Since H is a group we know that H contains an inverse h_H^{-1} for H. However, we cannot assume that h_H^{-1} is the same as h^{-1}, the inverse of h in G. But,

$$h_H^{-1} = h_H^{-1} e_G = h_H^{-1} \left(hh^{-1} \right) = \left(h_H^{-1} h \right) h^{-1} = e_H h^{-1} = e_G h^{-1} = h^{-1}.$$

So the inverse of h in H is the same as the inverse of H in G and condition (iii) is satisfied. For the converse, suppose that H is closed under the operation defined on G, that H contains the identity e_G from G, and that H contains an inverse for each of its elements. To show that H is a group, we only need to verify associativity. But associativity is a property of the operation and is thus inherited by H (as in part (a) of Preview Activity 19.1). Thus, H is a subgroup of G. ∎

The Center of a Group

Every group contains certain important subsets, one of which is its *center*.

Activity 19.6. In Activity 16.5 (see page 250), we determined the group of symmetries of a square. The operation table for this group (which we call D_4 for reasons we will see in Investigation 21) is reproduced in Table 19.1 for convenience.

(a) The group D_4 is not an Abelian group, but there are some elements in D_4 that commute with all the other elements. Find one such element. Are there any others?

(b) Let $Z(D_4)$ be the set of all elements in D_4 that commute with *every* element in D_4.

(i) Create an operation table for $Z(D_4)$ using the operation from D_4.

(ii) Is $Z(D_4)$ a group? Explain.

Activity 19.6 motivates the following definition:

Definition 19.7. Let G be a group. The **center** of G is the set

$$Z(G) = \{a \in G : ab = ba \text{ for all } b \in G\}.$$

In other words, the center of G is the set of all elements in G that commute with every element in G.

Activity 19.8. Find the center of each of the following groups:

(a) $\mathbb{Z}$.

(b) The group G of symmetries of an equilateral triangle.

As we saw in Activity 19.6, in at least one case, the center of G is a subgroup of G. The next activity explores whether this relationship holds in general.

Activity 19.9. Let G be a group with identity element e.

(a) Is e in $Z(G)$? Explain.

(b) Is $Z(G)$ closed under the operation in G? Prove your answer.

(c) If $a \in Z(G)$, is $a^{-1} \in Z(G)$? Prove your answer.

(d) Is $Z(G)$ a subgroup of G? Explain. If $Z(G)$ is a subgroup of G, is $Z(G)$ an Abelian subgroup of G?

(e) Complete the statement of the following theorem as specifically and precisely as you can.

> **Theorem 19.10.** *Let G be a group. The center of G is a(n)* _____ _____ *of G.*

So every group has at least three defined subgroups: the trivial subgroup consisting of just the identity, the entire group itself, and the center of the group. It is important to note that these subgroups need not be distinct. For example, the center of every Abelian group is the entire group. It is also possible for the center of a group to contain only the identity.

The Subgroup Generated by an Element

Given a group G and an element $a \in G$, it is natural to look for the smallest subgroup of G that contains a. The next activity demonstrates what this subgroup looks like for one particular example.

Activity 19.11. Let H be the smallest subgroup of $\mathbb{Z}$ containing 5.

(a) Find three other elements that must be in H.

(b) Explain why the set $H' = \{5m : m \in \mathbb{Z}\}$ is a subset of H.

(c) Explain why H' must equal H.

Let's now generalize what we saw in Activity 19.11. From this point on, we will assume the operation in G is written multiplicatively.

Activity 19.12. Denote by $\langle a \rangle$ the smallest subgroup of G containing the element a.

(a) Since $\langle a \rangle$ is a group, $\langle a \rangle$ must be closed under the group operation. Given $a \in \langle a \rangle$, list five other elements that must be in $\langle a \rangle$.

(b) Since $\langle a \rangle$ is a group, $\langle a \rangle$ must contain the inverse of each of its elements. Given $a \in \langle a \rangle$, list five more elements that must be in $\langle a \rangle$.

(c) Explain why $\langle a \rangle$ must contain the set $\{a^n \mid n \in \mathbb{Z}\}$. What is a^0?

(d) Let $H = \{a^n \mid n \in \mathbb{Z}\}$. If H is a group, then the previous results tell us that H must be the smallest group containing a. That is, $\langle a \rangle = H$. Prove that $H = \{a^n \mid n \in \mathbb{Z}\}$ is a subgroup of G.

The subgroup $\langle a \rangle$ of G is called the *subgroup generated by* a or the *cyclic subgroup* generated by a. The element a is called a *generator* of the group $\langle a \rangle$.

Definition 19.13. Let G be a group, and let $a \in G$. The **cyclic subgroup generated by** a, denoted $\langle a \rangle$, is defined by

$$\langle a \rangle = \{a^n \mid n \in \mathbb{Z}\}$$

if the group operation is written in multiplicative notation, or

$$\langle a \rangle = \{na \mid n \in \mathbb{Z}\}$$

if the group operation is written in additive notation.

As stated in the next definition, any group that is generated by a single element is said to be *cyclic*.

Definition 19.14. A group G is a **cyclic group** if $G = \langle a \rangle$ for some $a \in G$.

Activity 19.15. For the given group G and element a, determine the elements in the cyclic subgroup generated by a.

(a) $G = \mathbb{Z}_{10}$, $a = [3]$.

(b) $G = U_{15}$, $a = [8]$. (Recall that U_n is the group of units in $\mathbb{Z}_n$.)

Activity 19.16. Which of the following groups is a cyclic group?

(a) $\mathbb{Z}$

(b) $\mathbb{Z}_n$ for a positive integer n

(c) $\mathbb{R}$

(d) U_{10}

Just as we defined the order of a group in Investigation 17, we can use cyclic groups to define the order of an *element* in a group.

Definition 19.17. Let G be a group and $a \in G$. If $\langle a \rangle$ is a finite group, then the element a has **finite order**. In this case, the **order** of a is equal to the order of the subgroup generated by a. If $\langle a \rangle$ is an infinite group, the element a has **infinite order**.

The notation we use for the order of an element a in a group G is $|a|$ (not to be confused with absolute value; the meaning of the symbol should be clear from the context). Definition 19.17 tells us $|a| = |\langle a \rangle|$. For example, in $\mathbb{Z}_{10}$, we have $||[4]|| = |\langle [4] \rangle| = |\{[4], [8], [2], [6], [0]\}| = 5$.

Every group has a cyclic subgroup for each of its elements, although they may not all be different. Cyclic groups are very important in group theory, as we will learn throughout our investigations of group theory. As one example, we will see later that cyclic groups are the building blocks of all finite Abelian groups.

Concluding Activities

Activity 19.18. Let G be a group with identity e, and let H be a subset of G. To use the Subgroup Test to show that H is a subgroup of G, we need to verify three things: H is closed under the operation in G, $e \in H$, and $h^{-1} \in H$ whenever $h \in H$. In this activity, we will see that, with a little bit of thought, we can reduce the number of things we need to show from three down to two. Assume that G is a group and H is a nonempty subset of G.

(a) Show that if $hk^{-1} \in H$ for all $h, k \in H$, then H is closed.

(b) Show that if $hk^{-1} \in H$ for all $h, k \in H$, then H contains e.

(c) Show that if $hk^{-1} \in H$ for all $h, k \in H$, then H contains an inverse for each of its elements.

(d) Explain why if $hk^{-1} \in H$ for all $h, k \in H$, then H is a subgroup of G.

(e) Use the results from parts (a) – (d) to complete the following alternative form of the Subgroup Test. (Note that your statement should include two conditions.)

A subset H of a group G is a subgroup of G if ...

Activity 19.19. In this activity, we will explore a simple relationship between the subgroup generated by an element and the subgroup generated by its inverse.

(a) Determine the elements in the group $\langle [2] \rangle$ in $\mathbb{Z}_6$. What is the inverse of $[2]$ in $\mathbb{Z}_6$? Now determine the elements in the group $\langle -[2] \rangle$. What do you notice?

(b) Let $\alpha = \begin{pmatrix} 1 & 2 & 3 & 4 \\ 2 & 3 & 4 & 1 \end{pmatrix}$ in the group D_4 of symmetries of a square. Determine the elements in the group $\langle \alpha \rangle$. What is the inverse of α in D_4? Now determine the elements in the group $\langle \alpha^{-1} \rangle$. What do you notice?

(c) Let G be a group, and let a be an element of G. Based on your observations in parts (a) and (b), what relationship do you think exists between $\langle a \rangle$ and $\langle a^{-1} \rangle$? (Although it is dangerous to generalize from a small number of examples, in this case you should see a fairly clear relationship.) Prove this relationship.

(d) Let G be a group, and let a be an element of G. Using the relationship you found in part (c), what relationship do you think exists between the order of a and the order of a^{-1}? Explain. (Hint: Consider both the finite and infinite cases.)

Exercises

(1) Determine the orders of each of the indicated elements.

    (a) 2 in $\mathbb{Z}$

    (b) $[10]$ in $\mathbb{Z}_{18}$

    (c) $\begin{pmatrix} 1 & 2 & 3 & 4 \\ 2 & 3 & 4 & 1 \end{pmatrix}$ in D_4, the group of symmetries of the square

(2) Let H denote the set of all 2×2 matrices of the form

$$\begin{bmatrix} x & 0 \\ y & 0 \end{bmatrix},$$

    where $x, y \in \mathbb{R}$. Is H a subgroup of $\mathcal{M}_{2\times2}(\mathbb{R})$? Prove your answer.

(3) Let H denote the set of all 2×2 matrices of the form

$$\begin{bmatrix} x & y \\ -y & x \end{bmatrix},$$

    where $x, y \in \mathbb{R}$. Is H a subgroup of $\mathcal{M}_{2\times2}(\mathbb{R})$? Prove your answer.

(4) Let G be a group and H a subgroup of G. Which of the following conjectures do you think are true, and which do you think are false? Provide brief arguments or examples to justify your answers.

    (a) If G is finite, then H is finite.

    (b) If H is finite, then G is finite.

    (c) If G is Abelian, then H is Abelian.

    (d) If H is Abelian, then G is Abelian.

* (5) Let G be an Abelian group, and let H and K be subgroups of G. Let

$$HK = \{hk : h \in H \text{ and } k \in K\}.$$

    (a) Let $H = \langle [6] \rangle$ and $K = \langle [8] \rangle$ in $G = \mathbb{Z}_{12}$. Find the elements of HK in this example.

    (b) Prove that HK is a subgroup of G.

    (c) Is HK a subgroup of G if G is non-Abelian? Verify your answer.

* (6) In this exercise, we will explore some special subsets of certain groups.

  (a) Show that the set $H = \{n \in \mathbb{Z} : |n| = 1\}$ is a group under multiplication. Draw a picture to geometrically illustrate this group. Is H a subgroup of $\mathbb{Z}$?

  (b) Show that the set $K = \{q \in \mathbb{Q} : |q| = 1\}$ is a group under multiplication. Draw a picture to geometrically illustrate this group. Is K a subgroup of $U(\mathbb{Q})$, the group of units in $\mathbb{Q}$?

  (c) Show that the set $M = \{x \in \mathbb{R} : |x| = 1\}$ is a group under multiplication. Draw a picture to geometrically illustrate this group. Is M a subgroup of $U(\mathbb{R})$, the group of units in $\mathbb{R}$?

  (d) Show that the set $S = \{z \in \mathbb{C} : |z| = 1\}$ is a group under multiplication. (Here $|z|$ is the norm or modulus of the complex number z—that is, if $z = a + bi$, then $|z| = \sqrt{a^2 + b^2}$.) Draw a picture to geometrically illustrate this group. Is S a subgroup of $U(\mathbb{C})$, the group of units in $\mathbb{C}$?

  (e) Show that the set $\mathrm{SL}_n(\mathbb{R}) = \{A \in \mathcal{M}_{n \times n}(\mathbb{R}) : |A| = 1\}$ is a group under multiplication. (Here $|A|$ is the determinant of the matrix A.) This group is called the **special linear group**. Is $\mathrm{SL}_n(R)$ a subgroup of $\mathrm{GL}_n(\mathbb{R}) = U(\mathcal{M}_{n \times n}(\mathbb{R}))$, the group of units in $\mathcal{M}_{n \times n}(\mathbb{R})$? (Note that the group $\mathrm{GL}_n(\mathbb{R})$ is called the **general linear group**.)

(7) **Euler's formula**,
$$e^{ix} = \cos(x) + i\sin(x),$$
relates complex exponentials to the polar form of a complex number. Using Euler's formula we can see that if $z = e^{i(2\pi k)/n}$ for integers k between 0 and $n - 1$, then
$$z^n = \cos(2\pi k) + i\sin(2\pi k) = 1.$$

The numbers of the form $e^{i(2\pi k)/n}$ are called the **complex roots of unity**. Let n be a positive integer and let
$$H = \left\{ e^{i(2\pi k)/n} : 0 \le k < n, k \in \mathbb{Z} \right\}.$$

Show that H is a subgroup of the group of complex numbers with norm 1 (see Exercise (6)). Draw a picture of H with $n = 6$ to illustrate.

* (8) **Intersections of subgroups.** Let G be a group with subgroups H and K.

  (a) Is $H \cap K$ a subgroup of G? Prove your answer.

  (b) Can we generalize? That is, if $\{H_\alpha\}$ is a collection of subgroups of G indexed by α in an indexing set I, is it the case that $\bigcap_{\alpha \in I} H_\alpha$ is a subgroup of G? Prove your answer.

(9) **Unions of subgroups.** Let G be a group with subgroups H and K.

  (a) Is $H \cup K$ necessarily a subgroup of G? Prove your answer.

  (b) Under what conditions is $H \cup K$ a subgroup of G? Prove your answer.

(10) Let G be a group, and let $a \in G$. The **centralizer** of a is defined to be the set of all $g \in G$ such that $ga = ag$. In other words, the centralizer of a is the set of all elements that commute with a.

  (a) Find the centralizer of the 180° rotation R_2 in D_4, the group of symmetries of a square. (See Activity 19.6 for the notation and operation table for this group.)

  (b) Find the centralizer of the reflection r_1 around the vertical axis in D_4.

Table 19.2

Symmetries of a regular hexagon.

$\circ$	I	r_1	r_2	r_3	$\overline{r}_1$	$\overline{r}_2$	$\overline{r}_3$	R_1	R_2	R_3	R_4	R_5
I	I	r_1	r_2	r_3	$\overline{r}_1$	$\overline{r}_2$	$\overline{r}_3$	R_1	R_2	R_3	R_4	R_5
r_1	r_1	I	R_4	R_2	R_5	R_3	R_1	$\overline{r}_3$	r_3	$\overline{r}_2$	r_2	$\overline{r}_1$
r_2	r_2	R_2	I	R_4	R_1	R_5	R_3	$\overline{r}_1$	r_1	$\overline{r}_3$	r_3	$\overline{r}_2$
r_3	r_3	R_4	R_2	I	R_3	R_1	R_5	$\overline{r}_2$	r_2	$\overline{r}_1$	r_1	$\overline{r}_3$
$\overline{r}_1$	$\overline{r}_1$	R_1	R_5	R_3	I	R_4	R_2	r_1	$\overline{r}_3$	r_3	$\overline{r}_2$	r_2
$\overline{r}_2$	$\overline{r}_2$	R_3	R_1	R_5	R_2	I	R_4	r_2	$\overline{r}_1$	r_1	$\overline{r}_3$	r_3
$\overline{r}_3$	$\overline{r}_3$	R_5	R_3	R_1	R_4	R_2	I	r_3	$\overline{r}_2$	r_2	$\overline{r}_1$	r_1
R_1	R_1	$\overline{r}_1$	$\overline{r}_2$	$\overline{r}_3$	r_2	r_3	r_1	R_2	R_3	R_4	R_5	I
R_2	R_2	r_2	r_3	r_1	$\overline{r}_2$	$\overline{r}_3$	$\overline{r}_1$	R_3	R_4	R_5	I	R_1
R_3	R_3	$\overline{r}_2$	$\overline{r}_3$	$\overline{r}_1$	r_3	r_1	r_2	R_4	R_5	I	R_1	R_2
R_4	R_4	r_3	r_1	r_2	$\overline{r}_3$	$\overline{r}_1$	$\overline{r}_2$	R_5	I	R_1	R_2	R_3
R_5	R_5	$\overline{r}_3$	$\overline{r}_1$	$\overline{r}_2$	r_1	r_2	r_3	I	R_1	R_2	R_3	R_4

(c) How is the centralizer of an element different than the center of the group? Explain.

(d) Prove that the centralizer of $a \in G$ is a subgroup of G.

(e) Is the centralizer of $a \in G$ necessarily an Abelian group? Give a proof or counterexample to justify your answer.

(f) Show that $Z(G)$, the center of G, is equal to the intersection of all the centralizers of elements in G—that is,

$$Z(G) = \cap_{a \in G} C(a).$$

(11) If G is a group and H a subgroup of G, the **centralizer of the subgroup** H is the set

$$C(H) = \{a \in G : gh = hg \text{ for all } h \in H\}.$$

(a) We encountered the group of symmetries of a regular hexagon in Activity 16.5. (See page 250.) The operation table for this group, which we will label as D_6, is reproduced in Table 19.2. Find $C(H)$ if $H = \{I, R^2, R^4\}$.

(b) Let G be an arbitrary group and H a subgroup of G. Is $C(H)$ always a subgroup of G? Prove your answer.

(12) Determine whether H is a subgroup of G.

(a) $G = \mathbb{Z}_{20}$ under addition, $H = \{[0], [3], [6], [9], [12], [15], [18]\}$.

(b) $G = U_7$ under multiplication, $H = \{[1], [2], [4]\}$.

(c) $G = U_{16}$ and $H = \{[1], [7], [9], [15]\}$.

* (13) **Subgroups of $\mathbb{Z}$.** In this exercise, we will show that the only subgroups of $\mathbb{Z}$ are the subgroups of the form $n\mathbb{Z} = \{nk : k \in \mathbb{Z}\}$ for some $n \in \mathbb{Z}$.

(a) First show that $n\mathbb{Z}$ is a subgroup of $\mathbb{Z}$ for any $n \in \mathbb{Z}$.

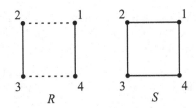

Figure 19.1
Rigid motions in the plane.

(b) Now follow the steps below to show that any subgroup of $\mathbb{Z}$ is equal to $n\mathbb{Z}$ for some $n \in \mathbb{Z}$.

  (i) Let H be a subgroup of $\mathbb{Z}$. Explain why there are two cases to consider: $H = \{0\}$ and $H \neq \{0\}$. Complete the case where $H = \{0\}$.

  (ii) Assume $H \neq \{0\}$. Show that H must contain a positive integer.

  (iii) Let $T = \{h \in H : h > 0\}$. Explain why T contains a smallest positive element. (Hint: Think of the Well-Ordering Principle.)

  (iv) Let a be the smallest positive element in T. Prove that $H = a\mathbb{Z}$. (Hint: Use the Division Algorithm to show $a \mid h$ for each $h \in H$.)

(14) If m and n are positive integers, what can we say about the relationship (as groups) between $m\mathbb{Z}$ and $n\mathbb{Z}$? (See Exercise (13).) Be as specific as possible, and prove your answer.

* (15) Prove that any group of order 2 or 3 is cyclic.

(16) Find the order of each element of the group U_{11}. Also find all subgroups of U_{11}.

(17) (a) Find all elements that generate $\mathbb{Z}_6$, $\mathbb{Z}_8$, and $\mathbb{Z}_{15}$.

   (b) Let $\langle a \rangle$ be a cyclic group of order 6, $\langle b \rangle$ a cyclic group of order 8, and $\langle c \rangle$ a cyclic group of order 15. Find all elements that generate $\langle a \rangle$, $\langle b \rangle$, and $\langle c \rangle$. Can you see a pattern here that characterizes the elements that generate a cyclic group?

(18) Is every cyclic group Abelian? Is every Abelian group cyclic? Justify your answers.

(19) Recall that U_{21} is the group of units in the ring $\mathbb{Z}_{21}$. List the elements in U_{21}. Then find the subgroup of U_{21} generated by the element $[5]$.

(20) Consider the two objects shown in Figure 19.1: a square S on the right and a square R with both solid and dashed sides on the left. We have already determined the group D_4 of symmetries of S (see Activity 19.6), but any symmetry of R will have to preserve the shading of the sides.

  (a) Explain why any symmetry of R is also a symmetry of S.

  (b) Determine the set H of symmetries of the object R.

  (c) Which of the group axioms (using the operation of composition of symmetries) does the set of symmetries of R satisfy?

  (d) Create the operation table for the set of symmetries of R.

  (e) Explain why H is a subgroup of D_4.

(21) **The Subgroup Test for finite groups.** We can reduce the number of items we need to prove in the Subgroup Test if we are working in a finite group. Assume that G is a finite group with identity e and that the order of G is n.

    (a) Let H be a nonempty subset of G that is closed under the operation in G, and let $h \in H$. Explain why the set of elements $S = \{h, h^2, h^3, \ldots\}$ must be finite.

    (b) Use the fact that the elements in S repeat to show that $e \in H$.

    (c) Again use the fact that the elements in S repeat to show that $h^{-1} \in H$.

    (d) Write a formal proof of the following theorem:

        Theorem. *Let G be a finite group. A nonempty subset H of G is a subgroup of G if and only if H is closed under the operation in G.*

* (22) **Subgroups of order 2.** In general, it is a difficult task to determine which subgroups a group has. In subsequent investigations, we will develop some powerful tools to help in this regard. For now, however, we will begin with a relatively straightforward argument. Show that any group of even order must contain an element (and hence a subgroup) or order 2. (Hint: Count elements and their inverses.)

(23) In this exercise, we will prove **Wilson's Theorem**:

    Theorem (Wilson's Theorem). *A positive integer p is prime if and only if*

$$(p-1)! \equiv -1 \pmod{p}.$$

Wilson's Theorem* gives a test (although not a practical one) for determining if a positive integer is prime. Wilson's Theorem also shows that $n! + 1$ is composite for infinitely many different positive integers n. It is unknown if $n! + 1$ is prime for infinitely many different integers n.

    (a) Let p be a prime. Show that $[p-1]$ is the only element in U_p of order 2.

    (b) Prove that if p is prime, then $(p-1)! \equiv -1 \pmod{p}$. (Hint: If p is an odd prime, pair each element in U_p with its inverse.)

    (c) Show that if m is composite, then $(m-1)! \not\equiv -1 \pmod{m}$.

    (d) Explain why parts (b) and (c) prove Wilson's Theorem.

Connections

In this investigation, we studied subgroups of groups. If you studied ring theory before group theory, you should notice connections between the topics in this investigation and those in Investigation 6. The idea of a subring is the same is that of a subgroup. In particular, if R is a ring, then a subring

---

*This theorem was conjectured by John Wilson in the 18th century, but appears to have been known to Ibn al-Haytham (according to Oystein Ore in *Number Theory and Its History*, Dover, 1988) as early as 1000 AD. Lagrange appears to have given the first proof in 1771.

of R is just a subset of R that is also a ring under the same operations as R. The only significant difference is that a group comes with one operation and a ring comes with two, so it is a bit more work to determine if a subset of a ring is a subring than if a subset of a group is a subgroup. Because of the simpler structure of groups, there are special types of subgroups that we identify (e.g., cyclic groups) that do not have a direct counterpart in rings. We will be able to exploit these special subgroups to classify some important families of groups—a task that is much more difficult to do with rings.

Investigation 20

Subgroups of Cyclic Groups

Focus Questions

By the end of this investigation, you should be able to give precise and thorough answers to the questions listed below. You may want to keep these questions in mind to focus your thoughts as you complete the investigation.

- What property must be satisfied by all subgroups of a cyclic group?

- What is the subgroup structure of a finite cyclic group?

- What is the subgroup structure of an infinite cyclic group?

- What important properties does the order of an element in a group have?

Preview Activity 20.1. The family of groups denoted by $\mathbb{Z}_n$ (where n is a positive integer) are the canonical examples of finite cyclic groups. In fact, it can be shown that, for each $n \in \mathbb{Z}^+$, $\mathbb{Z}_n$ is the only cyclic group of order n. Recall that every element in $\mathbb{Z}_n$ has the form $[k]_n = k[1]_n$ for some integer k, so $\mathbb{Z}_n$ is generated by $[1]_n$—that is, $\mathbb{Z}_n = \langle [1]_n \rangle$. (Note that, when the context is clear, we will typically omit subscripts and simply write $\mathbb{Z}_n = \langle [1] \rangle$.) One question we will address in this investigation is what the subgroup structure of $\mathbb{Z}_n$ looks like. In other words, if H is a subgroup of $\mathbb{Z}_n$, what kind of things can we say about H? In this activity, we will consider the specific example where H is a nontrivial subgroup of $\mathbb{Z}_{100}$.

(a) Suppose $[20] \in H$. List at least three other elements that are also in H. What is the smallest number of elements H can have? Explain.

(b) If $[20] \in H$, must it be true that H contains $\langle [20] \rangle$? Why or why not?

(c) If $[20] \in H$, is it possible that $[k] \in H$ for some integer k with $1 < k < 20$? If so, in what situations could that happen? If not, why is it impossible?

(d) Let $S = \{k \in \mathbb{Z}^+ : k[1] \in H\}$.

   (i) Why is S nonempty?

   (ii) What important conclusion can we draw about S as a nonempty subset of $\mathbb{Z}^+$? Explain.

   (iii) How do you think this important conclusion is related to H?

Introduction

Which subgroups a group contains can tell us a lot about the group. In general, it is difficult to explicitly describe all subgroups of a given group, but for some groups the subgroup structure is more accessible. The simplest type of group is a cyclic group, and the subgroup structure of a cyclic group is fairly simple, as we will see in this investigation.

Subgroups of Cyclic Groups

Throughout this section, we will let G be a cyclic group generated by an element $a \in G$. Recall that this means every element in G can be written as an integer power of a, or

$$G = \langle a \rangle = \{a^k : k \in \mathbb{Z}\}.$$

Our goal in this investigation is to determine all subgroups of G. Since every element in G is a power of a generator a, it follows that every element in any subgroup of G is also a power of a. This might lead us to conjecture the following theorem.

Theorem 20.2. *Every subgroup of a cyclic group is cyclic.*

The details of a proof of this theorem are presented in Activity 20.3, but a few words might be in order before we proceed. Let H be a subgroup of a cyclic group $G = \langle a \rangle$. Since $H \subseteq G$, every element of H is a power of a. The problem we encounter, though, is that it is not obvious exactly which powers of a are in H. This is the detail we need to address, and the main idea we will use to do so was introduced in Preview Activity 20.1.

Activity 20.3. Let $G = \langle a \rangle$ be a group.

(a) We want to prove that every subgroup of G is cyclic. How is this statement quantified? What should we assume to begin our proof?

(b) Let H be a subgroup of G. There is one subgroup of every group that is cyclic. What subgroup is that? If H is this subgroup, we are done. So we can assume H is not this subgroup. What additional assumption does that allow us to make?

(c) Let $h \in H$. Since $h \in G$, the element h must have a specific form. What form does h have?

(d) Let H be a nontrivial subgroup of G, and let $S = \{k \in \mathbb{Z}^+ : a^k \in H\}$. Why is S nonempty?

(e) What result tells us that S contains a smallest element?

(f) Let m be the smallest positive integer in S. That is, m is the smallest positive integer such that $a^m \in H$. This element a^m is a candidate for a generator for H. What must we do to show $\langle a^m \rangle = H$?

(g) Suppose $b \in H$. Why does b equal a^l for some integer l?

Table 20.1

Orders of elements in $\mathbb{Z}_4$ through $\mathbb{Z}_9$.

k	$\|[k]\|$ in $\mathbb{Z}_4$	$\|[k]\|$ in $\mathbb{Z}_5$	$\|[k]\|$ in $\mathbb{Z}_6$	$\|[k]\|$ in $\mathbb{Z}_7$	$\|[k]\|$ in $\mathbb{Z}_8$	$\|[k]\|$ in $\mathbb{Z}_9$
0	1	1	1	1		
1	4	5	6	7		
2	2	5	3	7		
3	4	5	2	7		
4	1	5	3	7		
5	4	1	6	7		
6	2	5	1	7		
7	4	5	6	1		
8	1	5	3	7		

(h) Since m is positive, we can divide m into l and obtain a unique quotient q and remainder r. What result tells us we can do this? What is true about the remainder? What is the specific relationship between m and l?

(i) Why can we say $a^l = a^{qm+r}$? How can we use this equation to show $r = 0$? What conclusion can we then draw?

Theorem 20.2 tells us an important fact about every subgroup of a cyclic group, but there is much more that we can say. First, we will need to know more about the orders of the elements in a group.

Properties of the Order of an Element

Preview Activity 20.4. The orders of the elements in $\mathbb{Z}_4$, $\mathbb{Z}_5$, $\mathbb{Z}_6$, and $\mathbb{Z}_7$ are given in Table 20.1.

(a) Complete Table 20.1 by calculating the orders of the elements in $\mathbb{Z}_8$ and $\mathbb{Z}_9$.

(b) For any positive integer n and any element $[k] \in \mathbb{Z}_n$, there is an explicit relationship between n and the order of $[k]$. Review the entries in Table 20.1, and make a conjecture about this relationship.

Preview Activity 20.4 illustrates the relationship between the order of a power of a generator in a cyclic group (such as $k[1] = [k]$ in $\mathbb{Z}_n$) and the order of the generator itself ($[1]$ in $\mathbb{Z}_n$). In what follows, we will prove that relationship and use it to completely determine the subgroup structure of finite and infinite cyclic groups. We will first determine when a power of an element of finite order in G can equal the identity.

Theorem 20.5. *Let G be a group with identity e, and let a be an element of G of order n. Then:*

(i) $a^n = e$ *and, moreover, n is the smallest positive integer so that* $a^n = e$.

(ii) *If s is an integer so that* $a^s = e$, *then n divides s.*

Proof. Let G be a group with identity e, and let a be an element of G of order n.

To prove (i), note that since $a \in \langle a \rangle$ and $\langle a \rangle$ is closed under the group operation, the elements $a, a^2, a^3, \ldots, a^n \ldots$ are all in $\langle a \rangle$. Because the order of $\langle a \rangle$ is finite, these powers cannot all be distinct. So there must exist $0 < i < j$ so that

$$a^j = a^i.$$

Multiplying both sides of this equation by a^{-i} gives us

$$a^{j-i} = e.$$

So there is at least one positive power of a that is equal to the identity. The Well-Ordering Principle tells us that there is a smallest positive power k of a so that $a^k = e$. To complete the proof, we will show that $k = n$.

First, we will show that $k \leq n$. We proceed by contradiction and assume $k > n$. Consider the elements $a, a^2, a^3, \ldots, a^k$ in $\langle a \rangle$. Since $|\langle a \rangle| = n$, these powers cannot all be distinct. So there exist $0 < i < j \leq k$ so that $a^j = a^i$. This implies $a^{j-i} = e$. However, $0 < j - i < j \leq k$, which contradicts the fact that k is the smallest positive power of a that equals the identity. We can thus conclude that $k \leq n$.

Next, we will show that $k \geq n$. Again we proceed by contradiction and assume $k < n$. Let $t \in \mathbb{Z}$. By the Division Algorithm, there are integers q and r such that

$$t = qk + r \quad \text{with} \ \ 0 \leq r < k.$$

Then

$$a^t = a^{qk+r} = (a^k)^q a^r = e^q a^r = a^r.$$

So any power of a is equal to a^r for some $0 \leq r < k$. This means that there are only $k < n$ distinct powers of a, and $|\langle a \rangle| \leq k < n$, a contradiction. Therefore, $k \geq n$.

Combining the inequalities $k \leq n$ and $k \geq n$ yields $k = n$ as desired.

To prove (ii), assume $a^s = e$ for some integer s. By the Division Algorithm, there are integers q and r such that

$$s = qn + r \quad \text{with} \ \ 0 \leq r < n. \tag{20.1}$$

Then

$$e = a^s = a^{qn+r} = (a^n)^q a^r = a^r.$$

But n is the smallest positive integer such that $a^n = e$. Therefore, we must have $r = 0$, and (20.1) shows us that n divides s. ∎

Now let's turn our attention to the result indicated in Preview Activity 20.4 about orders of elements in finite cyclic groups.

Theorem 20.6. *Let $G = \langle a \rangle$ be a cyclic group of order n, and let $k \in \mathbb{Z}$. Then*

(i) $\langle a^k \rangle = \langle a^{\gcd(k,n)} \rangle$

(ii) $|\langle a^k \rangle| = \frac{n}{\gcd(k,n)}$.

Proof. Let $G = \langle a \rangle$ be a cyclic group of order $n \in \mathbb{Z}^+$, and let $k \in \mathbb{Z}$. Let $d = \gcd(k, n)$. Then there exist integers u, v so that $du = k$ and $dv = n$.

To prove (i), we will show that $\langle a^k \rangle \subseteq \langle a^d \rangle$ and $\langle a^d \rangle \subseteq \langle a^k \rangle$. First, note that $d \mid k$, and so $a^k = a^{du} = (a^d)^u$, which implies that $a^k \in \langle a^d \rangle$. By closure, it follows that $\langle a^k \rangle \subseteq \langle a^d \rangle$. To prove containment in the other direction, we will show that $a^d \in \langle a^k \rangle$.

By Bezout's Identity, there exist integers x, y so that $d = kx + ny$. Using part (i) of Theorem 20.5 we see that

$$a^d = a^{kx+ny} = (a^k)^x (a^n)^y = (a^k)^x \in \langle a^k \rangle.$$

Therefore, $\langle a^d \rangle \subseteq \langle a^k \rangle$ by closure. Combining the two containments gives us $\langle a^k \rangle = \langle a^d \rangle$.

To prove (ii), we must show that the order of a^d is $\frac{n}{d}$. Part (i) of Theorem 20.5 implies that

$$(a^d)^{\frac{n}{d}} = a^n = e.$$

Then part (ii) of Theorem 20.5 tells us that $|a^d| \leq \frac{n}{d}$. To show $|a^d| \geq \frac{n}{d}$, we need to show that no smaller positive power of a^d is equal to e. If $0 < i < \frac{n}{d}$, then $0 < di < n$. Since n is the smallest positive power of a that is equal to e (by part (ii) of Theorem 20.5), we see that $(a^d)^i \neq e$. Therefore, $|a^d| \geq \frac{n}{d}$. The two inequalities ($|a^d| \leq \frac{n}{d}$ and $|a^d| \geq \frac{n}{d}$) show that $|a^d| = \frac{n}{d}$, and part (i) allows us to conclude that $|a^k| = |a^d| = \frac{n}{d}$. ∎

In the next section, we will use Theorem 20.6 to completely determine the subgroup structure of all finite cyclic groups.

Finite Cyclic Groups

Let $G = \langle a \rangle$ be a cyclic group of order n. Our previous theorems in this investigation allow us to completely classify the subgroup structure of G, as stated in the next theorem.

Theorem 20.7. *Let G be a finite cyclic group of order n. For each positive divisor m of n, there is exactly one subgroup of G of order m, and these are the only subgroups of G.*

An outline of a proof of Theorem 20.7 is contained in the next activity.

Activity 20.8. Let $G = \langle a \rangle$ be a finite cyclic group of order n, and let m be a positive divisor of n.

(a) Use Theorem 20.6 to find a subgroup of G of order m. This shows that G contains at least one subgroup of order m.

(b) Now suppose that G contains subgroups H and K of order m.

    (i) How can we conclude that $H = \langle a^s \rangle$ and $K = \langle a^t \rangle$ for some integers s and t?

    (ii) Explain how Theorem 20.6 tells us that $H = K$. What can we conclude about the number of subgroups of G of order m?

(c) Explain how we have completely classified the subgroup structure of finite cyclic groups.

As one final note, Theorem 20.6 also shows that a^k is a generator for G if and only if $\gcd(k, n) = 1$.

Activity 20.9.

(a) Find all the generators of $\mathbb{Z}_{30}$.

(b) Let a be an element of a group with $|a| = 15$. Find the orders of a^2, a^6, and a^{10}.

(c) Let G be a group, and let $a \in G$ be an element of order n. Explain why $\langle a^k \rangle = \langle a^{n-k} \rangle$. Use this result to prove that $\gcd(n, k) = \gcd(n - k, n)$.

Infinite Cyclic Groups

The subgroup structure of infinite cyclic groups is also known. If $G = \langle a \rangle$ is an infinite cyclic group, Theorem 20.2 tells us that every subgroup of G is also cyclic. Note that G will contain at least one finite cyclic subgroup. (Which subgroup must this be?) The structure of all other subgroups of G is described in the next theorem, whose proof is left as Exercise (9).

Theorem 20.10. *Let $G = \langle a \rangle$ be an infinite cyclic group with identity e, and let $b \neq e$ be an element in G. Then $\langle b \rangle$ is an infinite cyclic group.*

Concluding Activities

Activity 20.11. Use the steps outlined in Activity 20.3 to write a formal proof of Theorem 20.2.

Activity 20.12. Use the steps outlined in Activity 20.8 to write a formal proof of Theorem 20.7.

Activity 20.13. Let G be a group, and let $a \in G$ be an element of order n. If m is a positive integer and $|a^m| = k$, is it true that $|a^k| = m$? If yes, then prove your answer. If no, what can we say about $|a^k|$?

Exercises

(1) Let $G = \mathbb{Z}_{24} = \langle [1] \rangle$.

(a) Explain why $\langle [2] \rangle = \langle [10] \rangle$ in G.

(b) Find all elements $[m] \in G$ so that $\langle [m] \rangle = \langle [2] \rangle$.

(c) Find $\|[6]\|$ in G.

(2) Explicitly verify Theorem 20.6 for the group U_{26}.

(3) What is the order of $[10]$ in $\mathbb{Z}_{18}$?

(4) (a) Show that U_{22} is cyclic.

    (b) Find all generators of U_{22}. Explain how you know that each element is a generator.

(5) Let $A = \begin{bmatrix} 1 & 0 \\ 0 & -1 \end{bmatrix}$ and $B = \begin{bmatrix} 1 & 1 \\ 0 & -1 \end{bmatrix}$ in $GL_2(\mathbb{R})$.

    (a) Find $|A|$ and $|B|$.

    (b) Determine $|AB|$. Does your answer surprise you? Explain.

(6) Is it possible for a group G to contain a nonidentity element of finite order and also an element of infinite order? If yes, illustrate with an example. If no, give a convincing explanation for why it is not possible.

(7) Suppose $G = \langle a \rangle$ is a cyclic group of order 12.

    (a) Find all the generators of G. Explain your reasoning.

    (b) Find all proper subgroups of G, and list their elements. Find all the generators of each subgroup. Explain your reasoning.

(8) Let G be a group with identity element e, and let a be an element in G. Label each of the following statements as either true or false. Justify your answers.

    (a) If n is a positive integer and $a^n = e$, then $|a| = n$.

    (b) If $a^n \neq e$ for every nonzero integer n, then $G = \langle a \rangle$.

(9) Prove Theorem 20.10. (Hint: Let $G = \langle a \rangle$ be an infinite cyclic group with identity e, and let $b \neq e$ be an element in G. Show that the positive integer powers of b are all distinct.)

(10) Let G be a group with identity e, and let $a \in G$. If $a^6 = e$ but $a^2 \neq e$, what can we say about $|a|$? Prove your answer.

(11) Suppose that the only distinct subgroups a cyclic group G with identity e are G, $\{e\}$, and a subgroup of order 23. What is $|G|$?

(12) Let G be a group, and let a be an element of G of order n. Prove that $|a^k| = |a^{n-k}|$ for any integer k.

(13) Find two distinct subgroups of order 2 of the group D_3 of symmetries of an equilateral triangle. Explain why this fact alone shows that D_3 is not a cyclic group.

(14) Suppose G is a group of order n so that every proper subgroup of G is cyclic. Must G be cyclic? Prove your answer.

(15) Let $n \in \mathbb{Z}^+$, and let d be a positive divisor of n. Theorem 20.7 tells us that $\mathbb{Z}_n$ contains exactly one subgroup of order d, but not how many *elements* $\mathbb{Z}_n$ has of order d. We will determine that number in this exercise.

    (a) Determine the number of elements in $\mathbb{Z}_{12}$ of each order d. Fill in the table below to compare your answers to the number of integers between 1 and d that are relatively prime to d.

Divisor d	Number of elements of order d	Number of integers between 1 and d relatively prime to d
1		
2		
3		
4		
6		
12		

(b) Part (a) appears to indicate that there is a relationship between the number of elements in $\mathbb{Z}_n$ of a given order and the number of integers relatively prime to that order. There is a useful function that describes this number. The *Euler phi function* φ is defined to give the number of positive integers less than or equal to and relatively prime to a given positive integer. For example, $\varphi(2) = 1$ since 1 is the only positive integer less than 2 and relatively prime to 2. Similarly, $\varphi(3) = 2$, $\varphi(4) = 2$, $\varphi(5) = 4$, etc. Show that the number of elements in $\mathbb{Z}_n$ of order d is $\varphi(d)$.

(16) Let G be a finite group.

  (a) If G is cyclic, can G be the union of proper subgroups of G? Prove your answer.

  (b) If G is not cyclic, is it always true that G can be written as a union of proper subgroups of G?

  (c) Can an infinite cyclic group G be written as a union of proper subgroups of G?

(17) Let G be an Abelian group, let k be a positive integer, and let $O(k) = \{a \in G : |a| \text{ divides } k\}$.

  (a) Find the elements in $O(6)$ if $G = \mathbb{Z}_{12}$. Construct an operation table for $O(6)$ in this case.

  (b) Prove that $O(k)$ is a subgroup of G. (Hint: In G, why is $(ab)^n = a^n b^n$ for any $a, b \in G$ and any $n \in \mathbb{Z}^+$?)

* (18) Let G be a group, and let $a, b \in G$ with $|a| = n$ and $|b| = m$.

  (a) Is it necessarily true that $|ab| = mn$? Prove your answer.

  (b) If $ab = ba$, is it necessarily true that $|ab| = mn$? Prove your answer.

  (c) Prove that if $ab = ba$ and $\gcd(m, n) = 1$, then the order of ab is mn.

Connections

In this investigation, we continued our investigation subgroups that we began in Investigation 19. We were able to completely classify the subgroup structure of cyclic groups. This also provides some information about the additive structure of the ring $\mathbb{Z}$ as considered in Investigation 6.

Investigation 21

The Dihedral Groups

Focus Questions

By the end of this investigation, you should be able to give precise and thorough answers to the questions listed below. You may want to keep these questions in mind to focus your thoughts as you complete the investigation.

- How can we describe the symmetries of a regular polygon? What similarities do all such sets of symmetries have?

- What are the dihedral groups, and what are some of their important properties?

- What are generators and relations, and how can they be used to describe a group's structure? What is a presentation of a group?

Preview Activity 21.1. In Investigation 16, we determined the group D_3 of symmetries of an equilateral triangle. For convenience, the operation table for D_3 is reproduced in Table 21.1.

(a) Use Table 21.1 to determine if each of the following is a rotation or reflection in D_3:

    (i) The composition of two rotations.

    (ii) The composition of a rotation and a reflection.

    (iii) The composition of two reflections.

(b) Which elements in D_3 can be written in the form $r_1^i R_1^j$, where i and j are integers?

(c) If t is an integer, can $R_1^t r_1$ be written in the form $r_1 R_1^j$ for some j? If yes, for which j?

(d) Let $r = r_1$ and $R = R_1$. Rewrite Table 21.1 so that every element is in the form $r^i R^j$ for some i and j.

Introduction

In Investigation 16, we saw that the symmetries of certain regular polygons (e.g., an equilateral triangle, a square, a regular pentagon, and a hexagon) consist of rotations around the center and reflections through various axes. Now we will consider the general case of a regular polygon with

299

Table 21.1

D_3, the group of symmetries of an equilateral triangle.

$\circ$	I	r_1	r_2	r_3	R_1	R_2
I	I	r_1	r_2	r_3	R_1	R_2
r_1	r_1	I	R_1	R_2	r_2	r_3
r_2	r_2	R_2	I	R_1	r_3	r_1
r_3	r_3	R_1	R_2	I	r_1	r_2
R_1	R_1	r_3	r_1	r_2	R_2	I
R_2	R_2	r_2	r_3	r_1	I	R_1

$n \geq 3$ sides, also called a regular n-gon (an n-sided figure whose sides all have the same length and in which the angles formed by adjacent sides are all congruent).

First, note that there can be at most $2n$ symmetries of a regular n-gon. To see why, pick a vertex of our regular n-gon and label it as 1. Label the vertex adjacent to 1 counterclockwise from it as 2. A symmetry will need to send vertices to vertices, so there are n vertices to which vertex 1 can be sent. Once the image of vertex 1 is determined, then the image of vertex 2 must be adjacent to the image of vertex 1. So there are two choices for the image of vertex 2. It is not difficult to see that the images of vertices 1 and 2 completely determine the images of all other points on the n-gon. Thus, there are at most $2n$ symmetries of a regular n-gon.

Could a regular n-gon have fewer than n symmetries? The answer is no; in fact, we can explicitly describe all the $2n$ symmetries that a regular n-gon must have. To begin, let θ be the central angle of a regular n-gon—that is, the angle made by one side and the center. Then there are n rotational symmetries of angles $i\theta$ around the center of the n-gon for $0 \leq i \leq n - 1$. If n is odd, then there are also n reflections around the lines passing through each vertex and the midpoint of the opposite side. If n is even, there are still n reflections: $\frac{n}{2}$ through lines connecting the midpoints of opposite sides and $\frac{n}{2}$ through lines connecting pairs of antipodal (opposite) vertices. In each case, the set of symmetries of a regular n-gon contains n rotations and n reflections, for a total of $2n$ elements. We will denote the set of symmetries of a regular n-gon as D_n and call this set the *dihedral group* of order $2n$.*

Definition 21.2. Let n be an integer with $n \geq 3$. The **dihedral group** of order $2n$ is the group of symmetries of a regular n-gon.

Before we proceed, a word of caution about notation is in order. Some texts denote the dihedral group of order $2n$ as D_{2n}, but we will use the notation D_n. In the remainder of this investigation, we will examine the structure of D_n in more detail.

Relationships between Elements in D_n

Let n be an integer with $n \geq 3$. Activity 21.1 provides a framework from which we can succinctly and efficiently represent all the elements of D_n. Here we will establish the notation that we will use in this and subsequent investigations to represent elements in D_n.

---

*The term *dihedral* seems to have come from Felix Klein's study of symmetry groups as subgroups of motions in $\mathbb{R}^3$. Using a degenerate polyhedron, he invented and studied the *dihedron* (from the Greek "di," meaning two, and "hedron," meaning face or surface), which is a figure obtained by gluing together two congruent regular polygons of zero thickness.

Let $R_0, R_1, \ldots, R_{n-1}$ be the rotations in D_n, where R_i is a counterclockwise rotation of $i\theta$ around the center of the polygon. Note that $|R_1| = n$, so $\langle R_1 \rangle = \{R_0, R_1, \ldots, R_{n-1}\}$. For ease of notation. we will let $R = R_1$. Since we will only work with rotations in the plane, we know that R^j can be represented by the matrix transformation $\begin{bmatrix} \cos(j\theta) & -\sin(j\theta) \\ \sin(j\theta) & \cos(j\theta) \end{bmatrix}$. (See Exercise (7) on page 254 of Investigation 16.)

Now label one vertex of the regular n-gon with 1 and the others in order proceeding counterclockwise. Without loss of generality, we can rotate the n-gon so that vertex 1 is on the positive y-axis. (An example is illustrated by the pentagon to the right.) Let r be the reflection of the n-gon around the line through vertex 1 and the origin. In this case, r can be represented by the matrix transformation $\begin{bmatrix} -1 & 0 \\ 0 & 1 \end{bmatrix}$.

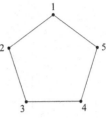

Activity 21.3. With the notation established above, answer the following.

(a) What is the order of r? Why?

(b) Explain why $R^i \neq r$ for any $0 \leq i \leq n - 1$.

(c) Now explain why rR^i cannot equal rR^j for $0 \leq i, j \leq n - 1$ unless $i = j$.

(d) Explain why

$$D_n = \{I, R, R^2, R^3, \ldots, R^{n-1}, r, rR, rR^2, \ldots, rR^{n-1}\}.$$

Activity 21.3 gives us a succinct way of representing all the elements of the dihedral group D_n; in particular, every element of D_n can be written in the form $r^i R^j$, where $0 \leq i \leq 1$ and $0 \leq j \leq n - 1$. This is very convenient and useful. Next we will see how this representation allows us to quickly generate the operation table for D_n.

Since we already know how to combine powers of the rotation R, to completely determine the structure of D_n we just need to know how to write $R^i r$ in the form rR^j for some j. Let's begin with Rr. Notice that

$$Rr = \begin{bmatrix} \cos(\theta) & -\sin(\theta) \\ \sin(\theta) & \cos(\theta) \end{bmatrix} \begin{bmatrix} -1 & 0 \\ 0 & 1 \end{bmatrix} = \begin{bmatrix} -\cos(\theta) & -\sin(\theta) \\ -\sin(\theta) & \cos(\theta) \end{bmatrix}$$

is the reflection around the line through the origin making an angle of $\left(\frac{\theta}{2} + 90\right)^\circ$ with the positive x-axis. (See Exercise (7) on page 254 of Investigation 16.) Thus, Rr is a reflection and $(Rr)^{-1} = Rr$. This can also be seen by taking the square of the matrix form for Rr. Hence,

$$(Rr)^{-1} = Rr$$
$$r^{-1}R^{-1} = Rr$$
$$rR^{-1} = Rr.$$

Therefore, $Rr = rR^{-1} = rR^{n-1}$. Note that this last inequality shows that D_n is a non-Abelian group for $n \geq 3$.

Activity 21.4. Now that we can write the product Rr in the form rR^j, to complete the operation table for D_n we need to determine $R^i r$ for $i > 1$. We will do so in this activity.

(a) Use the fact that $Rr = rR^{-1}$ to explain why $R^2 r = rR^{-2}$.

Table 21.2

Operation table for D_4.

$\circ$	I	R	R^2	R^3	r	rR	rR^2	rR^3
I	I	R	R^2	R^3	r	rR	rR^2	rR^3
R	R	R^2	R^3	I	rR^3	r	rR	rR^2
R^2	R^2	R^3	I	R	rR^2	rR^3	r	rR
R^3	R^3	I	R	R^2	rR	rR^2	rR^3	r
r	r	rR	rR^2	rR^3	I	R	R^2	R^3
rR	rR	rR^2	rR^3	r	R^3	I	R	R^2
rR^2	rR^2	rR^3	r	rR	R^2	R^3	I	R
rR^3	rR^3	r	rR	rR^2	R	R^2	R^3	I

(b) Assume $R^i r = rR^{-i}$ for some $i \geq 1$. Use this fact to show that $R^{i+1}r = rR^{-(i+1)}$.

(c) Explain how we have shown that $R^i r = rR^{-i}$ for all positive integers i.

Now that we know $R^i r = rR^{-i}$ for each i, we can compute every possible product in D_n in an efficient manner. For example, in D_4 we have

$$(rR^3)(rR^2) = r(R^3 r)R^2 = r(rR^{-3})R^2 = r^2 R^{-1} = IR^3 = R^3.$$

We can perform similar calculations to complete the operation table for D_4 (Table 21.2) in a format that is different from the one in Investigation 16 (Table 16.1 on page 249).

Generators and Group Presentations

We have seen that in a cyclic group, every element can be written as a power of a single element. That is, if G is a cyclic group, then $G = \{a^n : n \in \mathbb{Z}\} = \langle a \rangle$ for some $a \in G$. In other words, we say that the single element a generates the group G. Something similar, but a bit more complicated, happens with the dihedral group D_n. In the previous section, we found that every element in D_n can be written in terms of a single rotation and a single reflection. In other words,

$$D_n = \{r^i R^j : i, j \in \mathbb{Z}\},$$

with the stipulations that $r^2 = I$, $R^n = I$, and $rR = R^{-1}r$. In this situation, we can say that r and R generate D_n, or that D_n is generated by r and R. The next definition generalizes this idea.

Definition 21.5. A subset S of a group G **generates** G if every element in G can be written as a finite product of the elements of S (or a finite sum if the group operation is written additively).

The elements of the set S are called *generators* of G. Unless G is a single-element group, we assume that the identity is not an element of S. We can produce all the elements of the group G from generators for G, but the structure of G is also determined by how the generators interact with each other. For example, in D_4 the rotation R has order 4 ($R^4 = I$), the reflection r has order 2 ($r^2 = I$), and we have the relation that $rR = R^{-1}r$. These relations among the generators, along with the generators themselves, completely determine the group. We call the pair consisting of the generating subset S and the set of relations among these generators a *presentation* of the group G. We denote a group presentation by

$$\langle S \mid Relations \rangle.$$

For example, D_n has the following presentation:

$$\langle r, R \mid r^2 = 1, R^n = 1, rR = R^{-1}r \rangle.$$

As another example, a cyclic group of order n has the presentation $\langle a \mid a^n = 1 \rangle$, where 1 denotes the identity element in the group.[†]

Activity 21.6.

(a) Create the operation table for the group $\mathcal{V}$ with presentation

$$\langle a, b \mid a^2 = b^2 = (ab)^2 = 1 \rangle.$$

(b) Let $a = \begin{bmatrix} -1 & 0 \\ 0 & 1 \end{bmatrix}$ and $b = \begin{bmatrix} 1 & 0 \\ 0 & -1 \end{bmatrix}$. Demonstrate that the group generated by a and b in $\mathrm{GL}_2(\mathbb{R})$ (using matrix multiplication) is an example of a group of order 4 with presentation

$$\langle a, b \mid a^2 = b^2 = (ab)^2 = 1 \rangle.$$

Although we won't show it, every group has a presentation. Moreover, a group can have more than one presentation. Presentations provide a concise and simple way to describe many groups, but there are subtle issues that can make working with group presentations more complicated. For instance, it can be difficult to determine the size of a group from a presentation, or to identify when two seemingly different products of generators are equal. It can also be a challenge to determine when two different presentations yield the same group.

Concluding Activities

Activity 21.7. Use the notation from this investigation to create the operation tables for D_3, D_5, and D_6.

Activity 21.8. We defined the groups D_n for $n \geq 3$. You may wonder what we can do if $n = 1$ or $n = 2$.

(a) We can consider the group D_1 as the group of symmetries of a line segment. Find the elements of D_1, and write an operation table for D_1.

(b) We can define D_2 as the group of symmetries of a nonsquare rectangle. Find the elements of D_2, and write an operation table for D_2.

(c) The two groups D_1 and D_2 differ from the groups D_n for $n \geq 3$ in some important ways. Describe at least two of these differences.

[†]Although we often denote the identity in a group by e, it is standard notation for group presentations to use 1 instead.

Exercises

(1) Determine if each of the following is a rotation or reflection in D_n:

    (a) The composition of two rotations

    (b) The composition of a rotation and a reflection

    (c) The composition of a reflection and a rotation

    (d) The composition of two reflections

(2) Let n be an integer with $n \geq 3$.

    (a) If n is even, show that the center of D_n is not trivial. Then find all the elements in $Z(D_n)$.

    (b) If n is odd, find all elements in $Z(D_n)$.

(3) The Connected Mathematics Project (CMP, a collection of instructional materials for school mathematics) has, as part of its materials, various investigations in geometry for middle school students. One particular investigation deals with symmetries. As part of this investigation, students are asked to construct an operation table for the symmetries of various figures.

    (a) A figure similar to one presented to students in the CMP investigation is shown in Figure 21.1. Find all the symmetries of this figure, and construct an operation table for it. Clearly describe and label your symmetries. Place this group of symmetries in the context of group theory. Where have we seen something like this before?

At one point in the CMP exercise, students are asked to make a table of the symmetries of an equilateral triangle. They are then asked to, "Make an operation table for multiplication of the whole numbers 1, 2, 3, 4, 5, and 6. Compare the patterns in your multiplication table with the patterns in your table of transformation combinations (symmetries). Describe any interesting similarities and differences you discover." Parts (b) and (c) refer to this task.

Figure 21.1
A Connected Mathematics Project figure.

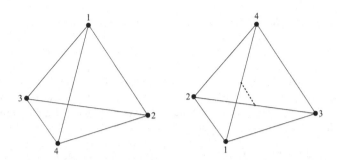

Figure 21.2
A tetrahedron and one of its symmetries.

    (b) What possible responses could students give to this question? Use appropriate terminology from group theory. Your response should include the terms "closed," "identity," and a discussion of operations.

    (c) Provide a more complete response to this CMP problem from the point of view of group theory, using appropriate terminology and notation from this course. Your discussion here should include the words "group," "modulus," "units," and "operations."

(4) **Symmetries of a tetrahedron.** We can also define symmetries in three-dimensions. For example, there are 12 rotational symmetries of the regular tetrahedron. (A regular tetrahedron has four vertices and four faces, all of which are equilateral triangles, as shown in Figure 21.2.) Describe these symmetries, and then represent each in permutation notation. Finally, construct the operation table for this group of symmetries. (We will call this group *Tetra*. As a hint, note that nine symmetries are fairly easy to find, and the other three are more difficult to see. One of these more difficult symmetries is shown in Figure 21.2 (on the right) where the dashed line connects the midpoints of two nonadjacent sides.)

(5) Find a presentation for the group $\mathbb{Z}$.

$\ast$ (6) Let $\mathbf{Q}$ be the group with presentation $\langle a, b \mid a^4 = 1, a^2 = b^2, ab = b^{-1}a \rangle$. (The group $\mathbf{Q}$ is called the *quaternion* group or the group of *quaternions*.)

    (a) Show that the elements $1, a, a^2, a^3, b, ab, a^2b$, and a^3b are all distinct.

    (b) Create the operation table for the set $\{1, a, a^2, a^3, b, ab, a^2b, a^3b\}$ to show that this set is a group. This is the group $\mathbf{Q}$.

    (c) Let $a = \begin{bmatrix} 0 & 1 \\ -1 & 0 \end{bmatrix}$, and $b = \begin{bmatrix} 0 & i \\ i & 0 \end{bmatrix}$, where $i^2 = -1$. Demonstrate that the group generated by a and b in $\mathrm{GL}_2(\mathbb{C})$ (using matrix multiplication) is an example of a group of order 8 with presentation $\langle a, b \mid a^4 = 1, a^2 = b^2, ab = b^{-1}a \rangle$.

(7) (a) Show that the quaternion group $\mathbf{Q}$ (see Exercise (6) in this investigation) contains exactly one subgroup of order 2.

    (b) Show that the quaternion group $\mathbf{Q}$ contains more than one cyclic subgroup of order 4. How many distinct subgroups of order 4 does $\mathbf{Q}$ have? Verify your answer.

(8) As we have seen, the group D_n has the presentation

$$\langle r, R \mid r^2 = 1, R^n = 1, rR = R^{-1}r \rangle.$$

Let $a = r$ and $b = rR$. Show that D_n also has the presentation

$$\langle a, b \mid a^2 = b^2 = (ab)^n = 1 \rangle.$$

This will show that D_n can be generated by elements of order 2. (You could construct the group with presentation $\langle a, b \mid a^2 = b^2 = (ab)^n = 1 \rangle$ and then compare to D_n, but it might be easier to show that the generators and relations from each presentation can be obtained from the other.)

⋆ (9) Let G be a group with identity e. If $a, b \in G$, the element $[a, b] = a^{-1}b^{-1}ab$ is called the **commutator** of the pair a, b.

    (a) Calculate $[a, b]$ if $G = D_4$, $a = rR$, and $b = R^3$.

    (b) Let $a, b \in G$. Prove that $[a, b] = e$ if and only if a and b commute.

    (c) Let G' be the subgroup (called the **commutator** subgroup) of G generated by all the commutators in G, so that every element of G' can be written as a finite product of elements of the form $[a_i, b_i]$, where a_i, $b_i \in G$. Prove that $G' = \{e\}$ if and only if G is Abelian.

    (d) Determine G' if $G = D_n$. How does your answer depend on n? Explain.

Connections

The dihedral groups are special examples of the groups of symmetries that we introduced in Investigation 16. The dihedral groups have a well-defined structure that can be described in a concise manner, and so these groups are a good source of examples in mathematics. As symmetry groups of fundamental geometric objects, the dihedral groups play an important role in geometry. The dihedral groups also have applications outside of mathematics. For example, dihedral groups can be used to create a check digit scheme (see the online supplemental materials for an example). They also have applications in chemistry where the structure of molecules can often be modeled with regular polygons, and in other areas where an idea depends on the geometric properties of some underlying structure—for example, in the design of vision filters.

Investigation 22

The Symmetric Groups

Focus Questions

By the end of this investigation, you should be able to give precise and thorough answers to the questions listed below. You may want to keep these questions in mind to focus your thoughts as you complete the investigation.

- What is a permutation of a set? Under what operation does the collection of permutations of a set form a group?

- What are the symmetric groups? What notation can be used to represent elements in symmetric groups?

- What is the cycle decomposition of a permutation, and how can we find it?

- What is a transposition, and why are transpositions important?

- What is the alternating group, and what important properties does it satisfy?

Preview Activity 22.1. Functions are important throughout mathematics, and permutations are special types of functions that will allow us to form an important family of groups. A permutation of a set is simply a rearrangement of the elements of the set, defined formally as follows:

Definition 22.2. A **permutation** of a set S is a bijection $f : S \to S$.

(a) Let S be a set. If we compose two permutations from S to S, is the resulting composite function a permutation? Why or why not?

(b) Let $f : S \to S$ be a permutation.

(i) How can we define the inverse function f^{-1} of f? (Hint: Define a function $g : S \to S$ by completing the following statement: $g(s) = t$ if $f(\underline{\quad}) = \underline{\quad}$. This function g will be our candidate for the inverse of f.)

(ii) Is the function you defined in part (i) a permutation of S? Verify your answer.

(iii) Show that the function you defined in part (i) is actually the inverse of f.

Introduction

We have seen that the set of symmetries of an equilateral triangle, a square, a regular pentagon, and a regular hexagon form groups under the operation of composition. These groups are actually subgroups of much larger groups called the *symmetric groups*, which we will study in this investigation. The symmetric groups are groups of permutations. As we will see later, symmetric groups are important in mathematics in the sense that every group can be viewed as a subgroup of a symmetric group. Symmetric groups also have applications to molecular chemistry, quantum mechanics, games, and many other areas.

The Symmetric Group of a Set

Preview Activity 22.1 showed that the collection of permutations of a set satisfies some of the group axioms using the operation of composition of functions. To show that the collection of permutations of a given set actually does form a group under composition, there are just a few more details we need to verify.

Activity 22.3. Let S be a set.

   (a) Why is the identity function from S to S a permutation of S?

   (b) Why is the composition of permutations on S an associative operation?

   (c) Explain how the results in this activity and Preview Activity 22.1 establish the following theorem:

   Theorem 22.4. *Let S be a set, and let $P(S)$ denote the collection of permutations of S. Then $P(S)$ is a group under the operation of composition of functions.*

   The group $P(S)$ is called the *permutation group of S*. It is not difficult to see that if S is a set with n elements, then $P(S)$ is basically the same as $P(\{1, 2, \ldots, n\})$. For this reason, and to make the notation easier, we typically focus on studying the permutations of the set $\{1, 2, \ldots, n\}$. This group of permutations is given a special name:

Definition 22.5. The **symmetric group** of degree n is the group

$$S_n = P(\{1, 2, \ldots, n\}).$$

In the next sections, we will investigate ways to represent elements of S_n.

Permutation Notation and Cycles

Let n be a positive integer. As we did with the symmetries of regular polygons we examined in Investigation 16, we can represent elements of S_n using permutation notation. If σ is a permutation in S_n such that $\sigma(i) = a_i$ for each i, then the permutation notation for σ is

$$\begin{pmatrix} 1 & 2 & 3 & \cdots & n \\ a_1 & a_2 & a_3 & \cdots & a_n \end{pmatrix}.$$

There is an alternate, more elegant, way to represent a permutation in S_n using *cycle notation*. To illustrate, note that the permutation $\sigma = \begin{pmatrix} 1 & 2 & 3 & 4 & 5 \\ 2 & 3 & 5 & 4 & 1 \end{pmatrix}$ has a special form. This permutation fixes 4 but sends 1 to 2, then 2 to 3, 3 to 5, and then *cycles* 5 back to 1. We could visualize the action of this permutation as follows:

$$\circlearrowleft 1 \longrightarrow 2 \longrightarrow 3 \longrightarrow 5 \circlearrowright$$

Such a permutation is called a *cycle*. More generally, a permutation $\sigma \in S_n$ is a *cycle* if for some distinct integers $a_1, a_2, \ldots, a_k$ between 1 and n we have

$$\sigma(a_1) = a_2$$
$$\sigma(a_2) = a_3$$
$$\sigma(a_3) = a_4$$
$$\vdots$$
$$\sigma(a_{k-1}) = a_k$$
$$\sigma(a_k) = a_1$$

and $\sigma(i) = i$ for all other i. In other words, a cycle has the form

$$\begin{pmatrix} a_1 & a_2 & a_3 & \cdots & a_k & b_1 & b_2 & \cdots & b_{n-k} \\ a_2 & a_3 & a_4 & \cdots & a_k & b_1 & b_2 & \cdots & b_{n-k} \end{pmatrix} \tag{22.1}$$

where $b_1, b_2, \ldots, b_{n-k}$ are the remaining integers between 1 and n that are not in $\{a_1, a_2, \ldots, a_k\}$. For example, the permutation

$$\begin{pmatrix} 1 & 2 & 3 & 4 & 5 & 6 \\ 1 & 4 & 3 & 6 & 2 & 5 \end{pmatrix}$$

can be rewritten as

$$\begin{pmatrix} 2 & 4 & 6 & 5 & 1 & 3 \\ 4 & 6 & 5 & 2 & 1 & 3 \end{pmatrix}$$

and has the form of a cycle. We can represent this cycle more concisely using the following *cycle notation*:

$$(2\ 4\ 6\ 5).$$

We can represent the general cycle from (22.1) in cycle notation as

$$(a_1\ a_2\ a_3\ \ldots\ a_k).$$

The next definition formalizes these ideas.

Definition 22.6. Let k be a positive integer. The k-cycle

$$\alpha = (a_1 \; a_2 \; a_3 \; \cdots \; a_{k-1} \; a_k)$$

is the permutation satisfying $\alpha(a_i) = a_{i+1}$ for $1 \le i \le k-1$, $\alpha(a_k) = a_1$, and $\alpha(j) = j$ for all $j \notin \{a_1, a_2, \ldots, a_{k-1}, a_k\}$.

Activity 22.7. Which of the following permutations is a cycle? Write each cycle in cycle notation.

(a) $\begin{pmatrix} 1 & 2 & 3 & 4 & 5 & 6 & 7 \\ 6 & 2 & 5 & 3 & 1 & 4 & 7 \end{pmatrix}$

(b) $\begin{pmatrix} 1 & 2 & 3 & 4 & 5 & 6 & 7 \\ 1 & 2 & 4 & 6 & 5 & 7 & 3 \end{pmatrix}$

(c) $\begin{pmatrix} 1 & 2 & 3 & 4 & 5 & 6 & 7 \\ 2 & 1 & 4 & 3 & 6 & 7 & 5 \end{pmatrix}$

The Cycle Decomposition of a Permutation

Not every permutation in S_n is a cycle, but it is not difficult to see that every permutation can be written as a product of cycles. Moreover, it turns out that each permutation can be written in some unique way as a product of disjoint cycles. To see how this works, consider the following example.

Let

$$\sigma = \begin{pmatrix} 1 & 2 & 3 & 4 & 5 & 6 & 7 & 8 \\ 2 & 3 & 1 & 7 & 5 & 6 & 4 & 8 \end{pmatrix}$$

in S_8. We can rewrite σ as

$$\sigma = \begin{pmatrix} 1 & 2 & 3 & 4 & 7 & 5 & 6 & 8 \\ 2 & 3 & 1 & 7 & 4 & 5 & 6 & 8 \end{pmatrix}.$$

Notice that σ does not have the form indicated in (22.1) and is therefore not a cycle. However, σ is made up of two cycles: $\sigma_1 = (1 \; 2 \; 3)$ and $\sigma_2 = (4 \; 7)$. If we consider the composition of σ_1 and σ_2, we find the following:

$$\sigma_1 \sigma_2(1) = \sigma_1(\sigma_2(1)) = \sigma_1(1) = 2$$
$$\sigma_1 \sigma_2(2) = \sigma_1(\sigma_2(2)) = \sigma_1(2) = 3$$
$$\sigma_1 \sigma_2(3) = \sigma_1(\sigma_2(3)) = \sigma_1(3) = 1$$
$$\sigma_1 \sigma_2(4) = \sigma_1(\sigma_2(4)) = \sigma_1(7) = 7$$
$$\sigma_1 \sigma_2(5) = \sigma_1(\sigma_2(5)) = \sigma_1(5) = 5$$
$$\sigma_1 \sigma_2(6) = \sigma_1(\sigma_2(6)) = \sigma_1(6) = 6$$
$$\sigma_1 \sigma_2(7) = \sigma_1(\sigma_2(7)) = \sigma_1(4) = 4$$
$$\sigma_1 \sigma_2(8) = \sigma_1(\sigma_2(8)) = \sigma_1(8) = 8$$

So $\sigma_1 \sigma_2 = \sigma$, and we call $\sigma_1 \sigma_2$ the *cycle decomposition* of σ.

There is an easy way to find the cycle decomposition of an arbitrary permutation $\sigma \in S_n$. We can begin by finding the image of 1 under σ (any starting place will do, but it is nice to have some consistency) and then follow the permutation along until we return to 1. This forms the first cycle. We can then repeat this process beginning with the smallest integer not in the first cycle, and continue until we have included each integer between 1 and n in some cycle. To illustrate, consider the permutation

$$\sigma = \begin{pmatrix} 1 & 2 & 3 & 4 & 5 & 6 & 7 & 8 & 9 \\ 3 & 6 & 4 & 2 & 7 & 1 & 9 & 8 & 5 \end{pmatrix}.$$

Note that

$$\sigma(1) = 3, \sigma(3) = 4, \sigma(4) = 2, \sigma(2) = 6, \text{ and } \sigma(6) = 1,$$

so one cycle in the cycle decomposition of σ is (1 3 4 2 6). Next we see that

$$\sigma(5) = 7, \sigma(7) = 9, \text{ and } \sigma(9) = 5,$$

so another cycle in the cycle decomposition of σ is (5 7 9). Since $\sigma(8) = 8$, the last cycle in the cycle decomposition of σ is simply (8). However, the cycle (8) is just the identity cycle, so we will omit it from the cycle decomposition. Therefore, the cycle decomposition of σ is

$$\sigma = (1\ 3\ 4\ 2\ 6)(5\ 7\ 9).$$

Activity 22.8. Find the cycle decomposition of each of the indicated permutations.

(a) $\begin{pmatrix} 1 & 2 & 3 & 4 & 5 & 6 & 7 \\ 2 & 4 & 3 & 1 & 7 & 6 & 5 \end{pmatrix}$

(b) $\begin{pmatrix} 1 & 2 & 3 & 4 & 5 & 6 & 7 & 8 \\ 1 & 2 & 5 & 3 & 4 & 6 & 8 & 7 \end{pmatrix}$

(c) $\begin{pmatrix} 1 & 2 & 3 & 4 & 5 & 6 & 7 & 8 & 9 & 10 \\ 6 & 10 & 9 & 4 & 7 & 1 & 5 & 8 & 2 & 3 \end{pmatrix}$

Activity 22.9. Write each product of cycles in permutation notation.

(a) $\sigma = (1\ 3\ 5\ 6)(2\ 4)$ in S_6

(b) $\sigma = (2\ 5\ 3)(7\ 4)(6\ 9)$ in S_{10}

One thing to note in each of our cycle decompositions is that every permutation was written as a product of nonoverlapping (that is, *disjoint*) cycles. This observation motivates our next definition.

Definition 22.10. Two cycles $\sigma = (a_1\ a_2\ \ldots\ a_k)$ and $\tau = (b_1\ b_2\ \ldots\ b_m)$ are **disjoint** if $a_i \neq b_j$ for all $1 \leq i \leq k$ and $1 \leq j \leq m$.

That we can always decompose a permutation as a product of disjoint cycles is the subject of the following theorem.

Theorem 22.11. *Let n be a positive integer. Every permutation in S_n is either a cycle or can be written as a product of disjoint cycles.*

Proof. We will show via strong induction that any permutation of a finite collection of numbers is either a cycle or can be written as a product of disjoint cycles. Let α be a permutation of m numbers. If $m = 1$, then α must be the identity permutation, and $\alpha = (1)$ is a cycle. For our induction hypothesis, we assume, for some integer $m \geq 1$, that any permutation of m or fewer numbers is a cycle or a product of disjoint cycles. Now suppose α permutes $m+1$ numbers. Since α is not the identity, there is some number that is not fixed by α. For the sake of convenience, label this number as a_1. For each positive integer s, consider the sequence $\alpha(a_1), \alpha^2(a_1), \alpha^3(a_1), \cdots$, where

$$\alpha^s(a_1) = \alpha(\alpha(\alpha(\cdots(\alpha(a_1))))),$$

with α appearing s times in this iterated composition. Since $m+1$ is finite, this sequence must repeat at some point. Let i and j be positive integers with $i < j$ such that $\alpha^j(a_1) = \alpha^i(a_1)$. Then $\alpha^{j-i}(a_1) = a_1$, and there is a positive power of α that fixes a_1. By the Well-Ordering Principle, there is a smallest positive integer k such that $\alpha^k(a_1) = a_1$. Since a_1 is not fixed by α, it must be that $k \geq 2$. For each integer t with $2 \leq t \leq k$, let $a_t = \alpha^{t-1}(a_1)$, and let $\sigma = (a_1 \ a_2 \ a_3 \ \cdots \ a_k)$. Note that $\sigma(a_i) = \alpha(a_i)$ for each i between 1 and k.

Now let α' be the permutation in S_n defined by

$$\alpha'(s) = \begin{cases} a_i, & \text{if } s = a_i \text{ for some } i \\ \alpha(s), & \text{otherwise.} \end{cases}$$

We will show that $\alpha = \sigma\alpha'$. To do so, choose $q \in \{1, 2, \ldots, m+1\}$. We will show that $(\sigma\alpha')(q) = \alpha(q)$ by considering cases.

For the first case, assume $q = a_i$ for some $1 \leq i \leq k$. In this case, we have

$$(\sigma\alpha')(q) = \sigma(\alpha'(q)) = \sigma(a_i) = \alpha(a_i).$$

For the remaining case, suppose $q \neq a_i$ for all i. To apply σ to $\alpha'(q)$, we need to know if $\alpha'(q)$ can be one of the a_i for i between 1 and k. The following claim answers this question.

Claim. $\alpha'(q) \neq a_i$ for all i.

Proof of Claim. We will prove the claim by contradiction. Suppose $\alpha'(q) = a_i$ for some $1 \leq i \leq k$. By definition, $\alpha'(q) = \alpha(q) = a_i$. Since α is an injection, this then implies $q = a_{i-1 \pmod k}$, a contradiction to the fact that q is not equal to one of the a_i. ∎

Since our claim establishes that $q \neq a_i$ for all i, it follows that

$$(\sigma\alpha')(q) = \sigma(\alpha'(q)) = \alpha'(q) = \alpha(q).$$

We can therefore conclude that $(\sigma\alpha')(q) = \alpha(q)$ for all q. Hence, $\sigma\alpha' = \alpha$.

Now α' is a permutation of $(m+1) - k$ integers, so by our induction hypothesis, we know α' is either a cycle or a product of disjoint cycles. Moreover, by our claim, these cycles are all disjoint from σ. Therefore, $\alpha = \sigma\alpha'$ is a cycle or a product of disjoint cycles. ∎

Activity 22.12.

(a) List the elements in S_3 as cycles or products of disjoint cycles.

(b) Construct the operation table for S_3.

(c) Is S_3 an Abelian group? Explain.

(d) List all the elements of S_4 as cycles or products of disjoint cycles.

(e) Is S_4 an Abelian group?

As you may suspect, S_n (for $n \geq 3$) is never an Abelian group. The proof of this fact is left as an exercise (see Exercise (6)), although it is not much more difficult than what you were asked to show in parts (c) and (e) of Activity 22.12.

Transpositions

Just like integers, permutations can be classified as being either even or odd. This classification is a useful tool in group theory and will ultimately allow us to define a special class of groups called the *alternating groups*. The parity of a permutation—that is, whether it is even or odd—is determined by first writing the permutation as a product of 2-cycles or *transpositions*. A straightforward computation shows that any cycle can be written as a product of transpositions as follows:

$$\sigma = (a_1\ a_2\ \ldots\ a_k) = (a_1\ a_2)(a_2\ a_3)(a_3\ a_4)\cdots(a_{k-2}\ a_{k-1})(a_{k-1}\ a_k). \qquad (22.2)$$

The price we pay for this 2-cycle decomposition is that our 2-cycles are not disjoint and do NOT commute. However, since every permutation can be written as a product of cycles and every cycle can be written as a product of 2-cycles, then every permutation can be written as a product of 2-cycles. We should also note that the decomposition in (22.2) is not the only way to decompose a cycle into a product of transpositions. For example, you should convince yourself that

$$(a_1\ a_2\ \ldots\ a_k) = (a_1\ a_k)(a_1\ a_{k-1})(a_1\ a_{k-2})\cdots(a_1\ a_2)$$

is another 2-cycle decomposition of $(a_1\ a_2\ \ldots\ a_k)$.

Activity 22.13. Write each permutation as a product of 2-cycles.

(a) $\begin{pmatrix} 1 & 2 & 3 & 4 & 5 & 6 & 7 \\ 2 & 4 & 3 & 1 & 7 & 6 & 5 \end{pmatrix}$

(b) $\begin{pmatrix} 1 & 2 & 3 & 4 & 5 & 6 & 7 & 8 \\ 1 & 2 & 5 & 3 & 4 & 6 & 8 & 7 \end{pmatrix}$

(c) $\begin{pmatrix} 1 & 2 & 3 & 4 & 5 & 6 & 7 & 8 & 9 & 10 \\ 6 & 10 & 9 & 4 & 7 & 1 & 5 & 8 & 2 & 3 \end{pmatrix}$

Why should we care about writing a permutation as a product of transpositions? Since the transpositions in the resulting product are not usually disjoint, this doesn't seem to be an improvement over the disjoint cycle decomposition we developed in Theorem 22.11. In fact, as we noted earlier, there are many different ways we can write a permutation as a product of transpositions. For example,

$$(1\ 2)(1\ 3)(1\ 4) = (1\ 4\ 3\ 2) = (1\ 4)(4\ 3)(3\ 2) = (1\ 4)(4\ 3)(3\ 2)(2\ 4)(4\ 2).$$

Notice that all of these decompositions are made up of an odd number of transpositions. This important observation holds in general, as stated in the next theorem.

Theorem 22.14. *Let $n \in \mathbb{Z}^+$, and let $\sigma \in S_n$. No matter how σ is written as a product of transpositions, the number of transpositions in the product will always have the same parity.*

In order to prove Theorem 22.14, we will first need to make some conclusions about the parity of the identity permutation, I.

Lemma 22.15. *Let $n \in \mathbb{Z}^+$, and let I be the identity permutation in S_n. If*

$$I = \tau_1 \tau_2 \cdots \tau_k,$$

where $\tau_1, \tau_2, \ldots, \tau_k$ are transpositions, then k is even.

Proof. Let $n \in \mathbb{Z}^+$, and let I be the identity permutation in S_n. Suppose

$$I = \tau_1 \tau_2 \cdots \tau_k$$

for some transpositions $\tau_1, \tau_2, \ldots, \tau_k$. First, note that the identity permutation is not itself a transposition, so $k > 1$. We will proceed by induction on k. For the base case, note that if $k = 2$, then k is even and we are done. For the inductive step, let $k \geq 2$, and assume that if I is written as a product of m transpositions with $m \leq k$, then m is even. Now suppose

$$I = \tau_1 \tau_2 \cdots \tau_k \tau_{k+1},$$

where $\tau_1, \tau_2, \ldots, \tau_k, \tau_{k+1}$ are transpositions. Let $\tau_{k+1} = (a\ b)$ and $\tau_k = (r\ s)$. There are four different possibilities for the product of these two transpositions:

(1) If $(r\ s) = (a\ b)$, then $\tau_k \tau_{k+1} = (a\ b)(a\ b) = I$.

(2) If $r = a$ and $s \neq b$, then

$$\tau_k \tau_{k+1} = (r\ s)(a\ b) = (a\ s)(a\ b) = (a\ b\ s) = (a\ b)(s\ b).$$

(3) If $s = b$ and $r \neq a$, then

$$\tau_k \tau_{k+1} = (r\ s)(a\ b) = (r\ b)(a\ b) = (a\ r\ b) = (a\ r)(r\ b).$$

(4) If τ_k and τ_{k+1} are disjoint cycles, then $(r\ s)(a\ b) = (a\ b)(r\ s)$.

In cases (2) – (4), we can rewrite $\tau_k \tau_{k+1}$ in the form $\alpha_1 \tau_k'$ for some transpositions τ_k' and α_1, where $\tau_k'(a) = a$ and $\alpha_1(a) \neq a$. Then

$$I = \tau_1 \tau_2 \cdots \tau_{k-1} \alpha_1 \tau_k'.$$

As long as $\alpha_1 \neq \tau_{k-1}$, we can repeat this process, interchanging α_1 with τ_{k-1}. In this case, $\tau_{k-1} \alpha_1 = \alpha_2 \tau_{k-1}'$, where $\tau_{k-1}'(a) = a$ and $\alpha_2(a) \neq a$, and

$$I = \tau_1 \tau_2 \cdots \tau_{k-2} \alpha_2 \tau_{k-1}' \tau_k'.$$

Continuing in this same manner, if we never encounter case (1)—that is, if we never find i and j so that $\tau_j = \alpha_i$—then we can interchange each α_i with each τ_j, leaving τ_j' and α_{i+1} so that $\tau_j'(a) = a$ and $\alpha_{i+1}(a) \neq a$, as shown in cases (2), (3), and (4). In this situation, we will ultimately have

$$I = \alpha_k \tau_1' \tau_2' \cdots \tau_{k-1}' \tau_k',$$

where $(\tau_1' \tau_2' \cdots \tau_{k-1}' \tau_k')(a) = a$ and $\alpha_k(a) \neq a$. Then

$$I(a) = (\alpha_k \tau_1' \tau_2' \cdots \tau_{k-1}' \tau_k')(a) = \alpha_k(\tau_1' \tau_2' \cdots \tau_{k-1}' \tau_k'(a)) = \alpha_k(a) \neq a,$$

which is impossible.

Therefore, as we interchange the α_i with the τ_j, we must encounter a j so that $\tau_j = \alpha_i$. As in case (1), the product $\tau_j \alpha_i$ will equal the identity and will drop out of the product. Then

$$I = \tau_1 \tau_2 \cdots \tau_{j-1} \tau_{j+1}' \tau_{j+2}' \cdots \tau_k',$$

and I is the product of $k - 1$ transpositions. By our induction hypothesis, it then follows that $k - 1$ is even, and therefore $k + 1$ is also even, as desired.   ∎

We are now ready to prove Theorem 22.14, which establishes that the parity of the number of transpositions in the 2-cycle decomposition of a permutation is an invariant.

Proof of Theorem 22.14. Let $n \in \mathbb{Z}^+$, and let $\sigma \in S_n$. Assume $\tau_1, \tau_2, \ldots, \tau_k$ and $\alpha_1, \alpha_2, \ldots, \alpha_m$ are transpositions such that

$$\sigma = \tau_1 \tau_2 \cdots \tau_k = \alpha_1 \alpha_2 \cdots \alpha_m. \tag{22.3}$$

We need to show that $k \equiv m \pmod 2$ or, equivalently, that $k + m$ is even. Multiplying the two decompositions in (22.3) on the right by $\alpha_m \alpha_{m-1} \cdots \alpha_2 \alpha_1$, keeping in mind that a transposition is its own inverse, we obtain

$$\tau_1 \tau_2 \cdots \tau_k \alpha_m \alpha_{m-1} \cdots \alpha_2 \alpha_1 = I.$$

We have thus written the identity permutation I as a product of $k + m$ transpositions. Lemma 22.15 now implies that $k + m$ must be even, as desired. ∎

Even and Odd Permutations and the Alternating Group

Theorem 22.14 shows that the number of transpositions in the 2-cycle decomposition of a permutation is always either even or odd. We can use this result to define the parity of a permutation.

Definition 22.16. Let $n \in \mathbb{Z}^+$. A permutation $\sigma \in S_n$ is **even** if σ can be written as a product of an even number of transpositions. A permutation $\sigma \in S_n$ is **odd** if σ is not even.

So, for example, the permutation $(1\ 2)(2\ 4)(3\ 5)(4\ 1)$ in S_5 is an even permutation.

Activity 22.17. Determine whether the indicated permutation is even or odd.

(a) $\begin{pmatrix} 1 & 2 & 3 & 4 & 5 & 6 & 7 \\ 2 & 4 & 3 & 1 & 7 & 6 & 5 \end{pmatrix}$

(b) $\begin{pmatrix} 1 & 2 & 3 & 4 & 5 & 6 & 7 & 8 \\ 1 & 2 & 5 & 3 & 4 & 6 & 8 & 7 \end{pmatrix}$

(c) $\begin{pmatrix} 1 & 2 & 3 & 4 & 5 & 6 & 7 & 8 & 9 & 10 \\ 6 & 10 & 9 & 4 & 7 & 1 & 5 & 8 & 2 & 3 \end{pmatrix}$

An interesting question is the following: If we collect all of even (or odd) permutations together, does the resulting set form a subgroup of S_n?

Activity 22.18.

(a) Explain why the collection of odd permutations is not a subgroup of S_n.

(b) Is the product of two even permutations always an even permutation? Explain.

(c) Is the inverse of an even permutation always an even permutation? Explain.

(d) Let A_n be the set of all even permutations in S_n. Explain why A_n is a subgroup of S_n.

The subgroup A_n defined in part (d) of Activity 22.18 is an important one and is given a special name.

Definition 22.19. For $n \geq 2$, the **alternating group** A_n is the subgroup of S_n consisting of the even permutations in S_n.

Activity 22.20.

(a) Find all the elements in A_3.

(b) Find all the elements in A_4.

You will show in Exercise (11) that the order of S_n is $n!$. But what is the order of A_n? Activity 22.20 seems to indicate that the order of A_n is half that of the order of S_n. To see that this is true in general, let $\alpha, \beta \in S_n$. If α is odd, then $(1\ 2)\alpha$ is even. Also, if $(1\ 2)\alpha = (1\ 2)\beta$, then $\alpha = \beta$. So the number of even permutations in S_n is greater than or equal to the number of odd permutations in S_n. If α is even, then $(1\ 2)\alpha$ is odd. Also, if $(1\ 2)\alpha = (1\ 2)\beta$, then $\alpha = \beta$. So the number of odd permutations in S_n is greater than or equal to the number of even permutations in S_n. Thus, the permutations in S_n are evenly divided between even and odd permutations, and $|A_n| = \frac{|S_n|}{2} = \frac{n!}{2}$.

Concluding Activities

Activity 22.21. Let n be an integer with $n \geq 3$. What is the connection between D_n and S_n? Explain.

Activity 22.22. One advantage of decomposing a permutation as a product of disjoint cycles is that disjoint cycles commute. To verify this property, let $\alpha = (a_1\ a_2\ \ldots\ a_k)$ and $\beta = (b_1\ b_2\ \ldots\ b_m)$ be disjoint cycles in S_n.

(a) What can we say about the sets $A = \{a_1, a_2, \ldots, a_k\}$ and $B = \{b_1, b_2, \ldots, b_m\}$?

(b) What is $\alpha(b_j)$ for any $1 \leq j \leq m$? What is $\beta(a_i)$ for any $1 \leq i \leq k$? Explain.

(c) If $r \in \{1, 2, \ldots, n\} \setminus (A \cup B)$, what are $\alpha(r)$ and $\beta(r)$? Explain.

(d) Show that $(\alpha\beta)(t) = (\beta\alpha)(t)$ for every $t \in \{1, 2, \ldots, n\}$ by considering cases.

Exercises

(1) Write $(3\ 1)(4\ 2)$ in S_5 in permutation notation.

(2) Write the permutation $\begin{pmatrix} 1 & 2 & 3 & 4 & 5 & 6 \\ 4 & 5 & 6 & 1 & 3 & 2 \end{pmatrix}$ as a product of disjoint cycles.

(3) Write the cycle $(1\ 2\ 4\ 3)$ as a product of transpositions.

(4) Write the given cycle as a product of transpositions. Identify if the permutation is even or odd.

    (a) $(1\ 2\ 4\ 3)$

    (b) $(2\ 4\ 8\ 9\ 3)$

(5) A friend of yours gives you the permutation $\begin{pmatrix} 1 & 2 & 3 & 4 & 5 & 6 & 7 & 8 & 9 \\ 7 & 5 & 3 & & 9 & & 8 & 1 & 2 \end{pmatrix}$ in S_9 and tells you that the images of 4 and 6 are lost. Your friend recalls, however, that this permutation is even. What are the images of 4 and 6? Explain.

* (6) Prove that for $n \geq 3$, S_n is non-Abelian.

(7) Find a subgroup of S_4 of order 8, or show that one does not exist.

(8) Find a subgroup of S_4 of order 6, or show that one does not exist.

(9) Let $n \in \mathbb{Z}^+$, and let k be a positive integer with $k \leq n$. What is the order of a k-cycle in S_n? Prove your result.

(10) Let n be an integer greater than 2. Prove that the center of S_n is $\{I\}$, where I is the identity permutation in S_n. (Hint: If α is a nonidentity element in S_n, then there exist distinct integers i and j between 1 and n (inclusive) such that $\alpha(i) = j$.)

* (11) Prove that the number of permutations of a set with n elements is $n!$. Based on this fact, what is the order of S_n?

(12) When is the cycle $(a_1\ a_2\ \cdots\ a_k)$ in S_n even, and when is it odd? Prove your answer.

(13) A fixed point of a function f is an input p so that $f(p) = p$. As functions, permutations can have fixed points.

    (a) Determine all permutations in S_4 that have 2 as a fixed point. Let G_2 be this set of permutations. Is G_2 a subgroup of S_4?

    (b) Let $n \in \mathbb{Z}^+$, and let k be a fixed integer between 1 and n. Define

$$G_k = \{\sigma \in S_n : \sigma(k) = k\}.$$

    Prove that G_k is a subgroup of S_n. (This group G_k is called a **stabilizer subgroup** of S_n.)

    (c) Can we generalize the result of the previous part to the following?

    Let T be a nonempty subset of $\{1, 2, \ldots, n\}$ and define G_T as

$$G_T = \{\sigma \in S_n : \sigma(k) = k \text{ for all } k \in T\}.$$

    Then G_T is a subgroup of S_n.

    Give a proof or counterexample to justify your answer.

    (d) Suppose $T \subset \{1, 2, \ldots, n\}$ contains m numbers. What do you expect the order of G_T to be? Explain.

(14) Let $n \geq 2$.

    (a) Let α and β be disjoint cycles in S_n. Determine $|\alpha\beta|$ in terms of $|\alpha|$ and $|\beta|$.

(b) Let $\alpha_1, \alpha_2, \ldots, \alpha_m$ be disjoint cycles in S_n for some integer $m \geq 2$. Find and prove a formula similar to that from part (a) for $|\alpha_1 \alpha_2 \cdots \alpha_m|$. (Hint: Exercise (4) on page 275 of Investigation 18 might be useful.)

(15) Let G be an Abelian group, and let $H = \{x^2 : x \in G\}$.

(a) Is H a subgroup of G? Prove your answer.

(b) In your proof in part (a), did you use the assumption that G is Abelian? Do you think it would be possible to prove this result without the assumption that G is Abelian? If not, then give a counterexample where G is non-Abelian and H is not a subgroup of G.

$\star$ (16) The permutation α in S_n is **conjugate** to $\beta \in S_n$ if $\alpha = \sigma \beta \sigma^{-1}$ for some $\sigma \in S_n$.

(a) Find all permutations in S_3 that are conjugate to $(1\ 2\ 3)$.

(b) Let $\beta = (b_1\ b_2\ \ldots\ b_k)$ be a k-cycle in S_n, and let $\sigma \in S_n$. Show that

$$\sigma \beta \sigma^{-1} = (\sigma(b_1)\ \sigma(b_2)\ \cdots\ \sigma(b_k)).$$

Conclude that the conjugate of a k-cycle is a k-cycle.

(c) Let $\pi \in S_n$ and suppose π is written as the product

$$\pi = \pi_1 \pi_2 \cdots \pi_m$$

of disjoint cycles. Let $\sigma \in S_n$. Show that

$$\sigma \pi \sigma^{-1} = \tau_1 \tau_2 \cdots \tau_m,$$

where the τ_i are disjoint cycles and, for each i, τ_i is a cycle of the same length as π_i. Conclude that conjugation preserves the cycle structure of a permutation.

(17) Write the operation table for A_4. (See Activity 22.20.)

$\star$ (18) Let $T = \langle \alpha, \beta \rangle$, where

$$\alpha = (1\ 2\ 3\ 4\ 5\ 6)(7\ 8\ 9\ 10\ 11\ 12) \quad \text{and} \quad \beta = (1\ 7\ 4\ 10)(2\ 12\ 5\ 9)(3\ 11\ 6\ 8)$$

in S_{12}. This group T has presentation $\langle s, t \mid s^6 = 1, s^3 = t^2, sts = t \rangle$.

(a) Find all the elements of T, and construct the operation table for T.

(b) Show that T also has the presentation

$$\langle x, y \mid x^4 = y^3 = 1, yxy = x \rangle.$$

(19) Theorem 22.11 shows that every permutation in S_n is either a cycle or can be written as a product of disjoint cycles. Equation (22.2) then demonstrates how every cycle can be written as a product of transpositions. Thus, the transpositions generate S_n. However, it turns out that we do not need all the transpositions to generate S_n. To see why this is the case, prove that the transpositions $(1\ 2)$, $(1\ 3)$, $(1\ 4)$, $\ldots$, $(1\ n)$ generate S_n.

(20) **Card shuffling.** Many magicians excel at card tricks. One important skill in card tricks is the perfect shuffle. A *shuffle* of a deck of $2n$ cards is obtained when the deck is split into two piles and then the cards from each pile are rearranged into one pile. A *perfect shuffle* occurs when the deck is split into two piles A and B of equal size and the cards are rearranged into

one pile by alternating cards from piles A and B. There are two types of perfect shuffles: the in-shuffle and the out-shuffle. The difference between the two shuffles is their effect on the top card; after an out-shuffle, the top card of the pre-shuffled deck remains on top, but an in-shuffle moves the top card of the pre-shuffled deck to the second position.

Label the top position in a deck as #1, the second card from the top as #2, and so on. A perfect in-shuffle can then be represented by the permutation

$$\begin{pmatrix} 1 & 2 & 3 & \cdots & n & n+1 & n+2 & \cdots & 2n \\ 2 & 4 & 6 & \cdots & 2n & 1 & 3 & \cdots & 2n-1 \end{pmatrix}.$$

(a) Determine the fewest number of perfect shuffles required to return a deck of 14 cards to its original position using only in-shuffles. Clearly explain your reasoning.

(b) Determine the fewest number of perfect shuffles required to return a deck of 52 cards to its original position using only in-shuffles. Explain.

(c) Represent a perfect out-shuffle in permutation notation.

(d) Determine the fewest number of perfect shuffles required to return a deck of 52 cards to its original position using only out-shuffles. Explain.

(21) **The Futurama Theorem.** In the Season 6 episode "The Prisoner of Benda" of the animated TV series *Futurama*, a machine is created that allows beings to swap minds. However, due to "cerebral immune response," the machine does not allow them to reverse the process. To solve the problem, the professor states, "I'm afraid we need to use math!". The characters in the episode spend their time figuring out a way to get all minds back with their original bodies.

The solution to the problem faced in this episode is contained in the following theorem (due to *Futurama* writer Ken Keeler, who holds a Ph.D. in applied mathematics), which states that all mind switches can be undone by introducing two more characters into the situation.

Theorem (The Futurama Theorem). *Let A be a finite set, and let x and y be distinct objects that do not belong to A. Any permutation of A can be reduced to the identity permutation by applying a sequence of distinct transpositions of $A \cup \{x, y\}$, each of which includes just one of x, y.*

(a) Since the group of permutations of any set with n elements can be identified with the permutations of the set $\{1, 2, \ldots, n\}$, we can assume that our set A is the set $\{1, 2, \ldots, n\}$. We will first prove the Futurama Theorem for cycles.

   (i) Assume we have a k-cycle σ that indicates mind swaps between k individuals (that is, individual i's mind is in individual $\sigma(i)$'s body). We can relabel to assume, without loss of generality, that

$$\sigma = (1\,2\,3\,\cdots\,k).$$

Explain how the Futurama Theorem states that we can find a permutation $\gamma \in S_{n+2}$ that is a product of transpositions, each of which permutes just one of $n+1$ or $n+2$, so that $\sigma\gamma = I$.

   (ii) Suppose we introduce two new characters x and y. Without loss of generality, we can let $x = n+1$ and $y = n+2$. For each i from 1 to k, let

$$\tau_i = (n+1\ \ 1)(n+1\ \ 2)\cdots(n+1\ \ i)$$
$$(n+2\ \ i+1)(n+2\ \ i+2)\cdots(n+2\ \ k)$$
$$(n+1\,i+1)(n+2\,1).$$

Consider σ as a permutation in S_{n+2}. What permutation is $\sigma\tau_i$? Explain.

(iii) Explain how we can reverse the mind swaps defined by the k-cycle σ so that all minds are back in their original bodies.

(b) Now prove the Futurama Theorem for any permutation.

Connections

A symmetry of a set is a bijection from the set to itself. Every symmetry of an object (as described in Investigation 16) is also a bijection from the object to itself. If we think of an object as a set of points, then every symmetry of the object corresponds to a symmetry of the set of points that represents it. In this sense, we can identify the collection of symmetries of an object with a subgroup of the symmetries of the set of points of the object. In particular, we can identify the dihedral group D_n, the set of symmetries of a regular n-gon, with a subset of S_n corresponding to permutations of the vertices of the n-gon. (We will make this identification more specific after Investigation 26 on group isomorphisms.) It is important to note, however, that not every symmetry of a set that defines an object is a symmetry of the object itself. By noting the difference between the orders of D_n and S_n, you should be able to come up with an example to convince yourself of this.

Investigation 23

Cosets and Lagrange's Theorem

Focus Questions

By the end of this investigation, you should be able to give precise and thorough answers to the questions listed below. You may want to keep these questions in mind to focus your thoughts as you complete the investigation.

- What are left and right cosets?

- What are some important properties of cosets? How do cosets partition the groups on which they are defined?

- What important fact does Lagrange's Theorem tell us about subgroups of finite groups?

- What are some of the consequences and corollaries of Lagrange's Theorem?

Preview Activity 23.1. In Investigation 20, we classified all the subgroups of finite cyclic groups. In particular, we learned in Theorem 20.7 (see page 295) that in a finite cyclic group of order n, there is exactly one subgroup of order m for each positive divisor m of n, and these are the only subgroups. It is natural to ask if Theorem 20.7 generalizes to *all* finite groups. As we will see, part of it does, and part of it does not.

(a) Find all the nontrivial, proper subgroups of D_3. How are the orders of these subgroups related to the order of D_3?

(b) For each divisor m of $|D_6|$, does D_6 contain a subgroup of order m? If yes, exhibit a subgroup for each divisor. If no, explain why.

Introduction

The subgroup structure of a group tells us a lot about the group. Theorem 20.7 completely classified the subgroup structure of finite cyclic groups, but Activity 23.1 shows us that this theorem does not generalize to arbitrary finite groups. For one thing, the group D_3 contains more than one subgroup of order 2. On the other hand, the order of every subgroup of D_3 does divide the order of D_3, and both D_3 and D_4 contain subgroups of each order that divides the order of the larger group. But what can we say in general? In other words, exactly what we can say about the subgroups of an arbitrary

321

finite group? This is an important question, and one that we will begin to answer throughout this investigation.

To uncover the underlying subgroup structure of a group, we often use counting arguments. Such arguments usually involve arranging the elements of a group into disjoint subsets so that we can count the elements without overlap. Throughout this and subsequent investigations, we will use counting arguments to prove important theorems (such as Lagrange's Theorem and the Sylow theorems) and to derive other critical tools (such as the Class Equation). In this investigation, we will introduce one of the more basic counting tools: the coset decomposition of a group. We will use cosets to prove Lagrange's Theorem, which is one of the most important theorems in finite group theory. In the next investigation (Investigation 24), we will also see how different types of subgroups lead to different coset structures, an observation that will motivate our study of normal subgroups and quotient groups.

A Relation in Groups

In Investigation 2, we constructed the set $\mathbb{Z}_n$, which we defined in terms of the congruence modulo n relation on $\mathbb{Z}$. Recall that if n is a positive integer and $a, b \in \mathbb{Z}$, we say that a is congruent to b modulo n if $b - a$ is a multiple of n.[*] We can rephrase this idea in terms of the subgroup $n\mathbb{Z} = \{nk : k \in \mathbb{Z}\}$ and say that $a \equiv b \pmod{n}$ if $b - a \in n\mathbb{Z}$. The advantage of this perspective is that we can apply it to other groups, as the next activity suggests.

Activity 23.2. Let H be the subset of $\mathbb{Z}_8$ defined by $H = \langle [4] \rangle = \{[0], [4]\}$. Define a relation $\sim_H$ on $\mathbb{Z}_8$ by

$$a \sim_H b \text{ if and only if } b - a \in H.$$

(a) Find all elements in $\mathbb{Z}_8$ that are related to $[2]$.

(b) For each $g \in \mathbb{Z}_8$, find all elements in $\mathbb{Z}_8$ related to g using the relation $\sim_H$.

(c) Let $[g]$ denote the set of elements in $\mathbb{Z}_8$ related to g. List three different things you notice about the collection of sets $[g]$ for $g \in \mathbb{Z}_8$.

The relations we defined on $\mathbb{Z}$ in Investigation 2, and on $\mathbb{Z}_8$ in Activity 23.2, can be extended to arbitrary groups and subgroups. The result of Activity 23.2 suggests that we might expect to be able to partition any group into a disjoint union of classes using a similar relation, in much the same way we partitioned $\mathbb{Z}$ into a disjoint union of congruence classes. Partitioning groups into classes will turn out to be a very important tool for us. Before we proceed to the general definition, however, we need to stop for a moment to talk about notation. Note that since the group operation for both $\mathbb{Z}$ and $\mathbb{Z}_8$ is addition, we used additive notation to define a congruence relation on these groups. We usually assume that an additive operation is commutative, so we could write $b - a$ as $(-a) + b$, which will be more convenient in what follows. When we use multiplicative notation, we replace $(-a) + b$ with $a^{-1}b$. This brings us to the next definition.

---

[*]In Investigation 2, we defined a to be congruent to b modulo n if $a - b$ (rather than $b - a$) was a multiple of n. Because of the symmetry of the congruence relation, these two definitions are equivalent. However, the version that uses $b - a$ will be more convenient for the purposes of working with cosets.

Definition 23.3. Let G be a group and H a subgroup of G. Let $\sim_H$ be the relation on G such that for all $a, b \in G$,

$$a \sim_H b \quad \text{if and only if} \quad a^{-1}b \in H.$$

For example, let $G = S_3$ and $H = \langle (1\ 2) \rangle$. Since $(1\ 3)^{-1}(1\ 2\ 3) = (1\ 2)$, the definition shows us that $(1\ 3) \sim_H (1\ 2\ 3)$. The groups G and H will usually be clear from the context, so we will normally suppress the subscript and just write $\sim$ instead of $\sim_H$.

Activity 23.4. Let $G = S_3$ and $H = \langle (1\ 2) \rangle$. Find all elements α in S_3 satisfying $\alpha \sim (1\ 2\ 3)$.

Whenever we have a relation, it is natural to ask what properties the relation satisfies.

Activity 23.5. Let G be a group with identity e and H a subgroup of G.

(a) Prove that $\sim$ is a reflexive relation on G.

(b) Prove that $\sim$ is a symmetric relation on G.

(c) Prove that $\sim$ is a transitive relation on G.

Recall that any relation that is reflexive, symmetric, and transitive is an *equivalence relation*. In Investigation 2, we learned that every equivalence relation partitions the underlying set into disjoint equivalence classes. (See Theorem 2.6 on page 27.) We will exploit this important property throughout the next sections.

Cosets

Let G be a group, let H be a subgroup of G, and let $\sim$ be the relation defined in the preceding section. It turns out that the equivalence classes corresponding to $\sim$ have a special form and can be easily described. Let $a \in G$, and choose $b \in [a]$ (where $[a]$ denotes the equivalence class of a under $\sim$). Then $a \sim b$. So $a^{-1}b \in H$ and $a^{-1}b = h$ for some $h \in H$. Thus, $b = ah$. If we let $aH = \{ah : h \in H\}$, then it follows that $[a] \subseteq aH$. Conversely, let $b \in aH$. Then $b = ah'$ for some $h' \in H$. So $a^{-1}b = h' \in H$ and $a \sim b$, which implies $b \in [a]$. Thus, $aH \subseteq [a]$, and so we have shown that $[a] = aH$. We can therefore recognize the equivalence classes under the relation $\sim$ as the sets of the form aH. These sets play an important role in counting techniques in finite group theory and are given a special name.

Definition 23.6. Let G be a group and H a subgroup of G. Let $g \in G$. The **left coset**[†] of H in G containing g is the set

$$gH = \{gh : h \in H\}.$$

Similarly, the **right coset** of H in G containing g is the set

$$Hg = \{hg : h \in H\}.$$

The element g is called the coset representative of gH (or Hg).

[†]The idea of a coset seems to have been first used by Evariste Galois in his paper *Mémoire sur les conditions de résolubilité des équations par redicaux* in 1830. The term *coset* was first applied by G.A. Miller in the *Quarterly Journal of Mathematics* in 1910. The word coset appears to have literally meant "co-set" to Miller (in the same vein that a word like co-pilot is used) and replaced the previous term *Nebengruppen* for this idea.

The sets in Definition 23.6 are written using multiplicative notation. If we were using additive notation instead, we would write $a + H$ for a left coset rather than aH.

Since the equivalence classes under $\sim$ are exactly the left cosets of H in G, we have the following result.

Theorem 23.7. *Let G be a finite group and H a subgroup of G.*

(i) *If a and b are in G, then $aH = bH$ or $aH \cap bH = \emptyset$.*

(ii) *The group G can be written as a disjoint union of left cosets of H.*

Theorem 23.7 is important in that once we have partitioned a finite group into a disjoint union of left cosets, we can count the elements in the group by simply summing the elements in the cosets, without having to worry about repeated elements. We will see several applications of this type of partition in what follows. It should be noted at this point that everything we have done with left cosets can be replicated for right cosets as well.

Some additional examples are in order.

Activity 23.8.

(a) Find all the left and right cosets of $H = \langle (1\ 2) \rangle$ in $G = S_3$.

(b) Find all the left and right cosets of $H = \langle (1\ 2\ 3) \rangle$ in $G = S_3$.

(c) If G is any group and H is a subgroup of G, must the left coset of H containing $g \in G$ be the same as the right coset of H in G containing g? Explain.

Note that, in general, a left coset of G is NOT a subgroup of G.

Activity 23.9. In each of the cases below, write the group G as a disjoint union of left cosets of the subgroup H.

(a) $G = \mathbb{Z}_{16}$, $H = \langle [4] \rangle$

(b) $G = S_3$, $H = \langle (1\ 2) \rangle$

Lagrange's Theorem

Lagrange's Theorem, which was hinted at in Preview Activity 23.1, is one of the most important theorems in all of finite group theory. In this section, we will discover what Lagrange's Theorem tells us about the subgroups of a finite group. Throughout this section, we will let G be a group of finite order and H a subgroup of G. For any finite set X, we will denote by $|X|$ the number of elements in X.

Preview Activity 23.10. To prove Lagrange's Theorem, we need to formalize an observation we have made about the number of elements in each left coset. Let $a \in G$. There is a natural function $\varphi : aH \to H$ defined by $\varphi(ah) = h$.

(a) Is φ an injection? Prove your answer.

(b) Is φ a surjection? Prove your answer.

(c) Is φ a bijection?

(d) What do your answers to parts (a) – (c) tell us about the number of elements in aH and in H? Explain.

You have probably noticed by now that in all the examples we have seen, when G has been a finite group and H a subgroup of G, the order of H has been a divisor of the order of G. Lagrange's Theorem says that this is always true.

Theorem 23.11 (Lagrange's Theorem). *If G is a finite group and H is a subgroup of G, then the order of H divides the order of G.*

To understand why Lagrange's Theorem holds, we need to connect the result of Preview Activity 23.10 to the partition of G determined by the relation $\sim$. Recall that the relation $\sim$ on G defined by $a \sim b$ if and only if $a^{-1}b \in H$ is an equivalence relation on G that partitions G into a disjoint union of left cosets.

Activity 23.12.

(a) Suppose G is partitioned into m distinct equivalence classes with representatives $a_1, a_2, \ldots, a_{m-1}, a_m$. Fill in the blanks to write G as a disjoint union of left cosets.

$$G = \underline{\quad} \cup \underline{\quad} \cup \underline{\quad} \cup \underline{\quad} \quad \cdots \quad \cup \underline{\quad} \tag{23.1}$$

(b) What does Preview Activity 23.10 tell us about how the number of elements in each left coset is related to $|H|$?

(c) You may have learned at one point that if X and Y are disjoint finite sets, then $|X \cup Y| = |X| + |Y|$. Use this fact and your response to parts (a) and (b) to write an equation that relates $|G|$ to $|H|$.

(d) Explain how your response to part (c) proves Lagrange's Theorem.

Lagrange's Theorem provides important information about the subgroup structure of a group and tells us that a finite group of order n cannot have a subgroup whose order does not divide n. We must be careful though, since **Lagrange's Theorem does NOT say that if m is a divisor of $|G|$, then G contains a subgroup of order m.** This conclusion is FALSE. (See Activity 23.20.) Lagrange's Theorem tells us only about the *possible* orders of subgroups of a finite group, but it does not tell us that subgroups of these possible orders must exist. For all we know at this point, it may be possible for a group of order 10 to have no subgroups of order 2 or 5. The first corollary to Lagrange's Theorem shows that that cannot happen. There are also other theorems that we will encounter in later investigations that will tell us something more about the existence of certain types of subgroups of any group.

Corollary 23.13. *Let G be a group of order n with $n > 1$. Then there is a prime integer p such that G contains a subgroup of order p.*

Proof. Let G be a group of order $n > 1$ with identity e. Choose an element $a \neq e$ in G. Then $|a| > 1$. Since $|G|$ is finite, Lagrange's Theorem tells us $|a| = |\langle a \rangle|$ divides n, and $|a|$ is finite. Let $m = |a|$. Let p be a prime factor of m and let $k \in \mathbb{Z}$ so that $pk = m$. Theorem 20.6 (see page 295) then shows that

$$|a^k| = \frac{m}{\gcd(m, k)} = \frac{m}{k} = p,$$

and so G contains an element a of order p. Therefore, $\langle a^k \rangle$ is a subgroup of G of order p. ∎

We should be cautious about Corollary 23.13. This corollary does not say that if a prime divides the order of a finite group G, then G contains a subgroup of order p—only that there is a subgroup of G of *some* prime order. For example, we can conclude that a group of order 10 contains either a subgroup of order 2 or a subgroup of order 5, but we can't say which one. Furthermore, such a group may contain subgroups of both orders (as in $\mathbb{Z}_{10}$), but Corollary 23.13 does not guarantee this.

Another simple, yet important corollary of Lagrange's Theorem is the following:

Corollary 23.14. *Let G be a finite group with identity element e. Then $a^{|G|} = e$ for every $a \in G$.*

Proof. Let G be a finite group of order n with identity element e and let $a \in G$. Lagrange's Theorem shows that $|a| = |\langle a \rangle|$ divides n, so there exists an integer m such that $|a|m = n$. Then

$$a^n = a^{m|a|} = \left(a^{|a|} \right)^m = e^m = e.$$

■

The exercises will provide several additional examples of the usefulness of Corollary 23.14.

Activity 23.12 shows us that if m is the number of left cosets of a subgroup H in a finite group G, then $m|H| = |G|$. This number of left cosets of H in G is given a name:

Definition 23.15. Let G be a group and H a subgroup of G. The **index** of H in G is the number of distinct left cosets of H in G.

We denote the index of H in G as $[G : H]$. If G is a finite group and H a subgroup of G, Activity 23.12 tells us that $[G : H] = \frac{|G|}{|H|}$. It should be noted though that Definition 23.15 also applies to infinite groups. For more details, see Exercise (17).

Activity 23.16. In each of the cases below, find $[G : H]$. Refer to Activity 23.9.

(a) $G = \mathbb{Z}_{16}$, $H = \langle [4] \rangle$

(b) $G = S_3$, $H = \langle (1\ 2) \rangle$

We will conclude this investigation with an application of Lagrange's Theorem to groups of prime order. In previous investigations, we have seen that any group of order 2 or 3 is cyclic (see Exercise (15) in Investigation 19), but that this is not true of groups of order 4. (Can you think of an example of a noncyclic group of order 4?) A major goal in finite group theory is to classify all groups of a given type. While this problem is far from completed, much progress has been made. Lagrange's Theorem is a key tool in the classification process, and the next activity illustrates a small portion of its power in this regard.

Activity 23.17.

(a) Let G be a group of order p, where p is a prime, and let H be a subgroup of G. What does Lagrange's Theorem tell us must be true of H? Explain. How many subgroups does G have?

(b) Let G be a group of order p, where p is a prime, and let $a \in G$. What must be true of $|a|$? Explain. Must G be Abelian? Must G be cyclic? Explain. What can we conclude about every group of prime order?

Concluding Activities

Activity 23.18. Write a formal proof of Lagrange's Theorem (Theorem 23.11).

Activity 23.19. You might be wondering why we focus on the relation $\sim_H$ instead of some other relation. In general, a relation just partitions the underlying set without concern for any structure the set possesses. With groups, however, there is a structure that we want to have preserved. An equivalence relation on a group that preserves the group structure as well is called a *congruence relation* (a generalization of the relation called congruence). In other words, an equivalence relation $\sim$ on a group G is a **congruence relation** if whenever $a, b, c, d \in G$ so that $a \sim b$ and $c \sim d$, then

(i) $ac \sim bd$ and

(ii) $a^{-1} \sim b^{-1}$.

As we will see in this activity, there is only one type of congruence relation on a group, and that is why we focus on the relation $\sim_H$. Let G be a group with identity e, and assume that $\sim$ is a congruence relation on G.

(a) Let $H = \{x \in G : x \sim e\}$. Show that H is a subgroup of G.

(b) Let $a, b \in G$. Prove that $a \sim b$ if and only if $a^{-1}b \in H$.

(c) Explain why $\sim_H$ is the only possible congruence relation on a group G. (Note that this does not say that $\sim_H$ is always a congruence relation on G. We will address that issue in Investigation 24.)

Activity 23.20. The Converse of Lagrange's Theorem. Lagrange's Theorem tells us that the order of any subgroup H of finite group G must divide the order of G. We did not prove the converse of Lagrange's Theorem for the simple reason that it is not true. In this activity, we'll see why this is the case.

(a) State the converse of Lagrange's Theorem. What do we need to do to show that the converse of Lagrange's Theorem is not true?

(b) Consider the group $G = A_4$. List the elements of A_4 in cycle notation, and determine the order of A_4.

(c) Assume that H is a subgroup of A_4 of order 6.

    (i) Explain why the nonidentity elements of H must have order 2 or 3.

    (ii) Explain why there must be an element α of A_4 of order 3 that is not in H.

    (iii) Explain why the left cosets H, αH and $\alpha^2 H$ cannot all be distinct.

    (iv) Show that it is not possible for any two of H, αH and $\alpha^2 H$ to be equal.

(d) Explain why the converse of Lagrange's Theorem is not true.

Exercises

(1) Suppose a classmate told you that there is a subgroup of Z_{15} of order 10. What would your response be? Explain as clearly and completely as possible, citing all relevant theorems.

(2) Let G be a group and H a subgroup of G.

    (a) If $a \in G$ and $aH = H$, what must be true about a? Prove your answer.

    (b) If $a \in H$, what coset is aH? Prove your answer.

    (c) Correctly complete the statement of the following theorem:

        Theorem. *Let G be a group and H a subgroup of G. If $a \in G$, then $aH = H$ if and only if* _____.

(3) Let G be a group of order n. If H is a subgroup of G and $|H| \geq \frac{n}{2} + 1$, what group must H be? Prove your answer.

(4) A group G contains elements of every order from 1 to 10. What is the smallest order G could have? Find a group G of that order that contains elements of every order from 1 through 10.

(5) For each of the examples given below, list the distinct cosets of H in G.

    (a) $G = \mathbb{Z}_{12}$, $H = \langle [4] \rangle$

    (b) $G = U_{15}$, $H = \langle [11] \rangle$

(6) Let $H = \{I, r\}$ in D_4.

    (a) Determine all distinct left cosets of H in D_4.

    (b) Determine all distinct right cosets of H in D_4.

(7) Let G be a group, and let $a \in G$ be an element of order 21. Let $H = \langle a \rangle$ and $K = \langle a^7 \rangle$.

    (a) How many left cosets are there of K in H?

    (b) Find all left cosets of K in H.

(8) Let G be a group of order 840, and suppose K is a subgroup of G of order 42. If H is a subgroup of G that contains K as a subgroup, what could the order of H be? Explain.

(9) Let $G = \langle a \rangle$ be a cyclic group of order n. If k is an integer between 1 and n, what is the index of $\langle a^k \rangle$ in G? Prove your answer.

(10) The subgroup structure of a group tells us much about the group. For example, there are groups that have no nontrivial proper subgroups (subgroups of order larger than 1 that are not the entire group), and these groups have a very simple structure.

    (a) Suppose G is a finite group that contains no nontrivial proper subgroups. What can you say about $|G|$? Prove your answer.

    (b) What conditions are there on the order of a finite group G that ensure that G contains a nontrivial proper subgroup? Prove your answer.

    (c) Correctly complete the statement of the following theorem:

Theorem. *A group G of order n has a nontrivial proper subgroup if and only if n is*

_____.

(11) **Subgroups of p-groups.** A group G is called a p-group if $|G| = p^k$ for some prime p and some integer $k \geq 1$. For example, the groups D_4, $\mathbb{Z}_8$, U_{15}, U_{24}, and $\mathbf{Q}$ from Exercise (6) in Investigation 21 (see page 305) are all 2-groups of order 8. In this exercise, we will examine subgroups of p-groups in more detail.

  (a) Exercise (22) of Investigation 19 (see page 288) shows that any 2-group must contain an element of order 2. Must a p-group of order p^k contain elements of order p^t for all $1 \leq t \leq k$? Explain.

  (b) Can we generalize Exercise (22) of Investigation 19 to show that a p-group must contain an element of order p? Prove your answer.

(12) Let G be a finite group with identity e, and let H and K be subgroups of G.

  (a) If $|H| \neq |K|$, must $H \cap K = \{e\}$? Prove your answer.

  (b) Is there any condition on $|H|$ and $|K|$ that ensures that $H \cap K = \{e\}$? Prove your conjecture. (Hint: If $a \in H \cap K$, what does Corollary 23.14 tell us?)

(13) Over the real numbers, we are able to solve all equations of the form $x^m = b$ for x if m is a positive integer. A solution to the equation $x^m = b$ is called an mth *root* of b.

  (a) If G is a finite group, $b \in G$, and $m \in \mathbb{Z}^+$, can we always solve the equation $x^m = b$ in G? That is, does every group contain all m^{th} roots of each of its elements? Prove your answer.

  (b) There are certain situations in which a finite group must contain some of the mth roots of its elements. We will explore that situation in this part of the exercise. Let G be a group of order n with identity e.

   (i) If $g \in G$ and $g^m = e$ for some $m \in \mathbb{Z}^+$ with $\gcd(m, n) = 1$, show that $g = e$. (Hint: Use Corollary 23.14 on page 326.)

   (ii) Let $m \in \mathbb{Z}^+$ so that $1 = rn + sm$ for some integers r and s. If $g \in G$, show that g^s is an mth root of g. Conclude that if $\gcd(m, n) = 1$, then G contains an mth root for each of its elements.

   (iii) Illustrate the process in the previous part to find a third root of the element $[4]$ in U_{11}.

  (c) A nontrivial Abelian group G is said to be **divisible** if G contains an mth root for each of its elements. That is, G is divisible if for each $a \in G$ and each nonzero $m \in \mathbb{Z}$, there exists $x \in G$ so that $x^m = a$.

   (i) Is the group $\mathbb{Q}$ of rational numbers divisible? Prove your answer.

   (ii) Is the group $U(\mathbb{Q})$ (the group of units in $\mathbb{Q}$, which consists of all nonzero rational numbers) divisible? Prove your answer.

   (iii) Can a finite Abelian group be divisible? Prove your answer.

(14) (a) Let G be a group of order 4 with identity e. Show that G is either cyclic or $a^2 = e$ for all $a \in G$.

  (b) Does the result of part (a) generalize to groups of order p^2 for any positive integer p? In other words, is it the case that if G is a group of order p^2 with identity e, then G is either cyclic or $a^p = e$ for every $a \in G$? Prove your answer.

(c) Is there any condition on p that will make the answer to the question in part (b) yes? If so, state and prove a conjecture. If no, explain why.

(d) Does the result of part (a) generalize to groups of order 2^k for any positive integer $k > 2$? That is, if G is a group of order 2^k for some $k \in \mathbb{Z}$ with $k > 2$, must it be the case that either G is cyclic or $a^2 = e$ for every $a \in G$? Prove your answer.

* (15) Show that every group of order 4 is Abelian. Must every group of order 4 be cyclic? Explain. (Hint: Use part (a) of Exercise (14) on page 329.)

(16) Let n be a positive integer. Let $H = n\mathbb{Z} = \{nk : k \in \mathbb{Z}\}$ (that is, H is the set of all integer multiples of n). Find all left and right cosets of H in $\mathbb{Z}$. What is $[\mathbb{Z} : n\mathbb{Z}]$?

* (17) If G is a finite group and H a subgroup of G, then Lagrange's Theorem tells us about $[G : H]$. But Definition 23.15 also applies to infinite groups.

(a) Is it possible to find an infinite group G and subgroup H of G so that $[G : H]$ is finite? If yes, find and explain such an example. If no, explain why.

(b) Is it possible to find an infinite group G and subgroup H of G so that there are infinitely many different left cosets of H in G? (In this case, we would say that $[G : H]$ is infinite.) If yes, find and explain such an example. If no, explain why.

(18) Let G be a group and H a subgroup of G. We have indicated that the number of left cosets of H in G is equal to the number of right cosets of H in G. Prove that statement. (Hint: Consider the function f that maps the left coset aH of H in G to the right coset Ha^{-1}.)

(19) Let $G = D_n$, the group of symmetries of a regular n-gon.

(a) Determine the distinct left cosets of $H = \langle R \rangle$ in G.

(b) Use the result of (a) to explain why $D_n = \{r^i R^j : 0 \le i \le 1, 0 \le j \le n - 1\}$.

* (20) A famous theorem in number theory is Fermat's Little Theorem. (One interesting application of this theorem is to prove the validity of RSA encryption. See the online supplemental materials for more details.)

Theorem (Fermat's Little Theorem). *Let p be a prime number. If $a \in \mathbb{Z}$, then*

$$a^p \equiv a \bmod p.$$

Prove Fermat's Little Theorem. (Hint: Use the result of Corollary 23.14 in the group U_p.)

(21) A collection $m_1, m_2, \ldots, m_n$ of integers is said to be *pairwise relatively prime* if $\gcd(m_i, m_j) = 1$ whenever $i \ne j$. For example, the integers 121, 122, and 123 are pairwise relatively prime even though each of the integers is composite.

(a) Let $m_1, m_2, \ldots, m_n$ be a collection of pairwise relatively prime integers. Let k be between 1 and n. Show that any product of the integers $m_1, m_2, \ldots, m_n$ that doesn't have m_k as a factor is relatively prime to m_k.

(b) Prove the following lemma.

Lemma 23.21. *Let $m_1, m_2, \ldots, m_n$ be pairwise relatively prime positive integers. If $x \equiv 1 \pmod{m_i}$ for each $1 \le i \le n$, then $x \equiv 1 \pmod{m_1 m_2 \cdots m_n}$.*

(22) We can use Fermat's Little Theorem (see Exercise (20)) to test if an integer is composite. If n is a positive integer and for some $a \in \mathbb{Z}^+$ with $\gcd(a, n) = 1$ we have $a^{n-1} \not\equiv 1 \pmod{n}$, then we know n is not prime. For example, if $n = 1121107$, $a = 2337$, then we can use some method like successive squaring to see that $a^{n-1} \equiv 325391 \pmod{n}$ and n is not prime.

A **pseudoprime** is an integer that passes some tests for primality, but is still not prime. Pseudoprimes are classified according to which property of primes they satisfy. For example, a Fermat pseudoprime to a base a, written psp(a), is a composite number m such that $a^{n-1} \equiv 1 \pmod{n}$. Alternatively, $a^n \equiv a \pmod{n}$. That is, n satisfies Fermat's little theorem for the base a.

In 1904, Cipolla gave a method for producing an infinite number of pseudoprimes base a for $a > 1$. [‡] We explore Cipolla's method in this exercise.

Let a be a positive integer and let p be any odd prime that does not divide $a(a^2 - 1)$. Let $A = \frac{a^p - 1}{a - 1}$ and let $B = \frac{a^p + 1}{a + 1}$. We will show that $n = AB$ is psp(a).

   (a) Explain why A and B are integers. Thus, n is an integer.

   (b) Show that $n = \frac{a^{2p} - 1}{a^2 - 1}$. Use this to show that

$$(n - 1)(a^2 - 1) = a(a^{p-1} - 1)(a^p + a).$$

   (c) Explain why $a^p + a$ is always even.

   (d) Show that $(a^2 - 1)p$ divides $a^{p-1} - 1$. (Hint: Use the fact that $p - 1$ is even to explain why $a^2 - 1$ divides $a^{p-1} - 1$. Then use Fermat's Little Theorem.)

   (e) Now show that $2p$ divides $n - 1$. From this, draw the conclusion that $a^{n-1} \equiv 1 \pmod{n}$.

   (f) Use Cipolla's method to find a Fermat pseudoprime to the base 2 and a Fermat pseudoprime to the base 4.

(23) It is possible to find composite numbers m so that $a^{n-1} \equiv 1 \pmod{n}$ for every a with $\gcd(a, n) = 1$. That is, there are integers that are Fermat pseudoprimes for *every* base (see Exercise (22)). So Fermat's Little Theorem cannot distinguish these numbers from primes. Such numbers are called **Carmichael numbers** (named for Robert Carmichael who found the smallest Carmichael number). In this exercise we will demonstrate that 561 is a Carmichael number (in fact, 561 is the smallest Carmichael number). We will also generalize the method to find other Carmichael numbers.

   (a) Let $n = 561$. Show that $a^{n-1} \equiv 1 \pmod{p}$ for all prime divisors p of n and all integers a with $\gcd(a, n) = 1$.

   (b) Use Lemma 23.21 to show that $a^{n-1} \equiv 1 \pmod{n}$.

   (c) Now we generalize the above argument. Let $n = p_1 p_2 \cdots p_k$, where $p_1 < p_2 < \cdots < p_k$ are prime and $p_i - 1$ divides $n - 1$ for each $1 \leq i \leq k$. We will show that m is a Carmichael number.
      (i) First, show that k must be at least 3. (Hint: What is q modulo $q - 1$? What does this say about n modulo $q - 1$?)
      (ii) Now show that n is a Carmichael number.

[‡] M. Cipolla, Sui numeri composti P, che verificanola congruenza di Fermat $a^{P-1} \equiv 1 \pmod{P}$. *Annali di Matematica* **9** (1904), 139–160.

(iii) Let $n = (6k + 1)(12k + 1)(18k + 1)$ for some positive integer k. Show that if $6k + 1, 12k + 1$, and $18k + 1$ are all prime, then n is a Carmichael number.

(iv) Find two Carmichael numbers using the ideas from part (iii.).

(24) The Euler phi function φ is defined to give the number of positive integers less than or equal to and relatively prime to a given positive integer. For example, $\varphi(2) = 1$ since 1 is the only positive integer less than 2 and relatively prime to 2. Similarly, $\varphi(3) = 2$, $\varphi(4) = 2$, $\varphi(5) = 4$, and so on. There is a generalization of Fermat's Little Theorem (from Exercise (20)) due to Euler:

Theorem. *Let n be a positive integer and $a \in \mathbb{Z}$ with $\gcd(a, n) = 1$. Then*

$$a^{\varphi(n)} \equiv 1 \pmod{n}.$$

Prove this theorem. Then explain why this theorem is a generalization of Fermat's Little Theorem. (Hint: Use the result of Corollary 23.14 in the group U_n.)

(25) If G is a finite group and H and K are subgroups of G with K a subgroup of H, then Lagrange's Theorem tells us that $[G : H] = \frac{|G|}{|H|}$, $[H : K] = \frac{|H|}{|K|}$, and $[G : K] = \frac{|G|}{|K|}$. So in this case we have $[G : K] = [G : H][H : K]$. If G is an infinite group, we cannot apply Lagrange's Theorem to obtain this index property. In this exercise, we will determine if $[G : K] = [G : H][H : K]$ when G is infinite and H and K are subgroups of a group G so that $K \subset H$ and $[G : H]$ and $[H : K]$ are finite.

(a) Let $G = \mathbb{Z}$, $H = 4\mathbb{Z}$ and $K = 12\mathbb{Z}$.

(i) Let $a_1 + H, a_2 + H, \ldots, a_n + H$ be the collection of distinct left cosets of H in G. Find n and a representative a_i for each coset. What is $[G : H]$?

(ii) Let $b_1 + K, b_2 + K, \ldots, b_m + K$ be the collection of distinct left cosets of K in H. Find m and a representative b_j for each coset. What is $[H : K]$?

(iii) Find all left cosets of K in G in terms of the a_i and b_j. What is $[G : K]$? Is $[G : K] = [G : H][H : K]$?

(b) Let G be an infinite group, and let H and K be subgroups of G such that $K \subset H$ and both $[G : H]$ and $[H : K]$ are finite. Is it true that $[G : K]$ is finite and $[G : K] = [G : H][H : K]$? Prove your answer.

Connections

Given a subgroup H of a group G, we showed in this investigation that congruence modulo H is an equivalence relation. As we will see in subsequent investigations, this relation will lead us to the useful construction of quotient structures. We are familiar with other notions of congruence in groups; in particular, congruence modulo an integer n led us to the group $\mathbb{Z}_n$ in Investigation 2. If you studied ring theory before group theory, you should notice connections between the topics in this investigation and those in Investigations 12 and 13. In particular, congruence modulo a subgroup is the same idea as congruence modulo a polynomial in a polynomial ring $F[x]$, or congruence modulo an ideal I in a ring R.

Investigation 24

Normal Subgroups and Quotient Groups

Focus Questions

By the end of this investigation, you should be able to give precise and thorough answers to the questions listed below. You may want to keep these questions in mind to focus your thoughts as you complete the investigation.

- What is a normal subgroup of a group? Why are normal subgroups important?

- Under what conditions is the quotient of a group by a subgroup a group itself?

- What does Cauchy's Theorem say about the subgroups that a finite Abelian group must have?

- What is a simple group, and what are some examples of simple groups?

Preview Activity 24.1. Recall that if G is a group and H is a subgroup of G, the cosets of H partition G into disjoint equivalence classes. In this activity, we will investigate what these cosets might look like.

(a) Table 24.1 presents the operation table for a group G of order 6. (Note that we have used the letters a through f to denote the elements of this group, but we are not saying that e is the identity of the group.) We are going to rearrange the elements of G so that the operation table for G allows us to observe more easily what some of the cosets in G look like. The table on the right will eventually be the rearranged table for G, but we will replace each of the labels g_1, g_2, g_3, g_4, g_5, and g_6 with the elements a, b, c, d, e, and f from the group G.

To begin, notice that f is an element of order 2 in this group. Let the element g_1 in the table on the right be f and let $H = \langle f \rangle$ be the subgroup of G generated by f. Let the element g_2 in the table on the right be the other element in $\langle f \rangle$ (which will be d). The table on the right now appears as shown in Table 24.2.

  (i) We know that H is one left coset of H in G, and now we will find the others. Choose any element $g_3 \in G$ so that $g_3 \notin H$. (That is, choose g_3 to be either a, b, c, or e.) Then replace g_3 in the table on the right with the element you chose. List the elements in $g_3 H$, and replace g_4 with the remaining element in $g_3 H$.

  (ii) There is an element $g_5 \in G$ with $g_5 \notin H \cup g_3 H$. Replace g_5 and g_6 in the table on the right with the elements in $g_5 H$. What is the relationship between H, $g_3 H$, and $g_5 H$?

  (iii) Now fill in the operation table for G writing the elements in the order you used when you relabeled the table on the right in parts (i) and (ii). Color the cells in this table that correspond to elements in H with the color blue (or any other color you choose).

Table 24.1

Operation table for a group G of order 6 (left) and a rearrangement of the table (right).

	a	b	c	d	e	f
a	b	d	f	a	c	e
b	d	a	e	b	f	c
c	f	e	a	c	d	b
d	a	b	c	d	e	f
e	c	f	d	e	b	a
f	e	c	b	f	a	d

	g_1	g_2	g_3	g_4	g_5	g_6
g_1						
g_2						
g_3						
g_4						
g_5						
g_6						

Table 24.2

Operation table for a group G of order 6 (left) and a rearrangement of the table (right).

	a	b	c	d	e	f
a	b	d	f	a	c	e
b	d	a	e	b	f	c
c	f	e	a	c	d	b
d	a	b	c	d	e	f
e	c	f	d	e	b	a
f	e	c	b	f	a	d

	f	d				
f	d	f				
d	f	d				

Table 24.3

Operation table for a second group G of order 6 (left) and a rearrangement of the table (right).

	a	b	c	d	e	f
a	a	b	c	d	e	f
b	b	a	e	f	c	d
c	c	f	a	e	d	b
d	d	e	f	a	b	c
e	e	d	b	c	f	a
f	f	c	d	b	a	e

	g_1	g_2	g_3	g_4	g_5	g_6
g_1						
g_2						
g_3						
g_4						
g_5						
g_6						

Then color the cells that correspond to elements in $g_3 H$ with a different color. Finally, select a third color, and color the cells that correspond to elements in $g_5 H$ with that third color. What do you notice about this table if you ignore the labels and just focus on the colors?

(b) Repeat part (a) with the group G whose operation table is shown in Table 24.3. That is, let $H = \langle g_1 \rangle$ be a subgroup of G of order 2, and then: (i) find the distinct left cosets H, $g_3 H$, and $g_5 H$ of H in G; (ii) complete Table 24.3 at right; and (iii) color-code the cells of the table corresponding to distinct cosets of H. What do you notice? Compare your observations to those from part (a).

Introduction

Let G be a group and H a subgroup of G. Preview Activity 24.1 seems to indicate that in some cases the collection of distinct cosets of H in G has a group structure, and in other cases it does not. In this investigation, we will explore the structure of the set of distinct left cosets of a subgroup of a group.

An Operation on Cosets

Let G be a group (with the operation written multiplicatively), H a subgroup of G, and $g \in G$. Recall that the left coset of H in G containing g is the set

$$gH = \{gh : h \in H\}.$$

We will denote by G/H the collection of all distinct left cosets of the subgroup H in the group G. We want to understand when the set G/H can be made into a group. To identify a group structure on G/H, we will first need to define an operation on G/H. There is a natural way to try to define such an operation. In particular, for any elements aH, bH in G/H, we could define the product $(aH)(bH)$ as

$$(aH)(bH) = (ab)H. \tag{24.1}$$

Activity 24.2. One question we must address with the operation on G/N proposed in (24.1) is whether or not the operation is well-defined.

(a) What would it mean for the operation from (24.1) to be well-defined on G/H? Be specific. In general, why is it important for operations to be well-defined?

(b) Let G be the 12 element group in Table 24.4. (You may assume that G is a group.) Let $H = \{I, a_9, a_{10}, a_{11}\}$ and $K = \{I, a_3, a_4\}$. It can be shown that H and K are subgroups of G.

   (i) Find the distinct left cosets of H in G. What is G/H?

   (ii) Find the right cosets of H in G. How are the left and right cosets of H in G related?

   (iii) Find the distinct left cosets of K in G. What is G/K?

   (iv) Find the right cosets of K in G. What do you notice about the left and right cosets of K in G?

   (v) Give an example to show that the operation proposed in (24.1) is not well-defined on G/K.

There is a substantive difference between the cosets of H in G and the cosets of K in G (from Activity 24.2) that makes the operation from (24.1) well-defined on G/H but not on G/K. As we saw, for each $a \in G$, we had $aH = Ha$, but this is not true for K. Note that $aH = Ha$ does not

Table 24.4

Operation table for a group G of order 12.

	I	a_1	a_2	a_3	a_4	a_5	a_6	a_7	a_8	a_9	a_{10}	a_{11}
I	I	a_1	a_2	a_3	a_4	a_5	a_6	a_7	a_8	a_9	a_{10}	a_{11}
a_1	a_1	a_2	I	a_{10}	a_6	a_7	a_{11}	a_9	a_3	a_5	a_8	a_4
a_2	a_2	I	a_1	a_8	a_{11}	a_9	a_4	a_5	a_{10}	a_7	a_3	a_6
a_3	a_3	a_{11}	a_5	a_4	I	a_{10}	a_8	a_1	a_9	a_6	a_2	a_7
a_4	a_4	a_7	a_{10}	I	a_3	a_2	a_9	a_{11}	a_6	a_8	a_5	a_1
a_5	a_5	a_3	a_{11}	a_9	a_7	a_6	I	a_{10}	a_2	a_1	a_4	a_8
a_6	a_6	a_9	a_8	a_1	a_{10}	I	a_5	a_4	a_{11}	a_3	a_7	a_2
a_7	a_7	a_{10}	a_4	a_5	a_9	a_{11}	a_1	a_8	I	a_2	a_6	a_3
a_8	a_8	a_6	a_9	a_{11}	a_2	a_3	a_{10}	I	a_7	a_4	a_1	a_5
a_9	a_9	a_8	a_6	a_7	a_5	a_4	a_2	a_3	a_1	I	a_{11}	a_{10}
a_{10}	a_{10}	a_4	a_7	a_6	a_1	a_8	a_3	a_2	a_5	a_{11}	I	a_9
a_{11}	a_{11}	a_5	a_3	a_2	a_8	a_1	a_7	a_6	a_4	a_{10}	a_9	I

mean $ah = ha$ for all $h \in H$; rather, for each $h \in H$ there is an element $h' \in H$ so that $ah = h'a$. Under this condition, we can show that the operation from (24.1) is well-defined on G/H.

Activity 24.3. Let G be a group, and let H be a subgroup of G such that $gH = Hg$ for all $g \in G$. Let a, a', b, b' be elements of G with $aH = a'H$ and $bH = b'H$.

(a) Explain why there must be elements $h_a, h_b \in H$ so that

$$a = a'h_a \quad \text{and} \quad b = b'h_b.$$

(b) Explain why there is an element $h' \in H$ so that $h_a b' = b'h'$. (Hint: Every left coset of H is a right coset of H.)

(c) Use the previous two results to show that ab is an element of $a'b'H$. What does this tell us about the relationship between abH and $a'b'H$? Explain.

(d) Explain why the operation from (24.1) is well-defined on G/H.

The condition on H that makes the operation from (24.1) well-defined is an important one, and we will study it in more detail in the next section.

Normal Subgroups

In the previous section we saw that if G is a group and H is a subgroup of G so that $aH = Ha$ for all $a \in G$, then we can construct a well-defined operation on the collection of left cosets of H in G. The subgroups that allow us to define such an operation are called *normal* subgroups.

Definition 24.4. Let G be a group. A subgroup N of G is **normal** in G (or is a **normal subgroup** of G) if $aN = Na$ for all $a \in G$.*

The normal subgroups of G are exactly the subgroups for which we have a well-defined operation on G/N, the collection of all distinct left cosets of N in G. When N is a normal subgroup of G, we write $N \triangleleft G$.

To show that a subgroup N is normal in a group G, we can use Definition 24.4, which involves proving two different containment statements—specifically, that $aN \subseteq Na$ and $Na \subseteq aN$ for all $a \in G$. However, we can simplify this process a bit. Informally, if $aN \subseteq Na$, then we might think that we could "multiply" both sides of this containment on the right by a^{-1} to obtain the equivalent statement that $aNa^{-1} \subseteq N$. Similarly, we might think that the statement $Na \subseteq aN$ is equivalent to $a^{-1}Na \subseteq N$. The next theorem formalizes these intuitive ideas.

Theorem 24.5. *Let G be a group and N a subgroup of G. Then N is normal in G if and only if $aNa^{-1} \subseteq N$ for all $a \in G$, where*

$$aNa^{-1} = \{ana^{-1} : n \in N\}.$$

Proof. The usefulness of this theorem is in the reverse implication, so we leave the proof of the forward implication to the exercises. (See Exercise (6).) Let G be a group and N a subgroup of G. Assume $aNa^{-1} \subseteq N$ for all $a \in G$. To show $N \triangleleft G$, we must show $aN = Na$ for all $a \in G$. Choose $a \in G$. To prove $aN = Na$, we will demonstrate containment in both directions. Let $x \in aN$. Then there exists $n \in N$ such that $x = an$. By hypothesis we know that $aNa^{-1} \subseteq N$, so $ana^{-1} \in N$. Let $n' = ana^{-1}$. Then

$$x = an = n'a \in Na.$$

Therefore, $aN \subseteq Na$. To prove containment in the opposite direction, let $x \in Na$. So there exists $n \in N$ such that $x = na$. By hypothesis, we know that

$$a^{-1}Na = (a)^{-1}N\left(a^{-1}\right)^{-1} \subseteq N,$$

so $a^{-1}na \in N$. Let $n' = a^{-1}na$. Then

$$x = na = an' \in aN.$$

The two containments we have proved show that $Na = aN$, and so N is a normal subgroup of G. ∎

A few examples are in order.

Activity 24.6. For each part below, decide if the given subgroup N is a normal subgroup of the group G. You may want to use Theorem 24.5.

(a) $G = \mathbb{Z}_8$, $N = \langle [2] \rangle$

(b) $G = \mathbb{Z}$, $N = \langle 5 \rangle$

(c) $G = S_3$, $N = \langle (1\ 2\ 3) \rangle$

---

*The term *normal* seems to be used here because in Galois theory (not discussed in these investigations), the Galois correspondence between subgroups and subfields has normal subgroups corresponding to subfields that are normal extensions of the base field. Normal subgroups are also called *invariant subgroups* because they are invariant under inner automorphisms.

Quotient Groups

Let G be a group and $N \triangleleft G$. Recall that the operation on G/N defined as

$$(aN)(bN) = (ab)N \tag{24.2}$$

is well-defined. It is natural to ask what kind of structure this operation imposes on G/N.

Activity 24.7. By definition, G/N is closed under the operation from (24.2). Here we will investigate some other properties of G/N under this same operation.

   (a) There is an identity element in G/N. What is it? Verify your answer.

   (b) Prove that the operation from (24.2) is associative on G/N.

   (c) Prove that G/N contains an inverse for each of its elements.

The responses to Activity 24.7 show that G/N is a group under the operation from (24.2). This group is called the *quotient group* (or *factor group*) of G by N.

Definition 24.8. Let G be a group and N a normal subgroup of G. The **quotient group** (or **factor group**) of G by N is the group

$$G/N = \{aN : a \in G\}$$

with the operation $(aN)(bN) = (ab)N$ for all $a, b \in G$.

At this point, it will be helpful for us to explore a few examples of quotient groups and their structures.

Activity 24.9. For each choice of G and N below, construct an operation table for the quotient group G/N.

   (a) $G = \mathbb{Z}_8, N = \langle [2] \rangle$

   (b) $G = \mathbb{Z}, N = \langle 5 \rangle$

   (c) $G = D_4, N = \langle R^2 \rangle$

You may have noticed that if G is a finite group, then the order of G/N is related to the orders of G and N. In fact, Lagrange's Theorem (see page 325) tells us that $|G/N| = \frac{|G|}{|N|}$ when G is finite. Notice that G/N can be a finite group even if G is an infinite group.

Quotient groups are very useful, in part because if G is a finite group and N a nontrivial normal subgroup of G, then G/N has smaller order than G. Consequently, it may be easier to prove properties of G/N than it is for G. Often, information about G/N tells us something important about G, as the next theorem shows. (Recall that for any group G, $Z(G)$ denotes the center of G—that is, the set of elements that commute with every element of G.)

Theorem 24.10. *If G is a group and $G/Z(G)$ is cyclic, then G is Abelian.*

Proof. Let G be a group such that $G/Z(G)$ is cyclic. So that the proof is easier to follow, let $N = Z(G)$. Since G/N is cyclic, we know there is an element $gN \in G/N$ that generates G/N.

That is, $G/N = \langle gN \rangle$. To show G is Abelian, we need to show $ab = ba$ for all $a, b \in G$. Let $a, b \in G$. Then $aN, bN \in G/N$. Moreover, $aN = (gN)^k = g^k N$ and $bN = (gN)^m = g^m N$ for some integers k, m. Since $a \in aN$, we have $a = g^k z_a$ for some $z_a \in N$. Since $b \in bN$, we also have $b = g^m z_b$ for some $z_b \in N$. Since z_a and z_b are in the center of G, we know z_a and z_b commute with all elements of G. Therefore,

$$
\begin{aligned}
ab &= (g^k z_a)(g^m z_b) \\
&= z_b(g^k g^m) z_a \\
&= z_b g^{k+m} z_a \\
&= z_b g^{m+k} z_a \\
&= z_b g^m g^k z_a \\
&= (z_b g^m)(z_a g^k) \\
&= ba.
\end{aligned}
$$

It follows that G is an Abelian group. ∎

Cauchy's Theorem for Finite Abelian Groups

In our study of groups so far, we have seen that it is important to understand the subgroup structure of a group. Lagrange's Theorem (see page 325) tells us what orders are possible for subgroups of a finite group, but it does not tell us anything about the existence of subgroups of any particular order. One result we have so far in that direction is Corollary 23.13 (see page 325), which shows that every finite group must contain a subgroup of prime order. However, the drawback of Corollary 23.13 is that it doesn't tell us *which* prime order subgroup must exist. A more general result, Cauchy's Theorem, does just that. We will focus here on Cauchy's Theorem for Finite Abelian Groups, and we will later extend the theorem to all finite groups.

Theorem 24.11 (Cauchy's Theorem for Finite Abelian Groups). *Let G be an Abelian group of finite order n. If p is a prime divisor of n, then G contains an element of order p.*

Preview Activity 24.12. In the proof of Cauchy's Theorem for Finite Abelian Groups, we will see that the construction of a quotient group allows us to reduce the order of a group and then apply the principle of mathematical induction. This is a powerful idea, and in this activity we will explore an example to see how it works. Let $G = U_{21} = \{[1], [2], [4], [5], [8], [10], [11], [13], [16], [17], [19], [20]\}$. Once proved, Cauchy's Theorem will tell us that if a prime p divides the order of G, then G must contain an element of order p. Thus, U_{21} must contain elements of order 2 and 3. Since U_{21} is a finite group of relatively small order, we could easily determine the orders of each of its elements, but our goal in this activity is to create a general process that will allow us to find such an element in any finite Abelian group. In this example, we will find an element of order 3 in U_{21}. To find such an element in U_{21}, we will start with Corollary 23.13, which tells us that G contains an element of some prime order q that divides $|G|$. Recall, however, that we don't know exactly what prime q is.

(a) Explain why we are done with this activity if $q = 3$.

(b) Now assume $q \neq 3$. For the purposes of this illustration, we will consider the case that $q = 2$. Recall that we are still looking for an element of order 3 in G. We are assuming there is an

Table 24.5

Operation table for $U_{21}/\langle[8]\rangle$.

	N	$[2]N$	$[4]N$	$[5]N$	$[10]N$	$[13]N$
N	N	$[2]N$	$[4]N$	$[5]N$	$[10]N$	$[13]N$
$[2]N$	$[2]N$	$[4]N$	N	$[10]N$	$[13]N$	$[5]N$
$[4]N$	$[4]N$	N	$[2]N$	$[13]N$	$[5]N$	$[10]N$
$[5]N$	$[5]N$	$[10]N$	$[13]N$	$[4]N$	N	$[2]N$
$[10]N$	$[10]N$	$[13]N$	$[5]N$	N	$[2]N$	$[4]N$
$[13]N$	$[13]N$	$[5]N$	$[10]N$	$[2]N$	$[4]N$	N

element of order 2 in G. In fact, there are three such elements, and $[8]$ happens to be one of them. Let $N = \langle[8]\rangle$. Since G is Abelian, we know that N is a normal subgroup of G, and so the quotient $\overline{G} = G/N$ is a group.

(i) Explain why $|\overline{G}| < |G|$. Since $|\overline{G}| < |G|$, we can apply the idea of induction. We will be more specific about the induction hypothesis a bit later, but the basic concept is that since $|\overline{G}| < |G|$, we can assume the result of Theorem 24.11 for $\overline{G}$. Explain why doing so allows us to assume that $\overline{G}$ contains an element of order 3.

(ii) Note that $\overline{G} = \{N, [2]N, [4]N, [5]N, [10]N, [13]N\}$. The operation table for $\overline{G}$ is given in Table 24.5. Show that $[2]N$ has order 3 in $\overline{G}$.

(iii) If $\overline{G} = G/N$ contains an element of order 3, it might be natural to expect that there is a corresponding element of order 3 in G. (See Exercise (16).) This, however, is not always the case. For instance, the element $[2]N$ has order 3 in $\overline{G}$, but $|[2]| \neq 3$ in G. Find $|[2]|$ in G. Then, using only what we know about finite cyclic groups and $|[2]|$ in G, find an element of order 3 in G. (This shows that G contains an element of order 3, as desired.)

If G is a finite Abelian group of order n and p is a prime that divides n, Theorem 24.11 tells us that G must contain an element of order p. Preview Activity 24.12 illustrates the method we will use to prove Theorem 24.11. First, we will find an element $a \in G$ of some prime order q that divides n. If $q \neq p$, then the quotient group $\overline{G} = G/\langle a\rangle$ has smaller order than G. If p divides $|\overline{G}|$, then we can apply induction and find an element $\overline{b} \in \overline{G}$ so that $|\overline{b}| = p$. We can associate the element $\overline{b}$ in $\overline{G}$ with an element $b \in G$ so that $\overline{b} = bN$. If $|b|$ is a multiple of p, then we can find an element of order p in G. The proof of Cauchy's Theorem presented below completes these steps in detail.

Proof of Cauchy's Theorem for Finite Abelian Groups. This proof is a bit more complicated than some others we have done, so you should be sure that you can provide the rationale for any steps that are omitted or labeled with the ⑦ symbol. To begin, let G be an Abelian group of finite order n with identity element e, and let p be a prime divisor of n. We will show that G contains an element of order p. Our proof will be by induction on the order of the group.

The base case will be when $n = 2$.⑦ When $|G| = 2$, the only prime p that divides $|G|$ is $p = 2$. Let $a \in G$ be the nonidentity element. By Lagrange's Theorem, $|a| = |\langle a\rangle| = 2$, so a is an element of order 2. Thus, our theorem is true when $n = 2$.

For the inductive step, assume that for any Abelian group G' of order $k < n$ and for any prime divisor q of k, G' contains an element of order q. ⊙ By Corollary 23.13 (see page 325), we know that G contains an element a of some prime order q, where q divides n. If $q = p$, then a has order p and we are done because we have found an element in G of order p. So assume that $q \neq p$. Let $N = \langle a \rangle$. Since $N \lhd G$, ⊙ the quotient $\overline{G} = G/N$ is a group. Since a has prime order, we know $|a| = |N|$ is greater than 1. Thus, $|\overline{G}| = |G/N| = \frac{|G|}{|N|} < |G|$, and so $\overline{G}$ has order smaller than $|G|$—that is, $|\overline{G}| < |G| = n$. Furthermore, since G is an Abelian group, so is $\overline{G}$. ⊙ Thus, we can apply the induction hypothesis to $\overline{G}$ and any prime divisor of $|\overline{G}|$. Now p divides n and p does not divide q, so p divides $|\overline{G}|$. ⊙ We can then apply our induction hypothesis to conclude there is an element $\overline{b}$ of order p in $\overline{G}$. Now all we need to do is translate this conclusion about $\overline{G}$ to one about G. To do so, note that since $\overline{b} \in \overline{G}$, there exists $b \in G$ such that $\overline{b} = bN$. Recall that $|\overline{b}| = p$, so bN is not the identity in $\overline{G}$, and $(bN)^p = N$. ⊙ Since $(bN)^p = b^p N$, it follows that $b^p \in N$. ⊙ Note that $b \neq e$, since that would imply $bN = N$ is the identity element in $\overline{G}$. Thus, b is a nonidentity element in G such that $b^p \in N$. We will consider two possibilities: $b^p = e$ or $b^p \neq e$.

If $b^p = e$ and $b \neq e$, then b must have order p, ⊙ and we have found an element of order p in G, as desired. So assume $b^p \neq e$. Recall that $|N| = |\langle a \rangle| = q$, and so the order of b^p must be q. ⊙ Thus, $b^{pq} = e$, and the order of b divides pq. So the possible orders of b are 1, p, q, and pq. We have already argued that $b \neq e$, and we have considered the case where $|b| = p$. Therefore, we are left to consider the cases where $|b| = q$ or $|b| = pq$. Now if $|b| = q$, then $\overline{b}^q = (bN)^q = b^q N = eN = N$, and $|\overline{b}|$ divides q. But p and q are distinct primes and $|\overline{b}| = p$, so this is impossible. ⊙ Thus, the only case left to consider is when $|b| = pq$. If $|b| = pq$, then Theorem 20.6 (see page 295) shows that $|b^q| = \frac{pq}{\gcd(pq,q)} = p$. Therefore, b^q is an element in G of order p. In each case, we have demonstrated that G contains an element of order p, as desired. ∎

One elementary consequence of Cauchy's Theorem for Finite Abelian Groups is a determination of the structure of all Abelian groups of order pq, where p and q are distinct primes.

Corollary 24.13. *Any Abelian group of order pq, where p and q are distinct primes, is cyclic.*

Proof. Let p and q be distinct primes, and let G be an Abelian group of order pq. Cauchy's Theorem for Finite Abelian Groups tells us that G contains an element a of order p and an element b of order q. Since $\gcd(|a|, |b|) = 1$ and G is Abelian, it follows (see Exercise (18) on page 298 of Investigation 20) that $|ab| = pq$, and so $G = \langle ab \rangle$ is a cyclic group. Hence, every Abelian group of order pq (where p and q are distinct primes) is a cyclic group. ∎

Corollary 24.13 tells us, for example, that the Abelian groups of order 6, 10, 15, 21, etc., are all cyclic.

Simple Groups and the Simplicity of A_n

While mathematicians have not been able to classify all the finite groups (that is, determine all the groups of any given order), it is a remarkable accomplishment that the classification of all the finite *simple* groups has been completed. Simple groups are important in group theory because they form the building blocks of all groups. (We'll explain this statement in more detail a little later.)

Definition 24.14. A group G is **simple** if G has no nontrivial proper normal subgroups.

Note that by *nontrivial proper normal subgroups*, we mean normal subgroups of more than one element that are not the entire group.

There is one straightforward piece of the classification of simple groups: the finite Abelian simple groups. (See Exercise (34).) The best known examples of non-Abelian simple groups are the alternating groups A_n. In fact, it is the simplicity of these groups that can be used to show that there is no general method for finding roots of polynomials with complex coefficients of degree 5 or higher. After proving a couple of preliminary results, we will be able to prove that A_n is simple for $n \geq 5$.

To begin, recall that A_n is the set of even permutations in S_n—that is, the permutations that can be written as a product of an even number of transpositions. The first fact we will prove is that every element in A_n can also be written as a product of 3-cycles. The general idea behind the proof is contained in the next activity.

Activity 24.15. Let $n \geq 3$ be an integer and let $a, b, c \in \{1, 2, 3, \ldots, n\}$.

(a) Write the product $(a\ b)(a\ c)$ as a cycle or a product of disjoint cycles. Why does this show that A_n contains every 3-cycle?

(b) Write each of the following products as a product of two transpositions.

   (i) $(1\ 3\ 2)(1\ 3\ 4)$

   (ii) $(1\ 2\ 3)$

   (iii) $(1\ 2\ 3)(1\ 3\ 2)$

Activity 24.15 illustrates how A_n contains all 3-cycles (for $n \geq 3$) and also how products of 3-cycles are related to products of even numbers of transpositions. This leads us to the next result.

Lemma 24.16. *Let n be an integer with $n \geq 3$. Any permutation in A_n can be written as a product of 3-cycles.*

Proof. Let n be an integer with $n \geq 3$. Note that each 3-cycle $(a\ b\ c) = (a\ c)(a\ b)$ is an even permutation. So A_n contains the 3-cycles.

Let $\sigma \in A_n$. Then there exist transpositions $\tau_1, \tau_2, \ldots, \tau_m$ so that

$$\sigma = \tau_m \cdots \tau_3 \tau_2 \tau_1.$$

Since $\sigma \in A_n$, we know that m is even. So there is an integer k such that $m = 2k$. We can then collect the τ_i into groups of 2 so that

$$\sigma = (\tau_{2k} \tau_{2k-1}) \cdots (\tau_4 \tau_3)(\tau_2 \tau_1).$$

If in any pair we have $\tau_{2i} = \tau_{2i-1}$, then $\tau_{2i} \tau_{2i-1}$ is the identity and can be removed from the product for σ. Therefore, we can assume that $\tau_{2i} \neq \tau_{2i-1}$ for each pair of transpositions $(\tau_{2i}\ \tau_{2i-1})$. It follows that each pair has the form $(a\ b)(c\ d)$ with $a \neq b$, $c \neq d$, and $(a\ b) \neq (c\ d)$. There are two cases that we must consider.

(1) The first case is when a, b, c, and d are all distinct. In this situation, we have $(a\ b)(c\ d) = (a\ c\ b)(a\ c\ d)$.

(2) The other case is when $b = c$ and $a \neq d$. (This case is equivalent to those in which $a = d$ and $b \neq c$, or $a = c$ and $b \neq d$. Therefore, we will assume, without loss of generality, that $b = c$ and $a \neq d$.) Then $(a\ b)(c\ d) = (a\ b)(b\ d) = (a\ b\ d)$.

In each case we see that we can write a product of two transpositions in terms of 3-cycles. Therefore, every even permutation—that is, every element of A_n—can be written as a product of 3-cycles. ∎

The fact that A_n is generated by the 3-cycles also tells us something about the normal subgroups of A_n.

Activity 24.17. Let $n \geq 3$ and suppose N is a normal subgroup of A_n that contains the 3-cycle $(1\ 2\ 3)$.

(a) Why must N contain the 3-cycle $(1\ 3\ 2)$?

(b) Let $r, s, t \in \{1, 2, 3, \ldots, n\}$. Write each of the following products in A_n as a cycle or a product of disjoint cycles. Why must each of these elements be in N?

    (i) $(1\ t\ 3)(1\ 2\ 3)(1\ 3\ t)$

    (ii) $(r\ 1\ t)(1\ t\ 2)(r\ t\ 1)$

    (iii) $(1\ 2\ s)(r\ 2\ t)(1\ s\ 2)$

(c) How does the result of part (b) show that $N = A_n$?

Activity 24.17 seems to provide a method for generating all 3-cycles in a normal subgroup of A_n that contains a 3-cycle. The next lemma shows this in general.

Lemma 24.18. *Let n be an integer with $n \geq 3$. If N is a normal subgroup of A_n and N contains a 3-cycle, then $N = A_n$.*

Proof. Let n be an integer with $n \geq 3$. Let N be a normal subgroup of A_n, and let $(a\ b\ c)$ be a 3-cycle in N. We will show that $N = A_n$ by showing that N contains all possible 3-cycles. Then Lemma 24.16 will establish that $N = A_n$. Let r, s, t be integers between 1 and n. Since N is normal in A_n, we know $\alpha(a\ b\ c)\alpha^{-1}$ is in N for every $\alpha \in A_n$. From Lemma 24.16, we know that A_n contains all 3-cycles. Thus,

- $(a\ t\ c)(a\ b\ c)(a\ c\ t) = (a\ t\ b) \in N$, which implies that
- $(r\ a\ t)(a\ t\ b)(r\ t\ a) = (r\ b\ t) \in N$, which implies that
- $(b\ s\ a)(r\ b\ t)(b\ a\ s) = (r\ s\ t) \in N$.

Since r, s, and t were chosen to be arbitrary integers between 1 and n, this shows that N contains every possible 3-cycle, and so $N = A_n$. ∎

We will now show that A_n is a simple group for all $n \geq 5$.

Theorem 24.19. *Let n be an integer with $n \geq 5$. Then A_n contains no nontrivial normal subgroups.*

Proof. Let n be an integer with $n \geq 5$. Suppose N is a nontrivial normal subgroup of A_n. Let σ be a nonidentity element in N that permutes the fewest number k of integers. If $k = 3$, then σ must be a 3-cycle and Lemma 24.18 shows $N = A_n$, a contradiction. Assume $k \geq 4$, and decompose σ into a product of disjoint cycles:

$$\sigma = \sigma_1 \sigma_2 \sigma_3 \cdots \sigma_m.$$

We will consider two cases:

- Case 1: For some i, σ_i is a q-cycle with $q \geq 3$.

  In this case, we can assume, without loss of generality, that $\sigma = (1\ 2\ 3\ \ldots)\tau$, where τ is disjoint from $(1\ 2\ 3\ \ldots)$. If σ permutes exactly four integers, then $\sigma = (1\ 2\ 3\ a)$ for some integer $a \neq 1, 2, 3$, and σ is odd. Thus, σ is not in N. Therefore, σ must permute at least five integers. Without loss of generality, we assume $\sigma(4) = a_4$ and $\sigma(5) = a_5$. Let $\beta = (3\ 4\ 5)$, and let

  $$\gamma = \sigma^{-1}\beta\sigma\beta^{-1}.$$

  Since N is normal in A_n, we know $\beta\sigma\beta^{-1} \in N$. Therefore, $\gamma \in N$. Now

  $$\gamma(1) = \sigma^{-1}\beta\sigma\beta^{-1}(1) = \sigma^{-1}\beta\sigma(1) = \sigma^{-1}\beta(2) = \sigma^{-1}(2) = 1,$$

  so $\gamma \neq \sigma$. In addition,

  $$\gamma(2) = \sigma^{-1}\beta\sigma\beta^{-1}(2) = \sigma^{-1}\beta\sigma(2) = \sigma^{-1}\beta(3) = \sigma^{-1}(4) \neq 2.$$

  Thus, γ is not the identity permutation. Also, if $t > 5$ and $\sigma(t) = t$, then $\gamma(t) = t$ as well. Since $\gamma(1) = 1$, γ permutes fewer elements than σ, a contradiction to our assumption that σ is the element of N that permutes the fewest number of integers.

- Case 2: For each i, σ_i is a transposition.

  In this case, we can assume, without loss of generality, that $\sigma = (1\ 2)(3\ 4) \cdots$. We again let $\beta = (3\ 4\ 5)$ and let

  $$\gamma = \sigma^{-1}\beta\sigma\beta^{-1}.$$

  Then $\gamma(1) = 1$ and $\gamma(2) = 2$. If $\sigma(t) = t$ for any $t > 5$, then $\gamma(t) = t$. Also,

  $$\gamma(3) = \sigma^{-1}\beta\sigma\beta^{-1}(4) = \sigma^{-1}\beta\sigma(3) = \sigma^{-1}\beta(4) = \sigma^{-1}(5) \neq 3,$$

  so γ is not the identity. Once again, γ permutes fewer elements than σ, a contradiction.

Since both cases lead to a contradiction, we can conclude that A_n contains no nontrivial normal subgroup group when $n \geq 5$. ∎

The simplicity of A_n will allow us to explicitly determine the normal subgroups of S_n. (See Exercise (37).)

We will close this section with a few comments about the importance of simple groups. Simply put (no pun intended!), simple groups form the building blocks of all groups. While we won't provide a rigorous formulation or proof of this statement, we can understand some of the ideas behind it.

Suppose G is a finite group with identity e. Let G_1 be a normal subgroup of G (other than G) of largest order. (Note that if G is simple, then $G_1 = \{e\}$.) Now G/G_1 must be a simple group. To see why, suppose M is a normal subgroup of G/G_1. Then there is a normal subgroup N of G with $G_1 \subset N \subset G$ so that $M = N/G_1$. This contradicts the maximal order of G_1.

If $G_1 \neq \{e\}$, then we can repeat this process on G_1 to obtain a second group $G_2 \subset G_1$, normal in G_1 (but not necessarily in G), so that G_1/G_2 is simple. Continuing this process, we obtain a sequence of groups

$$G \supseteq G_1 \supseteq G_2 \supseteq G_3 \cdots \supseteq G_n,$$

with $G_n = \{e\}$, such that for each i, G_{i+1} is a normal subgroup of G_i and G_i/G_{i+1} is simple. The quotient groups G_i/G_{i+1} are called the *composition factors* of G and turn out to be independent of the choices of the groups G_i. (In other words, we may have different choices for the normal subgroups G_i, but the quotient groups are always the same in some sense that we will make clear later.) A consequence of this result is that if we can classify all simple groups, then understand how the composition groups of an arbitrary group determine the group, we will be able to classify all finite groups.

In a remarkable collection of work, mathematicians have been able to classify all the finite simple groups. The final result is a combined effort of hundreds of researchers producing over 500 articles of more than 14,000 journal pages. In essence, the proof states that finite simple groups fall into two categories: several infinite families of groups for which there is an established pattern, and 26 other groups known as the sporadic groups:

(1) Cyclic groups of prime group order,

(2) Alternating groups of degree at least five,

(3) Lie-type Chevalley groups,

(4) Lie-type twisted Chevalley groups or the Tits group, and

(5) Sporadic groups:

- Mathieu groups $M_{11}, M_{12}, M_{22}, M_{23}, M_{24}$
- Janko groups J_1, J_2 (also known as the Hall-Janko group HJ), J_3, J_4
- Conway groups Co_1, Co_2, Co_3
- Fischer groups F_{22}, F_{23}, F_{24}
- Higman-Sims group HS
- McLaughlin group McL
- Held group He
- Rudvalis group Ru
- Suzuki sporadic group Suz
- O'Nan group $O'N$
- Harada-Norton group HN
- Lyons group Ly
- Thompson group Th
- Baby Monster group B
- Monster group M

Five of these sporadic groups were found in the 1860s and the other 21 were found between 1965 and 1975. The appropriately named Monster group is the largest of the sporadic groups and has order

$$808,017,424,794,512,875,886,459,904,961,710,757,005,754,368,000,000,000.$$

Quite a monster, indeed!

Concluding Activities

Activity 24.20. Let G be a group and H a subgroup of G. Theorem 24.5 allows us to tell if H is normal in G by determining if

$$aHa^{-1} = \{aha^{-1} : h \in H\}$$

is equal to H for every $a \in G$. As we will demonstrate in this activity, the set aHa^{-1} is a subgroup of G for any $a \in G$ and is called a **conjugate** of the subgroup H (or the conjugate of H by a).

(a) Let $H = \langle R^2 \rangle$ in the group D_6 of symmetries of a regular hexagon. Determine the elements in rHr^{-1}.

(b) Let $K = \langle r \rangle$ in D_6. Determine the elements in $(rR)K(rR)^{-1}$.

(c) Let $a \in G$. Prove that aHa^{-1} a subgroup of G.

Activity 24.21. In Activity 23.19 (see page 327) we showed that the only congruence relations on a group G with subgroup H are the relations $\sim_H$, where $a \sim_H b$ if $a^{-1}b \in H$. However, we did not show that $\sim_H$ is always a congruence relation. In this activity, we will show that the congruence relations on a group G are exactly the relations $\sim_N$, where N is a normal subgroup of G. This explains why we study only the relations $\sim_N$ and not others. Recall that an equivalence relation $\sim$ on a group G is a congruence relation if whenever $a, b, c, d \in G$ with $a \sim b$ and $c \sim d$ we also have

(i) $ac \sim bd$ and

(ii) $a^{-1} \sim b^{-1}$.

In other words, a congruence relation preserves the group structure.

(a) Let $\sim$ be a relation on a group G with identity e and let $N = \{x \in G : x \sim e\}$. Activity 23.19 shows that N is a subgroup of G. Prove that N is a normal subgroup of G. This establishes that every congruence relation on a group G corresponds to a normal subgroup of G.

(b) Now show that if N is a normal subgroup of a group G, then the relation $\sim_N$ is a congruence relation on G.

Activity 24.22. Let H and K be subgroups of the group G, with K normal in G. Let

$$KH = \{kh : k \in K \text{ and } h \in H\}.$$

(See Exercise (5) on page 284 in Investigation 19.)

(a) Let $G = D_4$, $K = \langle R^2 \rangle$, and $H = \langle r \rangle$. Assume K is a normal subgroup of G. Find the elements of KH.

(b) Let G be an arbitrary group, and let H and K be subgroups of G with K normal in G. Prove that KH is a subgroup of G.

Exercises

(1) Certain subgroups of every group are normal. Let G be any group, and let e be the identity in G.

   (a) Prove that $G \triangleleft G$.

   (b) Prove that $\{e\} \triangleleft G$.

   (c) Prove that if H is a subgroup of $Z(G)$, then $H \triangleleft G$. In particular, conclude that $Z(G) \triangleleft G$.

(2) Determine if the indicated subgroup H is a normal subgroup of G.

   (a) $G = A_4$ and $H = \{e, (1\ 2)(3\ 4), (1\ 3)(2\ 4), (1\ 4)(2\ 3)\}$.

   (b) $G = \mathrm{GL}_2(\mathbb{R})$ and $H = \left\{ \begin{bmatrix} a & 0 \\ 0 & a \end{bmatrix} : a \neq 0 \right\}$.

   (c) $G = \mathrm{GL}_2(\mathbb{R})$ and $H = \left\{ \begin{bmatrix} a & b \\ 0 & d \end{bmatrix} : a, b, d \in \mathbb{R}, ad \neq 0 \right\}$.

   (d) $G = D_4$ and H is the subgroup consisting of the rotations in D_4.

   (e) $G = \mathrm{GL}_n(\mathbb{R})$ and $H = \mathrm{SL}_n(\mathbb{R})$.

(3) If N is a normal subgroup of a group G and H is a subgroup of G that contains N, must N be normal in H? Prove your answer.

(4) (a) Give an example of a nontrivial normal subgroup of S_3, or explain why one doesn't exist.

   (b) Give an example of a nontrivial subgroup of S_3 that is not normal, or explain why one doesn't exist.

(5) Let G be a group and H a subgroup of G.

   (a) If $a, b \in G$ so that $aH = bH$, must it follow that $Ha = Hb$? Prove your answer.

   (b) For any $a, b \in G$, suppose that $Ha = Hb$ whenever $aH = bH$. Prove that $H = gHg^{-1}$ for any $g \in G$.

* (6) (a) Prove the forward implication of Theorem 24.5. That is, show that if N is a normal subgroup of a group G, then $aNa^{-1} \subseteq N$ for all $a \in G$.

   (b) Let G be a group and N a subgroup of G. To use Theorem 24.5 to show that N is a normal subgroup of G, we show that $aNa^{-1} \subseteq N$ for every $a \in G$. Is this the same as showing $a^{-1}Na \subseteq N$ for all $a \in G$? In other words, is the statement $aNa^{-1} \subseteq N$ for all $a \in G$ equivalent to $a^{-1}Na \subseteq N$ for all $a \in G$? Prove your answer.

   (c) Let N be a normal subgroup of a group G. Is the reverse containment in Theorem 24.5 true? That is, must $N \subseteq aNa^{-1}$ for every $a \in G$? Prove your answer.

(7) The result of Exercise (6) is that if N is a normal subgroup of a group G, then $aNa^{-1} = N = a^{-1}Na$ for every $a \in G$. Is the converse true—that is, if H is a subgroup of a group G and $aHa^{-1} = a^{-1}Ha$ for all $a \in G$, is $H \triangleleft G$? Prove your answer.

Table 24.6

Operation table for D_6

	I	R	R^2	R^3	R^4	R^5	r	rR	rR^2	rR^3	rR^4	rR^5
I	I	R	R^2	R^3	R^4	R^5	r	rR	rR^2	rR^3	rR^4	rR^5
R	R	R^2	R^3	R^4	R^5	I	rR^5	r	rR	rR^2	rR^3	rR^4
R^2	R^2	R^3	R^4	R^5	I	R	rR^4	rR^5	r	rR	rR^2	rR^3
R^3	R^3	R^4	R^5	I	R	R^2	rR^3	rR^4	rR^5	r	rR	rR^2
R^4	R^4	R^5	I	R	R^2	R^3	rR^2	rR^3	rR^4	rR^5	r	rR
R^5	R^5	I	R	R^2	R^3	R^4	rR	rR^2	rR^3	rR^4	rR^5	r
r	r	rR	rR^2	rR^3	rR^4	rR^5	I	R	R^2	R^3	R^4	R^5
rR	rR	rR^2	rR^3	rR^4	rR^5	r	R^5	I	R	R^2	R^3	R^4
rR^2	rR^2	rR^3	rR^4	rR^5	r	rR	R^4	R^5	I	R	R^2	R^3
rR^3	rR^3	rR^4	rR^5	r	rR	rR^2	R^3	R^4	R^5	I	R	R^2
rR^4	rR^4	rR^5	r	rR	rR^2	rR^3	R^2	R^3	R^4	R^5	I	R
rR^5	rR^5	r	rR	rR^2	rR^3	rR^4	R	R^2	R^3	R^4	R^5	I

(8) Write out an operation table for G/N for the given group G and subgroup N. Then explain how you know that N is normal in G.

    (a) $G = U_{13}$ and $N = \{[1], [3], [9]\}$.

    (b) $G = S_3$ and $N = \langle (1\ 2\ 3) \rangle$.

(9) (a) Find the order of the element $[10] + \langle [6] \rangle$ in the quotient group $\mathbb{Z}_{24}/\langle [6] \rangle$.

    (b) Find all the elements in $\mathbb{Z}_{24}/\langle [6] \rangle$, and determine the order of each.

(10) Let $G = U(\mathbb{R})$ (the group of units in $\mathbb{R}$) and let $N = \mathbb{R}^+$ (the subgroup of G consisting of all positive real numbers). Find the elements of G/N, and create the operation table for this quotient group. What kind of a group is G/N?

* (11) Let H and K be subgroups of the group G, with K normal in G. In Activity 24.22 we showed that

$$KH = \{kh : k \in K \text{ and } h \in H\}$$

is a subgroup of G.

    (a) Is the set

$$HK = \{hk : h \in H \text{ and } k \in K\}$$

also a subgroup of G? Prove your answer.

    (b) Does $KH = HK$? Prove your answer.

(12) Where in the proof of Theorem 24.11 did we use the fact that G is an Abelian group? Why doesn't our proof apply to non-Abelian groups?

(13) The operation table for D_6, the dihedral group of order 12, is given in Table 24.6.

    (a) Find the elements of the set $D_6/Z(D_6)$.

    (b) Write the operation table for the group $D_6/Z(D_6)$.

    (c) The examples of quotient groups we have seen so far have all been Abelian groups. Is it true that every quotient group is Abelian? Explain.

    (d) Give a necessary condition on a group G if G/N is a non-Abelian group. Is this necessary condition sufficient as well? Explain.

(14) Prove the following theorem:

Theorem 24.23. *Let G be a group and N a normal subgroup of G. If $a \in G$ and $n \in \mathbb{Z}$, then $(aN)^n = a^n N$.*

(15) (a) Can the order of a nonidentity element aN in a quotient group G/N be smaller than the order of a in G? If yes, provide an example to illustrate. If no, prove it.

(b) Can the order of an element aN in a quotient group G/N be greater than the order of a in G? If yes, provide an example to illustrate. If no, prove it.

(c) Can the order of a nonidentity element aN in a quotient group G/N be equal to the order of a in G? If yes, provide an example to illustrate. If no, prove it.

* (16) (a) If N is a normal subgroup of a finite group G and G contains an element of order n, must G/N contain an element of order n? If yes, provide an example to illustrate. If no, prove it.

(b) If N is a finite normal subgroup of a group G and G/N contains an element of order n, must G contain an element of order n? If yes, provide an example to illustrate. If no, prove it.

(17) (a) Is every quotient group of an Abelian group Abelian? Prove your answer.

(b) Is every quotient group of a cyclic group cyclic? Prove your answer.

(c) Is every quotient group of a non-Abelian group non-Abelian? Prove your answer.

* (18) Let G be a group and H a subgroup of G. Even though H may not be normal in G, we can consider in some sense how close H is to being normal by determining how many elements $g \in G$ have the property that $gHg^{-1} = H$. This leads us to define the **normalizer** of H in G as the set

$$N(H) = \{g \in G : gHg^{-1} = H\}.$$

(a) Find $N(\langle r \rangle)$ in D_4.

(b) Prove that $N(H)$ is always a subgroup of G. Must $N(H)$ be a normal subgroup of G? Prove your answer.

(c) Show that $H \subseteq N(H)$. Is H a normal subgroup of $N(H)$?

* (19) Determining when a subgroup is normal can be a challenge. In this exercise, we will consider a specific situation in which we can tell that a subgroup is normal.

(a) Prove that if G is a finite group and N is a subgroup of G so that $|G| = 2|N|$, then N is a normal subgroup of G.

(b) If H is a subgroup of G with $|G| = 2|H|$, then $[G : H] = 2$. Prove that, under this condition, if $x, y \in G$ and $x, y \notin H$, then $xy \in H$.

(c) Explain how the result of part (b) shows each of the following.
- The sum of two odd integers is an even integer.
- The product of two odd permutations is an even permutation.
- The product of two reflections is a rotation.

(20) Similar to Exercise (19), prove that if G is a group and has exactly one subgroup H of order n, then H is a normal subgroup of G. (Hint: Refer to Activity 24.20.)

Table 24.7

Operation table for the quaternions **Q**.

	1	i	j	k	$-i$	$-j$	$-k$	-1
1	1	i	j	k	$-i$	$-j$	$-k$	-1
i	i	-1	k	$-j$	1	$-k$	j	$-i$
j	j	$-k$	-1	i	k	1	$-i$	$-j$
k	k	j	$-i$	-1	$-j$	i	1	$-k$
$-i$	$-i$	1	$-k$	j	-1	k	$-j$	i
$-j$	$-j$	k	1	$-i$	$-k$	-1	i	j
$-k$	$-k$	$-j$	i	1	j	$-i$	-1	k
-1	-1	$-i$	$-j$	$-k$	i	j	k	1

(21) Let **Q** be the group of quaternions as introduced in Exercise (6) of Investigation 21. (See page 305.) If we let $i = a$, $j = b$, and $k = ab$, (with a and b as in Exercise (6) of Investigation 21), then the operation table for **Q** can be written as shown in Table 24.7. The structure of this group can be easily remembered if we think of i, j, and k as the standard unit vectors in $\mathbb{R}^3$ with the cross product as the operation and 1 as the identity.

  (a) Find all normal subgroups of **Q**. Make sure to verify that each subgroup you find is normal.

  (b) From Exercise (17), we know that every subgroup of an Abelian group is normal. Is it possible to have a non-Abelian group in which every subgroup is normal? Prove your answer.

(22) An operation table for the group **Q** of quaternions is given in Table 24.7. Determine the elements of $\mathbf{Q}/\langle -1\rangle$, and create the operation table for this group.

(23) Let G be a group and N a normal subgroup of G. If a and b are elements of G with $|aN| = |bN|$ in G/N, must it follow that $|a| = |b|$ in G? Prove your answer.

(24) Let N be a normal subgroup of G, and let $a \in G$. Suppose that the order of aN in G/N is 5 and the order of N is 12. What are the possible orders of a in G? Prove your answer, and provide examples that illustrate each possibility.

(25) (a) Let $G = U_{21}$ and $N = \langle [4]\rangle$. Since G is Abelian, we know that $N \triangleleft G$. Let $H = \langle [5]\rangle$ and let $\overline{H} = \{hN : h \in H\}$. Find the elements of $\overline{H}$. Is $\overline{H}$ a subgroup of G/N? If yes, write the operation table for $\overline{H}$. If no, explain why.

  (b) Let G be an arbitrary group and N a normal subgroup of G. If H is a subgroup of G, let $\overline{H} = \{hN : h \in H\}$. Prove that $\overline{H}$ a subgroup of G/N.

(26) Let p and q be primes (not necessarily distinct). Let G be a group of order pq. Prove that $|Z(G)| = 1$ or $|Z(G)| = pq$. Provide examples to illustrate each.

(27) It is possible to have a group that has elements of any order. Let $G = \mathbb{R}/\mathbb{Z}$.

  (a) Find the order of the element $\frac{1}{2} + \mathbb{Z}$ in G.

  (b) Let n be a positive integer. Does G contain an element of order n? Prove your answer.

  (c) Does every element of G have finite order? Prove your answer.

(d) Determine the conditions on $x \in \mathbb{R}$ so that $x + \mathbb{Z}$ has finite order in $\mathbb{R}/\mathbb{Z}$. In other words, correctly complete and prove the following theorem:

   Theorem. *An element $x+\mathbb{Z}$ in $\mathbb{R}/\mathbb{Z}$ has finite order if and only if* _____.

(e) Is it possible to find a group G so that every element in G has finite order, but we can find elements with any finite order? Explain.

(28) Let G be a group generated by a set S. Let N be a normal subgroup of G.

   (a) Find a set of generators for G/N. Prove your answer.

   (b) Recall that the transpositions $(1\ 2)$, $(1\ 3)$, and $(1\ 4)$ generate S_4 (see Exercise (19) in Investigation 22). Use this fact to find a set of generators for the group S_4/N, where $N = \langle I, (1\ 2)(3\ 4), (1\ 3)(2\ 4), (1\ 4)(2\ 3) \rangle$. (See Exercise (2).) Write the operation table for S_4/N.

(29) Let H and K be subgroups of a group G, with K normal in G. Let

$$KH = \{kh : k \in K \text{ and } h \in H\}$$

as in Activity 24.22

   (a) Can KH be a normal subgroup of G? Is it possible that KH is not a normal subgroup of G? Explain.

   (b) Prove that if K and H are normal subgroups of a group G, then KH is a normal subgroup of G. Is the converse true?

(30) Let N be a subgroup of a group G. We have seen that if every left coset aN of N in G is equal to the corresponding right coset Na, then N is normal in G. What if we change the assumption to something more general? Suppose that, for each $a \in G$, there exists $b \in G$ such that $aN = Nb$. In other words, every left coset of N in G is equal to *some* right coset of N in G. Must N then be a normal subgroup of G? Prove your answer.

(31) Let $N = \{e, n\}$ be a normal subgroup of order 2 in a group G, where e is the identity in G. Prove that n commutes with every element in G. Conclude that $N \subseteq Z(G)$.

(32) **Intersections of normal subgroups.** Is the intersection of two normal subgroups of a group G always a normal subgroup of G? Prove your answer.

(33) (a) Can we extend Corollary 24.13 to Abelian groups whose orders are the products of three distinct primes? That is, if G is an Abelian group of order pqr where p, q, and r are distinct primes, must G be a cyclic group? Prove your answer.

   (b) Can we extend Corollary 24.13 even more to Abelian groups whose orders are the products of any number of distinct primes? That is, if n is a positive integer and G is an Abelian group of order $p_1 p_2 \cdots p_n$ where $p_1, p_2, \ldots, p_n$ are distinct primes, must G be a cyclic group? Prove your answer.

$\star$ (34) **Finite Abelian simple groups.** Find all finite Abelian simple groups. Prove your answer.

(35) Theorem 24.19 shows that A_n is a simple group if $n \geq 5$. What if $n < 5$? Determine if A_2, A_3, and A_4 are simple.

$\star$ (36) **Normal subgroups of D_n.** Determining which subgroups of a given group are normal can be a difficult task. However, there are some groups whose normal subgroups can be completely classified. In this exercise, we will classify the normal subgroups of the dihedral groups. Let n be a positive integer.

(a) If i and j are integers, show that

$$(rR^j)(rR^i)(rR^j)^{-1} = rR^{2j-i} \quad \text{and} \quad (rR^j)(R^i)(rR^j)^{-1} = R^{-i}.$$

(b) Suppose N is a normal subgroup of D_n. Show that if N contains a reflection, then N contains R^2 and either r or rR.

(c) When are the groups $\langle r, R^2 \rangle$ and $\langle rR, R^2 \rangle$ proper subgroups of D_n? Your answers should depend on n. When $\langle r, R^2 \rangle$ and $\langle rR, R^2 \rangle$ are proper subgroups of D_n, what are their indices in D_n? Prove your answer.

(d) Prove the following theorem:

Theorem. *A nontrivial proper subgroup N of D_n is normal in D_n if and only if N is a subgroup of $\langle R \rangle$ or n is even and N is one of $\langle r, R^2 \rangle$ or $\langle rR, R^2 \rangle$.*

* (37) **Normal subgroups of S_n.** As with the dihedral groups (see Exercise (36)), the normal subgroups of S_n are also well known. We will investigate them in this exercise.

(a) Explain why $A_n \triangleleft S_n$ for $n \geq 2$.

(b) Let $n \geq 2$, and let N be a normal subgroup of S_n. Use Exercise (16) of Investigation 22 (see page 318) to show that if $\pi \in N$ can be written as the product $\pi_1 \pi_2 \cdots \pi_k$ of disjoint cycles, then N contains all elements of the form $\tau_1 \tau_2 \cdots \tau_k$, where the τ_i are disjoint cycles and τ_i is a cycle of the same length as π_i for each i.

(c) Determine the nontrivial proper normal subgroups of S_2 and S_3.

(d) Determine the normal subgroups of S_4.

(e) In the remainder of this exercise, we will prove the following theorem:

Theorem. *Let $n \geq 5$. Then A_n is the only nontrivial normal subgroup of S_n.*

Let $n \geq 5$, and let N be a nontrivial normal subgroup of S_n.
  (i) Explain why $N \cap A_n$ is a normal subgroup of A_n. Then explain why $N \cap A_n = A_n$ or $N \cap A_n = \{I\}$.

  (ii) What can be said about N if $N \cap A_n = A_n$?

  (iii) If $N \cap A_n = \{I\}$, why can N contain only odd permutations (in addition to I)? Show that N cannot contain more than one odd permutation. How does this show that $|N| = 2$?

  (iv) To complete our proof, show that if $N = \{I, \alpha\}$, where α is an odd permutation of order 2, then N is not normal in S_n.

(38) Let G be a group and N, K subgroups of G with K a subset of N. Prove or disprove: If N is a normal subgroup of G and K is a normal subgroup of N, then K is a normal subgroup of G. In other words, is normality transitive?

(39) Let G be a group, N a normal subgroup of G, and H any subgroup of G.

(a) Must $N \cap H$ be a normal subgroup of G? Prove your answer.

(b) Must $N \cap H$ be a normal subgroup of N? Prove your answer.

(c) Must $N \cap H$ be a normal subgroup of H? Prove your answer.

(40) The result of Exercise (38) shows that, in general, normality is not transitive. However, there are some special instances in which normality is transitive. In this exercise, we will investigate one such situation. Let G be a group and N a normal subgroup of G. Assume that N is cyclic with generator a. Let H be a subgroup of N.

    (a) Explain why we can write $H = \langle a^m \rangle$ for some integer m.

    (b) Let $y \in H$ and $x \in G$. Write y in terms of a. Then use the fact from Exercise (3) in Investigation 18 (see page 275) that

$$xy^k x^{-1} = (xyx^{-1})^k$$

for all $k \in \mathbb{Z}$ to show that H is a normal subgroup of G. Conclude that if N is a cyclic subgroup of G, $N \lhd G$, and H is a subgroup of N (and so H is normal in N because N is cyclic), then $H \lhd G$.

Connections

Given a normal subgroup N of a group G, we can define congruence modulo N and form the quotient group G/N. Quotient structures are useful in mathematics in that they often have a simpler structure than the original set and yet can provide important information about the original set. In Investigation 2, we studied congruence and quotient structures in the integers, constructing the set $\mathbb{Z}_n$. If you studied ring theory before group theory, you should also notice connections between the topics in this investigation and those in Investigations 12 and 13. In particular, the set G/N represents the set of distinct congruence classes modulo N in the same way that $F[x]/(f(x))$ represents the set of distinct congruence classes modulo $f(x)$, and R/I represents the distinct congruence classes modulo an ideal I. Since a ring R is an Abelian group under its addition operation, any ideal I of R is always a normal subgroup of R under addition. So, in the case of rings, we do not have to worry about normality as we do when constructing quotient groups.

Investigation 25

Products of Groups

Focus Questions

By the end of this investigation, you should be able to give precise and thorough answers to the questions listed below. You may want to keep these questions in mind to focus your thoughts as you complete the investigation.

- What is a direct product, and how can direct products be used to construct larger groups from smaller ones?

- What is the external direct product of two groups, and how is the operation in an external direct product defined?

- How can we find the order of an element in a direct product of groups?

- What is an internal direct product of groups? Under what conditions is the internal direct product defined?

- How are the external and internal direct products of groups similar, and how are they different?

Preview Activity 25.1. If we have two integers, say k and m, we can combine these integers in different ways (e.g., using addition or multiplication) to obtain another integer. In a similar manner, if we have two groups G and H, we can combine them together to make another group, called the *direct product* of G and H, that contains copies of both G and H as subgroups. We will soon define direct products formally, but before doing so, let's take a look at an example. An operation table for the direct product of $\mathbb{Z}_2$ and $\mathbb{Z}_3$, denoted $\mathbb{Z}_2 \oplus \mathbb{Z}_3$, is shown in Table 25.1. We use the notation $[a]_n$ to indicate the congruence class of a in $\mathbb{Z}_n$.

(a) Describe precisely how the elements of $\mathbb{Z}_2 \oplus \mathbb{Z}_3$ are related to the elements of $\mathbb{Z}_2$ and $\mathbb{Z}_3$.

(b) How does the operation in $\mathbb{Z}_2 \oplus \mathbb{Z}_3$ seem to be related to the operations in $\mathbb{Z}_2$ and $\mathbb{Z}_3$?

(c) Does $\mathbb{Z}_2 \oplus \mathbb{Z}_3$ seem to be a group under the operation given in Table 25.1? If so, what is the identity element of $\mathbb{Z}_2 \oplus \mathbb{Z}_3$? How could we find the inverse of an element $([a]_2, [b]_3)$ in $\mathbb{Z}_2 \oplus \mathbb{Z}_3$?

(d) Consider the set S defined by

$$S = \{([0]_2, [x]_3) : [x]_3 \in \mathbb{Z}_3\}.$$

Is S a subgroup of $\mathbb{Z}_2 \oplus \mathbb{Z}_3$? Why or why not?

(e) What is the relationship between $\mathbb{Z}_3$ and the set S defined in part (d)? Explain.

Table 25.1

Operation table for $\mathbb{Z}_2 \oplus \mathbb{Z}_3$.

+	$([0]_2, [0]_3)$	$([0]_2, [1]_3)$	$([0]_2, [2]_3)$	$([1]_2, [0]_3)$	$([1]_2, [1]_3)$	$([1]_2, [2]_3)$
$([0]_2, [0]_3)$	$([0]_2, [0]_3)$	$([0]_2, [1]_3)$	$([0]_2, [2]_3)$	$([1]_2, [0]_3)$	$([1]_2, [1]_3)$	$([1]_2, [2]_3)$
$([0]_2, [1]_3)$	$([0]_2, [1]_3)$	$([0]_2, [2]_3)$	$([0]_2, [0]_3)$	$([1]_2, [1]_3)$	$([1]_2, [2]_3)$	$([1]_2, [0]_3)$
$([0]_2, [2]_3)$	$([0]_2, [2]_3)$	$([0]_2, [0]_3)$	$([0]_2, [1]_3)$	$([1]_2, [2]_3)$	$([1]_2, [0]_3)$	$([1]_2, [1]_3)$
$([1]_2, [0]_3)$	$([1]_2, [0]_3)$	$([1]_2, [1]_3)$	$([1]_2, [2]_3)$	$([0]_2, [0]_3)$	$([0]_2, [1]_3)$	$([0]_2, [2]_3)$
$([1]_2, [1]_3)$	$([1]_2, [1]_3)$	$([1]_2, [2]_3)$	$([1]_2, [0]_3)$	$([0]_2, [1]_3)$	$([0]_2, [2]_3)$	$([0]_2, [0]_3)$
$([1]_2, [2]_3)$	$([1]_2, [2]_3)$	$([1]_2, [0]_3)$	$([1]_2, [1]_3)$	$([0]_2, [2]_3)$	$([0]_2, [0]_3)$	$([0]_2, [1]_3)$

External Direct Products of Groups

Preview Activity 25.1 introduces an operation on pairs of groups and indicates a way to construct new groups from given ones. The formal definition uses the Cartesian product of two sets. If R and S are sets, then the *Cartesian product* $R \times S$ of R and S is the set of ordered pairs of elements, where the first coordinate comes from R and the second coordinate comes from S. In other words,

$$R \times S = \{(r, s) : r \in R, s \in S\}.$$

Definition 25.2. Let G and H be groups with operations $\cdot_G$ and $\cdot_H$, respectively. The **external direct product** $G \oplus H$ of G and H is the Cartesian product $G \times H$, together with an operation defined componentwise—that is,

$$(g_1, h_1)(g_2, h_2) = (g_1 \cdot_G g_2, h_1 \cdot_H h_2).$$

It is important to note in Definition 25.2 that there are three distinct operations involved: the operation in G, the operation in H, and the operation in $G \oplus H$. In general, we will suppress any symbolism that indicates which operation is which and leave the context in which the notation is used to alleviate any potential ambiguities. For instance, when we write $(g_1, h_1)(g_2, h_2)$ with $(g_1, h_1), (g_2, h_2) \in G \oplus H$, it is understood that we are using the operation in $G \oplus H$. Likewise, when we write $g_1 g_2$ for a pair of elements $g_1, g_2 \in G$, then we know that we are using the operation from G.

A word about notation and terminology is in order here. Some authors write the external direct product as $G \times H$, and some just call it a direct product. We will use the word *external* to distinguish this product from the *internal* direct product that we will define later. The word external is used because there may be no relationship at all between the groups G and H, and the construction of $G \oplus H$ takes place outside of both groups. Some authors only use the notation $G \oplus H$ for the external direct product (and call it a direct sum) when both of the groups G and H are Abelian under additive operations. The definition of the internal direct product, and the connection between the external and internal direct products, will ultimately remove any conflict in the notation. We have adopted the $\oplus$ notation to distinguish between the external direct product and the internal direct product, and to be consistent with the notation we use for the direct sum in ring theory. (See Investigation 6.)

Activity 25.3.

(a) List all the elements in $\mathbb{Z}_2 \oplus \mathbb{Z}_2$.

(b) Construct the operation table for $\mathbb{Z}_2 \oplus \mathbb{Z}_2$ using the operation defined on the direct product.

You may have observed in Preview Activity 25.1 and Activity 25.3 that direct products appear to satisfy a number of important properties. We will explore some of these properties in the next activity.

Activity 25.4. Let G and H be groups.

(a) Why is $G \oplus H$ closed with respect to the operation defined on the direct product?

(b) Is the operation defined on $G \oplus H$ an associative operation? Prove your answer.

(c) Is there an identity element for the operation defined on $G \oplus H$? Explain.

(d) Does the element $(g, h) \in G \oplus H$ have an inverse in $G \oplus H$? Explain.

(e) Is $G \oplus H$ a group under the operation defined on $G \oplus H$? Explain.

Activity 25.4 provides the rationale for the following theorem.

Theorem 25.5. *Let G and H be groups. Then $G \oplus H$ is also a group.*

Finally, it is worth noting that $G \oplus H$ contains copies of both G and H as subgroups, as formalized in the next theorem. Its proof is left as an exercise. (See Exercise (3).)

Theorem 25.6. *Let G and H be groups with identities e_G and e_H, respectively. Then*

$$G \oplus \{e_H\} = \{(g, e_H) : g \in G\}$$

and

$$\{e_G\} \oplus H = \{(e_G, h) : h \in H\}$$

are both subgroups of $G \oplus H$.

A word of caution about Theorem 25.6: the group $G \oplus \{e_H\}$ is not equal to G; it just looks like G. The same is true for the groups $\{e_G\} \oplus H$ and H. However, you may have noticed that we can identify each element (g, e_H) of $G \oplus \{e_H\}$ with the element g in G. So while $G \oplus \{e_H\}$ is not equal to G, there is a natural identification of the elements of $G \oplus \{e_H\}$ with the elements of G in such a way that the two groups are virtually identical, both in their makeup and in the way they behave with respect to their operations. In that sense, the groups G and $G \oplus \{e_H\}$ are essentially the same, a notion that we will make more precise in Investigation 26 when we study group isomorphisms. Likewise, H and $\{e_G\} \oplus H$ can also be considered to be essentially the same group, which implies that $G \oplus H$ in some sense contains a copy of both G and H. Thus, direct products provide a way to construct a larger group that contains each of two smaller groups. This construction is useful in numerous examples, and it will play an essential role in the classification of finite Abelian groups, as we will see in Investigation 28.

Activity 25.7. If G is a group of order m and H a group of order n, what is $|G \oplus H|$? Explain.

Orders of Elements in Direct Products

As we have discussed, the subgroup structure of a group tells us much about the group. Each element in a group determines a subgroup—namely, the cyclic subgroup generated by that element—and we can learn a lot about a group by understanding its cyclic subgroups. In this section, we will examine the orders of elements in direct products, which will provide important insights about cyclic subgroups of direct products.

Let G and H be groups. If $g \in G$ has order m and $h \in H$ has order n, what can we say about the order of (g, h) in $G \oplus H$? It is natural at first to think that $|(g, h)| = |g| \, |h|$. In the next activity, we will see if this is actually the case.

Activity 25.8.

(a) Determine the orders of the elements $[4]_8 \in \mathbb{Z}_8$, $[3]_{12} \in \mathbb{Z}_{12}$, and $([4]_8, [3]_{12}) \in \mathbb{Z}_8 \oplus \mathbb{Z}_{12}$. Does $|([4]_8, [3]_{12})|$ equal $|[4]_8| \, |[3]_{12}|$?

(b) Determine the orders of the elements $[4]_8 \in \mathbb{Z}_8$, $[4]_{12} \in \mathbb{Z}_{12}$, and $([4]_8, [4]_{12}) \in \mathbb{Z}_8 \oplus \mathbb{Z}_{12}$. Does $|([4]_8, [4]_{12})|$ equal $|[4]_8| \, |[4]_{12}|$?

Activity 25.8 shows that there are times when $|(g, h)| = |g| \, |h|$ and times when $|(g, h)| \neq |g| \, |h|$. If we were to do enough examples, we would undoubtedly conjecture the formula for determining the order of an element (g, h) in $G \oplus H$ that is given in the next theorem. You may want to do enough other examples on your own to convince yourself that the theorem makes sense. In the proof of the theorem, one result we will use that we have not yet verified (although you will be asked to do so in Activity 25.18) is that for any element $(g, h) \in G \oplus H$ and any integer k, $(g, h)^k = (g^k, h^k)$.

Theorem 25.9. *Let G and H be groups with $g \in G$ and $h \in H$, both of finite order. Then*

$$|(g, h)| = \operatorname{lcm}(|g|, |h|).$$

Proof. Let e_G be the identity in G and e_H the identity in H. Let $m = |g|$ and $n = |h|$. First, we will show that $(g, h)^{\operatorname{lcm}(m,n)} = (e_G, e_H)$. Since $\operatorname{lcm}(m, n)$ is a common multiple of m and n, we know that $mr = \operatorname{lcm}(m, n) = ns$ for some integers r and s. Therefore,

$$(g, h)^{\operatorname{lcm}(m,n)} = \left(g^{\operatorname{lcm}(m,n)}, h^{\operatorname{lcm}(m,n)} \right) = ((g^m)^r, (h^n)^s) = (e_G, e_H).$$

So $|(g, h)| \leq \operatorname{lcm}(m, n)$.

Now suppose that $(g, h)^k = (e_G, e_H)$ for some positive integer k. Then

$$(e_G, e_H) = (g, h)^k = (g^k, h^k),$$

and so $g^k = e_G$ and $h^k = e_H$. Theorem 20.5 (see page 294) shows that m divides k and n divides k. By the definition of the least common multiple, we can therefore conclude that $\operatorname{lcm}(m, n) \leq k$. This, however, implies that $|(g, h)| \geq \operatorname{lcm}(m, n)$. Since we showed earlier that $|(g, h)| \leq \operatorname{lcm}(m, n)$, it follows that $|(g, h)| = \operatorname{lcm}(|g|, |h|)$. ∎

As the next activity demonstrates, Theorem 25.9 can be applied to yield important conclusions about the subgroup structure of direct products.

Activity 25.10.

(a) Is $\mathbb{Z}_9 \oplus \mathbb{Z}_{10}$ a cyclic group? Answer using Theorem 25.9.

(b) Does $\mathbb{Z}_9 \oplus \mathbb{Z}_{10}$ contain an element of order 6? If yes, find such an element. If no, explain why not.

Although we defined the external direct product for only two groups, we can extend the definition to any finite number of groups. The next activity illustrates how this can be done.

Activity 25.11. We can generalize the construction of the external direct product of two groups to any finite number of groups. Let $n \geq 2$ be an integer, and let $G_1, G_2, \ldots, G_n$ be groups. We define the elements in $G_1 \oplus G_2 \oplus \cdots \oplus G_n$ to be all ordered n-tuples of the form $(g_1, g_2, \ldots, g_n)$, where $g_i \in G_i$ for each i. We can then define an operation on $G_1 \oplus G_2 \oplus \cdots \oplus G_n$ by

$$(a_1, a_2, \ldots, a_n)(g_1, g_2, \ldots, g_n) = (a_1 g_1, a_2 g_2, \ldots, a_n g_n)$$

for any $(a_1, a_2, \ldots, a_n), (g_1, g_2, \ldots, g_n) \in G_1 \oplus G_2 \oplus \cdots \oplus G_n$.

(a) Show that $G_1 \oplus G_2 \oplus \cdots \oplus G_n$ is closed under its operation.

(b) For each i, let e_i be the identity in G_i. What is the identity element in $G_1 \oplus G_2 \oplus \cdots \oplus G_n$? Prove your answer.

(c) Show that the operation defined above is associative in $G_1 \oplus G_2 \oplus \cdots \oplus G_n$.

(d) What conclusion can we draw about the set $G_1 \oplus G_2 \oplus \cdots \oplus G_n$?

Internal Direct Products in Groups

We have seen that if G and H are groups, we can form the direct product $G \oplus H$ of G and H, and this direct product is a group under the obvious componentwise operation. The group $G \oplus H$ is an *external* direct product because it is constructed outside of both G and H.

We can take a different approach to the group $G \oplus H$ and view it as a product of subgroups—that is, an *internal* direct product. Recall that Theorem 25.6 shows that the groups $G' = G \oplus \{e_H\} = \{(g, e_H) : g \in G\}$ and $H' = \{e_G\} \oplus H = \{(e_G, h) : h \in H\}$ are subgroups of $G \oplus H$. Furthermore, both G' and H' are normal subgroups of $G \oplus H$. (See Exercise (3).) These two subgroups G' and H' generate all the elements of $G \oplus H$ in a natural way. We can write any element $(g, h) \in G \oplus H$ as $(g, h) = (g, e_H)(e_G, h)$, and so every element in $G \oplus H$ is of the form $g'h'$ with $g' \in G'$ and $h' \in H'$. In addition, this decomposition of an element $(g, h) \in G \oplus H$ into a product $g'h'$ with $g' \in G'$ and $h' \in H'$ is unique. As we will see, the uniqueness of the decomposition is due to the fact that $G' \cap H' = \{(e_G, e_H)\}$. So, to summarize, we have two normal subgroups G' and H' of a group $G \oplus H$, the intersection of G' and H' is trivial, and every element in $G \oplus H$ can be written uniquely as a product of an element in G' and an element in H'. Since G' and H' are subgroups of $G \oplus H$, we call this decomposition of $G \oplus H$ into products of elements from its subgroups G' and H' an *internal direct product*.

As it turns out, we can generalize this type of construction. Suppose that K and N are both normal subgroups of a group G and $K \cap N = \{e\}$ (where e is the identity in G). Then we can make the set

$$KN = \{kn : k \in K, n \in N\}$$

Table 25.2

Operation table for U_{28}.

	[1]	[3]	[5]	[9]	[11]	[13]	[15]	[17]	[19]	[23]	[25]	[27]
[1]	[1]	[3]	[5]	[9]	[11]	[13]	[15]	[17]	[19]	[23]	[25]	[27]
[3]	[3]	[9]	[15]	[27]	[5]	[11]	[17]	[23]	[1]	[13]	[19]	[25]
[5]	[5]	[15]	[25]	[17]	[27]	[9]	[19]	[1]	[11]	[3]	[13]	[23]
[9]	[9]	[27]	[17]	[25]	[15]	[5]	[23]	[13]	[3]	[11]	[1]	[19]
[11]	[11]	[5]	[27]	[15]	[9]	[3]	[25]	[19]	[13]	[1]	[23]	[17]
[13]	[13]	[11]	[9]	[5]	[3]	[1]	[27]	[25]	[23]	[19]	[17]	[15]
[15]	[15]	[17]	[19]	[23]	[25]	[27]	[1]	[3]	[5]	[9]	[11]	[13]
[17]	[17]	[23]	[1]	[13]	[19]	[25]	[3]	[9]	[15]	[27]	[5]	[11]
[19]	[19]	[1]	[11]	[3]	[13]	[23]	[5]	[15]	[25]	[17]	[27]	[9]
[23]	[23]	[13]	[3]	[11]	[1]	[19]	[9]	[27]	[17]	[25]	[15]	[5]
[25]	[25]	[19]	[13]	[1]	[23]	[17]	[11]	[5]	[27]	[15]	[9]	[3]
[27]	[27]	[25]	[23]	[19]	[17]	[15]	[13]	[11]	[9]	[5]	[3]	[1]

into a subgroup of G using the operation from G. (See Activity 24.22 on page 346.) The resulting group, denoted $K \times N$, is contained within G and is called an *internal* direct product of K and N.

Definition 25.12. Let G be a group with identity e, and let K and N be normal subgroups of G such that $K \cap N = \{e\}$. The **internal direct product** of K and N is the subgroup

$$K \times N = \{kn : k \in K, n \in N\}$$

of G.

When a subgroup H of a group G is decomposed as the product $H = K \times N$ for some normal subgroups K and N of G, these normal subgroups are called *factors* of H. The product $K \times N$ is called an *internal* product because all the products of group elements in $K \times N$ are performed within the group G.

Before we move on, a word of caution about notation is in order. Note that we use the $\times$ symbol for several different purposes, and you should be careful not to confuse the internal direct product $K \times N$ with the Cartesian product $K \times N$. The internal direct product $K \times N$ is a subgroup of the group G that contains K and N as normal subgroups, while the Cartesian product $K \times N$ is just a set of ordered pairs. The context should make clear which product is meant when the $\times$ symbol is used.

Activity 25.13. The operation table for

$$U_{28} = \{[1], [3], [5], [9], [11], [13], [15], [17], [19], [23], [25], [27]\}$$

is given in Table 25.2. Let $K = \langle [9] \rangle$ and $N = \langle [13] \rangle$. Note that $K \cap N = \{[1]\}$. Find the elements of the internal direct product $K \times N$, and then write the operation table for $K \times N$.

In some cases, we can write an entire group G as an internal direct product.

Activity 25.14. Let $G = U_{28}$, $K = \langle [3] \rangle$, and $N = \langle [13] \rangle$.

(a) Determine the elements in K and N. What is $K \cap N$?

(b) By completing the calculation of all the products in Table 25.3, show that every element in U_{28} can be written uniquely in the form kn for some $k \in K$ and $n \in N$.

Table 25.3
Products in $\langle [3] \rangle \times \langle [13] \rangle$ in U_{28}.

	[1]	[13]
[1]	[1]	[13]
[3]	[3]	
[9]	[9]	
[19]	[19]	
[25]	[25]	
[27]	[27]	

These last activities serve as a warm-up for a later investigation (Investigation 28) in which we will prove the Fundamental Theorem of Finite Abelian Groups, which will tell us that every finite Abelian group can be decomposed into an internal (and, in a sense, external) direct product of cyclic groups.

There are two important properties of internal direct products that deserve mention. First, if a group G is non-Abelian, we know that we cannot assume that elements in G commute. However, with an internal direct product, we can show that certain elements do commute. Theorem 25.15 formalizes this result. Second, each element in an internal direct product $K \times N$ can be written as a product kn where $k \in K$ and $n \in N$, but we don't know if such a representation is unique. Theorem 25.15 resolves this issue as well.

Theorem 25.15. *Let G be a group with identity e, and let K and N be normal subgroups of G such that $K \cap N = \{e\}$. Then*

(i) *$kn = nk$ for all $k \in K$ and $n \in N$; and*

(ii) *the representation of an element in $K \times N$ in the form kn, where $k \in K$ and $n \in N$, is unique.*

Proof. Let G be a group with identity e, and let K and N be normal subgroups of G such that $K \cap N = \{e\}$. To prove part (i), let $k \in K$ and $n \in N$. We know that $knk^{-1} \in N$ and $nk^{-1}n^{-1} \in K$, so $(knk^{-1})n^{-1} \in N$ and $k(nk^{-1}n^{-1}) \in K$. Therefore, $knk^{-1}n^{-1} \in K \cap N$ and so $knk^{-1}n^{-1} = e$. Now $e = knk^{-1}n^{-1} = (kn)(nk)^{-1}$, so right multiplication by nk yields $kn = nk$.

For part (ii), let $x \in K \times N$, and suppose that $x = k_1 n_1 = k_2 n_2$ for some $k_1, k_2 \in K$ and $n_1, n_2 \in N$. Then $k_2^{-1}k_1 = n_2 n_1^{-1} \in K \cap N$ and so $k_2^{-1}k_1 = n_2 n_1^{-1} = e$. Therefore, $k_1 = k_2$ and $n_1 = n_2$, which shows that the representation of the element x in $K \times N$ is unique. ∎

Note that part (i) of Theorem 25.15 shows that the elements of K and N commute with each other, but K and N themselves may not be Abelian groups.

Activity 25.16. We can generalize the construction of the internal direct product of two groups to any finite number of groups. Let G be a group with identity e, $m \geq 2$ an integer, and let N_1, N_2, ..., N_m be normal subgroups of G with

$$N_i \cap (N_1 \times N_2 \times \cdots N_{i-1} \times N_{i+1} \times \cdots \times N_m) = \{e\}$$

for each i. We define the elements in $N_1 \times N_2 \times \cdots \times N_m$ to be elements in G of the form $n_1 n_2 \cdots n_m$ where $n_i \in N_i$ for each i. The operation is the same as the operation in G.

(a) Explain why $N_1 \times N_2 \times \cdots \times N_m$ is closed under the operation from G.

(b) Explain why $N_1 \times N_2 \times \cdots \times N_m$ contains the identity from G.

(c) Explain why $N_1 \times N_2 \times \cdots \times N_m$ contains an inverse for each of its elements.

(d) What conclusion can we draw from parts (a) – (c)?

(e) You may be wondering why, in the above construction, we required that

$$N_i \cap (N_1 \times N_2 \times \cdots N_{i-1} \times N_{i+1} \times \cdots \times N_m) = \{e\}$$

for each i. This condition seems to be more complicated than in the two-subgroup case, but it is what we need in order to guarantee the same uniqueness of the representation of an element in $N_1 \times N_2 \times \cdots \times N_m$ that part (ii) of Theorem 25.15 gives in the $m = 2$ case. Prove that, with this condition, the representation of an element in $N_1 \times N_2 \times \cdots \times N_m$ in the form $n_1 n_2 \cdots n_m$, with $n_i \in N_i$ for each i, is in fact unique.

Concluding Activities

Activity 25.17. Write a formal proof of Theorem 25.5.

Activity 25.18. In this activity, we will verify a result that we used in the proof of Theorem 25.9.

(a) Let G and H be groups, and let $g \in G$ and $h \in H$. Prove that $(g, h)^n = (g^n, h^n)$ for each integer n.

(b) Let $n \geq 2$ be an integer, let $G_1, G_2, \ldots, G_n$ be groups, and let $(g_1, g_2, \ldots, g_n) \in G_1 \oplus G_2 \oplus \cdots \oplus G_n$. What element is $(g_1, g_2, \ldots, g_n)^k$, where k is an integer? Prove your answer.

Activity 25.19. Let G and H be groups.

(a) Under what conditions is $G \oplus H$ Abelian? State your answer in the form of a biconditional statement, and then prove it.

(b) Is $G \oplus H$ cyclic when G and H are both cyclic? If $G \oplus H$ is cyclic, must G and H both be cyclic? Prove your answers.

Exercises

(1) The operation table for the Klein 4-group $\mathcal{V}$ (see Activity 21.6 on page 303 of Investigation 21) is as follows:

	1	a	b	ab
1	1	a	b	ab
a	a	a	ab	b
b	b	ab	1	a
ab	ab	b	a	1

    (a) Construct the operation table for the group $\mathbb{Z}_2 \oplus \mathbb{Z}_2$. (See Activity 25.3.)

    (b) Compare the operation tables for $\mathbb{Z}_2 \oplus \mathbb{Z}_2$ and for $\mathcal{V}$. How are these groups similar? How are they different?

(2) Create the operation table for the group $(\mathbb{Z}_4 \oplus \mathbb{Z}_2)/\langle([2],[1])\rangle$.

* (3) Let G and H be groups with identities e_G and e_H, respectively.

    (a) Prove Theorem 25.6, which states that if G and H are groups with identities e_G and e_H, respectively, then
$$G \oplus \{e_H\} = \{(g, e_H) : g \in G\}$$
and
$$\{e_G\} \oplus H = \{(e_G, h) : h \in H\}$$
are both subgroups of $G \oplus H$.

    (b) Show that G' and H' are normal subgroups of $G \oplus H$.

(4) Let $G = \mathbb{Z}_4 \oplus \mathbb{Z}_6$.

    (a) Does G contain an element of order 12? If so, find one. If not, prove that no such element exists.

    (b) Calculate the orders of all elements in G.

(5) Let $n \geq 2$ be an integer, and let $G_1, G_2, \ldots, G_n$ be groups. For each i, let N_i be a subgroup of G_i.

    (a) Prove that $N_1 \oplus N_2 \oplus \cdots \oplus N_n$ is a subgroup of $G_1 \oplus G_2 \oplus \cdots \oplus G_n$.

    (b) Find necessary and sufficient conditions on the groups N_i so that $N_1 \oplus N_2 \oplus \cdots \oplus N_n$ is a normal subgroup of $G_1 \oplus G_2 \oplus \cdots \oplus G_n$.

    (c) Is it true that if M is a subgroup of $G_1 \oplus G_2 \oplus \cdots \oplus G_n$, then for each i there is a subgroup N_i of G_i so that $M = N_1 \oplus N_2 \oplus \cdots \oplus N_n$? Prove your answer.

(6) (a) Let G be a cyclic group of order m and H a cyclic group of order n. Find necessary and sufficient conditions m and n so that $G \oplus H$ is cyclic. Prove your answer.

    (b) Now extend part (a) to any finite number of cyclic groups. That is, let $k \geq 2$ be a positive integer, and let $n_1, n_2, \ldots, n_k$ be positive integers. For each i between 1 and k, let G_i be a cyclic group of order n_i. Find and prove necessary and sufficient conditions so that the group
$$G = G_1 \oplus G_2 \oplus \cdots \oplus G_k$$
is cyclic.

(7) Generalize Theorem 25.9. That is, let $n \geq 2$ be an integer, let $G_1, G_2, \ldots, G_n$ be groups, and, for each i, let $g_i \in G_i$ be an element of finite order in G_i. Determine $|(g_1, g_2, g_3, \ldots, g_n)|$ in $G_1 \oplus G_2 \oplus \cdots \oplus G_n$. Prove your answer.

(8) (a) Construct the operation table for the group $D_3 \oplus \mathbb{Z}_2$.

   (b) Find the orders of all elements in $D_3 \oplus \mathbb{Z}_2$.

   (c) Exercise (3) shows that the groups $D_3 \oplus \{[0]\}$ and $\{I\} \oplus \mathbb{Z}_2$ are both normal subgroups of $D_3 \oplus \mathbb{Z}_2$.

   (i) Find the elements of $(D_3 \oplus \mathbb{Z}_2)/(D_2 \oplus \{[0]\})$, and construct the operation table for this quotient group. What familiar group does $(D_3 \oplus \mathbb{Z}_2)/(D_3 \oplus \{[0]\})$ look like?

   (ii) Find the elements of $(D_3 \oplus \mathbb{Z}_2)/(\{I\} \oplus \mathbb{Z}_2)$, and construct the operation table for this quotient group. What familiar group does $(D_3 \oplus \mathbb{Z}_2)/(\{I\} \oplus \mathbb{Z}_2)$ look like?

(9) Let G be a finite group with identity e.

   (a) If K and N are normal subgroups of G with $K \cap N = \{e\}$ and $|K|\,|N| = |G|$, does $G = K \times N$? Prove your answer.

   (b) With the same conditions as in (a), does $G = N \times K$? Explain.

   (c) Does part (a) generalize to any number of normal subgroups? State your answer in the form of a conjecture, and then prove it.

(10) (a) Write the group U_{15} as an internal direct product of subgroups.

   (b) Is the representation of U_{15} as an internal direct product of groups (from part (a)) unique? That is, can you find normal subgroups K' and N' of U_{15} so that $U_{15} = K' \times N'$ but either $K' \neq K$ or $N' \neq N$?

(11) In this problem, we will determine when D_n can be written as an internal direct product of normal subgroups. Two previous results might be helpful. First, Exercise (36) of Investigation 24 (see page 351) tells us that a nontrivial proper subgroup N of D_n is normal in D_n if and only if N is a subgroup of $\langle R \rangle$ or n is even and N is one of $\langle r, R^2 \rangle$ or $\langle rR, R^2 \rangle$. Also, Exercise (2) of Investigation 21 (see page 304) shows that $Z(D_n) = \{I, R^{n/2}\}$ if n is even and $Z(D_n) = \{I\}$ if n is odd.

   (a) Explain why D_n cannot be written as an internal direct product of two normal subgroups if n is odd.

   (b) Can D_4 be written as an internal direct product of normal subgroups? Explain your answer.

   (c) Show that D_6 can be written as an internal direct product of normal subgroups.

   (d) State and prove a complete characterization of the groups D_n that can be written as an internal direct product of normal subgroups.

(12) In this investigation, we introduced the external direct product of a finite number of groups. We can also define an external direct product of an infinite number of groups. For each natural number i, let G_i be a group with identity e_i. Define the **infinite direct product**

$$\bigoplus_{i=1}^{\infty} G_i = G_1 \oplus G_2 \oplus G_3 \oplus \cdots$$

to be the set of all infinite tuples of the form

$$a = (a_1, a_2, a_3, \ldots),$$

where $a_i \in G_i$ for all i, with an operation defined componentwise—that is,

$$(a_1, a_2, a_3, \ldots)(b_1, b_2, b_3, \ldots) = (a_1 b_1, a_2 b_2, a_3 b_3, \ldots).$$

(a) Prove that the infinite direct product $\bigoplus_{i=1}^{\infty} G_i$ is a group.

(b) The **restricted direct product** (or **direct sum**) of the groups G_i, for $i \in \mathbb{Z}^+$, is the subset of $\bigoplus_{i=1}^{\infty} G_i$ consisting of the infinite tuples of the form $a = (a_1, a_2, a_3, \ldots)$, where $a_i \in G_i$ for all i, and $a_i = e_i$ for all but finitely many values of i. To illustrate, let $G_i = \mathbb{Z}_{i+1}$ for every natural number i, and define G to be the restricted direct product of the group

$$\bigoplus_{i=1}^{\infty} G_i = \bigoplus_{i=1}^{\infty} \mathbb{Z}_{i+1} = \mathbb{Z}_2 \oplus \mathbb{Z}_3 \oplus \mathbb{Z}_4 \oplus \cdots$$

Then

$$a = ([1]_2, [0]_3, [0]_4, [4]_5, [3]_6, [0]_7, [0]_8, [0]_9, \ldots) \in G,$$

since a_i is the zero element in $\mathbb{Z}_{i+1}$ for all but three values of i. In contrast,

$$b = ([1]_2, [1]_3, [1]_4, [1]_5, \ldots) \notin G,$$

since b_i is nonzero for infinitely many values of i.

    (i) Prove that the restricted direct product is a subgroup of the direct product.

    (ii) Is the restricted direct product a normal subgroup of the direct product? Prove your answer.

    (iii) Let $i \in \mathbb{Z}^+$, and let $G_i = \mathbb{Z}_{i+1}$ for each i. Show that $\bigoplus_{i=1}^{\infty} G_i$ has elements of infinite order, but every element of the corresponding restricted direct product has finite order.

Connections

In this investigation, we studied external and internal products of groups. If you studied ring theory before group theory, you should notice similarities between the external direct products of groups from this investigation and the direct sums of rings we considered in Investigation 6. Because of the simpler structure of groups, we can define more than one type of product of groups. Since there are two operations defined on a ring, rings are more complicated objects than groups and so we only defined one type of "product" (which we called a direct sum). The constructions of direct sums of rings and external direct products of groups are analogous. To create a direct sum/product of a pair of objects (either rings or groups), we make the Cartesian product of the two sets into the appropriate algebraic structure (either a ring or a group) using operations defined componentwise.

Investigation 26

Group Isomorphisms and Invariants

Focus Questions

By the end of this investigation, you should be able to give precise and thorough answers to the questions listed below. You may want to keep these questions in mind to focus your thoughts as you complete the investigation.

- Intuitively, what does it mean for two groups to be "essentially the same"?

- What does it mean for two groups to be isomorphic? How does the definition of isomorphism reflect the informal definition of "essentially the same"?

- What strategies can be used to prove that two groups are isomorphic? How are these strategies motivated by the definition of isomorphism?

- What is an invariant, and how does one prove that a property is an invariant?

- How can invariants be used to prove that two groups are not isomorphic? Can invariants be used to prove that two groups are isomorphic?

- What are isomorphism classes, and what is their role in the classification of finite groups?

- What does Cayley's Theorem say, and why is it important?

Preview Activity 26.1. The notion of "sameness" is very important in mathematics, for it allows us to identify when two objects should be considered indistinguishable, and thus treated identically. Identifying sameness also makes our analysis more efficient, since it allows us to consider entire classes of objects at the same time, instead of dealing with each object individually.

In this investigation, we will define precisely what it means for two groups to be essentially the same, or *isomorphic*. Before we do so, however, let's apply our intuitive ideas about sameness to a few examples.

The operation tables for four groups are shown below. Which of these groups would you consider to be essentially the same, and which would you consider to be different? Consider each possible pair of groups, and give a convincing argument to justify your answer for each.

G_1 :

	a	b	c	d	e	f	g	h
a	a	b	c	d	e	f	g	h
b	b	c	d	a	h	e	f	g
c	c	d	a	b	g	h	e	f
d	d	a	b	c	f	g	h	e
e	e	f	g	h	a	b	c	d
f	f	g	h	e	d	a	b	c
g	g	h	e	f	c	d	a	b
h	h	e	f	g	b	c	d	a

G_2 :

	α	β	γ	δ	ε	ζ	η	θ
α	α	β	γ	δ	ε	ζ	η	θ
β	β	γ	δ	ε	ζ	η	θ	α
γ	γ	δ	ε	ζ	η	θ	α	β
δ	δ	ε	ζ	η	θ	α	β	γ
ε	ε	ζ	η	θ	α	β	γ	δ
ζ	ζ	η	θ	α	β	γ	δ	ε
η	η	θ	α	β	γ	δ	ε	ζ
θ	θ	α	β	γ	δ	ε	ζ	η

G_3 :

	s	t	u	v	w	x	y	z
s	s	t	u	v	w	x	y	z
t	t	u	v	s	x	y	z	w
u	u	v	s	t	y	z	w	x
v	v	s	t	u	z	w	x	y
w	w	x	y	z	t	u	v	s
x	x	y	z	w	u	v	s	t
y	y	z	w	x	v	s	t	u
z	z	w	x	y	s	t	u	v

G_4 :

	♠	♡	◇	♣	✠	♯	†	℧
♠	♠	♡	◇	♣	✠	♯	†	℧
♡	♡	◇	♣	♠	♯	†	℧	✠
◇	◇	♣	♠	♡	†	℧	✠	♯
♣	♣	♠	♡	◇	℧	✠	♯	†
✠	✠	♯	†	℧	♠	♡	◇	♣
♯	♯	†	℧	✠	♡	◇	♣	♠
†	†	℧	✠	♯	◇	♣	♠	♡
℧	℧	✠	♯	†	♣	♠	♡	◇

Introduction

In Preview Activity 26.1, you were asked to decide which of the four groups shown were essentially the same, and which were not. At first glance, it would be easy to think that all the groups are different. After all, their elements are certainly different. But is this enough to conclude that the groups themselves are different?

To answer this question, let's consider the group $\mathbb{Z}_8$:

$\mathbb{Z}_8$:

	[0]	[1]	[2]	[3]	[4]	[5]	[6]	[7]
[0]	[0]	[1]	[2]	[3]	[4]	[5]	[6]	[7]
[1]	[1]	[2]	[3]	[4]	[5]	[6]	[7]	[0]
[2]	[2]	[3]	[4]	[5]	[6]	[7]	[0]	[1]
[3]	[3]	[4]	[5]	[6]	[7]	[0]	[1]	[2]
[4]	[4]	[5]	[6]	[7]	[0]	[1]	[2]	[3]
[5]	[5]	[6]	[7]	[0]	[1]	[2]	[3]	[4]
[6]	[6]	[7]	[0]	[1]	[2]	[3]	[4]	[5]
[7]	[7]	[0]	[1]	[2]	[3]	[4]	[5]	[6]

Let's suppose also that we decided to abbreviate the names of the equivalence classes in $\mathbb{Z}_8$ by assigning a variable to each one. In particular, we'll let $\alpha = [0]$, $\beta = [1]$, $\gamma = [2]$, $\delta = [3]$, $\epsilon = [4]$, $\zeta = [5]$, $\eta = [6]$, and $\theta = [7]$.

Activity 26.2. Substitute α, β, γ, δ, ϵ, ζ, η, and θ for $[0]$, $[1]$, $[2]$, $[3]$, $[4]$, $[5]$, $[6]$, and $[7]$ in the operation table for $\mathbb{Z}_8$. That is, each time $[0]$ appears in the tables, replace it with α. Do the same for the other classes as well, replacing $[1]$ with β, $[2]$ with γ, $[3]$ with δ, $[4]$ with ϵ, $[5]$ with ζ, $[6]$ with η, and $[7]$ with θ. What do you notice?

If you completed Activity 26.2 correctly, you probably observed that the operation table for $\mathbb{Z}_8$ can be made to look exactly like the operation table of G_2 in Preview Activity 26.1, simply by renaming the elements. In other words, the only differences between $\mathbb{Z}_8$ and G_2 are the names of the elements. As far as their structure as groups is concerned, $\mathbb{Z}_8$ and G_2 are basically the same group. It is the operation—the way the elements interact with each other—rather than the names of the elements themselves that determines a group.

As it turns out, we can carry out a similar renaming to show that $\mathbb{Z}_8$ and G_3 are also essentially the same. Here we have to be a bit more careful, however, since the elements of G_3 seem to be arranged in a different order than those of $\mathbb{Z}_8$ and G_2. To illustrate, let's see what would happen if we replaced the names of the elements of G_3 with the names of the elements of $\mathbb{Z}_8$, keeping the elements in the same order as they are listed in the tables. We would replace s with $[0]$, t with $[1]$, u with $[2]$, v with $[3]$, w with $[4]$, x with $[5]$, y with $[6]$, and z with $[7]$, which would yield the following operation table:

	[0]	[1]	[2]	[3]	[4]	[5]	[6]	[7]
[0]	[0]	[1]	[2]	[3]	[4]	[5]	[6]	[7]
[1]	[1]	[2]	[3]	[0]	[5]	[6]	[7]	[4]
[2]	[2]	[6]	[0]	[1]	[6]	[7]	[4]	[5]
[3]	[3]	[0]	[1]	[2]	[7]	[4]	[5]	[6]
[4]	[4]	[5]	[6]	[7]	[1]	[2]	[3]	[0]
[5]	[5]	[6]	[7]	[4]	[2]	[3]	[0]	[1]
[6]	[6]	[7]	[4]	[5]	[3]	[0]	[1]	[2]
[7]	[7]	[4]	[5]	[6]	[0]	[1]	[2]	[3]

This table does not look like the operation table for $\mathbb{Z}_8$; the renaming of the elements of G_3 that we used did not yield a group that seemed to be the same as $\mathbb{Z}_8$. So what can we conclude from this? Does our seemingly failed attempt at renaming the elements of G_3 necessarily imply that G_3 and $\mathbb{Z}_8$ are different groups? In fact, it does not. All we know at this point is that the particular renaming that we used yields a group that doesn't look like $\mathbb{Z}_8$. But what if we used a different renaming? For instance, notice that $w^2 = t$, $w^3 = x$, $w^4 = u$, $w^5 = y$, $w^6 = v$, $w^7 = z$, and $w^8 = s$, so w has order 8 and the group G_3 is cyclic. Because of this, if we want to identify elements of G_3 with the elements in $\mathbb{Z}_8$ so that the resulting group structures look identical, then we will need to identify w with an element in $\mathbb{Z}_8$ of order 8. This might lead us to try replacing w with $[1]$, $w^2 = t$ with $[2]$, $w^3 = x$ with $[3]$, $w^4 = u$ with $[4]$, $w^5 = y$ with $[5]$, $w^6 = v$ with $[6]$, $w^7 = z$ with $[7]$, and $w^8 = s$ with $[0]$. Doing so would yield the following operation table:

	[0]	[2]	[4]	[6]	[1]	[3]	[5]	[7]
[0]	[0]	[2]	[4]	[6]	[1]	[3]	[5]	[7]
[2]	[2]	[4]	[6]	[0]	[3]	[5]	[7]	[1]
[4]	[4]	[5]	[0]	[2]	[5]	[7]	[1]	[3]
[6]	[6]	[0]	[2]	[4]	[7]	[1]	[3]	[5]
[1]	[1]	[3]	[5]	[7]	[2]	[4]	[6]	[0]
[3]	[3]	[5]	[7]	[1]	[4]	[6]	[0]	[2]
[5]	[5]	[7]	[1]	[3]	[6]	[0]	[2]	[4]
[7]	[7]	[1]	[3]	[5]	[0]	[2]	[4]	[6]

Notice that, except for the order in which the elements are listed, this last operation table is identical to the operation table for $\mathbb{Z}_8$. Note also that the ordering of the elements affects only the way the table is displayed, and not the information that it contains. A simple rearrangement of the rows and columns puts the table in its more standard form. Thus, we can see that by renaming the elements of G_3, and possibly reordering the rows and columns of the resulting operation table, we are able to produce the operation table for $\mathbb{Z}_8$. Because of this, we say that G_3 and $\mathbb{Z}_8$ are basically the same group.

Now that we have seen a few examples, we are ready to state an informal definition, which we will use to formally define the notion of isomorphism in the next section.

Informal Definition 26.3. Let G and H be finite groups. The group G is said to be **essentially the same** as H if the operation table for G can be transformed into the operation table for H by doing nothing more than renaming the elements of G and/or reordering the rows and columns of G's operation table.

Note that this informal definition can easily be used to show that two groups are *not* essentially the same. For example, we saw that the group G_3 was cyclic, but neither G_1 nor G_4 contain an element of order 8. From these observations, it is clear that G_3 is a different group than either G_1 or G_3, since no matter how we rename and/or reorder the elements of G_3, we will still end up with a group that contains an element of order 8. Also, group G_1 is non-Abelian, so no matter how we relabel the elements of G_1, the resulting operation table will be that of a non-Abelian group. Since G_4 is an Abelian group, we can conclude that G_1 and G_4 are different groups.

To summarize, note that in order to show that two groups are the same, we must find a way to rename and/or reorder the elements of one group so that its operation table is identical to that of the other group. In order to show that two groups are different, however, it often suffices to identify a property that is different between the two groups—in particular, a property that could not possibly be different if one group had been obtained from the other by simply renaming and reordering elements.

Activity 26.4. We have already argued that groups G_1 and G_4 from Preview Activity 26.1 are different because G_1 is a non-Abelian group while G_4 is an Abelian group. Setting that difference aside, list at least two other properties that are different between G_1 and G_4 and would therefore show (by Informal Definition 26.3) that G_1 and G_4 are different groups.

Isomorphisms of Groups

Informal Definition 26.3 provides a helpful and intuitive way of thinking about what it means for two groups to be the same. This informal definition, however, has some significant limitations. First, it only works for finite groups. This is because it would be impossible to actually create the operation table for a group with infinitely many elements. Second, even for finite groups, the definition can be extremely cumbersome to work with, especially if the groups in question have more than a few elements. Can you imagine trying to create an operation table for a group with 50 or 1000 or even 50, 000 elements? The task would be daunting at best, and practically impossible at worst.

To deal with these difficulties, we will adopt a formal definition that captures the idea behind Informal Definition 26.3, but does so in a more precise manner. In order to motivate this definition, let's consider again the two main parts of Informal Definition 26.3.

Renaming Elements

When we argued that $\mathbb{Z}_8$ and G_2 were essentially the same group, we found a way to rename the elements of $\mathbb{Z}_8$ using the same names as the elements of G_2. This renaming was really just a bijective function (that is, a function that is both one-to-one and onto) from $\mathbb{Z}_8$ to G_2. Denoting this function by $\varphi : \mathbb{Z}_8 \to G_2$, we could write:

$$\varphi([0]) = \alpha, \quad \varphi([1]) = \beta, \quad \varphi([2]) = \gamma, \quad \varphi([3]) = \delta,$$
$$\varphi([4]) = \epsilon, \quad \varphi([5]) = \zeta, \quad \varphi([6]) = \eta, \quad \varphi([7]) = \theta.$$

Note that any function that actually corresponds to a valid renaming would have to be a bijection. This is because it wouldn't make sense to give two group elements the same name, or to leave a group element out. If two groups are truly the same, then the elements of one group should be able to be matched in a one-to-one correspondence with the elements of the other. As such, our formal definition of "sameness" will begin with a bijective function.

Preserving Operations

As we saw in our earlier example, just having a bijective function from one group to another is not enough to say that the two groups are the same. Indeed, this bijective function must also transform the operation table of the first group into the operation table of the second. To see exactly what this means, let's look at an example.

Consider two groups G and H, each having three elements. Suppose also that we have defined a bijective "renaming" function $\varphi : G \to H$. Let's consider the operation table for G, which we can write generically as follows (using a, b, and c to denote the elements of G):

	a	b	c
a	aa	ab	ac
b	ba	bb	bc
c	ca	cb	cc

If we simply replace each entry in this table with its new name (as given by φ), we obtain the following table:

	$\varphi(a)$	$\varphi(b)$	$\varphi(c)$
$\varphi(a)$	$\varphi(aa)$	$\varphi(ab)$	$\varphi(ac)$
$\varphi(b)$	$\varphi(ba)$	$\varphi(bb)$	$\varphi(bc)$
$\varphi(c)$	$\varphi(ca)$	$\varphi(cb)$	$\varphi(cc)$

Is this table the multiplication table for H? Its entries are certainly elements of H, since φ maps from G to H. But the actual multiplication table for H would be defined as follows:

$\cdot$	$\varphi(a)$	$\varphi(b)$	$\varphi(c)$
$\varphi(a)$	$\varphi(a)\varphi(a)$	$\varphi(a)\varphi(b)$	$\varphi(a)\varphi(c)$
$\varphi(b)$	$\varphi(b)\varphi(a)$	$\varphi(b)\varphi(b)$	$\varphi(b)\varphi(c)$
$\varphi(c)$	$\varphi(c)\varphi(a)$	$\varphi(c)\varphi(b)$	$\varphi(c)\varphi(c)$

If the notation seems confusing here, just keep in mind that $\varphi(a)$, $\varphi(b)$, and $\varphi(c)$ are the elements of H, and we have formed the multiplication table for H in the usual way. In particular, each entry in the table is the product of the corresponding row and column headers (so, for instance, the entry in the $\varphi(a)$ row and $\varphi(b)$ column is just $\varphi(a)\varphi(b)$).

So what can we conclude? Recall that we wanted the renamed G table to be equal to the H table. In order for this to happen, each entry of the renamed G table must be equal to the corresponding entry of the H table. Thus, it must be the case that $\varphi(aa) = \varphi(a)\varphi(a)$, $\varphi(ab) = \varphi(a)\varphi(b)$, $\varphi(ac) = \varphi(a)\varphi(c)$, $\varphi(ba) = \varphi(b)\varphi(a)$, and so on.

In other words, we want φ to *preserve the operation*, which means:

For all x, $y \in G$, $\varphi(xy) = \varphi(x)\varphi(y)$.

This condition helps us to state in a more precise way exactly what it means for two groups to have the same operation table—or, in other words, the same algebraic structure. Any bijective function φ that is both a bijection and that preserves the operation is called an *isomorphism*, defined formally below.

Definition 26.5. Let G and H be groups. An **isomorphism** is a bijective function $\varphi : G \to H$ such that for all $x, y \in G$,

$$\varphi(xy) = \varphi(x)\varphi(y).$$

If there is an isomorphism from the group G to the group H, we say that G is **isomorphic** to H.

The word *isomorphic* comes from two Greek words: *isos*, which means *equal* or *same*, and *morphe*, which means *form* or *structure*. Thus, when we say that two groups are isomorphic, we mean that they have the same structure. Likewise, we can think of an isomorphism as being an *operation-preserving bijection* or a *structure-preserving bijection*.

It is worth noting that the isomorphism relation is symmetric. (In fact, the isomorphism relation is an equivalence relation, as we will show later.) In particular, if $\varphi : G \to H$ is an isomorphism, then $\varphi^{-1} : H \to G$ is also an isomorphism. For this reason, we will often simply say that two groups G and H are isomorphic, rather than saying that G is isomorphic to H, or H is isomorphic to G. When G and H are isomorphic, we denote this relationship by writing $G \cong H$. So $G \cong H$ means that there is an isomorphism from G to H (or, equivalently, an isomorphism from H to G).

It's also important to note that the *function* in Definition 26.5 is called an *isomorphism*, while the *groups* G and H are said to be *isomorphic*. If the operations in G and H are written additively, then the operation-preserving condition for φ is written as

$$\varphi(x + y) = \varphi(x) + \varphi(y)$$

for all $x, y \in G$. If, on the other hand, the operations in G and H are written using different notations—for instance, suppose the operation in G is written multiplicatively and the operation in H is written additively—then the operation-preserving condition for φ must reflect this difference. In this case, we would write the condition as

$$\varphi(xy) = \varphi(x) + \varphi(y)$$

for all $x, y \in G$. Which notation is most appropriate should be clear from the context.

Activity 26.6.

(a) Use Definition 26.5 to explain why the function $\varphi : G_3 \to \mathbb{Z}_8$ (from Preview Activity 26.1) defined by

$$\varphi(w^k) = [k]$$

is an isomorphism.

(b) Is G_2 isomorphic to G_3? Use Definition 26.5 to justify your answer.

(c) Use Definition 26.5 to explain why the function $\varphi : G_2 \to G_4$ defined by

$$\varphi(\alpha) = \spadesuit, \quad \varphi(\beta) = \heartsuit, \quad \varphi(\gamma) = \diamondsuit, \quad \varphi(\delta) = \clubsuit,$$
$$\varphi(\epsilon) = \maltese, \quad \varphi(\zeta) = \sharp, \quad \varphi(\eta) = \dagger, \quad \varphi(\theta) = \mho$$

is **not** an isomorphism.

(d) What does your answer to part (c) allow you to conclude about whether G_2 is isomorphic to G_4?

Proving Isomorphism

Now that we have precisely defined what it means for one group to be isomorphic to another, let's consider how we might use this definition in the context of groups that have more than just a few elements. Let $G = \left\{ \begin{bmatrix} 1 & x \\ 0 & 1 \end{bmatrix} : x \in \mathbb{R} \right\}$.

Activity 26.7.

(a) Notice that G (as defined above) is a subset of $\mathrm{GL}_2(\mathbb{R})$ (the group of invertible 2×2 matrices under multiplication). Show that G is actually a *subgroup* of $\mathrm{GL}_2(\mathbb{R})$.

(b) To which familiar group do you think G is isomorphic? You don't have to prove your answer now, but you should make a reasonable conjecture with some solid reasoning to back it up.

Looking again at the definition of G, it appears that each element of G corresponds to a unique real number (and vice versa). Thus, it seems reasonable that we would try to prove that G is isomorphic to the group $\mathbb{R}$ under addition. Since both G and $\mathbb{R}$ have infinitely many elements, we will not be able to simply work with their operation tables. Instead, we must use Definition 26.5, which suggests the following steps:

(1) We must define an appropriate function $\varphi : G \to \mathbb{R}$.

(2) We must show that φ is bijective; that is, we must show that φ is both injective and surjective.

(3) We must show that φ preserves the operation in G.

Activity 26.8. Carefully read the following proof that G is isomorphic to $\mathbb{R}$, filling in all the missing details and providing additional explanations where appropriate. As you read the proof, try to identify where each of the three steps outlined above is taking place.

> **Theorem.** *Let* $G = \left\{ \begin{bmatrix} 1 & x \\ 0 & 1 \end{bmatrix} : x \in \mathbb{R} \right\}$. *Then G is isomorphic to $\mathbb{R}$.*

Proof. Let $\varphi : G \to \mathbb{R}$ be defined by

$$\varphi\left(\begin{bmatrix} 1 & x \\ 0 & 1 \end{bmatrix} \right) = x.$$

Suppose that

$$\varphi\left(\begin{bmatrix} 1 & x \\ 0 & 1 \end{bmatrix} \right) = \varphi\left(\begin{bmatrix} 1 & y \\ 0 & 1 \end{bmatrix} \right).$$

Then $x = y$, which implies that

$$\begin{bmatrix} 1 & x \\ 0 & 1 \end{bmatrix} = \begin{bmatrix} 1 & y \\ 0 & 1 \end{bmatrix}.$$

Also observe that for all $x \in \mathbb{R}$,

$$\varphi\left(\begin{bmatrix} 1 & x \\ 0 & 1 \end{bmatrix} \right) = x.$$

Finally,

$$\varphi\left(\begin{bmatrix} 1 & x \\ 0 & 1 \end{bmatrix}\begin{bmatrix} 1 & y \\ 0 & 1 \end{bmatrix}\right) = \varphi\left(\begin{bmatrix} 1 & x+y \\ 0 & 1 \end{bmatrix}\right)$$
$$= x + y$$
$$= \varphi\left(\begin{bmatrix} 1 & x \\ 0 & 1 \end{bmatrix}\right) + \varphi\left(\begin{bmatrix} 1 & y \\ 0 & 1 \end{bmatrix}\right).$$

Therefore, G is isomorphic to $\mathbb{R}$. ∎

Some Basic Properties of Isomorphisms

Since an isomorphism from a group G to a group H preserves the structure of G, there are certain things we should expect all isomorphisms to do. As an example, each group has exactly one identity element, so we should expect an isomorphism to map the identity to the identity. That property and a few others are given in the next activity.

Activity 26.9. Let φ be an isomorphism from a group G to a group H. Let e_G be the identity in G and e_H the identity in H.

(a) Prove that $\varphi(e_G) = e_H$. (Hint: It suffices to show that $h\varphi(e_G) = h$ for some $h \in H$. You may need to use the fact that φ is a surjection.)

(b) Prove that $\varphi\left(a^{-1}\right) = (\varphi(a))^{-1}$ for all $a \in G$. (Hint: It suffices to show that $\varphi(a)\varphi\left(a^{-1}\right) = e_H$.)

(c) Let $a \in G$. Prove that $\varphi(a^k) = (\varphi(a))^k$ for all positive integers k.

(d) Let $a \in G$. Use parts (b) and (c) to prove that $\varphi(a^k) = (\varphi(a))^k$ for all negative integers k. This will complete the proof of the following theorem.

   Theorem. *Let φ be an isomorphism from a group G to a group H. Then $\varphi(a^k) = (\varphi(a))^k$ for all $a \in G$ and $k \in \mathbb{Z}$.*

We will discuss other properties of isomorphisms later.

Well-Defined Functions

Preview Activity 26.10. Let f assign to each element $[a]_3$ in $\mathbb{Z}_3$ the element $[a]_6$ in $\mathbb{Z}_6$.

(a) Does f preserve the group operation in $\mathbb{Z}_3$? Explain.

(b) Consider the following proof that f is an injection:

   Let $[a]_3$ and $[b]_3$ be in $\mathbb{Z}_3$, and assume $f([a]_3) = f([b]_3)$. Then $[a]_6 = [b]_6$, and so 6 divides $b - a$. Thus, 3 divides $b - a$, which implies that $[a]_3 = [b]_3$.

This proof might seem to imply that $\mathbb{Z}_3$ is isomorphic to the set $f(\mathbb{Z}_3) = \{f([a]_3) : [a]_3 \in \mathbb{Z}_3\} = \{[0]_6, [1]_6, [2]_6\}$. What do you think about this conclusion? Explain your answer in detail.

There is one additional consideration we need to keep in mind when proving isomorphism. Activity 26.10 shows that we can define a map that preserves a group's structure and seems to behave like an isomorphism, but if the map treats equal elements with different representations in different ways, then whatever conclusions we might draw will make little sense. We saw this same idea in Investigation 2 when we discussed well-defined operations. To emphasize the point, any time we have multiple ways to represent the elements in a set (like in $\mathbb{Z}_n$ or $\mathbb{Q}$), we need to be sure that anything that acts on the elements of that set (like an operation or a function) is well-defined. The next definition formalizes this idea.

Definition 26.11. Let S and T be sets. A mapping $f : S \to T$ is **well-defined** if $f(a) = f(b)$ whenever $a = b$ in S.

When we use the word *function*, we always mean a well-defined mapping. Well-defined mappings or functions are also called *single-valued*. In many cases, we do not need to worry about a function begin well-defined; in particular, if there is only one way to represent an element in the domain, then there is nothing to show. If, however, there are multiple ways to represent elements in the domain (like in $\mathbb{Z}_n$ or $\mathbb{Q}$), then we need to verify that any mapping we consider is well-defined before we worry about any other properties the mapping might possess.

Activity 26.12. Let f be the mapping from $\mathbb{Q}$ to $\mathbb{Z}$ defined by $f\left(\frac{a}{b}\right) = a + b$. Is f well-defined? Why or why not?

Disproving Isomorphism

In Activity 26.8, we saw an example of how Definition 26.5 can be used to show that two groups are isomorphic. Although other examples may require more sophisticated arguments, the basic structure will often be the same: we first define a particular function, and then we show that this function is bijective and operation-preserving.

What should we do, however, if we want to prove that a group G is *not* isomorphic to a group H? For instance, consider the groups $\mathbb{Z}$ and $\mathbb{Q}$ under addition. These sets are quite different under multiplication (we can divide by nonzero elements in $\mathbb{Q}$, but not in $\mathbb{Z}$), but is the additive structure of $\mathbb{Z}$ different from the additive structure of $\mathbb{Q}$?

If we wanted to use the definition of isomorphism to prove that $\mathbb{Z}$ is not isomorphic to $\mathbb{Q}$, we would have to show that there does not exist an isomorphism $\varphi : \mathbb{Z} \to \mathbb{Q}$. In other words, we would have to show that every function that we could possibly define from $\mathbb{Z}$ to $\mathbb{Q}$ would violate at least one of the conditions that define isomorphisms. To show this directly seems daunting, if not impossible. The next activity provides one doable example.

Activity 26.13. To show that $\mathbb{Z}$ is not isomorphic to $\mathbb{Q}$, suppose, to the contrary, that there does exist a function $\varphi : \mathbb{Z} \to \mathbb{Q}$ that is both bijective and operation-preserving. Since 1 generates $\mathbb{Z}$, we will consider the rational number $q = \varphi(1)$.

(a) Explain why $\varphi(n) = nq$ for every integer n.

Table 26.1
Some common group invariants.

A Partial List of Invariants of Group Isomorphism

- Number of elements
- Commutativity
- Being cyclic
- Number of elements of each order
- Number of subgroups of each order
- Being simple

(b) Is φ an injection? Prove your answer.

(c) Is φ a surjection? Prove your answer.

(d) Deduce that $\mathbb{Z}$ cannot be isomorphic to $\mathbb{Q}$. (While it may seem difficult to believe, there are bijections from $\mathbb{Z}$ to $\mathbb{Q}$, but they cannot preserve the structure of $\mathbb{Z}$.)

Invariants

In Activity 26.9, we identified a few properties (e.g., mapping inverses to inverses) that isomorphisms must satisfy. We then used those properties in Activity 26.13 to show that the group $\mathbb{Z}$ is not isomorphic to the group $\mathbb{Q}$.

Any property that isomorphic groups must share is known as an *invariant*, or technically, an invariant of group isomorphisms. Invariants are properties that must be preserved by any isomorphism. Thus, if P is an invariant and a group G satisfies P, then every group to which G is isomorphic must also satisfy P. Consequently, if two groups differ with respect to an established invariant, then they cannot be isomorphic.

As an example, being cyclic is an invariant of a group isomorphism. That is, any group isomorphic to a cyclic group must also be cyclic. This was the basic idea in Activity 26.13, where we showed that $\mathbb{Z}$ is not isomorphic to $\mathbb{Q}$. Showing that two groups are different with respect to a particular invariant (and *any* invariant will do) is sufficient to establish that the two groups are not isomorphic. However, even if two groups agree with respect to every invariant we can think of, this does not prove that they are isomorphic. **Although invariants can be used to prove that two groups are different, they cannot be used to prove that two groups are the same. In order to prove that two groups are isomorphic, we must find an appropriate function from one group to the other, and then prove that this function is in fact an isomorphism.**

Table 26.1 lists some of the more common and useful invariants. This list, however, is far from complete, and we will add to it as needed throughout the remainder of our study of groups.

Isomorphism Classes

When we introduced the notation $G \cong H$ to denote the isomorphism relation between groups G and H, we claimed that this relation was in fact an equivalence relation on the set of all groups. The next theorem formalizes this result.

Theorem 26.14. *The relation of group isomorphism ($\cong$) is an equivalence relation on the set of all groups.*

Proof. To show that $\cong$ is an equivalence relation on the set of all groups, we must show that $\cong$ is reflexive, symmetric, and transitive.

To show that $\cong$ is a reflexive relation, we need to show that every group G is isomorphic to itself. Let G be a group. The natural map to use in this case is the identity map, which is always an isomorphism. (See Exercise (1).) So $G \cong G$, and $\cong$ is a reflexive relation.

To show that $\cong$ is a symmetric relation, let G and H be groups, and suppose that G and H are isomorphic—that is, suppose there is an isomorphism $\varphi : G \to H$. We need to show that there is also a corresponding isomorphism from H to G. Define $\varphi^{-1} : H \to G$ by $\varphi^{-1}(y) = x$ whenever $\varphi(x) = y$. To show that φ^{-1} is a group isomorphism, we will first show that φ^{-1} is well-defined. Suppose there is an element $y \in H$ so that $\varphi^{-1}(y) = x$ and $\varphi^{-1}(y) = x'$ for some $x, x' \in G$. Then $y = \varphi(x)$ and $y = \varphi(x')$. Since φ is an injection, it must be the case that $x = x'$. Therefore, φ^{-1} is well-defined.

Next we will show that φ^{-1} is an injection. Suppose $\varphi^{-1}(y) = \varphi^{-1}(y')$ for some $y, y' \in H$. Let $x = \varphi^{-1}(y) = \varphi^{-1}(y')$. Then $\varphi(x) = y$ and $\varphi(x) = y'$. Since φ is a function, we know $y = y'$. Therefore, φ^{-1} is an injection.

To show φ^{-1} is a surjection, let $x \in G$. Then let $y = \varphi(x)$. By definition, we have $\varphi^{-1}(y) = x$, so φ^{-1} is surjective.

To complete the proof that φ^{-1} is an isomorphism, we must show that φ^{-1} preserves the group operation. Let $y, y' \in H$, $x = \varphi^{-1}(y)$, and $x' = \varphi^{-1}(y')$. Then $\varphi(x) = y$ and $\varphi(x') = y'$. So $\varphi(xx') = \varphi(x)\varphi(x') = yy'$ and, by definition, $\varphi^{-1}(yy') = xx'$. Therefore,

$$\varphi^{-1}(yy') = xx' = \varphi^{-1}(y)\varphi^{-1}(y'),$$

and so φ^{-1} is operation-preserving. Therefore, φ^{-1} is an isomorphism, which verifies that $\cong$ is a symmetric relation.

To show that $\cong$ is a transitive relation, let G, H, and K be groups, and let $\varphi : G \to H$ and $\theta : H \to K$ be group isomorphisms. We must show that there is an isomorphism from G to K. The natural candidate is $\theta \circ \varphi$. We know that the composite of two bijective functions is also bijective. Therefore, $\theta \circ \varphi$ is a bijection, and we only need to show that $\theta \circ \varphi$ preserves the group operation. This argument is left as an exercise for the reader. (See Exercise (2).) Once it is completed, we will have established that $\theta \circ \varphi$ is a group isomorphism from G to K, and therefore $\cong$ is a transitive relation. Since $\cong$ is reflexive, symmetric, and transitive, it follows that $\cong$ is an equivalence relation.   ∎

As an equivalence relation, isomorphism partitions the set of all groups into disjoint equivalence classes—or, as we call them in this context, *isomorphism classes*. (See Theorem 2.6 on page 27.) In particular, the set of all groups isomorphic to a group G is called the isomorphism class of G. An important problem in group theory is to determine which isomorphism class a given group fits into.

We have already learned (or will soon learn) a few things about isomorphism classes of groups of certain orders.

- Activity 23.17 (see page 326) shows that every group of prime order is cyclic, and Theorem 26.18 will show that every cyclic group of order n is isomorphic to $\mathbb{Z}_n$. Consequently, every group of prime order p is isomorphic to $\mathbb{Z}_p$, and there is exactly one isomorphism class of groups of each prime order. Thus, we have classified up to isomorphism all groups of order 2, 3, 5, 7, 11, and so on.

- Corollary 24.13 (see page 341) tells us that if p and q are distinct primes, then any Abelian group G of order pq is cyclic. Again, Theorem 26.18 will show that G is isomorphic to $\mathbb{Z}_{pq}$. This tells us that, for distinct primes p and q, there is only one isomorphism class of Abelian groups of order pq. Consequently, any Abelian group of order 6 is isomorphic to $\mathbb{Z}_6$, any Abelian group of order 10 is isomorphic to $\mathbb{Z}_{10}$, any Abelian group of order 14 is isomorphic to $\mathbb{Z}_{14}$, and so on. Note, however, that Corollary 24.13 tells us nothing about non-Abelian groups of order pq.

Informally, when we say that there are k groups of a given order n, we mean that there are k isomorphism classes of groups of order n. So we would say that there is only one group of each prime order. Another way of saying this is that there are k groups of order n, *up to isomorphism*. Thus, we have classified up to isomorphism all groups of prime order and all Abelian groups of order pq, where p and q are distinct primes.

These results only begin to solve the problem of classifying finite groups. For example, neither of them tells us anything about groups of order 4, where things are just a bit more complicated. We know of one group of order 4 that is not cyclic—namely, $\mathbb{Z}_2 \oplus \mathbb{Z}_2$. We do know, however, that every group of order 4 is Abelian. (See Exercise (15) on page 330 of Investigation 23.) In Exercise (17), we will show that, up to isomorphism, there are only two groups of order 4—namely, $\mathbb{Z}_4$ and $\mathbb{Z}_2 \oplus \mathbb{Z}_2$.

In a later investigation, we will actually classify all finite Abelian groups. However, the general classification of finite non-Abelian groups is a very difficult (and unsolved) problem. In spite of its difficulty, significant progress has been made on this problem. For instance, Table 26.2 provides the number of isomorphism classes of groups of order 1 through 100, which is only a small portion of the information we know about the classification of finite groups. We will study some specific cases of the classification problem in the exercises and in later investigations.

For now, however, there is one classification of finite non-Abelian groups that we can complete.

Activity 26.15. Let p be an odd prime, and let G be a non-Abelian group of order $2p$ with identity e. In this activity, we will determine the isomorphism class of G.

(a) We know of one non-Abelian group of order $2p$. What is it?

(b) Exercise (2) of Investigation 18 (see page 274) shows that if $a^2 = e$ for all $a \in G$, then G is Abelian. Consequently, we can assume that there is an element $b \in G$ such that $b^2 \neq e$. Explain why $|b| = p$.

(c) Let $N = \langle b \rangle$. Since $|G|$ is even, Exercise (22) in Investigation 19 (see page 288) shows that every group of even order contains an element of order 2, so G contains an element a of order 2. Explain why $a \notin N$.

(d) Now explain why $G = N \cup aN$, and then explain why $G = \{e, b, b^2, \ldots, b^{p-1}, a, ab, ab^2, \ldots, ab^{p-1}\}$. What familiar group does G look like?

Table 26.2

The number $N(n)$ of isomorphism classes for groups of order n.

n	$N(n)$	n	$N(n)$	n	$N(n)$	n	$N(n)$
1	1	26	2	51	1	76	4
2	1	27	5	52	5	77	1
3	1	28	4	53	1	78	6
4	2	29	1	54	15	79	1
5	1	30	4	55	2	80	52
6	2	31	1	56	13	81	15
7	1	32	51	57	2	82	2
8	5	33	1	58	2	83	1
9	2	34	2	59	1	84	15
10	2	35	1	60	13	85	1
11	1	36	14	61	1	86	2
12	5	37	1	62	2	87	1
13	1	38	2	63	4	88	12
14	2	39	2	64	267	89	1
15	1	40	14	65	1	90	10
16	14	41	1	66	4	91	1
17	1	42	6	67	1	92	4
18	5	43	1	68	5	93	2
19	1	44	4	69	1	94	2
20	5	45	2	70	4	95	1
21	2	46	2	71	1	96	230
22	2	47	1	72	50	97	1
23	1	48	52	73	1	98	5
24	15	49	2	74	2	99	2
25	2	50	5	75	3	100	16

(e) Exercise (19) of Investigation 24 (see page 349) shows that every subgroup of index 2 in a group is a normal subgroup. So $N \triangleleft G$. Explain why $aba \in N$ and why $aba = b^i$ for some integer $2 \le i \le p-1$.

(f) Explain why $a = b^i a b^{-1}$, and then use the fact that $a^2 = e$ to show that $b^{i-1}a = ab^{-i+1}$.

(g) Recall that D_p has presentation $\langle r, R \mid r^2 = 1, R^p = 1, rR = R^{-1}r \rangle$. (See Investigation 21.) In Exercise (31), we will show that two groups with the same presentation are isomorphic. Explain why G also has the same presentation as D_p. What elements play the role of r and R? (Hint: What is $|b^{-i+1}|$ in G?)

(h) Explain how we have proved the following theorem:

Theorem 26.16. *Let p be an odd prime and G a non-Abelian group of order $2p$. Then $G \cong D_p$.*

Table 26.3
Known (to us) isomorphism classes of groups of orders 1 through 15.

n	Distinct groups of order n
1	$\{1\}$
2	$\mathbb{Z}_2$
3	$\mathbb{Z}_3$
4	$\mathbb{Z}_4$ and $\mathbb{Z}_2 \oplus \mathbb{Z}_2$
5	$\mathbb{Z}_5$
6	$\mathbb{Z}_6$ and D_3
7	$\mathbb{Z}_7$
8	$\mathbb{Z}_8, \mathbb{Z}_4 \oplus \mathbb{Z}_2, \mathbb{Z}_2 \oplus \mathbb{Z}_2 \oplus \mathbb{Z}_2, D_4, \mathbf{Q}$, and ?
9	$\mathbb{Z}_9, \mathbb{Z}_3 \oplus \mathbb{Z}_3$, and ?
10	$\mathbb{Z}_{10}$ and D_5
11	$\mathbb{Z}_{11}$
12	$\mathbb{Z}_{12}, \mathbb{Z}_6 \oplus \mathbb{Z}_2, D_6, A_4, T$ and ?
13	$\mathbb{Z}_{13}$
14	$\mathbb{Z}_{14}$ and D_7
15	$\mathbb{Z}_{15}$ and ?

Using Theorem 26.16, we can complete our classification of the groups of order 6, 10, and 14. Earlier we argued that there is exactly one Abelian group of each of these orders. Theorem 26.16 now shows that there is exactly one non-Abelian group of each of these orders, namely D_3, D_{10}, and D_{14}, which verifies those entries in Table 26.2. We now have completely classified all groups of orders 1 through 7, 10, 11, 13, and 14, as shown in Table 26.3. Exercise (27) shows that there are at least five different isomorphism classes of groups of order 8, while Exercise (28) tells us that there are at least five different isomorphism classes of groups of order 12. We will complete the classification of groups of order 8, 12, and 15 in later investigations. Since there are 14 groups of order 16, we will stop at the groups of order 15.

Isomorphisms and Cyclic Groups

When we study groups, we are interested in the structure that the group operation imposes on the set. We don't really care what the elements of the group look like—only how they interact with each other. When two groups are isomorphic, corresponding elements behave the same way, so for all intents and purposes, the isomorphic groups are the same (except for how the elements are represented). We often talk about classifying a group G, which means determining to which set of groups G is isomorphic. In this section, we will show that any cyclic group of finite order n is isomorphic to $\mathbb{Z}_n$. As a result, there is only one finite cyclic group of each order, up to isomorphism. The infinite case is left for the exercises.

Activity 26.17. Let $G = \langle a \rangle$ be a cyclic group of order $n \in \mathbb{Z}^+$. To show G is isomorphic to $\mathbb{Z}_n$, we need to construct a function $\varphi : G \to \mathbb{Z}_n$ that is an isomorphism.

(a) What is the form of an arbitrary element in G?

(b) There is a natural mapping φ from G to $\mathbb{Z}_n$. Based on your answer to part (a), write down an equation to precisely define this mapping.

(c) Show that the mapping φ you defined in part (b) is a well-defined function. Why is this step necessary (and important)?

(d) Prove that φ is a surjection.

(e) Prove that φ is an injection. (Hint: First, review the definition of an injection. Then, consider what $[k] = [l]$ in $\mathbb{Z}_n$ tells us about the relationship between k and l.)

(f) Explain how parts (b) – (e) establish that φ is an isomorphism, and thus G is isomorphic to $\mathbb{Z}_n$.

The result of Activity 26.17 is the following:

Theorem 26.18. *Any finite cyclic group of order n is isomorphic to $\mathbb{Z}_n$.*

Just as in the finite case, it is also true that, up to isomorphism, there is only one infinite cyclic group—namely, $\mathbb{Z}$. The next theorem states this result formally.

Theorem 26.19. *If G is an infinite cyclic group, then $G \cong \mathbb{Z}$.*

The proof of Theorem 26.19 is left for the exercises. (See Exercise (18).)

Cayley's Theorem

Group theory originated as the study of permutations, but we did not begin *our* study of group theory that way. Although the permutations of a set form a group, the abstract definition of a group may seem somewhat removed from what we know about permutations. In this section, we will study Cayley's Theorem, which establishes that all group theory can in fact be considered as the study of permutations. Consequently, if we could completely understand permutation groups, then we would completely understand all groups. (Of course, we are currently far from such an understanding!)

Preview Activity 26.20. To understand Cayley's Theorem, we need to understand how a group element can act as a permutation on a group. Let G be a group with identity e, and let $a \in G$. Define

$$T_a : G \to G \quad \text{by} \quad T_a(g) = ag.$$

(a) Illustrate the definition of T_a by determining $T_a(x)$ for each $a, x \in U_8$.

(b) What properties does the function T_a seem to possess? For example, is T_a an injection? Does T_a preserve the structure of U_8? List and verify as many properties as you can.

Cayley's Theorem is named after the mathematician Arthur Cayley and provides the surprising result that *every* group can be viewed as a group of permutations. Preview Activity 26.20 illustrates a method by which we can identify each element of a group with a permutation of the elements of that group. We will now investigate this construction in general.

Activity 26.21. Let G be a group with identity e, and let $a \in G$. Define

$$T_a : G \to G \text{ by } T_a(g) = ag. \tag{26.1}$$

(a) Is T_a an injection? Verify your answer.

(b) Is T_a a surjection? Verify your answer.

(c) What specific function is T_e? Explain.

(d) If $b \in G$, the composite $T_a T_b$ has the form T_c for some $c \in G$. For which c is this true? Prove your answer.

(e) What is the relationship between T_a and $T_{a^{-1}}$? Prove your conjecture.

Activity 26.21 shows us that for each $a \in G$, the function T_a is a permutation of G. Now we can see how G itself can be viewed as a subgroup of a group of permutations.

Activity 26.22. Let $P(G)$ be the collection of all permutations of G. Let $\Pi(G) = \{T_a : a \in G\}$. Explain why $\Pi(G)$ is a subgroup of $P(G)$.

Finally, let $\Theta : G \to \Pi(G)$ be defined by $\Theta(a) = T_a$.

Activity 26.23. Illustrate the definition of Θ by describing $\Theta : U_8 \to \Pi(U_8)$. (Hint: One way to do this is to specify what $\Theta(g)$ is for each $g \in U_8$.)

The major result of this section is the following:

Theorem 26.24 (Cayley's Theorem). *Every group is a subgroup of a group of permutations.*

Proof. Let G be a group with identity e, and define $\Theta : G \to \Pi(G)$ as above. We will show that Θ is an isomorphism.

First, we will show that Θ is an injection. Suppose $\Theta(a) = \Theta(b)$ for some $a, b \in G$. Then $T_a = T_b$. This means $T_a(e) = T_b(e)$ or $ae = be$. Thus, $a = b$, and so Θ is injective. Next, we will demonstrate that Θ is a surjection. If $T_a \in \Pi(G)$, then $\Theta(a) = T_a$. Therefore, the range of Θ is $\Pi(G)$, and so Θ is surjective. To complete the proof, we must show that Θ preserves the operation in G. Let $a, b \in G$. In part (d) of Activity 26.21, we showed that $T_a T_b = T_{ab}$. Therefore,

$$\Theta(a)\Theta(b) = T_a T_b = T_{ab} = \Theta(ab).$$

It follows that Θ is an isomorphism of groups, and so $G \cong \Pi(G)$. Thus, we have shown that G is isomorphic to a group of permutations. ∎

One corollary of Cayley's Theorem is the following:

Corollary 26.25. *If G is a finite group of order n, then G is isomorphic to a subgroup of the symmetric group S_n.*

Proof. The proof of this corollary is really just a matter of demonstrating that if T is a set with n elements, then $P(T) \cong S_n$. (See Exercise (35).) Then G is isomorphic to a subgroup of $P(G)$, which is then isomorphic to a subgroup of S_n. The transitivity of the isomorphism relation then shows that G is isomorphic to a subgroup of S_n. ∎

Corollary 26.25 can be useful in that it allows us to concretely represent any abstract finite group as a subgroup of S_n. The next activity illustrates how this representation works.

Activity 26.26. Let G be the group of order 4 with presentation $\langle a, b \mid a^2 = b^2 = (ab)^2 = 1 \rangle$. (See Activity 21.6 on page 303.) This group has the operation table:

	1	a	b	ab
1	1	a	b	ab
a	a	1	ab	b
b	b	ab	1	a
ab	ab	b	a	1

Identify the elements in G with the set $\{1, 2, 3, 4\}$ by $1 \leftrightarrow 1$, $a \leftrightarrow 2$, $b \leftrightarrow 3$, and $ab \leftrightarrow 4$.

(a) Recall that each element $g \in G$ is identified with the permutation $T_g \in P(G)$, where T_g is defined as in (26.1). Complete the table below to determine the actions of T_g for each $g \in G$.

x	1	a	b	ab
$T_1(x)$	1	a	b	ab
$T_a(x)$				
$T_b(x)$				
$T_{ab}(x)$				

(b) Use the identifications $1 \leftrightarrow 1$, $a \leftrightarrow 2$, $b \leftrightarrow 3$, and $ab \leftrightarrow 4$ to identify each T_g with an element in S_4. Then find a subgroup of S_4 isomorphic to G.

(c) Let S be the subgroup you found in part (b). Find a specific isomorphism from G to S.

Concluding Activities

Activity 26.27. Let G be a group with identity element e, and let K and N be normal subgroups of G with $K \cap N = \{e\}$. As we will demonstrate in this activity, $(K \times N) \cong (K \oplus N)$. Because of this isomorphism, we can identify an internal direct product with the corresponding external direct product, and so it doesn't matter which product we use.

To show that $(K \times N) \cong (K \oplus N)$, we need to exhibit an isomorphism between the two groups. The natural mapping to try is $\varphi : (K \times N) \to (K \oplus N)$ defined by

$$\varphi(kn) = (k, n).$$

To complete this activity, you may want to recall that the representation of an element in $K \times N$ as kn for some $k \in K$ and $n \in N$ is unique, and that if $k \in K$ and $n \in N$ then $kn = nk$. (See Theorem 25.15 on page 361.)

(a) Explain why φ is well-defined.

(b) Prove that φ is an isomorphism.

Activity 26.28. In Activity 26.13, we showed that if there were an isomorphism $\varphi : \mathbb{Z} \to \mathbb{Q}$, then φ would be completely determined by its action on 1, a generator of $\mathbb{Z}$. In this activity, we will show that this property is true for arbitrary cyclic groups. Let $G = \langle a \rangle$ be a cyclic group, and let $\varphi : G \to H$ be a group isomorphism. Show that the element $\varphi(a)$ completely determines the isomorphism φ.

Exercises

* (1) If G is a group, prove that the identity map $I : G \to G$ defined by $I(a) = a$ for all $a \in G$ is a group isomorphism.

* (2) Let G, H, and K be groups, and let $\varphi : G \to H$ and $\theta : H \to K$ be group isomorphisms. Prove that $\theta \circ \varphi$ preserves structure.

(3) Let G be a group. In Exercise (1), we showed that the identity map $I : G \to G$ is an isomorphism. In this exercise, we will determine when the function $\varphi : G \to G$ defined by $\varphi(a) = a^{-1}$ is an isomorphism.

    (a) Is φ an isomorphism when $G = \mathbb{Z}_3$?

    (b) Is φ an isomorphism when $G = D_3$?

    (c) Determine necessary and sufficient conditions on G so that φ is an isomorphism. Prove your answer.

(4) Let G be a group. In this exercise, we will consider when the mapping $\varphi : G \to G$ defined by $\varphi(a) = a^2$ is an isomorphism.

    (a) Is φ an isomorphism if $G = \mathbb{Z}_3$? Explain.

    (b) Is φ an isomorphism if $G = \mathbb{Z}_4$? Explain.

    (c) Suppose G is a finite Abelian group and G has no element of order 2. Show that φ is an isomorphism. Then show by example that if G is infinite, then φ need not be an isomorphism.

    (d) If G is a finite Abelian group and φ is an isomorphism, must G contain no element of order 2? Prove your answer.

(5) Let G be a group. In this exercise, we will investigate isomorphisms $\varphi : G \to G$ of the form $\varphi(a) = a^n$ for some integer n.

    (a) Show that the function $\varphi : U_{22} \to U_{22}$ defined by $\varphi(a) = a^3$ is an isomorphism of groups.

(b) Show that the function $\varphi : U_{22} \to U_{22}$ defined by $\varphi(a) = a^4$ is not an isomorphism of groups.

(c) Let G be a finite Abelian group and n a positive integer that is relatively prime to $|G|$. Show that the mapping $\varphi : G \to G$ defined by $\varphi(a) = a^n$ is an isomorphism.

(d) Let G be a finite Abelian group, and suppose $\varphi : G \to G$ defined by $\varphi(a) = a^n$ for some integer n is an isomorphism. Must $\gcd(n, |G|) = 1$? Prove your answer.

(6) To what familiar group is the group G_1 from Preview Activity 26.1 isomorphic? Prove your answer.

(7) Prove that commutativity is an invariant. That is, prove that if G and H are isomorphic groups and G is Abelian, then H must also be Abelian.

(8) Prove that the order of each element in a group is an invariant. That is, prove that if $\varphi : G \to H$ is an isomorphism between the groups G and H and $a \in G$, then the order of $\varphi(a)$ in H is the same as the order of a in G.

* (9) In Exercise (8), we showed that the order of each element in a group is an invariant. In this exercise, we will show that the number of elements of a given order in a group is also an invariant. We will then explore the converse; that is, we will consider whether two groups can have the same number of elements of each order and yet not be isomorphic.

(a) Let G be a finite group, k a positive integer, and $\nu_G(k)$ the number of elements of order k in G. Prove that if G and G' are isomorphic groups, then $\nu_G(k) = \nu_{G'}(k)$ for every positive integer k.

(b) Now we want to determine if the converse is true. That is, if G and G' are groups of the same order and $\nu_G(k) = \nu_{G'}(k)$ for every positive integer k, must it be true that

$$G \equiv G'? \text{ Let } p \text{ be an odd prime, and let } G = \left\{ \begin{bmatrix} [1] & [a] & [b] \\ [0] & [1] & [c] \\ [0] & [0] & [1] \end{bmatrix} : [a], [b], [c] \in \mathbb{Z}_p \right\}.^*$$

(i) Show that G is a group of order p^3. You may assume that $\mathrm{GL}_3(\mathbb{Z}_p) = \{A \in M_{3\times 3}(\mathbb{Z}_p) : \det(A) \neq 0\}$ is a group under standard matrix multiplication.

(ii) Show that the order of every nonidentity element in G is p. (Hint: Show that

$$\begin{bmatrix} [1] & [a] & [b] \\ [0] & [1] & [c] \\ [0] & [0] & [1] \end{bmatrix}^n = \begin{bmatrix} [1] & n[a] & \frac{n(n-1)}{2}[ac] + n[b] \\ [0] & [1] & n[c] \\ [0] & [0] & [1] \end{bmatrix}$$

for every positive integer n.)

(iii) Is it true that if G and G' are groups of the same order and $\nu_G(k) = \nu_{G'}(k)$ for every positive integer k, then $G \cong G'$? Prove your answer.

(10) Prove that being cyclic is an invariant. That is, prove that if G and H are isomorphic groups and G is cyclic, then H must also be cyclic.

(11) **Direct sums of $\mathbb{Z}_m$ and $\mathbb{Z}_n$.** Under what circumstances is $\mathbb{Z}_n \oplus \mathbb{Z}_m$ isomorphic to $\mathbb{Z}_{mn}$? State (and prove) your answer in the form of a biconditional (if and only if) statement.

---

*This group G is called the Heisenberg group modulo p (named after the Nobel Prize winning physicist Werner Heisenberg). The Heisenberg group (with real entries) is related to the Heisenberg Uncertainty Principle in quantum physics.

(12) Explain why the indicated groups are *not* isomorphic.

(a) $\mathbb{Z}_6$ and S_3

(b) $\mathbb{Z}_4 \oplus \mathbb{Z}_2$ and D_4

(c) $\mathbb{Z}_4 \oplus \mathbb{Z}_2$ and $\mathbb{Z}_2 \oplus \mathbb{Z}_2 \oplus \mathbb{Z}_2$

(d) $\mathbb{Z}$ and $\mathbb{R}$

(e) $\mathbb{R}^* = U(\mathbb{R})$ and $\mathbb{R}$

(f) $\mathbb{Q}$ and $\mathbb{R}$ (Hint: If s and t are nonzero rational numbers, show that there are nonzero integers m, n such that $ms = nt$.)

(13) One of the following statements is true, and the other is false. Which is which? Prove your answer.

(a) $\mathbb{Q}$ is isomorphic to the multiplicative group $\mathbb{Q}^+$ of positive rational numbers.

(b) $\mathbb{R}$ is isomorphic to the multiplicative group $\mathbb{R}^+$ of positive real numbers.

(14) Let $\varphi : U_7 \to \mathbb{Z}_6$ be an isomorphism such that $\varphi([3]_7) = [5]_6$. Find $\varphi([a]_7)$ for every $[a]_7 \in U_7$.

(15) Let $H = \left\{ \begin{bmatrix} 1+n & -n \\ n & 1-n \end{bmatrix} : n \in \mathbb{Z} \right\}$.

(a) Prove that H is a group under matrix multiplication.

(b) To what familiar group is H isomorphic? Prove your answer.

(16) Determine if the given groups are isomorphic. Prove your answers.

(a) U_5 and U_{10}

(b) U_{20} and U_{24}

* (17) **Groups of order 4.** In this exercise, we will classify all groups of order 4. Let G be a group of order 4 with identity e, and let a be a nonidentity element in G.

(a) What must be true about G if $|a| = 4$?

(b) If $|a| \neq 4$, what are the possible values for $|a|$? Why?

(c) If $a^2 = e$, then there must be another nonidentity element b in G. What, then, is the fourth element in G?

(d) What can we say about the relationship between ab and ba? Explain. What kind of group is G in this case?

(e) Explain why there are (up to isomorphism) exactly two groups of order 4 ($\mathbb{Z}_4$ and $\mathbb{Z}_2 \oplus \mathbb{Z}_2$).

* (18) Prove Theorem 26.19.

(19) Explain why any two cyclic groups of the same order are isomorphic to each other.

(20) Is it possible for a group to be isomorphic to one of its proper subgroups? Prove your answer.

(21) Let k and n be positive integers with $n > 1$. Recall that

$$k\mathbb{Z}_n = \{k[x] : [x] \in \mathbb{Z}_n\}.$$

   (a) Find (and prove) a necessary and sufficient condition for $k\mathbb{Z}_n$ to be isomorphic to $\mathbb{Z}_n$.

   (b) In general, to what familiar group is $k\mathbb{Z}_n$ isomorphic? Prove your answer.

(22) (a) Let G_1 and G_2 be groups. Prove that $(G_1 \oplus G_2) \cong (G_2 \oplus G_1)$.

   (b) Generalize part (a) to any number of factors to show that we can rearrange the factors in a direct product in any order and still have a group isomorphic to the original. A rearrangement of the order of the factors is just a permutation of the factors, so we can restate this problem more formally as follows. Let G_1, G_2, ..., and G_n be groups for some integer $n \geq 2$. For any $\sigma \in S_n$, prove that

$$\left(G_1 \oplus G_2 \oplus \cdots G_n\right) \cong \left(G_{\sigma(1)} \oplus G_{\sigma(2)} \oplus \cdots G_{\sigma(n)}\right).$$

(23) In this exercise, we will generalize the result of Activity 26.27 to direct products with any finite number of factors. Let G be a group with identity element e, and let $N_1, N_2, \ldots, N_m$ be normal subgroups of G for some integer $m \geq 2$ such that

$$N_i \cap (N_1 \times N_2 \times \cdots \times N_{i-1} \times N_{i+1} \times \cdots \times N_m) = \{e\}$$

for each i. Prove that

$$(N_1 \times N_2 \times \cdots \times N_m) \cong (N_1 \oplus N_2 \oplus \cdots \oplus N_m).$$

(24) Let φ be the Euler phi function, as defined in Exercise (15) on page 297 and Exercise (24) on page 332. Recall that $\varphi(n)$ is equal to the number of positive integers less than or equal to n that are relatively prime to n.

   (a) Show by example that $\varphi(mn)$ is not always equal to $\varphi(m)\varphi(n)$.

   (b) Let s and t be positive integers with $\gcd(s,t) = 1$. Prove that $U_{st} \cong (U_s \oplus U_t)$.

   (c) Show that if m and n are positive integers with $\gcd(m,n) = 1$, then $\varphi(mn) = \varphi(m)\varphi(n)$. (Functions that have this property are called *multiplicative functions*.)

(25) Prove, by comparing orders of elements, that $\mathbb{Z}_8 \oplus \mathbb{Z}_2$ is not isomorphic to $\mathbb{Z}_4 \oplus \mathbb{Z}_4$.

(26) Let $n = 2k$ be a positive even integer. In Exercise (2) of Investigation 21 (see page 304), we showed that $Z(D_n) = \langle R^k \rangle$. To what familiar group is $D_n/Z(D_n)$ isomorphic? Prove your answer.

* (27) We have seen several groups of order 8: $\mathbb{Z}_8$, $\mathbb{Z}_4 \oplus \mathbb{Z}_2$, $\mathbb{Z}_2 \oplus \mathbb{Z}_2 \oplus \mathbb{Z}_2$, D_4, and the quaternions $\mathbf{Q}$. (See Exercise (21) on page 350 of Investigation 24 and Exercise (6) on page 333 of Investigation 21.)

   (a) Determine the number of elements of each order in each of these groups.

   (b) Based on your work in part (a), separate these groups into their potential isomorphism classes. Can you be completely sure of your identifications based only on the work from part (a)? If so, explain why. If not, why not? Which of your identifications can you be sure of, and why?

* (28) We have seen several groups of order 12: A_4, D_6, *Tetra* (see Exercise (4) in Investigation 21), T (see Exercise (18) of Investigation 21), $D_3 \oplus \mathbb{Z}_2$, $\mathbb{Z}_{12}$, and $\mathbb{Z}_6 \oplus \mathbb{Z}_2$.

    (a) Determine the number of elements of each order in each of these groups.

    (b) Based on your work in part (a), separate these groups into their potential isomorphism classes. Can you be completely sure of your identifications based only on the work from part (a)? If so, explain why. If not, why not? Which of your identifications can you be sure of, and why?

(29) (a) Let G_1, G_2, H_1, and H_2 be groups with $G_1 \cong H_1$ and $G_2 \cong H_2$. Prove that $(G_1 \oplus G_2) \cong (H_1 \oplus H_2)$.

    (b) Does part (a) generalize to any finite number of groups? If so, formally state a claim and then prove it. If not, give a counterexample.

(30) Let $H = \langle 3 \rangle$ and $K = \langle 12 \rangle$ in $\mathbb{Z}$.

    (a) To what familiar group is $\langle 3 \rangle / \langle 12 \rangle$ isomorphic? Prove your answer.

    (b) To what familiar group is $\langle 8 \rangle / \langle 48 \rangle$ isomorphic? Prove your answer.

    (c) Generalize your work in parts (a) and (b) to arbitrary integers k and m. To what familiar group is $\langle k \rangle / \langle m \rangle$ isomorphic if k divides m? Prove your answer.

* (31) Let G and H be two groups with presentations of the form

$$\langle a_1, a_2, \ldots, a_n : r_1 = r_2 = \cdots = r_m = 1 \rangle.$$

In other words, G is generated by elements g_1, g_2, ..., g_n satisfying the relations $r_1 = r_2 = \cdots = r_m = 1$ (replacing each a_i in these relations with g_i), and H is generated by elements h_1, h_2, ..., h_n satisfying the relations $r_1 = r_2 = \cdots = r_m = 1$ (replacing each a_i in these relations with h_i). Prove that $G \cong H$.

(32) Prove that a group G of order 4 with presentation $\langle a, b \mid a^2 = b^2 = (ab)^2 = 1 \rangle$ (see Activity 21.6 on page 303 and Activity 26.26) is isomorphic to the group $\mathbb{Z}_2 \oplus \mathbb{Z}_2$. (This group $\mathcal{V} = \mathbb{Z}_2 \oplus \mathbb{Z}_2$—and any group isomorphic to it—is called the Klein 4-group. The name comes from the German "Viergruppe," found in 1884 in the paper *Vorlesungen über das Ikosaeder und die Aufloesung der Gleichungen vom funften Grade* by Felix Klein.)

(33) Let G be a group. An isomorphism from G to G is called an **automorphism** and the set of automorphisms of G is denoted Aut(G). That is,

$$\text{Aut}(G) = \{ \varphi : G \to G : \varphi \text{ is an isomorphism} \}.$$

    (a) Prove that there are exactly two automorphisms of $\mathbb{Z}$. (Hint: Activity 26.9 should be helpful.)

    (b) Prove that Aut(G) is a group under the operation of composition of functions.

    (c) To what familiar group is Aut($\mathbb{Z}$) isomorphic? Explain.

* (34) Exercise (33) shows that Aut(G) is a group whenever G is a group. In this exercise, we will determine Aut($\mathbb{Z}_n$). Let n be a positive integer.

    (a) Suppose that $\varphi \in \text{Aut}(\mathbb{Z}_n)$. What is $|\varphi([1])|$? What, specifically, does this result tell us about which elements $\varphi([1])$ could be in $\mathbb{Z}_n$? Explain.

(b) For each of the possibilities you found for $\varphi([1])$ in part (a), which define an automorphism of $\mathbb{Z}_n$? Prove your answer.

(c) Prove that $\text{Aut}(\mathbb{Z}_n) \cong U_n$.

$\star$ (35) Prove that if T is a set with n elements, then the permutation group $P(T)$ of T is isomorphic to the symmetric group S_n. (Hint: Label the elements in T in some order as $t_1, t_2, \ldots, t_n$. Then connect a permutation of T to a permutation of the subscripts.)

(36) The corollary to Cayley's Theorem tells us that if G is a group of order 8, then G is isomorphic to a subgroup of S_8. For each of the following groups, find a subgroup of S_8 to which the group is isomorphic. (Recall that $\mathbf{Q}$ denotes the group of quaternions.)

$$D_4 \quad \mathbb{Z}_8 \quad \mathbb{Z}_4 \oplus \mathbb{Z}_2 \quad \mathbb{Z}_2 \oplus \mathbb{Z}_2 \oplus \mathbb{Z}_2 \quad \mathbf{Q}$$

Connections

In this investigation, we studied isomorphisms of groups. If you studied ring theory before group theory, you should notice connections between isomorphisms of groups in this investigation and isomorphisms of rings in Investigation 7. The idea is the same in both contexts; isomorphic groups (or rings) are essentially the same, and an isomorphism is a bijection that preserves the underlying algebraic structure. Since there is only one operation defined in a group but two operations in a ring, the major difference between isomorphisms of groups and isomorphisms of rings is that an isomorphism of groups must preserve one operation, whereas an isomorphism of rings must preserve two operations. The process of verifying an isomorphism is the same in both contexts, but there is an extra step required for isomorphisms of rings.

Investigation 27

Homomorphisms and Isomorphism Theorems

Focus Questions

By the end of this investigation, you should be able to give precise and thorough answers to the questions listed below. You may want to keep these questions in mind to focus your thoughts as you complete the investigation.

- What is a homomorphism of groups, and how is a homomorphism different than an isomorphism?

- What are monomorphisms and epimorphisms of groups?

- What are the kernel and image of a group homomorphism, and what properties do they satisfy?

- What are the isomorphism theorems for groups, and how do they use homomorphisms to establish relationships between groups?

Preview Activity 27.1. As we saw in Investigation 26, the notion of isomorphism formalizes what it means for two groups to be essentially the same. Recall that an isomorphism of groups is a bijective, structure-preserving function. In group theory, structure-preserving maps are important even if they are not bijections. In this activity, we will explore three different kinds of structure-preserving functions. (Throughout the activity, recall that we use the notation $[k]_n$ to denote the congruence class of k in $\mathbb{Z}_n$.)

(a) Is the function $\varphi : \mathbb{Z}_3 \to \mathbb{Z}_6$ defined by $\varphi([k]_3) = [4k]_6$ structure-preserving? Is φ an injection? Is φ a surjection? Verify your answers. (You may assume that φ is well-defined.)

(b) Is the function $\varphi : \mathbb{Z}_6 \to \mathbb{Z}_3$ defined by $\varphi([k]_6) = [k]_3$ structure-preserving? Is φ an injection? Is φ a surjection? Verify your answers. (You may assume that φ is well-defined.)

(c) Is the function $\varphi : \mathbb{Z}_6 \to \mathbb{Z}_4$ defined by $\varphi([k]_6) = [2k]_4$ structure-preserving? Is φ an injection? Is φ a surjection? Verify your answers. (You may assume that φ is well-defined.)

Homomorphisms

Preview Activity 27.1 illustrates that it is possible to have structure-preserving maps that are injections but not surjections, surjections but not injections, or neither surjections nor injections. When we study groups, we are mostly interested in maps that preserve the group structure or operation. Such maps—whether they are injective, surjective, neither, or both—are called *homomorphisms*, defined formally as follows:

Definition 27.2. Let G and H be groups. A function φ from G to H is a **homomorphism** of groups if

$$\varphi(ab) = \varphi(a)\varphi(b)$$

for all $a, b \in G$.

Just like isomorphism, the word *homomorphism* comes from two Greek words: *homos*, which means *similar* or *like*, and *morphe*, which means *form* or *structure*. Thus, when there is a homomorphism from one group to another, it means that there is some similarity of structure between the two groups. Just like an isomorphism, a homomorphism is an *operation-preserving* or a *structure-preserving* function, but not necessarily a bijection.

Although it is not a requirement, some homomorphisms are also injections, surjections, or bijections (as seen in Preview Activity 27.1). Homomorphisms that satisfy these additional properties are given special names. In particular,

- a **monomorphism** is an injective homomorphism;

- an **epimorphism** is a surjective homomorphism; and

- an **isomorphism** is a bijective homomorphism.

If $\varphi : G \to G'$ is an epimorphism, we call G' a **homomorphic image** of G.

Activity 27.3. Determine whether each of the following functions is a homomorphism from G to H. If a function is a homomorphism, decide whether it is a monomorphism, an epimorphism, an isomorphism, or none of these.

(a) $G = \mathbb{Z}$, $H = \mathbb{Z}_5$, and $\varphi(k) = [k]_5$

(b) $G = \mathbb{Z}_3$, $H = \mathbb{Z}_{18}$, and $\varphi([k]_3) = [6k]_{18}$

(c) $G = \mathbb{Z}$, $H = \mathbb{Z}_2 \oplus \mathbb{Z}_4$, and $\varphi(k) = ([k]_2, [k]_4)$

(d) $G = \mathbb{R}^+$, $H = \mathbb{R}^+$, and $\varphi(k) = \sqrt{k}$

(e) $G = U_{12}$, $H = \mathbb{Z}_6$, and $\varphi([k]_{12}) = [k]_6$

The Kernel of a Homomorphism

In Investigation 26, we showed that every isomorphism between two groups G and H maps the identity in G to the identity in H. We also saw that isomorphisms map inverses to inverses. The same proofs show that these properties are also true for homomorphisms in general.

Theorem 27.4. *Let G and H be groups with identities e_G and e_H, respectively, and let $\varphi : G \to H$ be a homomorphism. Then:*

(i) $\varphi(e_G) = e_H$.

(ii) *If $a \in G$, then $\varphi(a^{-1}) = (\varphi(a))^{-1}$.*

Proof. Let G and H be groups with identities e_G and e_H, respectively, and let $\varphi : G \to H$ be a homomorphism. Then

$$\varphi(e_G)\varphi(e_G) = \varphi(e_G e_G) = \varphi(e_G) = e_H \varphi(e_G),$$

and cancellation shows that $\varphi(e_G) = e_H$. So φ must map the identity element of G to the identity element of H.

Now let $a \in G$. To show that $\varphi(a^{-1}) = (\varphi(a))^{-1}$, note that

$$e_H = \varphi(e_G) = \varphi\left(aa^{-1}\right) = \varphi(a)\varphi\left(a^{-1}\right).$$

Therefore, $(\varphi(a))^{-1} = \varphi(a^{-1})$. ∎

Let $\varphi : G \to H$ be a homomorphism of groups. By part (i) of Theorem 27.4, we know that φ maps the identity in G to the identity in H. If φ is a monomorphism—that is, if φ is also injective—then φ maps *only* the identity to the identity. If φ is not a monomorphism, we can measure (in a sense) how close φ is to being a monomorphism by determining the number of elements φ maps to the identity. This important idea leads us to the next definition.

Definition 27.5. Let $\varphi : G \to H$ be a homomorphism of groups, and let e_H be the identity element in H. The **kernel** of φ is the set

$$\text{Ker}(\varphi) = \{a \in G : \varphi(a) = e_H\}.$$

You may have seen an idea related to the kernel of a homomorphism in linear algebra, where the set of objects that are sent to the zero vector under a matrix transformation is called the *null space* of the matrix. In this sense, the notion of the kernel of a homomorphism is not entirely new.

Activity 27.6. Find $\text{Ker}(\varphi)$ for each of the homomorphisms in Activity 27.3.

Since $\text{Ker}(\varphi)$ is a subset of G, it is natural to ask if $\text{Ker}(\varphi)$ is a subgroup of G.

Activity 27.7. Let G and H be groups with identities e_G and e_H, respectively, and let $\varphi : G \to H$ be a homomorphism.

(a) Is e_G in $\text{Ker}(\varphi)$? Explain.

(b) Is $\text{Ker}(\varphi)$ closed under the operation in G? Prove your answer.

(c) If $a \in \text{Ker}(\varphi)$, is $a^{-1} \in \text{Ker}(\varphi)$? Prove your answer.

(d) Is $\text{Ker}(\varphi)$ a subgroup of G? Explain.

The next theorem tells us an important fact about the kernel of a group homomorphism.

Theorem 27.8. *Let G and H be groups, and let $\varphi : G \to H$ be a group homomorphism. Then $\text{Ker}(\varphi)$ is a normal subgroup of G.*

Proof. Let G and H be groups, and let $\varphi : G \to H$ be a group homomorphism. Let e_G and e_H be the identities for G and H, respectively. Let $K = \text{Ker}(\varphi)$. To show that K is a normal subgroup of G, we will show that $aKa^{-1} \subseteq K$ for all $a \in G$. Let $a \in G$. We then need to show $aka^{-1} \in K$ for every $k \in K$. Now let $k \in K$. Then $\varphi(k) = e_H$. So

$$\varphi(aka^{-1}) = \varphi(a)\varphi(k)\varphi(a^{-1}) = \varphi(a)e_H(\varphi(a))^{-1} = \varphi(a)(\varphi(a))^{-1} = e_H,$$

and $aka^{-1} \in K$. Thus, $aKa^{-1} \subseteq K$ for every $a \in G$, and so $K = \text{Ker}(\varphi)$ is a normal subgroup of G. ∎

The Image of a Homomorphism

Let $\varphi : G \to H$ be a homomorphism of groups G and H with identities e_G and e_H, respectively. As we have seen, $\varphi(e_G) = e_H$, and so e_H is always the image of the identity element under any homomorphism. If φ is an epimorphism, then every element in H is the image under φ of some element from G. If φ is not an epimorphism, then some elements in H are "missed" by φ—that is, they are not the images of any elements from G. The size of the set of images of elements in G under φ measures (in some sense) how close φ is to being an epimorphism. This leads us to the next definition.

Definition 27.9. Let $\varphi : G \to H$ be a homomorphism of groups. The **image** of φ is the set

$$\text{Im}(\varphi) = \{\varphi(a) : a \in G\}.$$

Activity 27.10. Find $\text{Im}(\varphi)$ for each of the homomorphisms in Activity 27.3.

Since $\text{Im}(\varphi)$ is a subset of H, it is natural to ask if $\text{Im}(\varphi)$ is a subgroup of H.

Activity 27.11. Let G and H be groups with identities e_G and e_H, respectively, and let $\varphi : G \to H$ be a homomorphism.

(a) Is e_H in $\text{Im}(\varphi)$? Explain.

(b) Is $\text{Im}(\varphi)$ closed under the operation in H? Prove your answer.

(c) If $y \in \text{Im}(\varphi)$, is $y^{-1} \in \text{Im}(\varphi)$? Prove your answer.

(d) Is $\text{Im}(\varphi)$ a subgroup of H? Explain.

The Isomorphism Theorems for Groups

Sometimes homomorphisms can be used to recognize isomorphic groups, even if the homomorphisms themselves are not isomorphisms. Four theorems, called the *isomorphism theorems* for groups, formalize this idea. The First Isomorphism Theorem connects the kernel and image of a homomorphism. The Second Isomorphism Theorem connects intersections and products of subgroups. The Third Isomorphism Theorem provides a sort of "cancellation" rule for quotient groups. Finally, the Fourth Isomorphism Theorem establishes a correspondence between certain subgroups of a group and one of its quotients.

The First Isomorphism Theorem for Groups

Preview Activity 27.12. Let $G = \mathbb{Z}_{24}$, $H = \mathbb{Z}_8$, and let $\varphi : G \to H$ be defined by $\varphi([m]_{24}) = [6m]_8$.

(a) Show that φ is well-defined. Why is this necessary?

(b) Show that φ is a homomorphism of groups. Is φ a monomorphism? Is φ an epimorphism? Is φ an isomorphism? Explain.

(c) Let $K = \text{Ker}(\varphi)$. Find all the elements of K.

(d) Determine the elements of the group G/K. Is G/K Abelian? Is G/K cyclic? Explain.

(e) Let $R = \text{Im}(\varphi)$. Find all the elements of R.

(f) What specific relationship is there between the groups G/K and R? Explain.

Preview Activity 27.12 shows that in one example where $\varphi : G \to H$ is a group homomorphism, we have $G/\text{Ker}(\varphi) \cong \text{Im}(\varphi)$. The First Isomorphism Theorem tells us that this is always true.

Theorem 27.13 (The First Isomorphism Theorem). *Let G and H be groups, and let $\varphi : G \to H$ be a group homomorphism. Then $G/\text{Ker}(\varphi) \cong \text{Im}(\varphi)$.*

The proof of the First Isomorphism Theorem is outlined in the next activity.

Activity 27.14. Let G and H be groups, and let $\varphi : G \to H$ be a group homomorphism. For the sake of convenience, let $K = \text{Ker}(\varphi)$. We have already seen that $K = \text{Ker}(\varphi)$ is a normal subgroup of G, so G/K is a group. To prove Theorem 27.13, we first need to define a function from G/K to H. A natural choice for such a function is

$$\Phi : G/K \to H \quad \text{defined by} \quad \Phi(aK) = \varphi(a)$$

for all $aK \in G/K$.

(a) Prove that Φ is well-defined. Why do we need to do this?

(b) Prove that Φ is a homomorphism of groups.

(c) Prove that Φ is a monomorphism.

(d) Prove that $\mathrm{Im}(\Phi) = \mathrm{Im}(\varphi)$.

(e) Explain how the previous parts of this activity prove the First Isomorphism Theorem.

As an example of the use of the First Isomorphism Theorem, let n be a positive integer n. We know that the set $n\mathbb{Z} = \{nk : k \in \mathbb{Z}\}$ is a subgroup of $\mathbb{Z}$. It can be shown that the canonical map $\varphi : \mathbb{Z} \to \mathbb{Z}_n$ defined by $\varphi(k) = [k]$ is an epimorphism, and $\mathrm{Ker}(\varphi) = n\mathbb{Z}$. (See Exercise (10).) Therefore, $\mathbb{Z}/n\mathbb{Z} \cong \mathbb{Z}_n$, and $\mathbb{Z}_n$ is really a quotient group of the group of integers. We will explore additional applications of the First Isomorphism Theorem in the exercises.

The Second Isomorphism Theorem for Groups

Before stating the Second Isomorphism Theorem, we need to recall two prior results. Let G be a group, and let K and N be subgroups of G. The result of Exercise (8) of Investigation 19 (see page 285) shows that $K \cap N$ is a subgroup of G. Furthermore, Exercise (11) of Investigation 24 (see page 348) establishes that if N is a normal subgroup of G, then the set

$$KN = \{kn : k \in K, n \in N\}$$

is a subgroup of G.

The next activity illustrates the Second Isomorphism Theorem.

Activity 27.15. Let $G = D_6$, $K = \langle r, R^3 \rangle = \{I, R^3, r, rR^3\}$, and $N = \langle r, R^2 \rangle = \{I, R^2, R^4, r, rR^2, rR^4\}$. Exercise (19) or 36 of Investigation 24 (see page 349 or 351) shows that N is a normal subgroup of G.

(a) Find the elements of $K \cap N$.

(b) Find the elements of KN.

(c) What are the elements in $K/(K \cap N)$? To what familiar group is $K/(K \cap N)$ isomorphic?

(d) What are the elements in KN/N? To what familiar group is KN/N isomorphic?

(e) What is the relationship between $K/(K \cap N)$ and KN/N? Explain.

The result of Activity 27.15 is that, for the given groups, we have $K/(K \cap N) \cong KN/N$. The Second Isomorphism states that this is always true.

Theorem 27.16 (The Second Isomorphism Theorem). *Let G be a group, K a subgroup of G, and $N \triangleleft G$. Then $K/(K \cap N) \cong KN/N$.*

Proof. Let G be a group with identity e, K a subgroup of G, and $N \triangleleft G$. For each $k \in K$, we have $keN = kN \in KN/N$. Define a function $\varphi : K \to KN/N$ by

$$\varphi(k) = kN.$$

We will show that φ is an epimorphism with kernel $K \cap N$. The First Isomorphism Theorem will then allow us to conclude $K/(K \cap N) \cong KN/N$.

First we will show that φ is a homomorphism. Let $k_1, k_1 \in K$. Then

$$\varphi(k_1 k_2) = (k_1 k_2)N = (k_1 N)(k_2 N) = \varphi(k_1)\varphi(k_2).$$

Thus, φ is a homomorphism.

Next, we will show that φ is an epimorphism. Let $(kn)N \in KN/N$ for some $k \in K, n \in N$. Then $(kn)N = (kN)(nN) = (kN)N = kN$, since N is the identity element in KN/N. So $\varphi(k) = (kn)N$, and φ is an epimorphism.

Finally, we will show that $\text{Ker}(\varphi) = K \cap N$. Let $k \in \text{Ker}(\varphi)$. Then $\varphi(k) = kN = N$. Thus, $k \in N$. Since $\text{Ker}(\varphi) \subseteq K$, we also know $k \in K$. So $k \in K \cap N$. This shows $\text{Ker}(\varphi) \subseteq K \cap N$. Now let $a \in K \cap N$. Then $a \in N$, and so $\varphi(a) = aN = N$, and $a \in \text{Ker}(\varphi)$. Therefore, $K \cap N \subseteq \text{Ker}(\varphi)$. Since $\text{Ker}(\varphi) \subseteq K \cap N$ also, it follows that $K \cap N = \text{Ker}(\varphi)$. The First Isomorphism Theorem then allows us to conclude that $K/(K \cap N) \cong KN/N$. ∎

The Third Isomorphism Theorem for Groups

The Third Isomorphism Theorem establishes a cancellation law of sorts for quotient groups, as the next activity illustrates.

Activity 27.17. Let $G = D_6$, $N = \langle R \rangle = \{I, R, R^2, R^3, R^4, R^5\}$, and $K = \langle R^2 \rangle = \{I, R^2, R^4\}$. Exercise (36) of Investigation 24 (see page 351) shows that both N and K are normal subgroups of G.

(a) Find the elements of G/N. To what familiar group is G/N isomorphic?

(b) Find the elements of G/K and N/K. Why is N/K a normal subgroup of G/K?

(c) What are the elements of $(G/K)/(N/K)$? To what familiar group is $(G/K)/(N/K)$ isomorphic?

(d) What is the relationship between $(G/K)/(N/K)$ and G/N? Explain.

The result of Activity 27.17 is that, for these particular groups,

$$(G/K)/(N/K) \cong G/N.$$

In other words, we can "cancel" the Ks in this particular example. The Third Isomorphism Theorem tells us that this type of cancellation works in general.

Theorem 27.18 (The Third Isomorphism Theorem). *Let G be a group, and let K and N be normal subgroups of G with $K \subseteq N$. Then $(G/K)/(N/K) \cong G/N$.*

The proof of the Third Isomorphism Theorem is left as an exercise. (See Exercise (15).)

The Fourth Isomorphism Theorem for Groups

Let G be a group and K a normal subgroup of G. Among other things, the Fourth Isomorphism Theorem provides a correspondence between the subgroups of G that contain K and the subgroups of G/K. The next activity addresses this correspondence for a particular example.

Activity 27.19. Let $G = D_4$, the group of symmetries of a square. We know all the subgroups of G, and we can list them and indicate their containments in a *subgroup lattice*, as shown in Figure 27.1. The single lines indicate containment, and the meaning of the double lines should become clear shortly. Let $K = \langle R^2 \rangle$. Exercise (36) in Investigation 24 (see page 351) shows that K is a normal subgroup of G.

(a) Create the subgroup lattice for the quotient group $\overline{G} = G/K$.

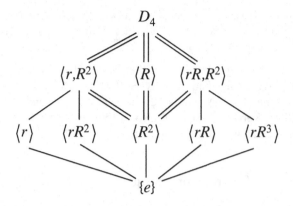

Figure 27.1
The subgroup lattice for D_4.

(b) Compare the subgroups of $\overline{G}$ to the subgroups of G. List three things you notice about the two sets of subgroups. Include in your discussion how the normal subgroups of $\overline{G}$ are related to subgroups of G. In particular, compare the subgroups of $\overline{G}$ to the subgroups of G that contain K.

Activity 27.19 shows that, in the case of $G = D_4$ and $K = \langle R^2 \rangle$, there is a one-to-one correspondence between the subgroups of G/K and the subgroups of G that contain K. Moreover, the normal subgroups of G/K correspond to normal subgroups of G that contain K. Although it is stated in terms of group epimorphisms (rather than normal subgroups), the Fourth Isomorphism Theorem shows that this is true in general. In particular, since every normal subgroup K of G can be viewed as the kernel of a group epimorphism—namely, the mapping $G \to G/K$ defined by $a \mapsto aK$ (see Exercise (11))—the Fourth Isomorphism Theorem, when applied to this map, yields the conclusions about normal subgroups that we noted above.

Theorem 27.20 (The Fourth Isomorphism Theorem). *Let $\varphi : G \to G'$ be a group epimorphism, and let $K = Ker(\varphi)$.*

(i) *If A is a subgroup of G, then $\varphi(A) = \{\varphi(a) : a \in A\}$ is a subgroup of G'.*

(ii) *If A' is a subgroup of G', then $A' = \varphi(H)$, where $H = \{a \in G : \varphi(a) \in A'\}$ is a subgroup of G containing K.*

(iii) *Let Φ be the function that assigns to each subgroup A of G that contains K the subgroup $\varphi(A)$. Then Φ is a bijection.*

(iv) *The function Φ preserves inclusion—that is $K \subseteq A \subseteq B$ if and only if $\varphi(A) \subseteq \varphi(B)$.*

(v) *The function Φ preserves normality—that is, N is a normal subgroup of G containing K if and only if $\varphi(N)$ is a normal subgroup of G'.*

(vi) *If N is a normal subgroup of G containing K, then the function $(G/N) \to (G'/\varphi(N))$ given by $aN \mapsto \varphi(a)\varphi(N)$ for all $a \in G$ is an isomorphism.*

Because the Fourth Isomorphism Theorem gives us a correspondence between the subgroups in the subgroup lattices for G and G', this theorem is also referred to as the **lattice isomorphism theorem** or the **correspondence theorem**. The proof of the Fourth Isomorphism Theorem is left as an exercise. (See Exercise (18).)

Concluding Activities

Activity 27.21. Let G and H be arbitrary groups, and let $\varphi : G \to H$ be a homomorphism of groups.

(a) If φ is a monomorphism, what is $\mathrm{Ker}(\varphi)$? Explain.

(b) Is the converse of part (a) true? That is, if $\mathrm{Ker}(\varphi) = \{e_G\}$, must φ be a monomorphism? Explain.

(c) Write a formal proof of the following theorem:

> **Theorem 27.22.** *Let $\varphi : G \to H$ be a homomorphism of groups, and let e_G be the identity element in G. Then φ is a monomorphism if and only if $\mathrm{Ker}(\varphi) = \{e_G\}$.*

Activity 27.23. Write a formal proof of the First Isomorphism Theorem.

Exercises

(1) Let $\varphi : G \to H$ be a group homomorphism, and let $a \in G$. If φ is an isomorphism, we know that $|\varphi(a)| = |a|$. Is the same result true if φ is not an isomorphism? If yes, prove it. If no, provide a counterexample and then determine and prove what is true about the relationship between $|\varphi(a)|$ and $|a|$.

(2) Let m and n be positive integers and define $\varphi : \mathbb{Z}_m \to \mathbb{Z}_n$ by $\varphi([a]_m) = [a]_n$.

    (a) Under what conditions on m and n is φ well-defined?

    (b) When φ is well-defined, is φ a homomorphism? If yes, under what conditions is φ a monomorphism or an epimorphism? Prove your answers.

(3) We can take Exercise (2) a step further. Let m and n be positive integers, and let k be any integer. Define $\varphi : \mathbb{Z}_m \to \mathbb{Z}_n$ by $\varphi([a]_m) = [ka]_n$.

    (a) Under what conditions on m, n, and k is φ well-defined?

    (b) When φ is well-defined, is φ a homomorphism? If yes, under what conditions is φ a monomorphism or an epimorphism? Prove your answers.

(4) Let $n \in \mathbb{Z}^+$, and let $\varphi : \mathrm{GL}_n(\mathbb{R}) \to \mathbb{R}^*$ be defined by $\varphi(A) = \det(A)$. (Here $\mathbb{R}^* = U(\mathbb{R})$ is the group of nonzero real numbers.)

    (a) Explain why φ is a homomorphism.

(b) Show that φ is an epimorphism.

(c) Find $\mathrm{Ker}(\varphi)$. Use the First Isomorphism Theorem to find a quotient of $GL_n(\mathbb{R})$ that is isomorphic to $\mathbb{R}^*$.

(5) The Alternating Group, A_4, is an important group because of its relation to Lagrange's Theorem.

(a) Explain why the group A_4 is important in this context.

(b) Now show that there is no homomorphism from A_4 onto a group of order 2 and show that there is no homomorphism from A_4 onto a group of order 4. (You do not need to use the operation table for A_4 for this problem.)

(6) Let n be a positive integer, and let $G_1, G_2, \ldots, G_n$ be groups with identities $e_1, e_2, \ldots, e_n$, respectively. For each $1 \leq i \leq n$, define

$$\varphi_i : G_i \to (G_1 \oplus G_2 \oplus \cdots \oplus G_n)$$

by

$$\varphi_i(g) = (e_1, e_2, \ldots, e_{i-1}, g, e_{i+1}, \ldots, e_n).$$

Show that each φ_i is a monomorphism and that

$$\mathrm{Im}(\varphi_i) = \{e_1\} \oplus \{e_2\} \oplus \cdots \oplus \{e_{i-1}\} \oplus G_i \oplus \{e_{i+1}\} \oplus \cdots \oplus \{e_n\}.$$

The map φ_i shows that $G_1 \oplus G_2 \oplus \cdots \oplus G_n$ contains an isomorphic copy of G_i.

(7) Table 26.1 (see page 377) lists a number of invariants of group isomorphisms. Activity 27.4 shows that group homomorphisms preserve the identity and inverses, while the result of Exercise (1) demonstrates that group homomorphisms do not necessarily preserve the orders of elements. In this exercise, we will examine a few more invariants of isomorphisms to see if they are also invariants of group homomorphisms. Let $\varphi : G \to H$ be a group homomorphism.

(a) Prove or disprove: $|G| = |H|$. If the statement is false, are there any conditions on φ that are weaker than requiring φ to be an isomorphism but that still make the statement true? Prove your answer.

(b) Prove or disprove: If G is Abelian, then H is Abelian. If the statement is false, are there any conditions on φ that are weaker than requiring φ to be an isomorphism but that still make the statement true? Prove your answer.

(c) Prove or disprove: If G is cyclic, then H is cyclic. If the statement is false, are there any conditions on φ that are weaker than requiring φ to be an isomorphism but that still make the statement true? Prove your answer.

(8) (a) Determine the number of homomorphisms from $\mathbb{Z}_{12}$ to $\mathbb{Z}_{42}$.

(b) Let n and m be positive integers. Determine as best you can the number of homomorphisms from $\mathbb{Z}_m$ to $\mathbb{Z}_n$. (Note that Exercise (15) on page 297 of Investigation 20 might be useful.) Does your result agree with the specific example in part (a)?

(c) Let G and G' be finite groups. If $\gcd(|G|, |G'|) = 1$, explain why there is only one homomorphism from G to G'.

(9) Find a homomorphism $\varphi : \mathbb{Z}_8 \to U_{10}$, if possible. Either prove that your choice of φ is a homomorphism or that no such homomorphism exists. If you can find a homomorphism φ, compute $\mathrm{Ker}(\varphi)$.

* (10) Let n be a positive integer. Define $\varphi : \mathbb{Z} \to \mathbb{Z}_n$ by $\varphi(k) = [k]$.

  (a) Show that φ is a homomorphism.

  (b) Find $\mathrm{Ker}(\varphi)$ and $\mathrm{Im}(\varphi)$.

  (c) To what familiar group is the quotient group $\mathbb{Z}/n\mathbb{Z}$ isomorphic? Explain.

* (11) Let G and H be groups, and let $\varphi : G \to H$ be a homomorphism. Theorem 27.8 shows that $\mathrm{Ker}(\varphi)$ is a normal subgroup of G. The converse of this statement is also true, as we will show in this exercise. Prove that if G is a group and N a normal subgroup of G, then the function $\varphi : G \to G/N$ defined by $\varphi(g) = gN$ is a homomorphism. Then show that $\mathrm{Ker}(\varphi) = N$.

(12) As we will see in later investigations, the number of normal subgroups of a given order in a group tells us something about the subgroup structure of the group.

  (a) Is it possible for a finite group G to have more than one normal subgroup of a given order? If yes, provide an example. If no, prove your answer.

  (b) Let G be a group. In Exercise (20) of Investigation 24 (see page 349), we showed that if G has exactly one subgroup N of a given order, then N is a normal subgroup of G. This problem does not, however, show that if N is normal in G, then there is only one subgroup of order $|N|$ in G. However, there are certain situations in which we can be sure that there is only one normal subgroup of a given order in a finite group. Show that if G is a finite group and K a normal subgroup of G with $\gcd(|K|, [G : K]) = 1$, then K is the only subgroup of G of order $|K|$. (Hint: Use Exercise (25) from Investigation 24.)

  (c) Let G be a group of order 15, and suppose G has a subgroup K of order 3 and a subgroup N of order 5. Explain why $G = K \times N$. Explain how this shows that G is cyclic.

(13) Let $G = \mathbb{Z}$ and $H = \mathbb{Z}_2 \oplus \mathbb{Z}_4$. Define $\varphi : \mathbb{Z} \to (\mathbb{Z}_2 \oplus \mathbb{Z}_4)$ by $\varphi(k) = ([k]_2, [k]_4)$.

  (a) Prove that φ is a homomorphism.

  (b) Find $\mathrm{Ker}(\varphi)$.

  (c) Find $\mathrm{Im}(\varphi)$.

  (d) Find the elements of $G/\mathrm{Ker}(\varphi)$. How is $G/\mathrm{Ker}(\varphi)$ related to $\mathrm{Im}(\varphi)$?

(14) Let G_1 and G_2 be groups with identities e_1 and g_2, respectively, and let $G = G_1 \oplus G_2$. Show that $G/(G_1 \oplus \{e_2\}) \cong G_2$ and that $G/(\{e_1\} \oplus G_2) \cong G_1$.

* (15) Prove the Third Isomorphism Theorem.

(16) Let a, b, and c be integers such that a divides b and b divides c. Then $c\mathbb{Z} \subseteq b\mathbb{Z} \subseteq a\mathbb{Z}$. Let $\varphi : a\mathbb{Z}/c\mathbb{Z} \to a\mathbb{Z}/b\mathbb{Z}$ be defined by $\varphi(at + c\mathbb{Z}) = at + b\mathbb{Z}$.

  (a) Show that φ is a well-defined epimorphism.

  (b) What is $\mathrm{Ker}(\varphi)$? (Hint: Use the Third Isomorphism Theorem.)

(17) If N and K are normal subgroups of a group G with K a subgroup of N, the Third Isomorphism Theorem tells us that $(G/K)/(N/K) \cong G/N$. A consequence of this is that

$$[G/K : N/K] = [G : N]. \tag{27.1}$$

Table 27.1

Orders of elements in A.

Element	a_1	a_2	a_3	a_4	a_5	a_6	a_7	a_8	a_9	a_{10}	a_{11}	a_{12}
Order	1	12	12	6	2	4	6	3	12	4	3	12

Element	a_{13}	a_{14}	a_{15}	a_{16}	a_{17}	a_{18}	a_{19}	a_{20}	a_{21}	a_{22}	a_{23}	a_{24}
Order	12	6	4	12	6	6	4	2	6	12	12	2

Recall that we can define $[G : N]$ as the number of distinct left cosets of N in G even if N is not normal in G. Is Equation (27.1) true even if N and K are not normal subgroups of G? In other words, if G is any group, N is any subgroup of G (not necessarily normal), and K is any subgroup of N (not necessarily normal), must $[G/K : N/K] = [G : N]$? Prove your answer. (Hint: One way to show that two sets X and Y have the same number of elements is to find a bijection $F : X \to Y$.)

$*$ (18) Prove the Fourth Isomorphism Theorem.

(19) Find all groups G' so that there is an epimorphism $\varphi : D_4 \to G'$. (Hint #1: The normal subgroups of D_4 are $\{I\}$, $\langle R^2 \rangle$, $\{I, r, rR^2, R^2\}$, $\langle R \rangle$, and D_4; see Exercise (36) on page 351 of Investigation 24. Hint #2: Use Theorem 27.8 and the First Isomorphism Theorem.)

(20) Let A be an Abelian group of order 24 whose elements have orders as shown in Table 27.1. Suppose $\varphi : A \to \mathbb{Z}_{22}$ is a nontrivial homomorphism. What can you say about $|\text{Ker}(\varphi)|$? What specific group is $\text{Im}(\varphi)$? Explain.

(21) Let $G = U_{44}$, $N = \langle [3] \rangle$, and $K = \langle [9] \rangle$.

(a) Find $G/K, G/N, N/K$, and $(G/K)/(N/K)$.

(b) Explain why $(G/K)/(N/K) \cong G/N$.

(22) We can use the Second Isomorphism Theorem to derive a relationship between the greatest common divisor and the least common multiple of two integers. Let a and b be nonzero integers. Let $K = a\mathbb{Z}$ and $N = b\mathbb{Z}$.

(a) What is $K \cap N$? Prove your answer.

(b) What is $K + N$? Prove your answer.

(c) Let r and s be nonzero integers such that r divides s. What is $[r\mathbb{Z} : s\mathbb{Z}]$? Prove your answer.

(d) Explain how the Second Isomorphism Theorem and the result of part (c) tell us that

$$\text{lcm}(a, b) = \frac{ab}{\gcd(a, b)}.$$

(23) In this exercise, we will prove an interesting fact about the subgroups of the symmetric groups. Let $n \geq 2$ be an integer. Show that every subgroup H of S_n contains either all even permutations or exactly one-half of the elements in H are even permutations. (Hint: If H is a subgroup of S_n, consider the function $\varphi : H \to \{1, -1\}$ defined by

$$\varphi(\alpha) = \begin{cases} 1, & \text{if } \alpha \text{ is even,} \\ -1, & \text{if } \alpha \text{ is odd} \end{cases},$$

and then apply the First Isomorphism Theorem.)

(24) Let G be a group. Let $a \in G$, and define $\varphi_a : G \to G$ by $\varphi_a(g) = aga^{-1}$.

  (a) Completely describe the mappings φ_a for all $a \in D_3$. Then construct an operation table for the set $\{\varphi_a : a \in D_3\}$, using the operation of function composition. What do you notice about this table?

  (b) Show that φ_a is an automorphism of G. (See Exercise (33) on page 389 of Investigation 26.) The automorphism φ_a is called the **inner automorphism of G induced by** a. All other automorphisms of a group are called **outer** automorphisms.

  (c) Let Inn(G) denote the set of all inner automorphisms of G. Prove that Inn(G) is a subgroup of Aut(G).

  (d) Let G be an Abelian group. Which specific subgroup of Aut(G) is Inn(G)? Explain.

(25) In this exercise, we will determine a specific fact about inner automorphisms (see Exercise (24)) and then find Inn(D_n).

  (a) Let G be a group. We can associate to any element $a \in G$ the inner automorphism φ_a defined by $\varphi_a(g) = aga^{-1}$ for each $g \in G$. Define a function $\Phi : G \to$ Inn(G) by

$$\Phi(a) = \varphi_a.$$

   (i) Show that Φ is a homomorphism.

   (ii) Is Φ an epimorphism? Prove your answer.

   (iii) Find Ker(Φ). Is Φ a monomorphism? Explain.

   (iv) Explain why the following theorem is true:
   Theorem 27.24. *If G is any group, then $G/Z(G) \cong Inn(G)$.*

  (b) To what familiar group is Inn(D_n) isomorphic? (Hint: The result of Exercise (2) on page 304 of Investigation 21 is useful.)

(26) Let $n \geq 2$ be an integer.

  (a) Find Inn(S_3). (Hint: Use the results of Exercise (25) in this investigation and Exercise (10) on page 304 of Investigation 22.)

  (b) Show that Aut$(S_3) \cong$ Inn(S_3). Note that this result—namely, that Aut$(S_n) \cong$ Inn(S_n)—is almost true in general, but fails when $n = 6$, providing another fascinating property of the number 6. (Other interesting properties of the number 6 are that 6 is the smallest order a non-Abelian group can have, 6 is the smallest perfect number, and 6 is the smallest positive integer that is the product of two distinct primes.) Segal* gives a reasonably accessible proof of the fact that Aut$(S_n) \cong$ Inn(S_n) for *every* n not equal to 6. (Segal states that the first proof is due to Hölder.[†]) A good project would be to find an outer automorphism of S_6.

*I. Segal,"The automorphisms of the symmetric group," *Bulletin of the American Mathematical Society*, **46**(6), p. 565.
[†]Otto Hölder, "Bildung zusammengesetzter Gruppen", *Mathematische Annalen*, **46**, 1895, pp. 321–422.

Connections

A homomorphism from a group G to a group H is a structure-preserving map. The isomorphism theorems for groups show us different ways that homomorphisms can determine isomorphisms. If you studied ring theory before group theory, you should notice the connections between group homomorphisms in this investigation and ring homomorphisms in Investigation 13. The major difference between a group homomorphism and a ring homomorphism is that a ring has two operations, and so there is a bit more structure to preserve. In spite of this difference, a careful perusal of the group isomorphism theorems shows that they are completely analogous to the corresponding ring isomorphism theorems.

Investigation 28

The Fundamental Theorem of Finite Abelian Groups

Focus Questions

By the end of this investigation, you should be able to give precise and thorough answers to the questions listed below. You may want to keep these questions in mind to focus your thoughts as you complete the investigation.

- What is a p-group?

- In what sense is each finite Abelian group made up of p-subgroups?

- What does the Fundamental Theorem of Finite Abelian Groups tell us, and why is the theorem considered "fundamental"?

Preview Activity 28.1. The orders of the elements of the group $U_{35} = \{[1], [2], [3], [4], [6], [8], [9], [11], [12], [13], [16], [17], [18], [19], [22], [23], [24], [26], [27], [29], [31], [32], [33], [34]\}$ are given in Table 28.1.

(a) The group U_{35} contains subgroups K and N of orders 12 and 2, respectively, so that U_{35} is the internal direct product $K \times N$. Find such groups and explain why $U_{35} = K \times N$.

(b) Find two cyclic groups $\mathbb{Z}_m$ and $\mathbb{Z}_n$ (for some integers m and n) so that U_{35} is isomorphic to the external direct product $\mathbb{Z}_m \oplus \mathbb{Z}_n$. Explain your reasoning.

(c) Notice that $|U_{35}| = 2^3 \times 3$. Let $G(2)$ be the subset of all elements a of U_{35} such that $|a|$ is a power of 2. Does $G(2)$ form a subgroup of U_{35}? Explain.

(d) Let $G(3)$ be the set of all elements of U_{35} whose order is a power of 3. Does $G(3)$ form a subgroup of U_{35}? Explain.

(e) Explain why $U_{35} = G(2) \times G(3)$.

(f) Explain why $G(2) \cong \mathbb{Z}_4 \oplus \mathbb{Z}_2$. (Hint: What is the maximal order an element in $G(2)$ has?) Why does this imply $U_{35} \cong \mathbb{Z}_4 \oplus \mathbb{Z}_2 \oplus \mathbb{Z}_3$? Does this contradict your answer to part (b)? Explain.

Table 28.1
Orders of elements in U_{35}.

$\|[1]\| = 1$	$\|[9]\| = 6$	$\|[18]\| = 12$	$\|[27]\| = 4$
$\|[2]\| = 12$	$\|[11]\| = 3$	$\|[19]\| = 6$	$\|[29]\| = 2$
$\|[3]\| = 12$	$\|[12]\| = 12$	$\|[22]\| = 4$	$\|[31]\| = 6$
$\|[4]\| = 6$	$\|[13]\| = 4$	$\|[23]\| = 12$	$\|[32]\| = 12$
$\|[6]\| = 2$	$\|[16]\| = 3$	$\|[24]\| = 6$	$\|[33]\| = 12$
$\|[8]\| = 4$	$\|[17]\| = 12$	$\|[26]\| = 6$	$\|[34]\| = 2$

Introduction

Classification theorems are among the most difficult in mathematics. To classify something is to understand it completely. As a result, it is a significant accomplishment that there is a classification theorem for all finite Abelian groups. It is that classification that we will study in this investigation.

The Components: p-Groups

Preview Activity 28.1 suggests that certain subgroups of a finite Abelian group—namely, those whose orders are powers of a prime number p—might provide factors of the group. This brings us to the important idea of what we call *p-groups*.

Definition 28.2. A *p*-**group**, where p is a prime, is a group whose order is a power of p.

The p-groups are important because they will form the building blocks of all finite Abelian groups. As we will see, a finite Abelian group can be broken down into a direct sum of p-subgroups. Preview Activity 28.1 gave us a hint about how we might begin with such a decomposition when we wrote the subgroup $G(2)$ as an internal direct product using an element of maximal order (as a power of 2) in the group. The next theorem shows us that we can always take this step for a p-group. Its proof is quite technical, so we will present it in its entirety and then consider an example to illustrate.

Theorem 28.3. *Let p be a prime, and let G be a finite Abelian p-group. Let $a \in G$ be an element of maximal order in G. Then there exists a subgroup K of G so that $G = \langle a \rangle \times K$.*

Proof. Let p be a prime, let G be a p-group with identity e, and let $a \in G$ have maximal order in G. To find a candidate for the subgroup K of G, we will look for a group K that is large (to account for the elements not in $\langle a \rangle$) and that also satisfies $\langle a \rangle \cap K = \{e\}$. There is at least one subgroup H of G that has the property that $\langle a \rangle \cap H = \{e\}$ (namely, $H = \{e\}$). Since G is a finite group, G must contain such a subgroup K of largest order.

To show that $G = \langle a \rangle \times K = \{a^m k : m \in \mathbb{Z}, k \in K\}$, we must show that any element $g \in G$ can be written in the form $a^m k$ for some nonnegative integer m and some element $k \in K$. We will proceed by contradiction and assume there is some element $b \in G$ so that b cannot be written in this form. Since $e = ee = a^0 e$, we know that $b \neq e$. Since G is a finite Abelian p-group, there is a positive integer r so that $b^{p^r} = e$ and, consequently, $b^{p^r} \in \langle a \rangle \times K$. Let s be the smallest positive integer so that $b^{p^s} \in \langle a \rangle \times K$. Let $c = b^{p^{s-1}}$. Then

$$c \notin \langle a \rangle \times K \quad \text{and} \quad c^p = \left(b^{p^{s-1}}\right)^p = b^{p^s} \in \langle a \rangle \times K. \qquad (28.1)$$

Let m be a nonnegative integer and $k \in K$ so that

$$c^p = a^m k. \qquad (28.2)$$

Since $a \in G$ and G is a p-group, there is a positive integer n so that $|a| = p^n$.

Claim 1. If $g \in G$, then $g^{p^n} = e$ for every $g \in G$.

To verify Claim 1, let $g \in G$. Since G is a p-group, we must have $|g| = p^t$ for some integer t. Since a has maximal order in G, we know $|g| \leq p^n$. It follows that $|g|$ divides $|a|$, and so $g^{p^n} = e$ for every $g \in G$.

By Claim 1, we know $c^{p^n} = e$, and so

$$e = c^{p^n} = (c^p)^{p^{n-1}} = (a^m k)^{p^{n-1}}.$$

Thus $(a^m)^{p^{n-1}} = k^{-p^{n-1}}$ is an element of both $\langle a \rangle$ and K. Since $\langle a \rangle \cap K = \{e\}$, it follows that

$$(a^m)^{p^{n-1}} = k^{-p^{n-1}} = e.$$

Therefore, the order of a must divide mp^{n-1}—that is, $p^n \mid mp^{n-1}$. Since $n > n - 1$, Euclid's Lemma (see page 10) implies that $p \mid m$. So there is an integer w such that $pw = m$. Thus,

$$c^p = a^m k = a^{pw} k,$$

or

$$k = c^p a^{-pw} = (ca^{-w})^p. \qquad (28.3)$$

Let

$$d = ca^{-w}. \qquad (28.4)$$

Claim 2. $d \notin K$, but $d^p \in K$.

To verify Claim 2, note that if $d = ca^{-w} = k_0 \in K$, then $c = a^w k_0 \in \langle a \rangle \times K$, which contradicts (28.1). Thus, $d \notin K$. However, (28.3) shows $d^p \in K$.

Claim 3. The subgroup $H = K\langle d \rangle$ of G contains K as a proper subgroup.

To verify Claim 3, note that $H = K\langle d \rangle = \{xd^y : x \in K, y \in \mathbb{Z}\}$ is a subgroup of G since K is a normal subgroup of G. (See Activity 24.22 on page 346.) Since $xd^0 = x$ is in H for every $x \in K$, we see that K is a subset of H. Since $d \notin K$ and $d = ed^1 \in H$, we see that K is a proper subset of H.

Table 28.2

Orders of elements in U_{60}.

$\lvert[1]\rvert = 1$	$\lvert[17]\rvert = 4$	$\lvert[31]\rvert = 2$	$\lvert[47]\rvert = 4$
$\lvert[7]\rvert = 4$	$\lvert[19]\rvert = 2$	$\lvert[37]\rvert = 4$	$\lvert[49]\rvert = 2$
$\lvert[11]\rvert = 2$	$\lvert[23]\rvert = 4$	$\lvert[41]\rvert = 2$	$\lvert[53]\rvert = 4$
$\lvert[13]\rvert = 4$	$\lvert[29]\rvert = 2$	$\lvert[43]\rvert = 4$	$\lvert[59]\rvert = 2$

Recall that K is the largest subgroup of G with the property that $\langle a \rangle \cap K = \{e\}$. Therefore, $\langle a \rangle \cap H \neq \{e\}$. Let z be a nonidentity element in $\langle a \rangle \cap H$. Then there exist a nonnegative integer j, an element k_1 in K, and an integer q so that

$$z = a^j = k_1 d^q. \tag{28.5}$$

Recall that $d^p \in K$ by Claim 2. So if $p \mid q$, then for some integer t, we have $z = k_1 d^q = k_1(d^p)^t \in K$. Thus, $z \in \langle a \rangle \cap K$. Since z is not the identity element, this cannot happen. So p cannot divide q. Therefore, $\gcd(p, q) = 1$, and Corollary 1.20 (see page 13) implies that there exist integers u, v so that

$$pu + qv = 1.$$

Substituting from (28.2) and (28.4) gives us

$$c = c^{pu+qv} = (c^p)^u(c^q)^v = (a^m k)^u((da^w)^q)^v = (a^m k)^u(d^q a^{wq})^v.$$

Substituting from (28.5) now yields

$$c = (a^m k)^u(d^q a^{wq})^v = (a^m k)^u(a^j k_1^{-1} a^{wq})^v = (a^{mu} a^{jv} a^{wqv})(k^u k_1^{-v}).$$

Now $c = (a^{mu} a^{jv} a^{wqv})(k^u k_1^{-v}) \in \langle a \rangle \times K$, which contradicts (28.1). Therefore, we can conclude that every element in G is in $\langle a \rangle \times K$. Since every element in $\langle a \rangle \times K$ is also in G, it follows that $G = \langle a \rangle \times K$, as desired. ∎

There are two key steps in applying Theorem 28.3:

- First, we must find an element $a \in G$ of maximal order.

- Next, we must find a maximal subgroup K of G satisfying $\langle a \rangle \cap K = \{e\}$. Then, for this K, we will have
$$G = \langle a \rangle K = \{a^m k : m \in \mathbb{Z}, k \in K\}.$$

Activity 28.4. We will illustrate the proof of Theorem 28.3 using $G = U_{60}$ as an example. Note that

$$U_{60} = \{[1], [7], [11], [13], [17], [19], [23], [29], [31], [37], [41], [43], [47], [49], [53], [59]\},$$

and so $|U_{60}| = 16 = 2^4$. The orders of the elements of U_{60} are given in Table 28.2.

(a) Find an element a in U_{60} of maximal order. Compute $\langle a \rangle$.

(b) Find a maximal subgroup K of G so that $\langle a \rangle \cap K = \{[1]\}$. (Hint: Recall that we want to have $\langle a \rangle \times K = G$, so what must the order of K be? Also note that $[7]^2 = [13]^2 = [17]^2 = [23]^2 = [37]^2 = [43]^2 = [47]^2 = [53]^2 = [49] \in U_{60}$.)

(c) It will necessarily be the case that every element in U_{60} can be written uniquely in the form $a^m k$ for some nonnegative integer m and some element $k \in K$. Verify that this does in fact happen for the choices of a and K you found in parts (a) and (b).

The group K in Theorem 28.3 is itself a p-group, so we can apply Theorem 28.3 to K. Continuing this process will allow us to write the original group as an internal direct product of cyclic subgroups.

Activity 28.5. Let p be a prime, and let G be a finite Abelian p-group with order p^n for some nonnegative integer n. In this activity, we will show that G can be written as an internal direct product of cyclic groups.

(a) Explain why we are done if $n = 1$.

(b) Now assume that any p-group of order less than p^n is an internal direct product of cyclic p-groups.

    (i) What does Theorem 28.3 tell us about G?

    (ii) If K is the subgroup as described in Theorem 28.3, what can we say about $|K|$?

    (iii) By considering cases and using Theorem 28.3, complete the proof of the following corollary to Theorem 28.3:

        Corollary 28.6. *Let p be a prime and G a finite Abelian p-group. Then G is an internal direct product of cyclic p-groups.*

Next we will apply Corollary 28.6 to an example to see a complete decomposition of a finite Abelian p-group into an internal direct product of cyclic subgroups.

Activity 28.7. Consider the group U_{60} whose decomposition as an internal direct product of cyclic groups we began in Activity 28.4. In the inductive step of Corollary 28.6, we applied Theorem 28.3 to the subgroup K. Illustrate this process by completing the steps below for the subgroup K you found in part (b) of Activity 28.4.

(a) Choose an element a_K of maximal order in K.

(b) Find a maximal subgroup K_1 of K so that $\langle a_K \rangle \cap K_1 = \{[1]\}$.

(c) Construct U_{60} from these cyclic subgroups.

Example 28.8. As an illustration of Corollary 28.6, we will determine all Abelian groups of order 8. Let G be an Abelian group of order $8 = 2^3$. By Corollary 28.6, we know that G is isomorphic to a direct product of cyclic p-groups for $p = 2$. The only possibilities for G are then

$$\mathbb{Z}_8, \qquad \mathbb{Z}_4 \oplus \mathbb{Z}_2, \qquad \text{or} \qquad \mathbb{Z}_2 \oplus \mathbb{Z}_2 \oplus \mathbb{Z}_2.$$

Since $\mathbb{Z}_8$ contains one element of order 2, $(\mathbb{Z}_4 \oplus \mathbb{Z}_2)$ contains three elements of order 2, and $\mathbb{Z}_2 \oplus \mathbb{Z}_2 \oplus \mathbb{Z}_2$ contains seven elements of order 2, we see that these three groups form three distinct isomorphism classes of groups of order 8. So there are exactly three Abelian groups of order 8, up to isomorphism.

An interesting question to ask is, given a prime power q^n, how many isomorphism classes are there of finite Abelian groups of order q^n? (The reason we are switching to use q as our prime instead of p should become clear shortly.) The answer is connected to integer partitions. An **integer**

Table 28.3

Isomorphism classes of Abelian groups of order q^n, for $1 \leq n \leq 5$.

n	Partitions of n	Isomorphism classes of groups of order q^n
1	1	$\mathbb{Z}_q$
2	2	$\mathbb{Z}_{q^2}$
	1+1	$\mathbb{Z}_q \oplus \mathbb{Z}_q$
3	3	$\mathbb{Z}_{q^3}$
	2+1	$\mathbb{Z}_{q^2} \oplus \mathbb{Z}_q$
	1+1+1	$\mathbb{Z}_q \oplus \mathbb{Z}_q \oplus \mathbb{Z}_q$
4	4	$\mathbb{Z}_{q^4}$
	3+1	$\mathbb{Z}_{q^3} \oplus \mathbb{Z}_q$
	2+2	$\mathbb{Z}_{q^2} \oplus \mathbb{Z}_{q^2}$
	2+1+1	$\mathbb{Z}_{q^2} \oplus \mathbb{Z}_q \oplus \mathbb{Z}_q$
	1+1+1+1	$\mathbb{Z}_q \oplus \mathbb{Z}_q \oplus \mathbb{Z}_q \oplus \mathbb{Z}_q$
5	5	$\mathbb{Z}_{q^5}$
	4+1	$\mathbb{Z}_{q^4} \oplus \mathbb{Z}_q$
	3+2	$\mathbb{Z}_{q^3} \oplus \mathbb{Z}_{q^2}$
	3+1+1	$\mathbb{Z}_{q^3} \oplus \mathbb{Z}_q \oplus \mathbb{Z}_q$
	2+2+1	$\mathbb{Z}_{q^2} \oplus \mathbb{Z}_{q^2} \oplus \mathbb{Z}_q$
	2+1+1+1	$\mathbb{Z}_{q^2} \oplus \mathbb{Z}_q \oplus \mathbb{Z}_q \oplus \mathbb{Z}_q$
	1+1+1+1+1	$\mathbb{Z}_q \oplus \mathbb{Z}_q \oplus \mathbb{Z}_q \oplus \mathbb{Z}_q \oplus \mathbb{Z}_q$

partition of a positive integer n is a nonincreasing sequence of positive integers whose sum is n. As an example,

$$
\begin{aligned}
4 &= 4 \\
&= 3 + 1 \\
&= 2 + 2 \\
&= 2 + 1 + 1 \\
&= 1 + 1 + 1 + 1,
\end{aligned}
$$

so there are five partitions of 4. The **integer partition function** is denoted p, where $p(n)$ is the number of partitions of the positive integer n. Our example above shows that $p(4) = 5$. Some values of $p(n)$, along with connections to various branches of mathematics, can be found in the *On-Line Encyclopedia of Integer Sequences* (OEIS).* Table 28.3 suggests that there is a connection between the number of isomorphism classes of finite Abelian groups of order q^n and the number $p(n)$ of partitions of the power n. You will formalize this connection in Exercise (5).

---

*This is sequence A000041 at `oeis.org`. The OEIS is a fascinating site and you can learn a lot of interesting mathematics there. We highly recommend that you visit the site.

Table 28.4

Orders of elements in U_{28}.

$\|[1]\| = 1$	$\|[11]\| = 6$	$\|[19]\| = 6$
$\|[3]\| = 6$	$\|[13]\| = 2$	$\|[23]\| = 6$
$\|[5]\| = 6$	$\|[15]\| = 2$	$\|[25]\| = 3$
$\|[9]\| = 3$	$\|[17]\| = 6$	$\|[27]\| = 2$

The Fundamental Theorem

Preview Activity 28.9. Let $G = U_{28}$ where

$$U_{28} = \{[1], [3], [5], [9], [11], [13], [15], [17], [19], [23], [25], [27]\}.$$

The orders of the elements in U_{28} are shown in Table 28.4.

(a) Let $G(2)$ be the subset of G consisting of the elements whose orders are powers of 2. Find the elements of $G(2)$.

(b) Let $G(3)$ be the subset of G consisting of the elements whose orders are powers of 3. Find the elements of $G(3)$.

(c) Are $G(2)$ and $G(3)$ subgroups of G? Explain.

(d) What group is $G(2) \times G(3)$?

We will now extend the results of the last section to all finite Abelian groups, using p-groups as our building blocks. Let G be a finite Abelian group of order n. By the Fundamental Theorem of Arithmetic, we can find distinct primes $p_1, p_2, \ldots, p_k$ and positive integers $m_1, m_2, \ldots, m_k$ so that

$$n = p_1^{m_1} p_2^{m_2} \cdots p_k^{m_k}.$$

In this section, we will show that we can break G up into an internal direct product of p-groups,

$$G = G(p_1) \times G(p_2) \times \cdots \times G(p_k), \tag{28.6}$$

where $|G(p_i)| = p_i^{m_i}$ for each i from 1 to k. We can then use Corollary 28.6 to write each $G(p_i)$ as a direct product of cyclic groups. When we put this all together, we have G as an internal direct product of cyclic groups as well, which will be isomorphic to an external direct product of cyclic groups.

In what follows, we will see how to decompose G into a direct product of p-groups as in (28.6). First, we will define the groups $G(p_i)$. Preview Activity 28.9 indicates that the groups we want are exactly those groups whose elements have orders that are powers of the primes. The next definition formalizes this idea.

Definition 28.10. Let G be a finite Abelian group of order n, and let p be a prime factor of n. The p-**primary component** of G is the set

$$G(p) = \{a \in G : |a| = p^t \text{ for some nonnegative integer } t\}.$$

In other words, $G(p)$ is the set of all elements $a \in G$ whose order is a power of p.

In our example in Preview Activity 28.9, the p-primary components of G were subgroups of G. In the next activity, we will determine if this is always the case.

Activity 28.11. Let G be a finite Abelian group with identity e, and let p be a prime divisor of $|G|$.

(a) Is $e \in G(p)$? Explain.

(b) Is $G(p)$ closed? Explain.

(c) If $a \in G(p)$, must a^{-1} be in $G(p)$? Explain.

(d) Is $G(p)$ a subgroup of G? Explain.

The next lemma formalizes the result of Activity 28.11 and also shows that $G(p)$ is a p-group.

Lemma 28.12. *If p is a prime factor of $|G|$, then $G(p)$ is a subgroup of G whose order is a power of p.*

Proof. Let p be a prime factor of $|G|$. The proof that $G(p)$ is a subgroup of G is contained in Activity 28.11. We will show that $G(p)$ has order a power of p. Suppose q is a prime, with $q \neq p$, such that q divides $|G(p)|$. By Cauchy's Theorem for Finite Abelian Groups, we know that $G(p)$ contains an element g of order q. This contradicts the fact that every element of $G(p)$ has order p^t for some positive integer t. Therefore, the only prime divisor of $|G(p)|$ is p, and so the order of $G(p)$ is a power of p. ∎

In other words, Lemma 28.12 shows that $G(p)$ is a p-group. It is these p-primary components of G that will form the factors of G, as in (28.6).

Theorem 28.13 (The Fundamental Theorem of Finite Abelian Groups). *Let G be a finite Abelian group. Then*

$$G = G(p_1) \times G(p_2) \times \cdots \times G(p_k), \tag{28.7}$$

where $p_1, p_2, \ldots, p_k$ are the distinct prime factors of $|G|$.

Proof. Let G be a finite Abelian group with identity e, and let $p_1, p_2, \ldots, p_k$ be the distinct prime factors of $|G|$. We will show that any element $a \in G$ can be written in the form

$$a = a_1 a_2 \cdots a_k, \tag{28.8}$$

with $a_i \in G(p_i)$ for each i.

Let $a \in G$. Since $|a|$ divides $|G|$ and $p_1, p_2, \ldots, p_k$ are the distinct prime factors of $|G|$, the order of a must have the form

$$|a| = q_1^{r_1} q_2^{r_2} \cdots q_s^{r_s},$$

where $Q = \{q_1, q_2, \ldots, q_s\}$ is a subset of $\{p_1, p_2, \ldots, p_k\}$ and $r_i > 0$ for each i. We will assume, without loss of generality, that $q_1 < q_2 < \cdots < q_s$. If we show that a can be factored as

$$a = b_1 b_2 \cdots b_s,$$

where $b_i \in G(q_i)$ for each i, then to verify (28.8) we can take $a_i = b_i$ when $p_i \in Q$ and $a_i = e$ when $p_i \notin Q$.

To show $a = b_1 b_2 \cdots b_s$, where $b_i \in G(q_i)$ for each i, we will proceed by induction on s, the number of distinct prime factors of $|a|$. If $s = 1$, then there is only one prime factor of $|a|$. Thus, we can choose $b_1 = a \in G(q_1)$ and we will be done. Now suppose that for any positive integer $t < s$, whenever $q_1, q_2, \ldots, q_t$ are the distinct prime factors of the order of an element $g \in G$, then we can write $g = b_1 b_2 \cdots b_t$, where $b_i \in G(q_i)$ for each i between 1 and t.

Now suppose $q_1, q_2, \ldots, q_s$ are the distinct prime factors of $|a|$. Let $|a| = q_1^{r_1} q_2^{r_2} \cdots q_s^{r_s}$ for some positive integers $r_1, r_2, \ldots, r_s$. Let $u = q_1^{r_1}$ and $v = q_2^{r_2} \cdots q_s^{r_s}$. Then $\gcd(u, v) = 1$, and so we can find integers x, y such that $xu + yv = 1$. (See Corollary 1.20 on page 13.) Now

$$a = a^{xu+yv} = (a^{xu})(a^{yv}). \tag{28.9}$$

Note that

$$(a^{xu})^v = (a^{uv})^x = (a^{|a|})^x = e,$$

and so $|a^{xu}|$ must be a divisor of v. (See Theorem 20.5 on page 294.) Therefore, the only primes dividing $|a^{xu}|$ are $q_2, q_3, \ldots, q_s$. Thus, we can apply the induction hypothesis to a^{xu} and write

$$a^{xu} = b_2 b_3 \cdots b_s, \tag{28.10}$$

where $b_i \in G(q_i)$ for each i between 2 and s.

Similarly,

$$(a^{yv})^u = (a^{uv})^y = (a^{|a|})^y = e,$$

and so the only prime factor of $|a^{yv}|$ is q_1. We can again apply the induction hypothesis, this time to a^{yv}, and write

$$a^{yv} = b_1, \tag{28.11}$$

where $b_1 \in G(q_1)$. Combining (28.9), (28.10), and (28.11), gives us

$$a = a^{yv} a^{xu} = b_1 b_2 \cdots b_s,$$

where $b_i \in G(q_i)$ for each i between 1 and s. This completes our proof that a has the form from (28.8).

Finally, to show that G is an internal direct product of $G(p_1), G(p_2), \ldots, G(p_k)$, we will show that the decomposition of an element in G into a product of elements in the p-primary components of G is unique. (Activity 28.16 establishes the validity of this approach.) Let $a \in G$, and suppose

$$a = a_1 a_2 \cdots a_k = b_1 b_2 \cdots b_k,$$

where $a_i, b_i \in G(p_i)$ for each i. Then, for each j between 1 and k we have

$$a_j b_j^{-1} = (b_1 a_1^{-1})(b_2 a_2^{-1}) \cdots (b_{j-1} a_{j-1}^{-1})(b_{j+1} a_{j+1}^{-1}) \cdots (b_k a_k^{-1}).$$

The only prime that can divide the order of $a_j b_j^{-1}$ is p_j. If p_j divides $|a_j b_j^{-1}|$, then p_j divides

$$|(b_1 a_1^{-1})(b_2 a_2^{-1}) \cdots (b_{j-1} a_{j-1}^{-1})(b_{j+1} a_{j+1}^{-1}) \cdots (b_k a_k^{-1})|. \tag{28.12}$$

But for each $i \neq j$, we have $b_i a_i^{-1} \in G(p_i)$. So the only primes that can divide (28.12) are $p_1, p_2, \ldots, p_{j-1}, p_{j+1}, \ldots, p_k$. Therefore, $|a_j b_j^{-1}| = 1$, and so $a_j = b_j$. Since j was chosen arbitrarily, we can conclude that $a_j = b_j$ for each j from 1 to k. Thus, the decomposition of the element a as $a = a_1 a_2 \cdots a_k$, where $a_i \in G(p_i)$ for each i, is unique.

It follows that if G is a finite Abelian group and $p_1, p_2, \ldots, p_k$ are the distinct prime divisors of $|G|$, then

$$G = G(p_1) \times G(p_2) \times \cdots \times G(p_k).$$

Table 28.5

U_{28} as an internal direct product of p-groups.

	[1]	[9]	[25]
[1]			
[13]			
[15]			
[27]			

Activity 28.14. In this activity, we will illustrate Theorem 28.13 with $G = U_{28}$.

(a) Determine specific groups $G(p_1), G(p_2), \ldots, G(p_k)$ so that $G = U_{28}$ is the internal direct product

$$G = G(p_1) \times G(p_2) \times \cdots \times G(p_k).$$

(b) Complete Table 28.5 to show that

$$U_{28} = G(p_1) \times G(p_2) \times \cdots \times G(p_k).$$

If these groups $G(p_i)$ are different than the ones you found in part (a), then explicitly identify $G(p_1), G(p_2), \ldots, G(p_k)$ in this case.

Example 28.15. To illustrate the Fundamental Theorem of Finite Abelian Groups, we will find all Abelian groups of order 360. If G is such a group, then since $360 = 2^3 \times 3^2 \times 5$, it follows that

$$G = G(2) \times G(3) \times G(5).$$

By Corollary 28.6, we know:

- $G(2)$ is isomorphic to $\mathbb{Z}_8$, $\mathbb{Z}_4 \oplus \mathbb{Z}_2$, or $\mathbb{Z}_2 \oplus \mathbb{Z}_2 \oplus \mathbb{Z}_2$.

- $G(3)$ is isomorphic to $\mathbb{Z}_9$ or $\mathbb{Z}_3 \oplus \mathbb{Z}_3$.

- $G(5)$ is isomorphic to $\mathbb{Z}_5$.

Therefore, G is isomorphic to one of the following:

$$\begin{array}{ll}
\mathbb{Z}_8 \oplus \mathbb{Z}_9 \oplus \mathbb{Z}_5 & \mathbb{Z}_8 \oplus \mathbb{Z}_3 \oplus \mathbb{Z}_3 \oplus \mathbb{Z}_5 \\
\mathbb{Z}_4 \oplus \mathbb{Z}_2 \oplus \mathbb{Z}_9 \oplus \mathbb{Z}_5 & \mathbb{Z}_4 \oplus \mathbb{Z}_2 \oplus \mathbb{Z}_3 \oplus \mathbb{Z}_3 \oplus \mathbb{Z}_5 \\
\mathbb{Z}_2 \oplus \mathbb{Z}_2 \oplus \mathbb{Z}_2 \oplus \mathbb{Z}_9 \oplus \mathbb{Z}_5 & \mathbb{Z}_2 \oplus \mathbb{Z}_2 \oplus \mathbb{Z}_2 \oplus \mathbb{Z}_3 \oplus \mathbb{Z}_3 \oplus \mathbb{Z}_5.
\end{array}$$

One question remains. Let G be a finite Abelian group whose order has prime power factorization

$$|G| = p_1^{m_1} p_2^{m_2} \cdots p_k^{m_k}.$$

How do we know that the groups we obtain using the method in Example 28.15 are all different? It seems obvious, but of course the obvious isn't always true. However, it is true in this case. Exercise (5) tells us how to determine all p_i-primary components, but it doesn't tell us what happens when we put them together. If we can show that different choices for $G(p_i)$ yield nonisomorphic groups G when combined as in (28.7), then we will have our answer. You are asked to supply the details in Exercise (10).

Concluding Activities

Activity 28.16. In the proof of the Fundamental Theorem of Finite Abelian Groups, we showed that the decomposition of an element in G into a product of elements in the p-primary components of G is unique. We then concluded that G is the internal direct product of the p-primary components of G. In this activity, we will verify the validity of this conclusion.

Assume that $N_1, N_2, \ldots, N_k$ are normal subgroups of a group G with $|N_1||N_2| \cdots |N_k| = |G|$ and that the decomposition of an element in $N_1 N_2 \cdots N_k$ is unique. That is, if $a_i, b_i \in N_i$ for each i and

$$a_1 a_2 \cdots a_k = b_1 b_2 \cdots b_k,$$

then it follows that $a_i = b_i$ for all $1 \leq i \leq k$.

(a) Let e be the identity in G. Prove that

$$N_i \cap (N_1 N_2 \cdots N_{i-1} N_{i+1} \cdots N_k) = \{e\}$$

for each i between 1 and k. (You may want to look back at Activity 25.16.) Conclude that

$$N_1 N_2 \cdots N_k = N_1 \times N_2 \times \cdots \times N_k.$$

(b) Use part (a) and the fact that $|N_1||N_2| \cdots |N_k| = |G|$ to explain why

$$G = N_1 \times N_2 \times \cdots \times N_k.$$

Activity 28.17. There is an alternate version of the Fundamental Theorem of Finite Abelian Groups that says that if G is a finite Abelian group, then G is isomorphic to a direct product of cyclic groups

$$\mathbb{Z}_{c_1} \oplus \mathbb{Z}_{c_2} \oplus \cdots \oplus \mathbb{Z}_{c_t}, \tag{28.13}$$

where $c_i \geq 2$ for all i and $c_t \mid c_{t-1} \mid c_{t-2} \mid \cdots \mid c_1$. The integers $c_1, c_2, \ldots, c_t$ are called the **invariant factors** of G because two finite Abelian groups with the same invariant factors are isomorphic. (See Exercise (13).) In this activity, we will prove the alternate, invariant factor version of the Fundamental Theorem of Finite Abelian Groups.

Let G be a finite Abelian group so that $|G| = p_1^{m_1} p_2^{m_2} \cdots p_k^{m_k}$ for some distinct primes $p_1, p_2, \ldots, p_k$ and some positive integers $m_1, m_2, \ldots m_k$. By the original version of the Fundamental Theorem of Finite Abelian Groups, we know that

$$G = G(p_1) \times G(p_2) \times \cdots \times G(p_k),$$

where $G(p_i)$ is a p_i group for each i. Also, for each i we have

$$G(p_i) \cong \mathbb{Z}_{p_i^{r_{i,1}}} \oplus \mathbb{Z}_{p_i^{r_{i,2}}} \oplus \cdots \oplus \mathbb{Z}_{p_i^{r_{i,t_i}}} \tag{28.14}$$

for some $t_i \geq 1$, where $r_{i,1} \geq r_{i,2} \geq \cdots \geq r_{i,t_i}$.

(a) The group U_{1001} has order $720 = 2^4 \times 3^2 \times 5$. Also,

$$U_{1001}(2) = \langle [34] \rangle \times \langle [274] \rangle \times \langle [428] \rangle$$
$$U_{1001}(3) = \langle [100] \rangle \times \langle [144] \rangle$$
$$U_{1001}(5) = \langle [92] \rangle.$$

(i) Identify all the variables in (28.14) for the group U_{1001}.

(ii) Find the invariant factors of U_{1001}.

(b) To prove this alternate formulation of the Fundamental Theorem of Finite Abelian Groups, we will need some additional notation. Let $t = \max\{t_i : 1 \leq i \leq k\}$. Note that $\mathbb{Z}_1 = \{[0]\}$, so we can extend $G(p_i)$ to

$$G(p_i) \cong \mathbb{Z}_{p_i^{r_{i,1}}} \oplus \mathbb{Z}_{p_i^{r_{i,2}}} \oplus \cdots \oplus \mathbb{Z}_{p_i^{r_{i,t_i}}} \oplus \cdots \oplus \mathbb{Z}_{p_i^{r_{i,t}}},$$

where $r_{i,j} = 0$ for $j > t_i$. Using this notation, determine the invariant factors of G, and show that G can be written in the form of Equation (28.13).

Exercises

(1) Find all isomorphism classes of finite Abelian groups of the following orders.

        (a) 12　　　　　　　(b) 16　　　　　　　(c) 30

        (d) 36　　　　　　　(e) 252　　　　　　(f) 8600

(2) Write the group U_{120} as an internal direct product of cyclic groups. (Hint: No element in U_{120} has order greater than 4.)

(3) For this exercise, let A be an Abelian group of order 24 whose elements have orders as shown in Table 28.6.

    (a) Find all isomorphism classes of Abelian groups of order 24. Use the Fundamental Theorem of Finite Abelian Groups to determine which class the group A belongs to. Explain. (Note that $a_6^2 \neq a_{20}$.)

    (b) It is known that there are 15 groups (up to isomorphism) of order 24. How many of these groups are non-Abelian? Explain.

    (c) List six potentially different non-Abelian groups of order 24. Explain why at least two of the non-Abelian groups that you listed are not isomorphic.

(4) Prove that the integer partition function p is a strictly increasing function.

* (5) Let p be prime. Table 28.3 indicates a connection between the number of distinct isomorphism classes of a finite Abelian group of order q^m and the partition function p. Find this connection

Table 28.6
Orders of elements in A.

Element	a_1	a_2	a_3	a_4	a_5	a_6	a_7	a_8	a_9	a_{10}	a_{11}	a_{12}
Order	1	12	12	6	2	4	6	3	12	4	3	12
Element	a_{13}	a_{14}	a_{15}	a_{16}	a_{17}	a_{18}	a_{19}	a_{20}	a_{21}	a_{22}	a_{23}	a_{24}
Order	12	6	4	12	6	6	4	2	6	12	12	2

and explain it. Be sure to provide an explanation for why there are *exactly* the number of isomorphism classes you claim.

(6) (a) Classify the orders of all finite Abelian groups for which there are exactly four distinct isomorphism classes. (Hint: Exercise (5) might be useful.)

   (b) Classify the orders of all finite Abelian groups for which there are exactly five distinct isomorphism classes.

(7) (a) Suppose the order of some finite Abelian group is divisible by 6. Prove that the group has a cyclic subgroup of order 6.

   (b) Is there something special about the number 6 in part (a)? That is, if k is a positive integer and the order of a finite Abelian group G is divisible by k, must G contain a cyclic subgroup of order k? If yes, prove it. If no, find a counterexample and then determine conditions on k that make the statement true.

(8) We know that a finite cyclic group of order n contains exactly one subgroup of order m for each divisor m of n. Is the converse true if our group is Abelian? That is, suppose G is a finite Abelian group that has exactly one subgroup for each divisor of $|G|$. Must G be cyclic? Prove your answer.

(9) Let $G = \mathbb{Z}_{720} \oplus \mathbb{Z}_{120} \oplus \mathbb{Z}_{15}$. To which groups are the p-primary components of G isomorphic? Explain.

* (10) Let G be a finite Abelian group of order $p_1^{m_1} p_2^{m_2} \cdots p_k^{m_k}$, where $p_1, p_2, \ldots, p_k$ are distinct primes and $m_1, m_2, \ldots, m_k$ are positive integers. The Fundamental Theorem of Finite Abelian Groups tells us that

$$G = G(p_1) \times G(p_2) \times \cdots \times G(p_k).$$

Since $|G(p_i)|$ is a power of p_i, the Fundamental Theorem of Arithmetic tells us that $|G(p_i)| = p_i^{m_i}$ for each i. In this exercise, we will show that different choices for $G(p_i)$ yield nonisomorphic groups G. To do so, let $H_1, H_2, \ldots, H_k$ and $Q_1, Q_2, \ldots, Q_k$ be groups with $|H_i| = |Q_i| = p_i^{m_i}$ for each i. If

$$G \cong (H_1 \oplus H_2 \oplus \cdots \oplus H_k) \cong (Q_1 \oplus Q_2 \oplus \cdots \oplus Q_k),$$

prove that $H_i \cong Q_i$ for each i.

(11) (a) Use the results of Exercises (5) and (10) to determine the number of isomorphism classes of Abelian groups of order $8318750000 = 2^4 \times 5^8 \times 11^3$. Use the fact that $p(8)$, the number of partitions of 8, is 22. Please do not attempt to write out all representatives of all the isomorphism classes.

   (b) Let G be an Abelian group of order $q_1^{m_1} q_2^{m_2} \cdots q_k^{m_k}$, where $q_1, q_2, \ldots, q_k$ are distinct primes and $m_1, m_2, \ldots, m_k$ are positive integers. Find a formula, in terms of the partition function, for the number of isomorphism classes of Abelian groups of order $|G|$.

(12) Let G be a finite Abelian group, and let a be an element of maximum order in G. Show that $|b|$ divides $|a|$ for any element b in G. (Hint: Activity 28.17 might be useful.)

* (13) Let G and H be finite Abelian groups, and let $\nu_G(k)$ and $\nu_H(k)$ denote the number of elements of order k in G and H, respectively.

   (a) Prove that if $\nu_G(k) = \nu_H(k)$ for all $k \in \mathbb{Z}^+$, then G and H have the same invariant factors. (See Activity 28.17.)

(b) Prove that $G \cong H$ if and only if G and H have the same invariant factors.

(c) In Exercise (9) of Investigation 26 (see page 386), we showed that there exists a finite Abelian group K and a finite non-Abelian group L so that $|K| = |L|$ and $\nu_K(k) = \nu_L(k)$ for all $k \in \mathbb{Z}^+$, but $K \not\cong L$. Is this same property possible if both G and H are finite Abelian groups? That is, if $\nu_G(k) = \nu_H(k)$ for all $k \in \mathbb{Z}^+$, must G and H be isomorphic? Prove your answer.

(14) Let G be a group, and define $t(G)$ to be the set of all elements of G of finite order.

(a) Is $t(G)$ always a subgroup of G? If the answer is yes, prove it. If the answer is no, illustrate with an example and then find at least two different types of groups for which the answer is yes. Explain. (Hint: Refer to Exercise (5) on page 297 of Investigation 20.)

(b) Let S be the set of all complex numbers with norm 1. (See Exercise (6) on page 285 of Investigation 19.) Determine the elements of $t(S)$.

(c) Let G be an Abelian group. Then $t(G)$ is a normal subgroup of G called the **torsion subgroup** of G. What can we say about the orders of the elements in $G/t(G)$? Prove your answer.

(d) Prove that if G is an Abelian group, then $G \cong t(G) \oplus G/t(G)$.

(e) Describe as best you can the elements of the group $S/t(S)$, where S is the group of complex numbers of norm 1. Your description must involve more than just the definition of a quotient group.

Connections

In this investigation, we completely classified all finite Abelian groups up to isomorphism, as discussed in Investigation 26. The Fundamental Theorem of Finite Abelian Groups provides information about the subgroup structure of finite Abelian groups that we began to study in Investigation 19. The construction of group products that was introduced in Investigation 25, cyclic groups as described in Investigation 19, and groups of units in $\mathbb{Z}_n$ as we saw in Investigation 17 provide important tools for this work.

Investigation 29

The First Sylow Theorem

Focus Questions

By the end of this investigation, you should be able to give precise and thorough answers to the questions listed below. You may want to keep these questions in mind to focus your thoughts as you complete the investigation.

- What does it mean for two elements to be conjugate in a group? What are some important properties of conjugacy?

- How is the conjugacy class of an element in a group G related to a special subgroup of G?

- How does the Class Equation work, and what are two important consequences of the Class Equation?

- What is the general formulation of Cauchy's Theorem, and why is it important?

- What are the three Sylow theorems, and what do they tell us about finite groups?

- What is a Sylow p-subgroup of a group G? Why are these subgroups important?

Preview Activity 29.1. In this investigation we will introduce three important theorems known as the Sylow theorems. In order to understand the Sylow theorems, we will first need to understand *conjugacy*, defined formally as follows:

Definition 29.2. Let G be a group. The element $a \in G$ is **conjugate** to $b \in G$ if there is an element $g \in G$ so that $a = gbg^{-1}$.

The element $b = gag^{-1}$ is called a *conjugate* of a or the *conjugate of a by g*.

(a) Find all conjugates of the element r in D_4.

(b) Note that conjugacy can be used to define a relation on a group. In particular, let $a \sim b$ if a is conjugate to b. It seems natural to ask what properties this conjugacy relation satisfies.

    (i) Is $\sim$ a reflexive relation? Prove your answer.

    (ii) Is $\sim$ a symmetric relation? Prove your answer.

    (iii) Is $\sim$ a transitive relation? Prove your answer.

(c) Is the relation $\sim$ defined in part (b) an equivalence relation? If so, find the distinct equivalence classes of this relation (also known as *conjugacy classes*) for the group D_4.

Introduction

After Lagrange's Theorem, the most important results in finite group theory are arguably the Sylow theorems. These theorems are named after Norwegian mathematician Peter Ludvig Mejdell Sylow (1832–1918). * The Sylow theorems are among the most important and profound results in the theory of finite groups.

As we have discussed, the subgroup structure of a group tells us a lot about the group. Our first big theorem in studying the subgroup structure of finite groups was Lagrange's Theorem. Recall that Lagrange's Theorem told us that if G is a finite group and H is a subgroup of G, then $|H|$ divides $|G|$. While this theorem tells us something about the subgroup structure of finite groups, it is not as strong a result as we might like. It would be ideal if the converse of Lagrange's Theorem were true—that is, if whenever a positive integer k divides the order of a finite group G, then G must have a subgroup of order k. Unfortunately, this is not the case, as we have seen that the alternating group A_4 has order 12 but has no subgroup of order 6. The Sylow theorems are important because they give us some information about which subgroups a finite group *must* have.

In this investigation, we will state the three Sylow theorems and prove the first one. The second and third Sylow theorems will be discussed in more detail in the next investigation. To begin, we will develop a counting technique that depends on conjugacy.

Conjugacy and the Class Equation

The results of Preview Activity 29.1 show that conjugacy is an equivalence relation on a group G. As such, the conjugacy relation partitions a group G into disjoint equivalence classes, called *conjugacy classes*.

Definition 29.3. Let G be a group and $a \in G$. The **conjugacy class** of a in G is the equivalence class of a under the conjugacy relation.

We denote the conjugacy class of an element a in a group G as $\mathrm{cl}(a)$. So

$$\mathrm{cl}(a) = \{gag^{-1} : g \in G\}.$$

Activity 29.4. Find $\mathrm{cl}(a)$ for each $a \in S_3$.

It is important to note that, in general, $\mathrm{cl}(a)$ is **NOT** a subgroup of G.

The notion of a relation on a group is not new to us. Recall that if G is a group and H a subgroup of G, then we can define an equivalence relation $\sim_H$ on G by letting $a \sim_H b$ if $b^{-1}a \in H$. The equivalence classes under $\sim_H$ are the left cosets of H in G. The $\sim_H$ relation was useful in that it allowed us to write any finite group G as a disjoint union of left cosets and then count the elements in the distinct left cosets to obtain Lagrange's Theorem. What made this argument work was that any two equivalence classes (left cosets of H in G) had the same number of elements. As our previous activities have shown, two conjugacy classes can contain different numbers of elements,

---

*The paper in which the Sylow theorems appeared was titled, "Théorèmes sur les groupes de substitutions" and was published in *Mathematische Annalen*, vol. 5 (pp. 584–594) in 1872.

so counting with conjugacy classes will be a bit more complicated than counting with left cosets. However, many useful results will follow from the conjugacy relation.

Since conjugacy is an equivalence relation on G, we know that G is the union of the disjoint conjugacy classes. So we can write

$$G = \bigcup_{a \in G} \text{cl}(a).$$

If we want to use conjugacy classes to count elements, we will need to know how to determine the number of elements in $\text{cl}(a)$ for a given $a \in G$. The answer, of course, depends on the element a. In some cases, it is easy to determine the number of elements in a conjugacy class, as the next activity demonstrates.

Activity 29.5. Let G be a group, and let $a \in G$.

(a) In Preview Activity 29.1 we saw that it is possible to have $\text{cl}(a) = \{a\}$. Determine specific conditions on a for this to happen.

(b) If $a \in Z(G)$, what is $\text{cl}(a)$? Explain.

(c) Complete the following lemma:

Lemma 29.6. *Let G be a group and $a \in G$. Then $\text{cl}(a) = \{a\}$ if and only if* _____.

Lemma 29.6 seems to imply that conjugacy classes are not very interesting for Abelian groups (where $G = Z(G)$). However, if G is non-Abelian, then conjugacy classes are very helpful in studying the subgroup structure of G.

Let G be a group, and let $a \in G$. Since $\text{cl}(a)$ is not in general a subgroup of G, there is no obvious relationship between the size of $\text{cl}(a)$ and the order of G. (For example, we can't use Lagrange's Theorem to make any conclusions about the relationship between $|\text{cl}(a)|$ and $|G|$.) If we dig a little deeper, however, we will find that there is a relationship that will make it easier for us to count elements using conjugacy classes. Suppose we have two conjugates, $x = gag^{-1}$ and $y = hah^{-1}$, of an element $a \in G$. In order to count only distinct conjugates, we would like to have a way to tell when these conjugates are the same. In other words, we would like to determine the circumstances in which

$$gag^{-1} = hah^{-1}?$$

If $gag^{-1} = hah^{-1}$, then we have

$$\left(g^{-1}h\right) a = a \left(g^{-1}h\right).$$

Thus, $g^{-1}h$ commutes with a. So, to count the number of distinct conjugates of an element a, we need to understand how many elements of G commute with a. This observation motivates the following definition.

Definition 29.7. Let G be a group and $a \in G$. The **centralizer** of $a \in G$ is the set $C(a) = \{g \in G : ga = ag\}$.

In other words, $C(a)$ is the set of elements in G that commute with a.

Activity 29.8. Let $G = S_3$.

(a) Find $C(a)$ for each $a \in G$.

(b) Compare the sets $C(a)$ in G for choices of a that are in the same conjugacy class. (See Activity 29.4.) What do you notice?

Our observations about the sets $C(a)$ in Activity 29.8 seem to suggest that, unlike the conjugacy class of an element, the centralizer of an element might always be a group. In the next activity, we will see if this is actually the case.

Activity 29.9. Let G be a group with identity e, and let $a \in G$.

(a) Show that $C(a)$ is not empty.

(b) Is $C(a)$ closed under the operation in G? Verify your answer.

(c) If $g \in C(a)$, must g^{-1} also be in $C(a)$? Verify your answer.

(d) Is $C(a)$ a subgroup of G?

Recall that the number of distinct conjugates of an element a in a finite group G is the number of elements that commute with a, or the order of $C(a)$. So we now have a relationship between conjugacy classes and subgroups. We can exploit this relationship to count the number of elements in a group. Recall that two conjugates gag^{-1} and hah^{-1} of a are equal when $g^{-1}h$ is in $C(a)$. We also know that $g^{-1}h \in C(a)$ implies $gC(a) = hC(a)$. So $gag^{-1} = hah^{-1}$ implies $gC(a) = hC(a)$. Conversely, if $gC(a) = hC(a)$, then $g^{-1}h$ is in $C(a)$ and $gag^{-1} = hah^{-1}$. Therefore, the number of elements in the conjugacy class of $a \in G$ is the number of distinct right (or left) cosets of $C(a)$ in G. These observations establish the following theorem:

Theorem 29.10. *Let G be a finite group, and let $a \in G$. The number of elements in the conjugacy class of $a \in G$ is*

$$[G : C(a)] = \frac{|G|}{|C(a)|}.$$

The next activity asks you to verify this theorem for a particular example.

Activity 29.11. Verify Theorem 29.10 for the elements in S_3.

The Class Equation

Counting techniques are important throughout finite group theory. We saw one example of the use of a counting technique when we proved Lagrange's Theorem. In this section, we will develop another counting tool known as the *Class Equation*.

As we observed earlier, we can write G as the union of the disjoint conjugacy classes of elements in G. In other words,

$$G = \bigcup_{a \in G} \mathrm{cl}(a).$$

By Theorem 29.10, we know how many elements are in each conjugacy class. This leads to the following result:

Theorem 29.12 (The Class Equation). *Let G be a finite group, and let $\mathrm{cl}(a_1), \mathrm{cl}(a_2), \ldots, \mathrm{cl}(a_k)$ be the distinct conjugacy classes of G. Then*

$$|G| = \sum_{i=1}^{k} \frac{|G|}{|C(a_i)|}. \tag{29.1}$$

The next activity uses our prior calculations and provides a quick example of the Class Equation.

Activity 29.13. Verify the Class Equation for the group S_3.

There are several variations on the Class Equation. For one variation, recall that the single element classes correspond exactly to those elements in the center of G. Therefore, if $cl(a_1), cl(a_2), \ldots, cl(a_r)$ are the distinct multi-element conjugacy classes of G, then

$$|G| = |Z(G)| + \sum_{i=1}^{r} \frac{|G|}{|C(a_i)|}. \tag{29.2}$$

Note also that $a \in Z(G)$ if and only if $C(a) = G$. So if $a \notin Z(G)$, then, since both the identity and a commute with a, we have $|C(a)| \geq 2$.

The Class Equation has many applications in group theory. Later on, we will use the Class Equation to prove the First Sylow Theorem. The Class Equation can also be used to prove other results about groups, such as the following:

Theorem 29.14. *Let p be a prime and n a positive integer. If G is a group of order p^n, then $Z(G)$ is nontrivial.*

Proof. Let p be a prime and n a positive integer. Let G be a group of order p^n. Let $z = |Z(G)|$. Since every element in $Z(G)$ has a one-element conjugacy class, there are z distinct one-element conjugacy classes. Let $a \in G$ such that $a \notin Z(G)$. Then $|C(a)|$ is larger than 1, less than $|G|$, and divides p^n. So $|C(a)| = p^{k_a}$ for some positive integer k_a less than n. Then $\frac{|G|}{|C(a)|} = p^{n-k_a} \geq p$. The Class Equation (29.2) then implies that

$$p^n = z + \sum_{a} \frac{|G|}{|C(a)|} = z + \sum_{a} p^{n-k_a}.$$

A little algebra shows that

$$z = p^n - \sum_{a} p^{n-k_a} = p\left(p^{n-1} - \sum_{a} p^{n-k_a-1}\right).$$

Therefore, $p \mid z$. Thus, z is larger than 1, and so $Z(G)$ is nontrivial. ∎

The next activity establishes another useful result.

Activity 29.15. Let p be a prime, and let G be a group of order p^2. In this activity, we will determine what type of group G must be.

(a) What does Theorem 29.14 tell us about $Z(G)$? What are the possible orders of $Z(G)$?

(b) What can we say about G if $|Z(G)| = p^2$?

(c) If $|Z(G)| \neq p^2$, then there is an element $g \in G \setminus Z(G)$. What can we say about $|C(g)|$? What conclusion can we draw?

(d) Explain how we have proved the following theorem:

Corollary 29.16. *Let p be a prime. If G is a group of order p^2, then G is Abelian.*

Corollary 29.16 allows us to classify all groups of order p^2, where p is prime. (See Exercise (3) for more details.)

Cauchy's Theorem

In Investigation 24, we proved Cauchy's Theorem for Abelian Groups, which states that if a prime p divides the order of a finite Abelian group G, then G contains a subgroup of order p. As it turns out, Cauchy's Theorem is true for any finite group G, and we can use the Class Equation to prove it. Cauchy's Theorem is an important result that we will use in our proofs of the Sylow theorems.

Theorem 29.17 (Cauchy's Theorem). *Let G be a finite group, and let p be a prime dividing the order of G. Then G has an element of order p.*

Proof. Let G be a finite group with identity element e. We will proceed by using strong induction on the order of G. Since we have completely classified the groups of order 1, 2, and 3, we know the theorem is true in those cases. Now let $n \in \mathbb{Z}^+$, and assume that, for any group of order less than n, if a prime p divides the order of the group, then the group has an element of order p.

Let G be a group of order n, and assume p is a prime factor of n. Then $n = pm$ for some positive integer m. We know that G is the union of finitely many conjugacy classes and that $Z(G)$ contains all single element conjugacy classes. Let $\mathrm{cl}(a_1)$, $\mathrm{cl}(a_2)$, ... $\mathrm{cl}(a_r)$ be the distinct multi-element conjugacy classes. The Class Equation tells us that

$$pm = |Z(G)| + \sum_{i=1}^{r} \frac{|G|}{|C(a_i)|}.$$

We will consider two cases: (1) if p divides $|C(a_i)|$ for some i; and (2) if p divides no $|C(a_i)|$.

Case 1: Suppose there is an integer i between 1 and r such that p divides $|C(a_i)|$. Since $1 < |C(a_i)| < |G|$, we can use the induction hypothesis to conclude that $C(a_i)$ contains an element of order p. Thus, G contains an element of order p.

Case 2: Now suppose there is no integer i such that p divides $|C(a_i)|$. Then $\frac{|G|}{|C(a_i)|} = \frac{pm}{|C(a_i)|}$ must be a multiple of p for each value of i. The Class Equation then tells us that p divides $pm - \sum_{i=1}^{r} \frac{|G|}{|C(a_i)|} = |Z(G)|$. Therefore, the order of $Z(G)$ is at least p. Since $Z(G)$ is an Abelian group whose order is a multiple of p, Cauchy's Theorem for Abelian Groups shows us that $Z(G)$ contains an element of order p.

In either case, we can conclude that G contains an element of order p. ∎

With these preliminary results out of the way, we can now turn our attention to the First Sylow Theorem.

The First Sylow Theorem

Preview Activity 29.18. We have seen that the group A_4 of order 12 does not contain a subgroup of order 6. Find all the nontrivial proper subgroups of A_4, and determine their orders. What can you say about the orders of these subgroups in relation to the order of A_4?

The result of Preview Activity 29.18 is that, while A_4 contains no subgroup of order 6, it does contain subgroups of every order p^n, where p is prime and p^n divides $|A_4|$. The First Sylow Theorem guarantees that this happens in general. In order to prove it, we will need one more piece of information about subgroups of quotient groups.

Lemma 29.19. *Let G be a group and N a normal subgroup of G. Every subgroup H' of G/N has the form H/N for some subgroup H of G containing N. Moreover, $H \triangleleft G$ if and only if $H' \triangleleft G/N$.*

The proof of Lemma 29.19 is left as Exercise (23).

We are now ready to prove the First Sylow Theorem.

Theorem 29.20 (First Sylow Theorem). *Let G be a finite group. If p is a prime number and $k \in \mathbb{Z}^+$ such that p^k divides $|G|$, then G has a subgroup of order p^k.*

Proof. Let G be a finite group. We will proceed by using strong induction on the order n of $|G|$. We have already classified all groups of orders 1 through 6, and so we know that the theorem is true in those cases. Now let $n \in \mathbb{Z}^+$, and assume that for any group H of order less than n and for any prime p and any positive integer k satisfying the property that p^k divides $|H|$, we know that H contains a subgroup of order p^k.

Let G be a group of order n. Let p be a prime number, and let $k \in \mathbb{Z}^+$ such that p^k divides $|G|$. Then there is a positive integer m such that $n = p^k m$. We will show that G contains a subgroup of order p^k.

We know that G is the union of finitely many conjugacy classes. We also know that $Z(G)$ contains all the single-element conjugacy classes. Let $\mathrm{cl}(a_1), \mathrm{cl}(a_2), \ldots, \mathrm{cl}(a_r)$ be the distinct multi-element conjugacy classes of G. The Class Equation tells us that

$$|G| = |Z(G)| + \sum_{i=1}^{r} \frac{|G|}{|C(a_i)|}.$$

As in the proof of Cauchy's Theorem, we will now consider two cases.

Case 1: Suppose there is an i between 1 and r such that p does not divide $\frac{|G|}{|C(a_i)|}$. Since $|C(a_i)|$ divides $|G| = p^k m$, we must then have that p^k divides $|C(a^i)|$. Recall that $1 < |C(a_i)| < |G|$, so we can apply our induction hypothesis to $C(a_i)$ to conclude that $C(a_i)$ contains a subgroup of order p^k. Hence, G contains a subgroup of order p^k.

Case 2: Suppose p divides $\frac{|G|}{|C(a_i)|}$ for each i. Then p divides

$$|G| - \sum_{i=1}^{r} \frac{|G|}{|C(a_i)|} = |Z(G)|.$$

So $Z(G)$ has order at least p. By Cauchy's theorem we know that $Z(G)$ has an element, x, of order p. Let $N = \langle x \rangle$. Since $x \in Z(G)$, we know $N \triangleleft G$. Now consider the group $K = G/N$. Since $|K| = \frac{|G|}{|N|} = p^{k-1} m < |G|$, by our induction hypothesis the group K has a subgroup X of order p^{k-1}. By Lemma 29.19, we know this subgroup X is of the form H/N for some subgroup H of G that contains N. We also know that

$$p^{k-1} = |X| = |H/N| = \frac{|H|}{|N|} = \frac{|H|}{p}.$$

Therefore, $|H| = p^k$.

In either case, we have found a subgroup of G of order p^k.   ∎

The First Sylow Theorem tells us that every finite group G contains subgroups of every prime power divisor of the order of G. This provides significant information about the subgroup structure of G. Recall that if G is an Abelian group with $p_1^{m_1} p_2^{m_2} \cdots p_k^{m_k}$ as the prime power factorization of $|G|$, then

$$G = G(p_1) \times G(p_2) \times \cdots \times G(p_k),$$

where $G(p_i)$ is the p_i-primary component of G and has order $p_i^{m_i}$. If G is a non-Abelian group with $p_1^{m_1} p_2^{m_2} \cdots p_k^{m_k}$ as the prime power factorization of $|G|$, the First Sylow Theorem tells us that G contains a subgroup of order $p_i^{m_i}$ for each i. These maximal p-subgroups are given a special name.

Definition 29.21. Let G be a finite group, and let p be a prime that divides $|G|$. Let $|G| = p^k m$ for some positive integers k and m with $\gcd(k, m) = 1$, so that p^k is the highest power of p that divides $|G|$. A subgroup S of G is a **Sylow p-subgroup** if $|S| = p^k$.

So a Sylow p-subgroup of an Abelian group is just the p-primary component of the group. A non-Abelian group may contain one or more Sylow p-subgroups for a given prime p, as the next activity illustrates.

Activity 29.22.

(a) Find a Sylow p-subgroup of D_6 for each prime divisor of $|D_6|$.

(b) Does D_6 contain more than one Sylow 2-subgroup? Does D_6 contain more than one Sylow 3-subgroup? Explain.

The second and third Sylow theorems provide additional information about the Sylow p-subgroups of a group. In particular, they tell us about the relationship between the different Sylow p-subgroups of a group, as well as the number of Sylow p-subgroups a group can have.

The Second and Third Sylow Theorems

In Activity 29.22, we showed that D_6 contains exactly one Sylow 3-subgroup, but it has more than one Sylow 2-subgroup. From Exercise (36) of Investigation 24 (see page 351), we know that the only normal subgroups of D_6 are the subgroups of $\{R\}$ along with $\langle r, R^2 \rangle$, $\langle rR, R^2 \rangle$, and D_6. So the sole Sylow 3-subgroup of D_6 is a normal subgroup. In general, this is one of the conclusions we can draw from the Second Sylow Theorem. In particular, the Second Sylow Theorem tells us that any two Sylow p-subgroups (for the same value of p) are *conjugate*. We first encountered conjugates of subgroups in Activity 24.20 (see page 346), and we will formally define them here.

Definition 29.23. Let G be a group and H a subgroup of G. The **conjugate** of H in G by $a \in G$ is the set

$$aHa^{-1} = \{aha^{-1} : h \in H\}.$$

Activity 24.20 shows that aHa^{-1} is always a subgroup of G. We say that the subgroup H is **conjugate** to the subgroup K if $H = aKa^{-1}$ for some $a \in G$. We will explore the conjugacy relation in more detail in the next investigation. For now, however, we can use it to formally state the Second Sylow Theorem.

Theorem 29.24 (Second Sylow Theorem). *Let G be a finite group, and let p be a prime divisor of $|G|$. If H and K are Sylow p-subgroups of G, then there exists $g \in G$ such that $H = gKg^{-1}$.*

One consequence of the Second Sylow Theorem is the following corollary:

Corollary 29.25. *Let G be a finite group, and let p be a prime divisor of $|G|$. A Sylow p-subgroup S of G is normal in G if and only if S is the only Sylow p-subgroup of G.*

The proof of Corollary 29.25 is outlined in Activity 29.30. The proof of the subsequent corollary is outlined in Activity 29.31.

The Third Sylow Theorem provides information—although it is not at all obvious—about the number of Sylow p-subgroups of a group G.

Theorem 29.26 (Third Sylow Theorem). *Let G be a finite group, and let p be a prime divisor of $|G|$. The number of Sylow p-subgroups of G divides $|G|$ and is of the form $1 + pk$ for some nonnegative integer k.*

Corollary 29.27. *Let G be a finite group, and let p be a prime such that $|G| = p^r m$ for some $r, m \in \mathbb{Z}^+$ with $\gcd(p, m) = 1$. Then the number of Sylow p-subgroups of G divides m and is of the form $1 + pk$ for some nonnegative integer k.*

Note that the First Sylow Theorem tells us about the *existence* of certain types of subgroups of a finite group, while the second and third Sylow theorems give us information about those subgroups. We will postpone the proofs of the second and third Sylow theorems to the next investigation, but the three theorems work together as a group (no pun intended), so we will freely use all three theorems from this point on. (You should do the same for the exercises.)

We will conclude this investigation with two examples that illustrate some of the power of the Sylow theorems.

Example 29.28. Let p be an odd prime, and let G be a non-Abelian group of order $2p$ with identity e. Then Theorem 26.16 (see page 380) shows that $G \cong D_p$. We can also prove this using the Sylow theorems. The Third Sylow Theorem tells us that the number n_p of Sylow p-subgroups of G is congruent to 1 mod p, and Corollary 29.27 shows that n_p divides 2. Since $p > 2$, the only way this can happen is if $n_p = 1$. So there is one Sylow p-subgroup N of G, and Corollary 29.25 tells us that N is normal in G. Since $|N| = p$, N is cyclic and is therefore generated by some element n. Let K be a Sylow 2-subgroup of G. By Activity 24.22 (see page 346), we know that NK is a subgroup of G. Furthermore, Exercise (2) shows that $N \cap K = \{e\}$. Now we will show that $G = NK$ by demonstrating that $|NK| = 2p$. By Exercise (7), we know that the representation of an element of NK in the form nk (where $n \in N$ and $k \in K$) is unique. Thus, $|NK| = |N||K| = 2p$, and so $G = NK$. If $K \triangleleft G$, then $G = (N \times K) \cong (N \oplus K) \cong Z_{2p}$. Therefore, we can assume that K is not normal in G.

Since $|K| = 2$, we know that K is cyclic and is generated by some element k. Since $G = NK$, every element in G is of the form $n^i k^j$ for some $0 \leq i < p$ and $0 \leq j < 2$. The operation table for G, and hence the structure of G, will be determined by the product kn. Since G is non-Abelian, it must be the case that $kn \neq nk$. Given that $N \triangleleft G$, we must have $knk = knk^{-1} \in N$. So $knk = n^t$

for some $1 \le t < p$. Therefore,

$$
\begin{aligned}
n &= k^2 n k^2 \\
&= k(knk)k \\
&= kn^t k \\
&= (knk)^t \\
&= \left(n^t\right)^t \\
&= n^{t^2}.
\end{aligned}
$$

Thus, $n^{t^2-1} = e$, and it follows that p divides $t^2 - 1 = (t+1)(t-1)$. So it must be the case that p divides $t + 1$ or $t - 1$. Recall that $1 \le t < p$, so if p divides $t + 1$, then $t = p - 1$. Also, if p divides $t - 1$, then $t = 1$. Since $kn \ne nk$, we can rule out $t = 1$, which leaves us with the conclusion that $t = p - 1$. This tells us that $knk = n^{p-1} = n^{-1}$. Therefore, G has presentation

$$
\langle n, k \mid n^p = k^2 = 1, knk^{-1} = n^{-1} \rangle \cong D_p.
$$

Example 29.29. As a second example, we will classify all groups of order 66. Note that $66 = 2 \times 3 \times 11$. Let G be a group of order 66, and let n_p be the number of Sylow p-subgroups of G for each prime divisor of $|G|$. The Third Sylow Theorem tells us that $n_{11} \equiv 1 \pmod{11}$ and that n_{11} divides 6. This can only happen if $n_{11} = 1$. So there is a unique Sylow 11-subgroup N of G. Let K be a Sylow 3-subgroup of G. Since $N \triangleleft G$ (by Corollary 29.25), we know that NK is a subgroup of G (by Activity 24.22 on page 346). Also, Exercise (2) shows that $K \cap N = \{e\}$, so $|NK| = 33$. The result of Activity 29.32 shows that NK is a cyclic group and therefore is generated by some element x. Note that $[G : NK] = 2$, so $NK \triangleleft G$ (by Exercise (19) on page 349 of Investigation 24).

Now let H be any Sylow 2-subgroup of G. Since NK is normal and $\gcd(|NK|, |H|) = 1$, Exercise (2) implies that $|NK \cap H| = \{e\}$. Exercise (7) then implies that $|NKH| = |NK||H|$. Since $|H|$ is a power of 2, it must be that $|H| = 2$ and $|NKH| = 66$. From this, we can conclude that $H = \langle h \rangle$ for some $h \in G$ and that $NKH = G$. Since NK is generated by x and H is generated by h, it follows that every element of G can be written as $x^t h^j$ for some integers t and j with $0 \le t < 33$ and $0 \le j < 2$. In particular, since hx is an element of G, $hx = x^t h^j$ for some $0 \le t < 33$ and $0 \le j < 2$. However, if $j = 0$, then $hx = x^t$, which implies that $h \in \langle x \rangle$, a contradiction. Thus, $j = 1$ and $hx = x^t h$—or, equivalently, $hxh^{-1} = x^t$. Furthermore, since $h \ne e$, we know that $1 \le t \le 32$. But since $|H| = 2$, we also know that $h^{-1} = h$, and so $hxh = x^t$. Therefore, by the same argument as in Example 29.28, $x^{t^2-1} = e$, and so 33 divides $t^2 - 1$. Since $1 \le t \le 32$, the possibilities are $t = 1$, $t = 10$, $t = 23$, and $t = 32$. Each of these four values of t determines a group with presentation

$$
\langle x, h \mid x^{33} = h^2 = 1, hx = x^t h \rangle.
$$

In Exercise (12), you will show that none of these four groups are isomorphic to each other. This establishes that, up to isomorphism, there are exactly four groups of order 66.

Concluding Activities

Activity 29.30. Use the Second Sylow Theorem to prove Corollary 29.25.

Activity 29.31. Use the Third Sylow Theorem to prove Corollary 29.27.

Activity 29.32. We have spent a significant amount of time classifying groups of various orders. The Sylow theorems are critical in this process except for very small groups. In addition, the Sylow theorems can help classify entire families of groups. In this activity, we will see how to explicitly determine the groups of order pq, where p and q are primes, $p < q$, and p does not divide $q - 1$. To begin, let G be such a group.

(a) How many Sylow q-subgroups must G have?

(b) How many Sylow p-subgroups must G have?

(c) Let K be a Sylow p-subgroup of G and N a Sylow q-subgroup of G. Explain why K and N are both normal subgroups of G.

(d) Prove that G is a cyclic group.

Exercises

(1) Describe the conjugacy classes of elements in an Abelian group G.

$\star$ (2) Let G be a group with identity e, and let H and K be finite subgroups of G.

    (a) Suppose that $\gcd(|H|, |K|) = 1$. Determine the elements of $H \cap K$.

    (b) Suppose that H and K are cyclic groups of the same prime order. What is $H \cap K$?

$\star$ (3) (a) Classify all groups of order p^2 when p is prime.

    (b) Is it true that every group of order n^2, where $n \in \mathbb{Z}^+$, is Abelian? Explain.

(4) Let A be an Abelian group of order 24 whose elements have orders as shown in Table 29.1. Find all the Sylow p-subgroups of A. Clearly explain your work.

(5) Recall that the Second Sylow Theorem tells us that any two Sylow p-subgroups are conjugate. Illustrate this theorem by finding a Sylow 2-subgroup H of S_3. Show specifically how all other Sylow 2-subgroups of S_3 are conjugate to H.

(6) Let G be a group of order 200. What are the sizes of the Sylow p-subgroups of G? Explain why at least one of the Sylow p-subgroups of G is normal in G.

$\star$ (7) Let G be a group with identity e, and let N and K be subgroups of G with $N \triangleleft G$. If $N \cap K = \{e\}$, show that the representation of an element of the form nk in the subgroup NK (where $n \in N$ and $k \in K$) is unique. Conclude that $|NK| = |N||K|$.

Table 29.1
Orders of elements in A.

Element	a_1	a_2	a_3	a_4	a_5	a_6	a_7	a_8	a_9	a_{10}	a_{11}	a_{12}
Order	1	12	12	6	2	4	6	3	12	4	3	12

Element	a_{13}	a_{14}	a_{15}	a_{16}	a_{17}	a_{18}	a_{19}	a_{20}	a_{21}	a_{22}	a_{23}	a_{24}
Order	12	6	4	12	6	6	4	2	6	12	12	2

(8) Let N be a normal subgroup of a finite group G, and let $a \in G$ such that $\gcd(|a|, |G/N|) = 1$. Show that $a \in N$.

* (9) Let G be a group and H a subgroup of G.

    (a) Show by example that not all conjugates of H must be the same.

    (b) Show that every conjugate of H in G is isomorphic to H.

(10) Consider the group $\mathbf{Q} = \{1, -1, i, -i, j, -j, k, -k\}$ of quaternions.

    (a) Determine the distinct conjugacy classes in $\mathbf{Q}$.

    (b) Let $\mathrm{cl}(a_1), \mathrm{cl}(a_2), \ldots, \mathrm{cl}(a_r)$ be the distinct conjugacy classes in $\mathbf{Q}$. Determine the elements in $C(a_i)$ for each i.

    (c) Find $Z(\mathbf{Q})$. Use the Class Equation to verify the order of $Z(\mathbf{Q})$.

(11) Consider the group $D_4 = \{r^i R^j : 0 \le i \le 1, 0 \le j \le 3\}$ of symmetries of the square.

    (a) Determine the distinct conjugacy classes in D_4.

    (b) Let $\mathrm{cl}(a_1), \mathrm{cl}(a_2), \ldots, \mathrm{cl}(a_r)$ be the distinct conjugacy classes in D_4. Determine the elements in $C(a_i)$ for each i.

    (c) Find $Z(D_4)$. Use the Class Equation to verify the order of $Z(D_4)$.

* (12) Demonstrate that the four groups $\mathbb{Z}_{66}$, D_{33}, $D_{11} \oplus \mathbb{Z}_3$, and $D_3 \oplus \mathbb{Z}_{11}$ form four different isomorphism classes of groups of order 66. Then determine the values of t from Example 29.29 that correspond to each group.

(13) We know that for each prime p there is exactly one group of order p. Is this true for any composite numbers? If so, find the smallest composite number n so that there is exactly one group of order n (up to isomorphism). Verify your result.

(14) Let G be a finite group and K a Sylow p-subgroup of G. If φ is an automorphism of G, must $\varphi(K) = K$? If not, what can we say about $\varphi(K)$?

(15) Let n and k be positive integers with $k \le n$, and let p be a prime number.

    (a) Determine the number of k-cycles in S_n.

    (b) Describe the Sylow p-subgroups of S_p. How many distinct Sylow p-subgroups does S_p contain?

(16) Let G be a group of order $p^m n$, where p is prime, $p > n$, and $\gcd(p, n) = 1$. Let S be a Sylow p-subgroup of G. Must S be normal in G? Prove your answer.

(17) Suppose G is a group of order pq, where p and q are distinct primes with $p < q$.

    (a) Must G contain a normal subgroup of order p? Explain.

    (b) Must G contain a normal subgroup of order q? Explain.

(18) (a) Classify all groups of order 45.

    (b) Let p and q be prime with $p < q$ and $q \not\equiv 1 \pmod{p}$. Classify all groups of order $p^2 q$.

(19) Find all Sylow p-subgroups of D_{12}.

(20) We have shown that normality is not transitive in general. However, let G be a finite group with subgroups P and N such that $P \triangleleft N$, $N \triangleleft G$, and P is a Sylow subgroup of N. Show that $P \triangleleft G$.

(21) Show that if all the Sylow subgroups of a finite group G are normal in G, then G is isomorphic to the direct sum of its Sylow subgroups. Is the converse true? Explain.

(22) In this exercise, we will classify all groups of order 21.

  (a) How many Abelian groups of order 21 are there? Explain.

  (b) Now let G be a non-Abelian group of order 21.

    (i) Determine the number of Sylow subgroups of each order in G.

    (ii) Let $N = \langle n \rangle$ be the Sylow 7-subgroup of G and let $K = \langle k \rangle$ be one Sylow 3-subgroup. Use the fact that $n = k^{-3}nk^3$ to show that $n^{t^3-1} = e$, where e is the identity in G. Conclude that 7 divides $t^3 - 1$.

    (iii) Determine the possible values of t from the previous part of this problem, and use your result to complete the classification of all groups of order 21.

* (23) Explain how the Fourth Isomorphism Theorem verifies Lemma 29.19.

(24) Classify all groups of order 30.

* (25) Let p be a prime, and let G be a p-group of order p^m. The First Sylow Theorem tells us that G contains subgroups of each order p^k for $0 \leq k \leq m$, demonstrating that the converse of Lagrange's Theorem is true for p-groups. We can show something more, though. Prove that G has *normal* subgroups of all orders p^k for $0 \leq k \leq m$.

Connections

The Sylow theorems provide a number of tools that we can use to make conclusions about the subgroup structure of an arbitrary group. In the case of a finite Abelian group G, the p-primary components of G that we defined in Investigation 28 are exactly the Sylow p-subgroups of G. While we can use the Fundamental Theorem of Finite Abelian Groups to completely classify all Abelian groups of a given order, the problem of classifying non-Abelian groups is much more difficult. The Sylow theorems play an important role in the classification of such groups.

Investigation 30

The Second and Third Sylow Theorems

Focus Questions

By the end of this investigation, you should be able to give precise and thorough answers to the questions listed below. You may want to keep these questions in mind to focus your thoughts as you complete the investigation.

- If G is a group and H is a subgroup of G, what does it mean for two subgroups of G to be H-conjugate? How is this different than two subgroups being conjugate?

- What properties does the H-conjugacy relation satisfy, and why are these properties important?

- What is the normalizer of a subgroup? Why are normalizers important, and what properties do they satisfy?

- What role do H-conjugacy and normalizers play in the proofs of the second and third Sylow theorems?

- How can the Sylow theorems be used to classify finite groups?

Preview Activity 30.1. In Investigation 29, we utilized conjugacy of elements in a group to develop an important counting tool, the Class Equation. In this investigation, we will take our work a step further and investigate the idea of conjugate subgroups. The definition is similar to that of conjugate elements.

Definition 30.2. Let G be a group and H a fixed subgroup of G. A subgroup K of G is H-**conjugate** to a subgroup N of G if there is an element $h \in H$ so that $K = hNh^{-1} = \{hnh^{-1} : n \in N\}$.

(a) Let $G = D_6$, and let $H = \langle R \rangle$. Is the subgroup $K = \langle rR \rangle$ H-conjugate to the subgroup $N = \langle r \rangle$? Find all subgroups N' of G that are H-conjugate to N.

(b) Conjugation is a relation on the collection of subgroups of a group. If H is a fixed subgroup of a group G, we will write $K \sim_H N$ to denote that the subgroup K of G is H-conjugate to the subgroup N of G. We will also say that two subgroups K and N of G are conjugate (without reference to a subgroup) if K and N are G-conjugate. It is natural at this point to ask what kind of relation $\sim_H$ is.

  (i) Show that $\sim_H$ is a reflexive relation.

  (ii) Show that $\sim_H$ is a symmetric relation.

(iii) Show that $\sim_H$ is a transitive relation.

(iv) What can you conclude about $\sim_H$ from your work in parts (i) – (iii)?

Introduction

Recall that the First Sylow Theorem tells us that if p is a prime and p^k divides the order of a finite group G, then G contains a subgroup of order p^k. In particular, if p is prime and the order of a finite group G is $p^k m$, where $k, m \in \mathbb{Z}+$ with $\gcd(p, m) = 1$, then any subgroup of G of order p^k is called a Sylow p-subgroup of G. In this investigation, we will prove the second and third Sylow theorems. These theorems tell us about the relationship between Sylow p-subgroups and the number of Sylow p-subgroups in a group.

Conjugate Subgroups and Normalizers

In Preview Activity 30.1, we began our exploration of conjugate subgroups. Recall that if G is a group and H is a subgroup of G, then two subgroups K and N of G are H-conjugate if $K = hNh^{-1}$ for some $h \in H$. As we saw in the activity, H-conjugacy defines an equivalence relation on the set of all subgroups of G, which we denote by $\sim_H$. Conjugates play an important role in group theory, and the lemma below provides some useful properties of conjugates.

Lemma 30.3. *Let G be a group, let K and N be subgroups of G, and let $a, b \in G$.*

(i) *If $K = aNa^{-1}$, then $N = a^{-1}Ka$.*

(ii) $(ab)K((ab)^{-1}) = a(bKb^{-1})a^{-1}$.

(iii) *If $a \in K$, then $aKa^{-1} = K$.*

We proved the first part of Lemma 30.3 in Preview Activity 30.1, and the proof of remaining parts of Lemma 30.3 are left for the exercises. (See Exercise (2).)

In order to prove the second and third Sylow theorems, we will also need to understand the normalizer of a subgroup. Recall that a subgroup N is normal in a group G if $aNa^{-1} = N$ for every $a \in G$. It is also possible to have $aNa^{-1} = N$ for just some elements $a \in G$, as the next activity demonstrates.

Activity 30.4. Let $K = \langle r \rangle$ in D_6.

(a) Is $RKR^{-1} = K$?

(b) Find all $a \in D_6$ so that $aKa^{-1} = K$. What kind of set do you obtain?

Activity 30.4 suggests that the set of elements a for which $aKa^{-1} = K$ might be interesting. This leads us to the following definition.

Definition 30.5. Let G be a group and K a subgroup of G. The **normalizer** of K in G is the set

$$N(K) = \{a \in G : aKa^{-1} = K\}.$$

You may have encountered the normalizer of a subgroup in Exercise (18) of Investigation 24. (See page 349.) If so, you already know that $N(K)$ is a subgroup of G, and you can omit parts (a) – (c) of the following activity.

Activity 30.6. Let G be a group with identity e, and let K be a subgroup of G.

(a) Can $N(K)$ be empty?

(b) Is $N(K)$ closed? You may use Lemma 30.3 if it helps.

(c) Does $N(K)$ contain the inverse of each of its elements? What conclusion can you draw about $N(K)$?

(d) Prove the following lemma. You may use Lemma 30.3 if it helps.

>   **Lemma 30.7.** *Let G be a group and K a subgroup of G. Then $N(K)$ is a subgroup of G and K is a normal subgroup of $N(K)$.*

As with Lagrange's Theorem and the relation that led us to left cosets, as well as the conjugacy relation on elements in a group that led us to the Class Equation, we will use the relation $\sim_H$ to count elements in finite groups. Doing so will allow us to prove the second and third Sylow theorems, but first we will need to learn a bit more about H-conjugacy. The next example suggests an important connection that will help us in this regard.

Activity 30.8. Let $K = \langle r \rangle$ in D_6, and let $H = \langle R \rangle$. We have seen that $N(K) = \{I, r, R^3, rR^3\}$ and that the distinct H-conjugates of K in D_6 are N, $\langle rR^2 \rangle$, and $\langle rR^4 \rangle$. There is a useful way to count the number of distinct H-conjugates of K in G. This number depends on $|H|$ and $|H \cap N(K)|$. Find $H \cap N(K)$ for this example, and use your work to conjecture a relationship between $|H|$, $|H \cap N(K)|$, and the number of distinct H-conjugates of K in G.

Although it is not wise to extrapolate from just one example, the result of Activity 30.8 might lead us to wonder if the relationship we observed there holds in general. The next lemma shows that it does.

Lemma 30.9. *Let G be a finite group and H a fixed subgroup of G. Let K be a subgroup of G. The number of distinct H-conjugates of K in G is the index of $H \cap N(K)$ in H. In particular, the number of distinct H-conjugates of K in G divides $|H|$.*

Proof. Let G be a group and H a fixed subgroup of G. Let K be a subgroup of G. To count the number of H conjugates of K, let aKa^{-1} and bKb^{-1} be H conjugates of K. Suppose $aKa^{-1} = bKb^{-1}$. Parts (a) and (b) of Lemma 30.3 imply that

$$K = b^{-1}(aKa^{-1})b = (b^{-1}a)K(b^{-1}a)^{-1},$$

so $b^{-1}a \in N(K)$. Since a and b are in H, we also have $b^{-1}a \in (H \cap N(K))$. Thus, $a(H \cap N(K)) = b(H \cap N(K))$. So different left cosets of $H \cap N(H)$ in G correspond to different H-conjugates of K in G. Therefore, the index of $H \cap N(K)$ in G is less than or equal to the number of distinct H-conjugates of K in G.

To prove equality, suppose $a(H \cap N(K)) = b(H \cap N(K))$ for some $a, b \in H$. Then $b^{-1}a \in (H \cap N(K))$ and

$$K = (b^{-1}a)K(b^{-1}a)^{-1} = b^{-1}(aKa^{-1})b,$$

from which it follows that

$$aKa^{-1} = bKb^{-1}.$$

Therefore, the number of left cosets of $H \cap N(K)$ in H is equal to the number of distinct H-conjugates of K in G. Since $H \cap N(K)$ is a subgroup of H, we know that the number of distinct H-conjugates of K in G divides $|H|$. ∎

There is one other result we will use in our proof of the Second Sylow Theorem. The H-conjugation relation is a tool that will allow us to count elements in a finite group. In order to do so, we will need to determine the circumstances in which a Sylow p-subgroup group can be H-conjugate to itself. The next lemma provides the details.

Lemma 30.10. *Let Q be a Sylow p-subgroup of a finite group G, and let $g \in G$. If $|g| = p^k$ for some nonnegative integer k and $gQg^{-1} = Q$, then $g \in Q$.*

Proof. Let Q be a Sylow p-subgroup of a finite group G with identity e. Since Q is a p-group, we know $|Q| = p^n$ for some positive integer n. Let $g \in G$ with $|g| = p^k$ for some nonnegative integer k such that $gQg^{-1} = Q$. Then $g \in N(Q)$. Lemma 30.7 shows that Q is a normal subgroup of $N(Q)$, so $N(Q)/Q$ is defined. Consider $gQ \in N(Q)/Q$. By the Fourth Isomorphism Theorem (see page 398), $\langle gQ \rangle = H/Q$ for some subgroup H of G that contains Q. Now $(gQ)^{p^k} = g^{p^k}Q = eQ = Q$, so $|gQ| = p^m$ for some nonnegative integer $m \leq k$. Therefore,

$$p^m = |\langle gQ \rangle| = |H/Q| = \frac{|H|}{|Q|} = \frac{|H|}{p^n},$$

so $|H| = p^{n+m}$. Since Q is a Sylow p-subgroup, n is the largest power of p that divides $|G|$. It follows that $m = 0$ and $H = Q$. Thus, $g \in H$ implies $g \in Q$, as desired. ∎

The Second Sylow Theorem

We are now in position to prove the Second Sylow Theorem, which tells us that any two Sylow p-subgroups of a group are conjugate. The next activity will help us to better understand the ideas behind the proof.

Preview Activity 30.11. Let $G = D_6$.

(a) Find all Sylow 2-subgroups of G. (You can use the result of the Second Sylow Theorem here.)

(b) Label the Sylow 2-subgroups of G as $K_1, K_2, \ldots, K_t$, and let $S = \{K_1, K_2, \ldots, K_t\}$. Find t and identify $K_1, K_2, \ldots, K_t$.

(c) Let H be one of the Sylow 2-subgroups of D_6. Let $[K_i]_H$ be the collection of all subgroups of G that are H-conjugate to K_i. Find $[K_i]_H$ for each i.

(d) Since $K_i \in [K_i]_H$ for each i, we can write S as a union of the sets $[K_i]_H$. However, there may be some overlap. To be more concise, we can write S as a disjoint union of the sets $[K_i]_H$, where we discard any repetition. Relabeling if necessary, identify s and the $K_1, K_2, \ldots, K_s$ so that

$$S = \bigcup_{v=1}^{s} [K_v]_H$$

is a disjoint union of sets of the form $[K_v]_H$.

The Second Sylow Theorem should not be all that surprising. Exercise (9) in Investigation 29 (see page 430) shows that if K is a subgroup of a group G, then any conjugate of K is isomorphic to K. This means that any conjugate of a Sylow p-subgroup is again a Sylow p-subgroup. The Second Sylow Theorem tells us that these are the *only* Sylow p-subgroups. Its proof uses H-conjugacy to count elements and is the first time we have used this technique.

Theorem 30.12 (Second Sylow Theorem). *Let G be a finite group, and let p be a prime divisor of $|G|$. If H and K are Sylow p-subgroups of G, then there is an element $g \in G$ so that $H = gKg^{-1}$.*

Proof. Let G be a finite group. Let p be a prime number, and let $k \in \mathbb{Z}^+$ such that $|G| = p^k m$ with $\gcd(p, m) = 1$. Let H and K be Sylow p-subgroups of G. Since K is a Sylow p-subgroup of G, we know $|K| = p^k$. Let $S = \{K_1, K_2, \ldots, K_t\}$ be the set of distinct conjugates of K in G. Lemma 30.9 shows that the number of conjugates of K in G is

$$t = [G : G \cap N(K)] = \frac{|G|}{|N(K)|}.$$

Since K is a subgroup of $N(K)$, we know p^k divides $|N(K)|$. Let $|N(K)| = p^k l$ for some positive integer l. So $t = \frac{p^k m}{p^k l} = \frac{m}{l}$. Since $\gcd(p, m) = 1$, it follows that $\gcd(p, t) = 1$. We will now show that $H = K_i$ for some i between 1 and t.

Let i and j be integers with $1 \le i, j \le t$. Note that both K_i and K_j are conjugate to K, and, by transitivity, K_i and K_j are conjugate to each other. Let $[K_i]_H$ be the collection of all subgroups of G that are H-conjugate to K_i. Since any H-conjugate of K_i is a G-conjugate of K_i, every H-conjugate of K_i is equal to K_r for some $1 \le r \le t$. Thus, $[K_i]_H \subseteq S$. Also, $K_i \in [K_i]_H$, so S is a union of the H-conjugacy classes of the form $[K_i]_H$. Since H-conjugacy is an equivalence relation, S can be written as a disjoint union of H-conjugacy classes with representatives from S. After relabeling, we can assume

$$S = \bigcup_{v=1}^{s} [K_v]_H, \tag{30.1}$$

for some integer s between 1 and t. By Lemma 30.9, the number of distinct H-conjugates of K_i in G is equal to $[H : H \cap N(K_i)] = \frac{|H|}{|H \cap N(K_i)|}$, which is a divisor of $|H| = p^k$. Let $\frac{|H|}{|H \cap N(K_i)|} = p^{k_i}$ for some positive integer $k_i \le k$. Then $|[K_i]_H| = p^{k_i}$.

We will now count the number of distinct conjugates of K in G in two different ways. On the one hand, S contains the distinct conjugates of K in G. On the other hand, we can add up the number of elements in each distinct conjugacy class in (30.1). Equating the two totals yields

$$t = |S| = \sum_{v=1}^{s} |[K_v]_H| = \sum_{v=1}^{s} p^{k_v}.$$

If all the k_v are positive, then $p \mid t$. This contradicts the fact that $\gcd(t, p) = 1$. Therefore, there is an index w so that $k_w = 0$. Then $\frac{|H|}{|H \cap N(K_w)|} = 1$, or $|H| = |H \cap N(K_w)|$. This means $H \subseteq N(K_w)$. So, if $x \in H$, then x has order p^m for some nonnegative integer m and $x K_w x^{-1} = K_w$. Lemma 30.10 then shows that $x \in K_w$. Thus, $H \subseteq K_w$. Since both K_w and H are Sylow p-subgroups of G, they have the same order. Thus, $H = K_w$, and by the transitivity of conjugacy, H and K are conjugate. ∎

The Third Sylow Theorem

The Third Sylow Theorem provides us with information about the number of Sylow p-subgroups in a group. More specifically:

Theorem 30.13 (Third Sylow Theorem). *Let G be a finite group, and let p be a prime dividing the order of G. The number of Sylow p-subgroups of G divides $|G|$ and is of the form $1 + pm$ for some nonnegative integer m.*

   Lemma 30.9 proves part of this result—namely, that the number of distinct conjugates of a Sylow p-subgroup in G divides $|G|$. The other conclusion, however, might seem mysterious. Activity 30.16 outlines a proof of the Third Sylow Theorem, and in doing so, sheds some light on why its second conclusion makes sense.

   To complete this investigation, we will consider two more examples that illustrate the use of the Sylow theorems.

Example 30.14. Let G be a group of order 36. We will show that the Sylow theorems guarantee that G contains a normal subgroup of order 3 or of order 9. If G is Abelian, then any subgroup of G is normal in G, so we will assume that G is non-Abelian. Since $36 = 2^2 \times 3^2$, the First Sylow Theorem tells us that G contains a Sylow 2-subgroup and a Sylow 3-subgroup. Let n_2 be the number of Sylow 2-subgroups of G and n_3 the number of Sylow 3-subgroups of G. The Third Sylow Theorem and its corollary (see Corollary 29.27 on page 427) tell us that $n_3 = 1 + 3t$ for some nonnegative integer t, and n_3 divides 4. These conditions can only be satisfied if $n_3 = 1$ or $n_3 = 4$. If $n_3 = 1$, then the Sylow 3-subgroup of G is normal in G. But what can we say if $n_3 = 4$?

   Let H be a Sylow 3-subgroup of G. Recall that $P(S)$ is the group of permutations of the set S and G/H is the set of left cosets of H in G. The mapping $\varphi : G \to P(G/H)$ defined by $\varphi(g) = \pi_g$ is a nontrivial homomorphism with $|\text{Im}(\varphi)| \geq [G : H]$, where $\pi_g(aH) = (ga)H$. (See Exercise (7).) Now $[G : H] = 4$, so $|P(G/H)| = 4! = 24$, and φ cannot be a monomorphism. Let $K = \text{Ker}(\varphi)$. Note that $|G/K| = \frac{|G|}{|K|} = \frac{36}{|K|}$ divides both $|G|$ and $|\text{Im}(\varphi)|$, so $|G/K| = 4$, $|G/K| = 6$, or $|G/K| = 12$. So $|K| = 3$, $|K| = 6$, or $|K| = 9$. If $|K| = 3$ or $|K| = 9$, then we are done. If $|K| = 6$, then the Third Sylow Theorem tells us that K contains a unique Sylow 3-subgroup N of order 3. Since $|N| = 3$, there is an element $n \in K$ such that $N = \langle n \rangle$. We will now show that N is normal in G. Let $g \in G$ and $m \in N$. Then $gmg^{-1} \in K$ since $K \triangleleft G$. If m is the identity e in K, then $gmg^{-1} = e \in N$. If m is not the identity, then $|gmg^{-1}| = 3$. Since N is the unique subgroup of K of order 3, N contains all the elements in K of order 3, and so $gmg^{-1} \in N$. Thus, N is a normal subgroup of G of order 3. In each case, we have that G contains a normal subgroup of order 9 or a normal subgroup of order 3.

Example 30.15. Let p and q be primes, and let G be a group of order p^2q^2. We will show that G is not a simple group—that is, G has a nontrivial proper normal subgroup. If G is Abelian, then any subgroup of G of order p is a nontrivial proper normal subgroup. So assume G is non-Abelian. Now consider the case where $p = q$. Then G is a p-group, and Theorem 29.14 shows that $Z(G)$ is nontrivial and is therefore a nontrivial proper normal subgroup of G. Thus, we can also assume that p and q are distinct primes with $p < q$. Let n_p be the number of Sylow p-subgroups of G, and let n_q be the number of Sylow q-subgroups of G. The Third Sylow Theorem and its corollary (Corollary 29.27) tell us that $n_q = 1 + qt$ for some nonnegative integer t, and n_q divides p^2. Since $p < q$, we cannot have $n_q = p$, so the only possibilities are $n_q = 1$ or $n_q = p^2$. If $n_q = 1$, then the Sylow q-subgroup of G is a nontrivial proper normal subgroup of G. So assume $n_q = p^2$. Then $1 + qt = p^2$, and $qt = p^2 - 1 = (p + 1)(p - 1)$. Since q is prime, it follows that q divides $p - 1$

or q divides $p + 1$. The former is impossible because $p < q$, so we conclude that q divides $p + 1$. The only way this can happen is if $p = 2$, $q = 3$, and $|G| = 36$. But Example 30.14 shows that any group of order 36 contains a normal subgroup of order 3 or 9, which completes the argument that G is not simple.

Concluding Activities

Activity 30.16. In this activity, we will provide the framework for a proof of the Third Sylow Theorem. Let G be a finite group, and let p be a prime divisor of $|G|$. Let $K_1 \ K_2$, ..., K_t be the distinct Sylow p-subgroups of G, and let $S = \{K_1, K_2, \ldots, K_t\}$. Let $H = K_1$, and for each i, let $[K_i]_H$ be the collection of all subgroups of G that are H-conjugate to K_i (as in our proof of the Second Sylow Theorem).

(a) In Activity 30.11, we saw that the sets $[K_i]_H$ did not all have the same number of elements. In particular, there was one case in which $[K_i]_H$ contained only one element. Use Lemma 30.10 to show that $|[K_i]_H| = 1$ if and only if $K_i = H$. (As we will see, this is where the 1 in $1 + pm$ comes from.)

(b) Suppose $K_i \neq H$. If we want to use H-conjugacy to count the elements in S (as we did in the proof of Theorem 30.12), then we will need to understand what $|[K_i]_H|$ can be. Review the proof of Theorem 30.12 and use part (a) of this activity to explain why p divides $|[K_i]_H|$. (This is where the pm part comes from.)

(c) As in the proof of Theorem 30.12, we can write S as a disjoint union of H-conjugacy classes—that is,

$$S = \bigcup_{v=1}^{s} [K_v]_H$$

for some integer s between 1 and t. Use this equation to explain why $t = 1 + pm$ for some nonnegative integer m.

(d) Combine the previous pieces of this activity into a complete and formal proof of the Third Sylow Theorem.

Activity 30.17. Let G be a finite group and p a prime divisor of $|G|$. Let H be a subgroup of G of order p^m for some m. Prove that H is a subgroup of a Sylow p-subgroup. (Hint: Refer to the proof of the Second Sylow Theorem.)

Exercises

(1) Let $G = S_3$.

    (a) For $H = \langle (1\ 2) \rangle$, find all H-conjugates of $K = \langle (1\ 2\ 3) \rangle$ in G.

    (b) For $H = \langle (1\ 2\ 3) \rangle$, find all H-conjugates of $K = \langle (1\ 2) \rangle$ in G.

$\star$ (2) Let G be a group, let H and K be subgroups of G, and let $a, b \in G$.

    (a) Prove part (b) of Lemma 30.3—that is, prove that

$$(ab)H(ab)^{-1} = a(bHb^{-1})a^{-1}.$$

    (b) Prove part (c) of Lemma 30.3—that is, prove that if $a \in H$, then $aHa^{-1} = H$.

(3) Find all Sylow p-subgroups of S_4. (Hint: See Exercise (37) on page 352 of Investigation 24.)

(4) In this exercise, we will investigate the question of whether subgroups inherit Sylow p-subgroups under intersections. Let G be a finite group, let K be a Sylow p-subgroup of G, and let N be any subgroup of G.

    (a) Must $K \cap N$ be a Sylow p-subgroup of N? (Hint: See Exercise (3).)

    (b) Must it be true that $K \cap N$ is a Sylow p-subgroup of N if N is normal in G? Prove your answer.

(5) Let p be a prime divisor of the order of a group G, and let N be a p-subgroup of G.

    (a) Must N be contained in every Sylow p-subgroup of G? Explain.

    (b) Are there any conditions on N so that N is contained in *every* Sylow p-subgroup of G? Prove your answer.

(6) **Classifying simple groups.** The Sylow theorems are especially useful in classifying simple groups, as we will explore in this exercise. We have already argued that the only simple Abelian groups are the groups $\mathbb{Z}_p$ for a prime p. Therefore, we will focus here on non-Abelian groups. Let G be a non-Abelian group of order n. Let p, q, and r be primes with $p < q < r$. At this point, we know the following:

    • One result of Activity 29.32 (see page 429) is that any group of order pq has a normal Sylow q-subgroup.

    • Exercise (25) of Investigation 29 (see page 431) shows that any non-Abelian p-group has nontrivial normal subgroups.

    • Exercise (18) of Investigation 29 (see page 430) shows that any group of order p^2q with $q \not\equiv 1 \pmod{p}$ has a normal Sylow q-subgroup.

    • Exercise (36) of Investigation 24 (see page 351) shows that D_n contains the nontrivial normal subgroup $\langle R \rangle$ of index 2 and also the normal subgroups $\langle r, R^2 \rangle$ and $\langle rR, r^2 \rangle$ when n is even.

In this exercise, we will show that the order of the smallest non-Abelian simple group is 60.

    (a) Show that if $|G| = pq^k$ with $p < q$ and $k \in \mathbb{Z}^+$, then G has a nontrivial proper normal subgroup.

    (b) Suppose $|G| = pqr$. Show that G has a nontrivial proper normal subgroup.

    (c) Explain why if G is a non-Abelian simple group of order less than or equal to 60, then $|G| = 60$.

$\star$ (7) This exercise was used in Example 30.15 and will be used in Exercise (8). Let G be a finite group and K a subgroup of G. Let G/K denote the collection of left cosets of K in G. (Note that we are not assuming the K is normal in G, so G/K may not be a group.) Recall that $P(S)$ denotes the collection of permutations of a set S.

(a) Let $a \in G$. Show that the function π_a defined by $\pi_a(gK) = (ag)K$ is a permutation of G/K.

(b) Define $\varphi : G \to P(G/K)$ by $\varphi(a) = \pi_a$. Show that φ is a homomorphism.

(c) If K is a proper subgroup of G, show that φ is not the trivial homomorphism.

(d) If K is a proper subgroup of G, show that $|\mathrm{Im}(\varphi)| \geq [G : K]$.

(8) **Simple groups of order 60.** In this exercise, we will use the result of Exercise (7) to show that the only simple group of order 60 is A_5. We know that there is no simple Abelian group of order 60, so for this exercise, assume that G is a non-Abelian simple group of order 60.

(a) For each prime divisor p of $|G|$, let n_p denote the number of Sylow p-subgroups of G. Determine the possible values of n_p for each p.

(b) Use Exercise (7) with K a Sylow 2-subgroup of G to show that $n_2 \neq 3$. (Hint: What can you say about $|\mathrm{Ker}(\varphi)|$?)

(c) If K and N are distinct Sylow 2-subgroups of G, what is the maximum value $|K \cap N|$ can have? Use this idea and count elements to show that $n_2 \neq 15$.

(d) Assume now that $n_2 = 5$, and let K be a Sylow 2-subgroup of G.
   (i) Let $\varphi : G \to P(G/K)$ be defined by $\varphi(a) = \pi_a$ (as in Exercise (7)). Explain why φ must be a monomorphism.
   (ii) Show that G is isomorphic to a subgroup G' of S_5.
   (iii) Show that $G' = A_5$, and conclude that $G \cong A_5$. (Hint: First, determine what group $G'A_5$ is if G' is not contained in A_5. Then use the Second Isomorphism Theorem to calculate $|G' \cap A_5|$.)

Connections

Lagrange's Theorem in Investigation 23 gives us some information about subgroups of finite groups. The First Sylow Theorem in Investigation 29 provides some additional information. The second and third Sylow theorems continue to give us even more information. In the process of understanding the second and third Sylow theorems, we developed the conjugacy relation, an equivalence relation as was discussed in Investigation 2. We also introduced a new subgroup, the normalizer, that adds to our knowledge of the subgroup structure of a group that began in Investigation 19. We saw further that the Class Equation from Investigation 29 is a powerful counting tool that is very useful in providing information about subgroups of finite groups.

Part VI

Fields and Galois Theory

Investigation 31

Finite Fields, the Group of Units in $\mathbb{Z}_n$, and Splitting Fields

---

Focus Questions

By the end of this investigation, you should be able to give precise and thorough answers to the questions listed below. You may want to keep these questions in mind to focus your thoughts as you complete the investigation.

- What is a finite field? How many elements can a finite field have?

- How are any two finite fields of the same order related?

- What is the structure of the group U_n of units in $\mathbb{Z}_n$?

- What is the splitting field of a polynomial? Why do we refer to "the" splitting field instead of "a" splitting field?

- How can one construct the splitting field of a polynomial?

Preview Activity 31.1. Let $p(x) = x^2 + x + [1]$ in $\mathbb{Z}_2[x]$.

(a) Find all the elements of $K = \mathbb{Z}_2[x]/p(x)$.

(b) Create the addition and multiplication tables for K.

(c) What special kind of ring is K? Explain.

(d) What kind of group is $U(K)$, the group of units in K? Explain.

(e) Explain why K contains all the roots of the polynomial $f(x) = x^4 - x$ over $\mathbb{Z}_2$. In fact, show that every element of K is a root of $f(x) = x^4 - x$ over $\mathbb{Z}_2$.

Introduction

The structure of the groups U_n, which consist of the units in $\mathbb{Z}_n$, is not obvious. However, with a bit of work we can determine the structure of these groups. To do so, we will first show that the group U_p is cyclic when p is prime. We will then use this result to describe every group U_n, whether n

445

is prime or not. In the process, we will classify all finite fields and also explore splitting fields of polynomials.

Finite Fields

When p is a prime, the ring $\mathbb{Z}_p$ is a field. The field $\mathbb{Z}_p$ is different than the more familiar fields like $\mathbb{Q}$ and $\mathbb{R}$ in that $\mathbb{Z}_p$ contains only a finite number of elements. The first question we will address in this investigation is whether there are finite fields other than $\mathbb{Z}_p$ for p prime. Preview Activity 31.1 shows that there are in fact such fields by demonstrating the existence of a field with four elements. As we will see, we can actually explicitly determine all the finite fields.

Activity 31.2. Let F be a finite field.

  (a) Every field has a characteristic. Explain why the characteristic of F must be prime. (Hint: In Exercise (7) on page 81 of Investigation 5, we showed that the characteristic of an integral domain is either zero or prime.)

  (b) Explain why F must contain a copy of $\mathbb{Z}_p$, and so, without loss of generality, we can consider F to be an extension of $\mathbb{Z}_p$.

  (c) Since F is finite, F must be a finite dimensional vector space over $\mathbb{Z}_p$.* Let $\{x_1, x_2, \ldots, x_n\}$ be a basis for F over $\mathbb{Z}_p$. Explain why F contains p^n elements.

The result of Activity 31.2 is that the order of any finite field is a power of a prime. We can arrive at this result in another way by observing that if F is a field of prime characteristic p, then as an Abelian group under addition every nonzero element has order p. Thus, by the Fundamental Theorem of Finite Abelian Groups (see page 412), F must be a p-group, and so $|F| = p^n$ for some integer n.

We now have the conditional statement that if F is a finite field, then the order of F is a power of a prime. It is natural to ask if the converse is true—that is, given a prime p and a positive integer n, is there a field of order p^n? This question is a little more complicated.

Let p be a prime, and let n be a positive integer. In Activity 31.1, we saw that the field F whose elements are the roots of the polynomial $f(x) = x^4 - x$ over $\mathbb{Z}_2$ is a field of order 4. As we will see, this construction works in general. If F is a finite field of order p^n, then the group $U(F)$ of units in F has order $p^n - 1$. So every nonzero element in F satisfies the polynomial equation $x^{p^n - 1} - 1 = 0$. If we toss in the 0 element, then every element in F is a root of the polynomial $x^{p^n} - x$. What we will ultimately show is that the set of all roots of the polynomial $x^{p^n} - x$ over $\mathbb{Z}_p$ is actually a field and has order p^n. Let $m = p^n$, and let $f(x) = x^m - x \in \mathbb{Z}_p[x]$. Kronecker's Theorem (see page 176) tells us that there is an extension field K of $\mathbb{Z}_p$ in which $f(x)$ *splits* into a product of linear factors. In other words, in $K[x]$, we have

$$f(x) = c(x - r_1)(x - r_2) \cdots (x - r_m)$$

for some $c, r_1, r_2, \ldots, r_m$ in K.

---

*If F is a field and K is a subfield of F, then F can be viewed as a vector space over K by simply taking scalar multiplication to be the same as multiplication within F, restricted to the set K of scalars.

We will show that the set $A = \{k \in K : f(k) = 0\}$ of roots of $f(x)$ in K is a subfield of K of order p^n, thus demonstrating that there is a field of order p^n. To do this, we first need to show that the set of roots of $f(x)$ is a subfield of K.

Activity 31.3. Let s and t be two roots of $f(x)$.

(a) Recall the Freshman's Dream from Exercise (11) of Investigation 5 (see page 81), which states that $(a + b)^p = a^p + b^p$ whenever a and b are elements of a commutative ring with prime characteristic p. A similar result is that $(a \pm b)^m = a^m \pm b^m$ for all $a \in K$. (See Exercise (4).) Use this fact to show that

$$f(s - t) = (s^m - s) - (t^m - t).$$

(b) Use the fact that s and t are roots of $f(x)$ to show that $s - t$ is a root of $f(x)$.

(c) Now assume that $t \neq 0$. Show that st^{-1} is a root of $f(x)$.

(d) Let $A = \{k \in K : f(k) = 0\}$. Explain why A is a subfield of K.

Activity 31.3 shows that the set A of roots of $f(x) = x^{p^n} - x$ in K is a subfield of K. By definition, all the elements of A are roots of $f(x)$. What remains is for us to determine the order of A. We know that A contains only $r_1, r_2, \ldots, r_m$, the roots of $f(x)$ in K. This means that $|A| \leq m = p^n$, but this inequality could be strict if $f(x)$ has a repeated root. So to show that $|A| = p^n$, we need to demonstrate that $f(x)$ has no repeated root in K. To do so, let r be a root of $f(x)$ in K. We need to show that $(x - r)^2$ does not divide $f(x)$. We will begin by rewriting $f(x)$ as follows:

$$\begin{aligned}
f(x - r) &= (x - r)^m - (x - r) \\
&= x^m - r^m - x + r \\
&= x^m - r - x + r \\
&= x^m - x \\
&= f(x).
\end{aligned}$$

Therefore,

$$f(x) = f(x - r) = (x - r)^m - (x - r) = (x - r)\left[(x - r)^{m-1} - 1\right].$$

But $(x - r)$ does not divide $(x - r)^{m-1} - 1$ (do you see why?) and so $(x - r)^2$ cannot divide $f(x)$. Thus, r is a root of $f(x)$ of multiplicity 1. Since our choice of r was arbitrary, we have shown that $f(x)$ does not have any repeated roots in K. Therefore, $|A| = p^n$, and A is the desired field.

We have now demonstrated that every finite field of characteristic p has order p^n for some n. We have also shown that there is a finite field of order p^n for each prime p and each positive integer n. The next theorem summarizes these results.

Theorem 31.4. *Every finite field has order p^n for some prime p and some positive integer n. Moreover, if p is a prime and n is a positive integer, then the set of all roots of the polynomial $x^{p^n} - x$ over $\mathbb{Z}_p$ is a field with p^n elements.*

One important question remains: How many finite fields are there of order p^n? We will answer this question later on in this investigation. In the meantime, we will return to one of the problems posed at the start of this investigation—namely, what can we say about the group U_p of units in $\mathbb{Z}_p$, where p is prime?

The Group of Units of a Finite Field

Preview Activity 31.5. Recall that $U_n = U(\mathbb{Z}_n)$ is the group of units in the ring $\mathbb{Z}_n$. There is a common structure to the groups U_p, where p is prime. Determine the structure of the groups U_2, U_3, U_5, U_7, and U_{11}. What property do they all have in common? Is this property satisfied even when p is not prime?

Preview Activity 31.5 indicates that there may be something special about the group of units in $\mathbb{Z}_p$, where p is prime. In fact, the observations we made about the units in $\mathbb{Z}_p$ are consequences of a more general result—namely, that if F is a finite field of order p^n for some prime p, then F contains an element with multiplicative order $p^n - 1$. A tool we will need to prove this result is the somewhat technical next lemma.

Lemma 31.6. *Let G be a group, and let $a, b \in G$ such that $ab = ba$. If $|a| = k$ and $|b| = n$, then there is an element $c \in G$ with $|c| = lcm(k, n)$.*

A note about Lemma 31.6: It is tempting to think that if $ab = ba$ in a finite group, then $|ab| =$ lcm$(|a|, |b|)$. Unfortunately, this is not true. For example, $|[4]| = 6$ and $|[16]| = 3$ in U_{35}, but $|[29]| = 2$. So there is some work to do to prove this lemma.

Proof of Lemma 31.6. Let G be a group, and let $a, b \in G$ such that $ab = ba$. Assume $|a| = k$ and $|b| = n$. Let $d = \gcd(k, n)$. Furthermore, let $p_1, p_2, \ldots p_m$ be the distinct primes that appear in either of the prime factorizations of k and n, and let

$$k = p_1^{\alpha_1} p_2^{\alpha_2} \cdots p_m^{\alpha_m} \quad \text{and} \quad n = p_1^{\beta_1} p_2^{\beta_2} \cdots p_m^{\beta_m},$$

where α_i and β_i are nonnegative integers for each i. Then

$$d = p_1^{\gamma_1} p_2^{\gamma_2} \cdots p_m^{\gamma_m},$$

where $\gamma_i = \min\{\alpha_i, \beta_i\}$ for each i. Break d into factors d_1 and d_2 as follows:

$$d_1 = p_1^{\sigma_1} p_2^{\sigma_2} \cdots p_m^{\sigma_m} \quad \text{where } \sigma_i = \begin{cases} \gamma_i & \text{if } \gamma_i = \alpha_i \text{ and } \alpha_i \neq 0 \\ 0 & \text{otherwise} \end{cases}$$

and

$$d_2 = p_1^{\mu_1} p_2^{\mu_2} \cdots p_m^{\mu_m} \quad \text{where } \mu_i = \begin{cases} \gamma_i & \text{if } \gamma_i = \beta_i \text{ and } \beta_i \neq \alpha_i \\ 0 & \text{otherwise.} \end{cases}$$

To see why $d = d_1 d_2$, we will examine $\sigma_i + \mu_i$. Consider two cases:

- If $\alpha_i \leq \beta_i$, then $\gamma_i = \alpha_i$ and $\sigma_i = \gamma_i$. So $\mu_i = 0$ and $\sigma_i + \mu_i = \gamma_i + 0 = \gamma_i$.

- If $\alpha_i > \beta_i$, then $\gamma_i = \beta_i$. So $\sigma_i = 0$ and $\mu_i = \gamma_i$. Again, $\sigma_i + \mu_i = 0 + \gamma_i = \gamma_i$.

It follows that

$$d_1 d_2 = p_1^{\sigma_1 + \mu_1} p_2^{\sigma_2 + \mu_2} \cdots p_m^{\sigma_m + \mu_m} = p_1^{\gamma_1} p_2^{\gamma_2} \cdots p_m^{\gamma_m} = d.$$

Next, we will compute $|a^{d_1}|$ and $|b^{d_2}|$. First, note that

$$|a^{d_1}| = \frac{k}{\gcd(k, d_1)} = \frac{k}{d_1} = p_1^{\alpha_1 - \sigma_1} p_2^{\alpha_2 - \sigma_2} \cdots p_m^{\alpha_m - \sigma_m}$$

and

$$|b^{d_2}| = \frac{n}{\gcd(n, d_2)} = \frac{n}{d_2} = p_1^{\beta_1 - \mu_1} p_2^{\beta_2 - \mu_2} \cdots p_m^{\beta_m - \mu_m}.$$

For each i, if $\alpha_i - \sigma_i > 0$, then $\alpha_i > 0$ and we must have $\sigma_i = 0$. Then $\gamma_i = \beta_i = \mu_i$, and so $\beta_i - \mu_i = 0$. Thus, $|a^{d_1}|$ and $|b^{d_2}|$ have no prime factors in common, and $\gcd(|a^{d_1}|, |b^{d_2}|) = 1$. Exercise (18) of Investigation 20 (see page 298; this result states that if a and b are elements in a group G with $ab = ba$ and $\gcd(|a|, |b|) = 1$, then $|ab| = |a||b|$) now implies that

$$
\begin{aligned}
|a^{d_1} b^{d_2}| &= |a^{d_1}||b^{d_2}| \\
&= (p_1^{\alpha_1 - \sigma_1} p_2^{\alpha_2 - \sigma_2} \cdots p_m^{\alpha_m - \sigma_m})(p_1^{\beta_1 - \mu_1} p_2^{\beta_2 - \mu_2} \cdots p_m^{\beta_m - \mu_m}) \\
&= p_1^{(\alpha_1 + \beta_1) - (\sigma_1 + \mu_1)} p_2^{(\alpha_2 + \beta_2) - (\sigma_2 + \mu_2)} \cdots p_m^{(\alpha_m + \beta_m) - (\sigma_m + \mu_m)} \\
&= \frac{p_1^{(\alpha_1 + \beta_1)} p_2^{(\alpha_2 + \beta_2)} \cdots p_m^{(\alpha_m + \beta_m)}}{p_1^{(\sigma_1 + \mu_1)} p_2^{(\sigma_2 + \mu_2)} \cdots p_m^{(\sigma_m + \mu_m)}} \\
&= \frac{kn}{p_1^{\gamma_1} p_2^{\gamma_2} \cdots p_m^{\gamma_m}} \\
&= \frac{kn}{d} \\
&= \mathrm{lcm}(k, n).
\end{aligned}
$$

Therefore, we have found an element in G of order $\mathrm{lcm}(k, n)$, as desired. ∎

The result of Lemma 31.6 is that, given commuting elements a and b of orders k and n in a group G, we will always be able to find an element in G of order $\mathrm{lcm}(a, b)$. We will use this fact in the next activity to prove that the group of units of any finite field is a cyclic group. Since $\mathbb{Z}_p$ is a field when p is prime, it naturally follows that the groups U_p are cyclic when p is prime.

Activity 31.7. Let F be a finite field, and let $G = U(F) = \{x \in F : x \neq 0_F\}$ be the group of units in F. Since F is finite, we can enumerate the elements in G. Let $G = \{a_1, a_2, \ldots, a_m\}$, and also let $|a_i| = n_i$ for each i.

(a) Explain why we can find an element $b_1 \in G$ with $|b_1| = \mathrm{lcm}(n_1, n_2)$.

(b) How can we show that for each j with $1 \leq j < m$, there exists an element b_j in G such that $|b_j| = \mathrm{lcm}(n_1, n_2, \ldots, n_{j+1})$?

(c) Let $k = |b_{m-1}|$. How is k related to each n_i? How is k related to $|G|$?

(d) What can we say about a_i^k for each i?

(e) How many roots can the polynomial $p(x) = x^k - 1$ in $F[x]$ have? What does this tell us about the relationship between k and m? Explain.

(f) Explain how we have proved the following theorem.

Theorem 31.8. *Let F be a finite field. The group $G = U(F)$ of units in F is a cyclic group.*

Now that we have proved Theorem 31.8, we can determine the groups U_n of the units in $\mathbb{Z}_n$, as we will see in the next section.

The Group of Units of $\mathbb{Z}_n$

Let n be a positive integer. The group U_n of units of $\mathbb{Z}_n$ is a finite Abelian group. The Fundamental Theorem of Finite Abelian Groups (see page 412) tells us that U_n must be isomorphic to a direct sum of finite cyclic groups. Recall that to find the summands in such a direct sum, we broke up the group in question into an internal direct product of its p-primary components. We then decomposed each p-primary component into internal direct products of cyclic subgroups of largest order. Even without knowing the order of U_n directly, we can do something similar here.

First, the result of Exercise (11) tells us that if $m = n_1 n_2 \cdots n_k$, where n_i and n_j are relatively prime for $i \neq j$, then

$$U_m \cong (U_{n_1} \oplus U_{n_2} \oplus \cdots \oplus U_{n_k}).$$

So once we decompose m into a product of primes, we have reduced our problem to that of determining the groups of the form U_{p^n} for each prime p.

To tackle this problem, we will first determine the order of U_{p^n}. Recall that the elements of U_{p^n} are the congruence classes $[a]$ such that $\gcd(a, p^n) = 1$. The only way $\gcd(a, p^n) > 1$ is if a is a multiple of p. The multiples of p between 1 and p^n are

$$p, 2p, 3p, \ldots, (p^{n-1})p,$$

so there are a total of p^{n-1} multiples of p between 1 and p^n. Thus, the number of integers a between 1 and p^n that are relatively prime to p^n is $p^n - p^{n-1}$. Therefore, $|U_{p^n}| = p^n - p^{n-1}$.

We will now proceed to the decomposition of U_{p^n} as a product of cyclic groups. As is often the case, the prime 2 exhibits different behavior than the odd primes. One can experiment with many odd primes and notice that U_{p^n} seems to always be a cyclic group. That conjecture is verified by the following theorem. The proof is a bit technical, so we present it in its entirety.

Theorem 31.9. *If p is an odd prime and n a positive integer, then $U_{p^n} \cong \mathbb{Z}_{p^n - p^{n-1}}$.*

Proof. Recall that $[a]_k$ denotes the class of the integer a in U_k. We already know that $|U_{p^n}| = p^n - p^{n-1} = (p-1)p^{n-1}$. To show that U_{p^n} is cyclic, we will demonstrate the existence of an element of order $p - 1$ and an element of order p^{n-1} in U_{p^n}. First, we will find an element of order $p - 1$.

Since U_p is cyclic, there is a generator $[a]_p$ for U_p with $|[a]_p| = p - 1$. Since $\gcd(a, p) = 1$, we also know that $\gcd(a, p^n) = 1$, and so $[a]_{p^n} \in U_{p^n}$. Suppose $|[a]_{p^n}| = k$. Then k divides $|U_{p^n}| = p^{n-1}(p-1)$, and so

$$kr = p^{n-1}(p - 1) \tag{31.1}$$

for some $r \in \mathbb{Z}^+$. Also, $([a]_{p^n})^k = [1]_{p^n}$, and so p^n divides $a^k - 1$, which implies that p divides $a^k - 1$. Thus, $[a]_p^k = [1]_p$ (in $\mathbb{Z}_p$), and so $|[a]_p| = p - 1$ divides k. Let $s \in \mathbb{Z}^+$ such that

$$k = (p - 1)s. \tag{31.2}$$

Equations (31.1) and (31.2) combine to give us $rs = p^{n-1}$, and so $s = p^j$ for some j with $0 \leq j \leq n - 1$. Thus,

$$|[a]_{p^n}| = k = (p - 1)s = (p - 1)p^j,$$

and

$$|[a]_{p^n}^{p^j}| = \frac{|[a]_{p^n}|}{\gcd(p^j, |[a]_{p^n}|)} = \frac{(p-1)p^j}{p^j} = p - 1.$$

Therefore, $[a]_{p^n}^{p^j}$ is an element of order $p - 1$ in U_{p^n}.

Next we will show that $|[p+1]_{p^n}| = p^{n-1}$. First, we will prove by induction that for each integer $k \geq 2$, $(p+1)^{p^{k-2}} - 1$ is divisible by p^{k-1} but not p^k. For our base case, note that $(p+1)^1 - 1 = p$ is divisible by p but not p^2, so our statement is true when $k = 2$. Now assume that, for some integer $k \geq 2$, $(p+1)^{p^{k-2}} - 1$ is divisible by p^{k-1} but not p^k. Then $(p+1)^{p^{k-2}} - 1 = qp^{k-1}$ for some integer q with $\gcd(q, p) = 1$. So $(p+1)^{p^{k-2}} = 1 + qp^{k-1}$, and raising both sides to the pth power gives us

$$(p+1)^{p^{k-1}} = (1 + qp^{k-1})^p$$

$$= 1 + p(qp^{k-1}) + \sum_{i=2}^{p} \binom{p}{i} (qp^{k-1})^i$$

$$= 1 + qp^k + p^{k+1}r$$

for some integer r. Therefore, p^k divides $(p+1)^{p^{k-1}} - 1$ but p^{k+1} does not (because $\gcd(q, p) = 1$). From this, we can conclude that $[p+1]_{p^n}^{p^{n-1}} = [1]_{p^n}$ but $[p+1]_{p^n}^{p^{k-1}} \neq [1]_{p^n}$ for any smaller positive value of k. This completes our proof that $|[p+1]_{p^n}| = p^{n-1}$.

Finally, since $|[a]_{p^n}^{p^j}| = p-1$ and $|[p+1]_{p^n}| = p^{n-1}$, it follows that $\gcd(|[a]_{p^n}^{p^j}|, |[p+1]_{p^n}|) = 1$. Thus,

$$|[a]_{p^n}^{p^j}[p+1]_{p^n}| = |[a]_{p^n}^{p^j}||[p+1]_{p^n}| = (p-1)p^{n-1} = |U_{p^n}|,$$

which implies that U_{p^n} is a cyclic group of order $(p-1)p^{n-1}$. We have therefore shown that $U_{p^n} \cong \mathbb{Z}_{p^n - p^{n-1}}$ when p is an odd prime. ∎

Next, we will turn to the case when $p = 2$. We know that $U_2 = \{[1]\}$ is a cyclic group of order 1 and that $U_4 = \{[1], [3]\} = \langle [3] \rangle$ is a cyclic group of order 2. Let us now consider the groups U_{2^n} for $n \geq 3$.

Recall that when we decomposed finite Abelian p-groups as internal direct products of cyclic groups (see Investigation 28), we began by finding an element of largest order. We have already shown that $|U_{2^n}| = 2^n - 2^{n-1} = 2^{n-1}$. Some routine calculations show that in U_8 the largest order of any element is 2, in U_{16} the largest order is 4, in U_{32} the largest order is 8, and the pattern seems to continue. This might lead us to conjecture that $U_{2^n} \cong (\mathbb{Z}_2 \oplus \mathbb{Z}_{2^{n-2}})$. It is also curious that, in each of the above examples, [5] appears to have order 2^{n-2}. This observation turns out to be a key idea in the proof of our conjecture, which is stated formally as Theorem 31.10.

Theorem 31.10. *Let $n \geq 3$ be an integer. Then*

$$U_{2^n} \cong (\mathbb{Z}_2 \oplus \mathbb{Z}_{2^{n-2}}).$$

Proof. First, we will demonstrate that [5] has order 2^{n-2} in U_{2^n}. Our proof will rely on the fact that $5^{2^{n-3}} \equiv 1 + 2^{n-1} \pmod{2^n}$ for each $n \geq 3$, which we will verify by induction. The case where $n = 3$ is straightforward, since $5^1 = 1 + 2^2$. Now assume that $5^{2^{n-3}} \equiv 1 + 2^{n-1} \pmod{2^n}$ for some $n \geq 3$. Then

$$5^{2^{n-3}} = 1 + 2^{n-1} + t2^n$$

for some integer t. Dividing t by 2, we obtain integers q and r with $0 \le r < 2$ and $t = 2q + r$. Then

$$5^{2^{n-3}} = 1 + 2^{n-1} + (2q + r)2^n = 1 + 2^{n-1} + r2^n + q2^{n+1}.$$

So

$$5^{2^{n-3}} \equiv 1 + 2^{n-1} + r2^n \pmod{2^{n+1}}.$$

We can now square both sides of this equation to obtain

$$\begin{aligned} 5^{2^{n-2}} &\equiv \left(1 + 2^{n-1} + r2^n\right)^2 \\ &\equiv 1 + 2^n + r2^{n+1} + 2^{2(n-1)} + r2^{2n} + r^2 2^{2n} \\ &\equiv 1 + 2^n \pmod{2^{n+1}}. \end{aligned}$$

So, we have $[5]^{2^{n-3}} = [1 + 2^{n-1}]$ in U_{2^n}. Now $[1 + 2^{n-1}]^2 = [1 + 2^n + 2^{2(n-1)}] = [1]$ in U_{2^n}, so $|[1 + 2^{n-1}]| = 2$ in U_{2^n}. This means $|[5]^{2^{n-3}}| = 2$ or that $|[5]| = 2^{n-2}$. Thus, in U_{2^n} we have that $\langle [5] \rangle \cong \mathbb{Z}_{2^{n-2}}$. Since $|U_{2^n}| = 2^{n-1}$, to complete our decomposition of U_{2^n} as an internal direct product of subgroups we only need to find an element of order 2 that is not in $\langle [5] \rangle$. We know that $[-1]^2 = [1]$ in U_{2^n}, and so $[-1]$ has order 2 in U_{2^n}. Recall that a cyclic group of order m has a unique subgroup of each order that divides m. (See Theorem 20.7 on page 295.) We have already seen that $\langle [5] \rangle$ has a subgroup $\langle [1 + 2^{n-1}] \rangle$ of order 2. Since $1 + 2^{n-1} \not\equiv -1 \pmod{2^n}$, we also see that $\langle [-1] \rangle \ne \langle [1 + 2^{n-1}] \rangle$. Therefore, $[-1] \notin \langle [5] \rangle$ and so

$$U_{2^n} = (\langle [-1] \rangle \times \langle [5] \rangle) \cong (\mathbb{Z}_2 \oplus \mathbb{Z}_{2^{n-2}})$$

when $n \ge 3$. ∎

Example 31.11. As an example, we will decompose U_{200} into a sum of cyclic groups. Note that $200 = 2^3 \times 5^2$, and so

$$U_{200} \cong (U_{2^3} \oplus U_{5^2}).$$

Theorem 31.10 shows that $U_{2^3} \cong (\mathbb{Z}_2 \oplus \mathbb{Z}_2)$, and Theorem 31.9 tells us that $U_{5^2} \cong \mathbb{Z}_{20}$. Combining these results shows that

$$U_{200} \cong (\mathbb{Z}_2 \oplus \mathbb{Z}_2 \oplus \mathbb{Z}_{20}).$$

Splitting Fields

In this section, we will return to the question of how many fields there are of a given order p^n. To address this question, we will in fact answer a much more general question about a special class of field extensions known as *splitting fields*.

Preview Activity 31.12. Splitting fields, which we will define shortly, are related to roots of polynomials. To begin our study of splitting fields, we will begin by investigating the structure of sets of polynomials with a common root. Let F be a field, and let K be an extension of F containing an element a that is a root of some nonzero polynomial in $F[x]$. Let $I = \{f(x) \in F[x] : f(a) = 0\}$.

(a) Show that I is an ideal of $F[x]$.

(b) We will now determine more specifically the structure of the ideal I.

(i) Explain why I must contain a polynomial $p(x)$ of smallest degree that has a as a root. (Hint: What principle have we used to find smallest elements?)

(ii) Show that if $f(x) \in I$, then $p(x)$ divides $f(x)$. Conclude that $I = \langle p(x) \rangle$. (Hint: How can we divide one polynomial into another?)

(iii) Show that $p(x)$ is an irreducible polynomial.

Elements that are roots of polynomials (as in Activity 31.12) are similar to the numbers we defined to be *algebraic* in Investigation 12. The next definition builds on this important similarity by extending the definition of an algebraic number.

Definition 31.13. Let F be a field, and let K be an extension field of F. An element $a \in K$ is **algebraic** over F if a is the root of some nonconstant polynomial in $F[x]$.

We have many familiar examples of algebraic elements. For example, $\sqrt{2}$ is algebraic over $\mathbb{Q}$ (being a root of $x^2 - 2$) and i is algebraic over $\mathbb{R}$ (as a root of $x^2 + 1$). Not all numbers are algebraic. Elements that are not algebraic are called **transcendental**.[†] Joseph Liouville[‡] was the first to prove the existence of transcendental numbers, while Hermite[§] showed that the number e is transcendental over $\mathbb{Q}$ and Lindemann[¶] proved that π is also transcendental over $\mathbb{Q}$. In general, it is very difficult to prove that a number is transcendental, and we won't say anything more about such numbers.

Let's now return to the question of how many fields there are of a given finite order. Let p be a prime, and let K and L be fields of order p^n for some positive integer n. As vector spaces over $\mathbb{Z}_p$, the two fields K and L have the same dimension. Therefore, K and L are isomorphic as vector spaces. In addition, $U(K)$ and $U(L)$ are both cyclic groups of the same order and are therefore isomorphic as groups. With this much structure in common, we might think that K and L are isomorphic as fields.

In fact, there is another very good reason to suspect that any two fields of order p^n are closely related. If K is any field of order p^n, then $U(K)$ is a cyclic group of order $p^n - 1$, and so $a^{p^n - 1} = 1$ for every nonzero $a \in K$. Thus, $a^{p^n} = a$ for every $a \in K$, which means that every element of K is a root of the polynomial $f(x) = x^{p^n} - x \in \mathbb{Z}_p[x]$. Since we also know that $|K| = p^n$ and that $f(x)$ has at most p^n roots (see Theorem 10.9 on page 147), it follows that K must be a field of smallest order that contains all the roots of $f(x)$. So any two fields of order p^n are fields of the same size that contain all the roots of the polynomial $f(x) = x^{p^n} - x$ over $\mathbb{Z}_p$. (Note the connection to Theorem 31.4, which states that the set of all roots of $f(x) = x^{p^n} - x$ is a field of order p^n. The above observations indicate that *every* field of order p^n contains this set of roots.) For these reasons, we might expect that any two fields of order p^n are isomorphic. We will verify this conjecture by proving a more general result.

As we have just discussed, a field K of order $m = p^n$ contains exactly the m roots of the polynomial $f(x) = x^m - x$ over $\mathbb{Z}_p$. Therefore, we can factor the polynomial $f(x)$ in $K[x]$ as a product of linear terms. In other words,

$$f(x) = c(x - r_1)(x - r_2) \cdots (x - r_n)$$

for some elements $c, r_1, r_2, \ldots, r_n$ in K. In this case, we say that $f(x)$ *splits* over the field K. Of

[†]Leibniz appears to have first used the word transcendental (as "transcendentes") in the fall of 1673 in *Progressio figurae segmentorum circuli aut ei sygnotae* when referring to the dimension of a sequence as being transcendent.
[‡]In the *Comptes-rendus*, **18**, 1844, p. 883, p. 910 (reproduced in *Journal de Mathématiques Pures et Appliquées*, **16**, 1851).
[§]C. Hermite, "Sur la fonction exponentielle," *C. R. Acad Sei.*, **77**, 1873, pp. 18–24, 74–79, 226–233, and 285–293; also in *Œuvres*, **3**, Gauthier-Villars, Paris, 1912, pp. 150–181.)
[¶]F. Lindemann, "Über die Zahl π," *Mathematische Annalen*, **20**, 1882, pp. 213–225. This also seems to be the place where the term "transcendental number" appears (as "transscendente Zahl").

course, this idea works in other fields as well. For example, $x^2 + 1 = (x + i)(x - i)$ in $\mathbb{C}[x]$. A smallest field that contains all the roots of a polynomial so that we can factor the polynomial into linear factors is called a *splitting field*.

Definition 31.14. Let F be a field, and let $f(x) \in F[x]$.

(i) The polynomial $f(x)$ **splits** in an extension K of F if

$$f(x) = c(x - r_1)(x - r_2) \cdots (x - r_n)$$

for some elements $c, r_1, r_2, \ldots, r_n$ in K.

(ii) An extension field K of F is a **splitting field** for $f(x)$ over F if $f(x)$ splits in K but does not split in any proper subfield of K.

In other words, a splitting field for a polynomial $f(x)$ in $F[x]$ is an extension of F of smallest order that contains all the roots of $f(x)$. It may not be surprising that any two splitting fields of a polynomial $f(x)$ over a field F are isomorphic, and we will finish this investigation by verifying this fact. Since any fields of order p^n for a prime p and a positive integer n are splitting fields for $f(x) = x^{p^n} - x$, this will show that any two finite fields of the same order are isomorphic.

We will need some results about extensions and irreducible polynomials to arrive at our final result. Recall that irreducible polynomials in polynomial rings behave like the primes do in the integers. In particular, irreducible polynomials form the building blocks of all polynomials in the sense that every polynomial in a polynomial ring over a field can be factored in some unique way as a product of irreducible polynomials. Since irreducible polynomials cannot be factored in a nontrivial way, if an element a is a root of a polynomial $f(x)$ in some polynomial ring $F[x]$, then a must be a root of some irreducible polynomial over $F[x]$. We can say a bit more than this, as the result of Activity 31.12 shows. The next theorem states that result formally.

Theorem 31.15. *Let F be a field, and let K be an extension of F containing an element a that is algebraic over F. Then there exists an irreducible polynomial $p(x) \in F[x]$ that has a as a root. Moreover, any polynomial in $F[x]$ that has a as a root is divisible by $p(x)$.*

The irreducible polynomial $p(x)$ in Theorem 31.15 need not be unique. In fact, any nonzero scalar multiple of $p(x)$ will work. If we want to specify a specific polynomial to use, we often select a monic polynomial—that is, one with a leading coefficient of 1. This polynomial is given a special name.

Definition 31.16. The **minimal polynomial** for an algebraic element a over a field F is the monic irreducible polynomial in $F[x]$ having a as a root.

So, for example, $x^2 + 1$ is the minimal polynomial for i over $\mathbb{R}$.

In Investigation 12 (specifically, Theorem 12.11 on page 175), we showed that the quotient ring $F[x]/\langle p(x) \rangle$ is a field if and only if $p(x)$ is irreducible in $F[x]$. There is something even more we can say about this quotient field—namely, that it is (up to isomorphism) the smallest extension of F that contains a. We will explore this idea in the next activity.

Activity 31.17. Let F be a field, let a be an element that is algebraic over F, and let $p(x)$ be an irreducible polynomial in $F[x]$ having a as a root. Denote by $F(a)$ the smallest extension of F that contains a. Define the function $\varphi : F[x] \to F(a)$ by $\varphi(f(x)) = f(a)$.

(a) Explain why φ is a ring homomorphism.

(b) Show that $\text{Ker}(\varphi) = \langle p(x) \rangle$.

(c) Show that $\text{Im}(\varphi) = F(a)$.

(d) Use the First Isomorphism Theorem for rings to conclude the following theorem:

> **Theorem 31.18.** *Let F be a field, let a be an algebraic element over F, and let $p(x) \in F[x]$ be an irreducible polynomial having a as a root. Then*
>
> $$F(a) \cong F[x]/\langle p(x) \rangle.$$

A useful corollary of Theorem 31.18 is the following:

Corollary 31.19. *Let F be a field, and let $p(x) \in F[x]$ be an irreducible polynomial. If a is a root of $p(x)$ in some extension E of F and b is a root of $p(x)$ in some extension K of F, then $F(a) \cong F(b)$.*

Proof. Let a be a root of $p(x)$ in some extension E of F, and let b be a root of $p(x)$ in some extension K of F. Theorem 31.18 shows that $F(a) \cong F[x]/\langle p(x) \rangle$ and also that $F[x]/\langle p(x) \rangle \cong F(b)$. Therefore, $F(a) \cong F(b)$. ∎

As a reminder, our goal in this section is to show that any two splitting fields for a polynomial are isomorphic. Corollary 31.19 seems to have us heading in the right direction. However, Corollary 31.19 only helps us if our polynomial is irreducible. Since any polynomial is the product of irreducible polynomials, we will need to do a bit more work to reach our destination. Our next step is to generalize Corollary 31.19 to the case where we have isomorphic—but not necessarily equal—base fields. (In Corollary 31.19, F is the base field.) Before we can do that, though, we need to discuss how to extend a field isomorphism to an isomorphism of polynomial rings.

Suppose F and E are fields and $\varphi : F \to E$ is an isomorphism. There is a natural function that maps $F[x]$ into $E[x]$ as follows: assign to the polynomial

$$f(x) = a_n x^n + a_{n-1} x^{n-1} + \cdots + a_1 x + a_0 \in F[x]$$

the polynomial

$$\varphi(a_n) x^n + \varphi(a_{n-1}) x^{n-1} + \cdots + \varphi(a_1) x + \varphi(a_0) \in E[x].$$

Although it is a bit of an abuse of notation, we will denote this image of $f(x)$ as $\varphi(f(x))$. Note that $\varphi : F[x] \to E[x]$ has the property that $\varphi(c) = c$ for any $c \in F$. So this mapping between the polynomial rings agrees with the corresponding isomorphism between the coefficient fields. For this reason, we say that $\varphi : F[x] \to E[x]$ **extends** the function φ, and it makes sense to use the same notation for both maps. Since φ is an isomorphism, we might expect that the extension of φ to the corresponding polynomial rings preserves much of the structure of the polynomial rings. In fact, in Exercise (5), you are asked to show that $\varphi : F[x] \to E[x]$ is a ring isomorphism. Knowing this, we can now proceed to the major result that will ultimately allow us to prove that any two splitting fields for a polynomial are isomorphic.

Theorem 31.20. *Let F be a field, and let $p(x) \in F[x]$ be an irreducible polynomial. Let a be a root of $p(x)$ in some extension F' of F. Let E be a field, and let $\varphi : F \to E$ be a field isomorphism. If b is a root of $\varphi(p(x))$ in some extension E' of E, then there is an isomorphism Φ from $F(a)$ to $E(b)$ so that*

(i) $\Phi(c) = \varphi(c)$ *for all $c \in F$, and*

(ii) $\Phi(a) = b$.

Proof. Let b be a root of $\varphi(p(x))$ in some extension E' of E. Since φ is an isomorphism, the polynomial $\varphi(p(x))$ is irreducible in $E[x]$. Let $\beta : E[x]/\langle\varphi(p(x))\rangle \to E(b)$ be the isomorphism whose existence was demonstrated in Theorem 31.18. Recall that $\beta(g(x) + \langle\varphi(p(x))\rangle) = g(b) \in E(b)$ and so $\beta(x + \langle\varphi(p(x))\rangle) = b$. Let α be the inverse of the corresponding isomorphism from $F[x]/\langle p(x)\rangle$ to $F(a)$. If we can show that there is an isomorphism $\psi : F[x]/\langle p(x)\rangle \to E[x]/\langle\varphi(p(x))\rangle$, then we can combine these three isomorphisms

$$F(a) \xrightarrow{\alpha} F[x]/\langle p(x)\rangle \xrightarrow{\psi} E[x]/\langle\varphi(p(x))\rangle \xrightarrow{\beta} E(b)$$

to obtain an isomorphism $\Psi : F(a) \to E(b)$.

Define $\psi : F[x]/\langle p(x)\rangle \to E[x]/\langle\varphi(p(x))\rangle$ by

$$\psi(f(x) + \langle p(x)\rangle) = \varphi(f(x)) + \langle\varphi(p(x))\rangle.$$

That ψ is a well-defined isomorphism is left for you to prove in Exercise (3). What remains is to demonstrate that $\Phi = \beta\psi\alpha$ satisfies conditions (i) and (ii). Let $c \in F$. Then $\alpha^{-1}(c + \langle p(x)\rangle) = c$, and so $\alpha(c) = c + \langle p(x)\rangle$. Thus,

$$\Phi(c) = (\beta\psi)(\alpha(c)) = \beta(\psi(c + \langle p(x)\rangle)) = \beta(\varphi(c) + \langle\varphi(p(x))\rangle) = \varphi(c),$$

and Φ satisfies condition (i). Moreover, $\alpha^{-1}(x + \langle p(x)\rangle) = a$, and so $\alpha(a) = x + \langle p(x)\rangle$. That, and the fact that $\varphi(x) = x$, gives us

$$\begin{aligned} \Phi(a) &= (\beta\psi)(\alpha(a)) \\ &= \beta(\psi(x + \langle p(x)\rangle)) \\ &= \beta(\varphi(x) + \langle\varphi(p(x))\rangle) \\ &= \beta(x + \langle\varphi(p(x))\rangle) \\ &= b, \end{aligned}$$

and so Φ satisfies condition (ii) as well. Therefore, Φ is the desired isomorphism. ∎

We now need to extend the result of Theorem 31.20 to any polynomial, and then we can show that any two splitting fields for a polynomial are isomorphic.

Corollary 31.21. *Let φ be an isomorphism from a field F to a field E, and let $f(x) \in F[x]$. Let F' be a splitting field for $f(x)$ over F, and let E' be a splitting field of $\varphi(f(x))$ over E. Then there is an isomorphism Φ from F' to E' so that $\Phi(c) = \varphi(c)$ for all $c \in F$.*

Since we don't know that $f(x)$ is irreducible, we cannot directly use Theorem 31.20 to prove Corollary 31.21. However, we can split $f(x)$ into irreducible factors and use Theorem 31.20 on each factor. This will allow us to build up extension fields with the desired properties from polynomials of smaller degree than $f(x)$. Mathematical induction is the tool that formalizes this idea in the proof that follows.

Proof of Corollary 31.21. We will induct on $n = \deg(f(x))$. If $n = 1$, then $F' = F$, $E' = E$, and $\Phi = \varphi$ is the desired isomorphism. Now assume $n > 1$. For our inductive hypothesis, assume that Corollary 31.21 is true for any polynomial of degree less than n.

Let $p(x)$ be an irreducible factor of $f(x)$, a a root of $p(x)$ in F', and b a root of $\varphi(p(x))$ in E'. By Theorem 31.20, there is an isomorphism α from $F(a)$ to $E(b)$ that agrees with φ on F and maps a to b. By the Division Algorithm,

$$f(x) = (x - a)g(x)$$

for some $g(x) \in F(a)[x]$ with $\deg(g(x)) < \deg(f(x))$. It follows that F' is a splitting field for $f(x)$ over $F(a)$ and E' is a splitting field for $\varphi(f(x))$ over $E(b)$. By our induction hypothesis (using $g(x)$ as our polynomial in $F(a)[x]$, and α an isomorphism from $F(a)$ to $E(b)$), there is an isomorphism Φ from F' to E' so that $\Phi(r) = \alpha(r)$ for all $r \in F(a)$. Note that if $c \in F$, then

$$\Phi(c) = \alpha(c) = \varphi(c),$$

and so Φ is the desired isomorphism. ∎

We can now establish our main result that splitting fields are unique.

Activity 31.22. Explain how Corollary 31.21 proves the following theorem:

Theorem 31.23. *Let F be a field, and let $f(x) \in F[x]$. If K and L are splitting fields for $f(x)$ over F, then $K \cong L$.*

Because any two splitting fields of a polynomial are isomorphic, we usually just refer to "the" splitting field of a polynomial. In particular, since a finite field of order p^n is the splitting field of $x^{p^n} - x$ over $\mathbb{Z}_p$, we have just proved the next result.

Corollary 31.24. *Let K and L be finite fields of order p^n for some prime p and some positive integer n. Then $K \cong L$.*

In other words, there is exactly one finite field (up to isomorphism) of any given order p^n, where p is prime and n is a positive integer. Different notations are used to denote finite fields, but we will denote the field of order p^n as $\mathbb{F}_{p^n}$. This field is called the *Galois field* of order p^n in honor of the mathematician Evariste Galois, whose work on the solvability of polynomial equations formed the basis of much of group theory.

Concluding Activities

Activity 31.25. Theorem 6.10 showed us the form of a simple or primitive quadratic extension of a field. Prove the following generalization of this theorem:

Theorem 31.26. *Let F be a field and $p(x)$ an irreducible polynomial of degree n in $F[x]$. Let E be an extension of F containing a root r of $p(x)$. The smallest extension of F containing field containing r has the form*

$$F(r) = \{a_0 + a_1 r + a_2 r^2 + \cdots + a^{n-1} r^{n-1} : a_0, a_1, \ldots, a_{n-1} \in F\}. \tag{31.3}$$

The field $F(r)$ is called a **simple** or **primitive** extension of F and is obtained from F by attaching the root r to F. Note that $F(r)$ is essentially the field $F[x]/\langle p(x) \rangle$.

Activity 31.27. Although we have studied the properties of splitting fields, we have not yet discussed a general method for finding the splitting field of a polynomial. In this activity, we will illustrate such a method with the example of $f(x) = x^4 - 5x^2 + 6 \in \mathbb{Q}[x]$.

(a) Show that $(x^2-2)(x^2-3)$ is a factorization of $f(x)$ into a product of irreducible polynomials in $\mathbb{Q}[x]$.

(b) The extension $\mathbb{Q}(\sqrt{2})$ contains both $\sqrt{2}$ and $-\sqrt{2}$, so $\mathbb{Q}(\sqrt{2})$ is the splitting field for $x^2 - 2$ over $\mathbb{Q}$. But is $\mathbb{Q}(\sqrt{2})$ the splitting field for $x^2 - 3$ over $\mathbb{Q}$? If yes, then we have found the splitting field for $f(x)$ over $\mathbb{Q}$. If no, then we will need to do more work. Show that $\sqrt{3} \notin \mathbb{Q}(\sqrt{2})$. How does that answer our question about whether $\mathbb{Q}(\sqrt{2})$ is the splitting field for $f(x)$ over $\mathbb{Q}$? (Hint: Use Theorem 31.26.)

(c) To completely factor $f(x)$ over $\mathbb{Q}$, we need to add a root of $x^2 - 3$ to $\mathbb{Q}(\sqrt{2})$. Denote by $\mathbb{Q}(\sqrt{2}, \sqrt{3})$ the field $\mathbb{Q}(\sqrt{2})(\sqrt{3})$. Explain why $\mathbb{Q}(\sqrt{2}, \sqrt{3})$ is the splitting field for $f(x)$ over $\mathbb{Q}$.

(d) Since $\sqrt{2}\sqrt{3} = \sqrt{6}$, it is tempting to think that $\mathbb{Q}(\sqrt{6}) = \mathbb{Q}(\sqrt{2}, \sqrt{3})$. Is this the case? Prove your answer.

As this activity illustrates, to find the splitting field of a polynomial $f(x)$ over a field F, we first factor $f(x)$ into a product of irreducible polynomials. If $p(x)$ is an irreducible factor of $f(x)$, then we know that the field $F[x]/\langle p(x) \rangle$ contains roots of $p(x)$. We attach these roots to F to obtain an extension field E_1. Then we divide $f(x)$ by $p(x)$ in $E_1[x]$ and repeat the process on the quotient in $E_1[x]$. Eventually, we arrive at an extension K of F that is obtained by adding roots of $f(x)$ to F so that K contains all the roots of $f(x)$. This field K is the splitting field for $f(x)$ over F.

Exercises

(1) The splitting field of the polynomial $f(x) = x^9 - x$ over $\mathbb{Z}_3$ is a field of order 9. We will explicitly construct this field in this exercise.

   (a) Note that $f(x)$ is not irreducible in $\mathbb{Z}_3[x]$. Factor $f(x)$ into a product of irreducible polynomials in $\mathbb{Z}_3[x]$.

   (b) Let i be a root of $x^2 + [1]$ in the extension $\mathbb{Z}_3(i)$ of $\mathbb{Z}_3$. Show that $f(x)$ factors completely over $\mathbb{Z}_3(i)$.

   (c) Write the operation tables for $\mathbb{Z}_3(i)$.

   (d) Find a generator for $U(\mathbb{Z}_3(i))$.

(2) Explicitly construct a field with 16 elements.

* (3) Let F be a field, and let $p(x) \in F[x]$ an irreducible polynomial. Let a be a root of $p(x)$ in some extension F' of F. Let E be a field and $\varphi : F \to E$ a field isomorphism. Define $\psi : F[x]/\langle p(x) \rangle \to E[x]/\langle \varphi(p(x)) \rangle$ by

$$\psi(f(x) + \langle p(x) \rangle) = \varphi(f(x)) + \langle \varphi(p(x)) \rangle.$$

Show that ψ is a well-defined isomorphism.

* (4) Let K be a field of characteristic p, and let n be a positive integer.

   (a) Prove that $(a \pm b)^p = a^p \pm b^p$ for any $a, b \in K$.

   (b) Prove that $(a \pm b)^{p^n} = a^{p^n} \pm b^{p^n}$ for any $a, b \in K$ and any positive integer n.

* (5) Let F and E be fields with $\varphi : F \to E$ an isomorphism. If $f(x) = a_n x^n + a_{n-1} x^{n-1} + \cdots + a_1 x + a_0 \in F[x]$, define $\varphi(f(x))$ to be the polynomial

$$\varphi(a_n)x^n + \varphi(a_{n-1})x^{n-1} + \cdots + \varphi(a_1)x + \varphi(a_0) \in E[x].$$

    (a) Prove that $\varphi : F[x] \to E[x]$, as defined above, is an isomorphism.

    (b) Show that $p(x)$ is irreducible in $F[x]$ if and only if $\varphi(p(x))$ is irreducible in $E[x]$.

(6) Find the splitting field of $x^2 + 1$ over $\mathbb{Q}$. Explain why this splitting field is not $\mathbb{C}$.

(7) Show that $x^2 + x + 1$ and $x^2 + 3$ have the same splitting fields over $\mathbb{Q}$.

(8) What are the possible splitting fields of a polynomial in $\mathbb{R}[x]$? Explain.

(9) Find the splitting field of $x^4 + 1$ over $\mathbb{Q}$. Would your answer be different if you had found the splitting field over $\mathbb{R}$ instead?

(10) Find the splitting field of $x^2 + x + 2$ over $\mathbb{Z}_3$. Then find all the roots of $f(x)$ in this splitting field.

* (11)   (a) Let s and t be positive integers with $\gcd(s, t) = 1$. Prove that $U_{st} \cong (U_s \oplus U_t)$.

    (b) Let $m = n_1 n_2 \cdots n_k$, where $\gcd(n_i, n_j) = 1$ for all $i \neq j$. Prove that

$$U_m \cong (U_{n_1} \oplus U_{n_2} \oplus \cdots \oplus U_{n_k}).$$

(12) Write each of U_{4752} and U_{114244} as a direct sum of cyclic groups.

(13) Let n be an integer with $n \geq 2$.

    (a) Use the result of Exercise (11) to calculate $|U_n|$ in terms of the prime factors of n.

    (b) Recall that the Euler phi function φ gives us the number of positive integers less than a given integer that are relatively prime to that integer. In Exercise 24 on page 388 we are asked to show that $\varphi(mn) = \varphi(m)\varphi(n)$ if m and n are positive integers and $\gcd(m, n) = 1$. (Such functions are called **multiplicative** functions and are important in number theory.) Find a formula for $\varphi(n)$ in terms of the prime factors of n. Use this formula to find $\varphi(6860)$.

Connections

In this investigation, we used material from both ring theory and group theory to study the structure of finite fields. Of course, any finite field is a ring, as introduced in Investigation 4, and we saw that facts about polynomials and polynomial rings (Investigation 8), field extensions (Investigation 6), ring isomorphisms (Investigation 7), roots of polynomials (Investigation 10), irreducible polynomials (Investigations 10 and 11), quotients of polynomial rings (Investigation 12), and ideals (Investigation 13) all came up in our classification of finite fields. Moreover, when determining the group of units of a finite field, we also used information about groups (Investigation 17), cyclic groups (Investigation 19), direct products of groups (Investigation 25), and the classification of finite Abelian groups (Investigation 28). As should be obvious from this lengthy list of investigations—all of which were necessary to answer the questions in this investigation—there are many important connections between ring theory and group theory.

Investigation 32

Extensions of Fields

Focus Questions

By the end of this investigation, you should be able to give precise and thorough answers to the questions listed below. You may want to keep these questions in mind to focus your thoughts as you complete the investigation.

- What is an extension of a field?

- What is the difference between a simple extension and a multiple extension of a field?

- What are some connections between simple and multiple extensions?

- What is the degree of a field extension?

- What does it mean to extend an isomorphism?

- What is an automorphism of a field?

- How do we determine automorphisms of splitting fields?

Introduction

One motivation for studying extensions of fields is to understand solutions to polynomial equations. For example, the polynomial equation $x^2 - 2x - 3 = 0$ has two solutions in the field of rational numbers, while the polynomial equation $x^2 - 2 = 0$ has no rational solutions. However, $x^2 - 2 = 0$ has two solutions in the field of real numbers. The equation $x^2 + 1 = 0$ has no real solutions, but does have two solutions in the field of complex numbers. To understand where polynomials have roots, we need to understand the fields in which we consider the roots.

An advantage to working in field extensions is the ability to do exact arithmetic. For example, if we consider $\sqrt{2}$ as an infinite decimal in the real number system, then a computer is only able to make approximate calculations to $\sqrt{2}$. On the other hand, if we consider $\sqrt{2}$ as an element of the field extension $\mathbb{Q}(\sqrt{2})$ of $\mathbb{Q}$, we can consider $\sqrt{2}$ as an ordered pair $(0, 1)$ that satisfies certain algebraic properties, then we can program a computer to perform exact arithmetic with $\sqrt{2}$.

To work effectively with field extensions, we will need a reminder of some background material from linear algebra. We begin this investigation with the necessary background, and then we continue our study of field extensions. One major goal of this investigation is to provide results that will be needed for the next investigation on Galois theory.

A Quick Review of Linear Algebra

A background in linear algebra is needed for this investigation – specifically, basic knowledge of vector spaces, spanning sets, linear independence and dependence, bases, and dimension. We provide some of the results we will need from linear algebra, but we proceed under the assumption that the reader is familiar with this information.

The main idea in linear algebra is that of a vector space.

Definition 32.1. A set V is a **vector space** over a field F if there are two operations, addition and a multiplication by scalars, defined on V such that for all elements $\mathbf{u}$, $\mathbf{v}$, and $\mathbf{w}$ in V and all elements a and b in F:

(1) $\mathbf{u} + \mathbf{v}$ is an element of V

(2) $\mathbf{u} + \mathbf{v} = \mathbf{v} + \mathbf{u}$

(3) $(\mathbf{u} + \mathbf{v}) + \mathbf{w} = \mathbf{u} + (\mathbf{v} + \mathbf{w})$

(4) there is a zero vector $\mathbf{0}$ in V so that $\mathbf{u} + \mathbf{0} = \mathbf{u}$

(5) for each $\mathbf{x}$ in V there is an element $\mathbf{y}$ in V so that $\mathbf{x} + \mathbf{y} = \mathbf{0}$

(6) $a\mathbf{u}$ is an element of V

(7) $(a + b)\mathbf{u} = a\mathbf{u} + b\mathbf{u}$

(8) $a(\mathbf{u} + \mathbf{v}) = a\mathbf{u} + a\mathbf{v}$

(9) $(ab)\mathbf{u} = a(b\mathbf{u})$

(10) $1\mathbf{u} = \mathbf{u}$

The elements in V are called *vectors*, and we refer to the elements of the field F as *scalars*. In an arbitrary vector space, we distinguish between vectors and scalars by writing vectors in bold font.

For example, $\mathbb{R}^n$, the set of all n-dimensional vectors with real components, is a vector space over $\mathbb{R}$ under component wise addition and multiplication by scalars. A more pertinent example is that the field $\mathbb{C}$ of complex numbers is a vector space over the field $\mathbb{R}$ of real numbers using the standard addition and multiplication in C. Recall that whenever one field is contained in another we call the smaller field a *subfield* of the larger field and the larger field an *extension* of the smaller one (see Definition 6.9 on page 89).

Another example of a vector space is the set $\{0\}$, where $0+0 = 0$ and $k \cdot 0 = 0$ for every $k \in \mathbb{R}$. This space is the simplest vector space possible and is called the *trivial* vector space. Every other vector space is called *nontrivial*.

If a field E is an extension of a field F, then E satisfies all the properties of the definition to make E a vector space over F. So, for example, $\mathbb{R}$ is an extension of $\mathbb{Q}$ and $\mathbb{R}$ is also a vector space over $\mathbb{Q}$.

An important idea in the study of vector spaces is that of a *linear combination* of vectors. Recall that the set of complex numbers is defined as $\mathbb{C} = \{a + bi : a, b \in \mathbb{R}\}$. So every element in $\mathbb{C}$ can be written in terms of 1 and i. We say that every element in $\mathbb{C}$ is a *linear combination* of 1 and i.

Definition 32.2. A **linear combination** of vectors $\mathbf{v}_1, \mathbf{v}_2, \ldots, \mathbf{v}_m$ in a vector space V over a field F is any vector of the form

$$c_1\mathbf{v}_1 + c_2\mathbf{v}_2 + \cdots + c_m\mathbf{v}_m, \tag{32.1}$$

where $c_1, c_2, \ldots, c_m$ are elements of F.

The notion of a span of a set of vectors is connected to linear combinations. Because every element in $\mathbb{C}$ a linear combination of the elements 1 and i, we say that 1 and i *span* $\mathbb{C}$.

Definition 32.3. Let V be a vector space over a field F. The **span** of the vectors $\mathbf{v}_1, \mathbf{v}_2, \ldots, \mathbf{v}_m$ in V is the collection of all linear combinations of the vectors $\mathbf{v}_1, \mathbf{v}_2, \ldots, \mathbf{v}_m$.

If the span of a set S of vectors in V is the entire vector space V, we say that S *spans* V. We denote the span of a set S of vectors as $\text{Span}(S)$. Spanning sets for vector spaces are not unique. For example, the sets $\{1, -i\}$ and $\{1 + i, 1 - i\}$ both span $\mathbb{C}$ over $\mathbb{R}$.

It may be the case that a vector can be written in a number of different ways as a linear combination of a given collection of vectors. For example, consider the set $\{1, 1 + i, 1 - i\}$ in $\mathbb{C}$. A little algebra shows that

$$0 = 0 + 0i = (-2t)(1) + (t)(1 + i) + (t)(1 - i)$$

for any real number t. So the complex number 0 can be written in infinitely many different ways as a linear combination of 1, $1 + i$, and $1 - i$. Consider instead the situation if we use the set $\{1, i\}$. Suppose x_1 and x_2 are real numbers and

$$0 = x_1 + x_2 i$$

If $x_2 \neq 0$, then $i = -\frac{x_1}{x_2}$ and i is a real number. But this is not the case, so $x_2 = 0$. Then $x_1 = 0$ as well. So the only way $0 = x_1 + x_2 i$ is if both x_1 and x_2 are 0. In this case, there is exactly one way to write 0 as a linear combination of 1 and i. This uniqueness is encapsulated in the notation of *linear independence*.

Definition 32.4. A set $\{\mathbf{v}_1, \mathbf{v}_2, \ldots, \mathbf{v}_k\}$ of vectors in a vector space V is **linearly independent** if the vector equation

$$x_1\mathbf{v}_1 + x_2\mathbf{v}_2 + \cdots + x_k\mathbf{v}_k = \mathbf{0}$$

for scalars $x_1, x_2, \ldots, x_k$ has only the trivial solution

$$x_1 = x_2 = x_3 = \cdots = x_k = 0.$$

If a set of vectors is not linearly independent, then the set is **linearly dependent**.

Alternatively, we say that the vectors $\mathbf{v}_1, \mathbf{v}_2, \ldots, \mathbf{v}_k$ are linearly independent (or dependent) if the set $\{\mathbf{v}_1, \mathbf{v}_2, \ldots, \mathbf{v}_k\}$ is linearly independent (or dependent).

When a set of vectors spans a space, then every vector is a linear combination of those vectors. If a set S spans a space and is linearly independent, then every vector in the space has exactly one representation as a linear combination of the vectors in S. This uniqueness of representation is important in that it reduces redundancy. Such a set is called a *basis* of the vector space.

Definition 32.5. A **basis** for a vector space V is a subset S of V such that

(1) $\text{Span}(S) = V$ and

(2) S is a linearly independent set.

Examples of bases for $\mathbb{C}$ over $\mathbb{R}$ are $\{1, i\}$, $\{1 + i, 1 - i\}$, and $\{2 + 3i, 4 - 7i\}$. Note that these bases all contain exactly the same number of elements. This is an important property of a vector space – that any two bases for the space contain the same number of elements, assuming that a basis for the space is a finite set (these will be the only bases that concern us).

Theorem 32.6. *If a nontrivial vector space V has a basis of n vectors, then every basis of V contains exactly n vectors.*

The key idea to understand this theorem is linear independence. Suppose there are two different bases $B_1 = \{\mathbf{v}_1, \mathbf{v}_2, \ldots, \mathbf{v}_n\}$ and $B_2 = \{\mathbf{w}_1, \mathbf{w}_2, \ldots, \mathbf{w}_m\}$ for a vector space V over a field F with $m > n$. Since B_1 is linearly independent, we should expect that B_1 contains the largest number of linearly independent vectors possible. So we will proceed to argue that B_2 is a linearly dependent set by demonstrating that the set $\{\mathbf{w}_1, \mathbf{w}_2, \ldots, \mathbf{w}_m\}$ is linearly dependent. Let $x_1, x_2, \ldots, x_m$ be scalars so that

$$x_1\mathbf{w}_1 + x_2\mathbf{w}_2 + \cdots + x_m\mathbf{w}_m = \mathbf{0}. \tag{32.2}$$

Since B_1 spans V, for each i there exist scalars a_{ij} so that

$$\mathbf{w}_i = a_{1i}\mathbf{v}_1 + a_{2i}\mathbf{v}_2 + \cdots + a_{ni}\mathbf{v}_n.$$

Substituting into (32.2) yields

$$\begin{aligned}
\mathbf{0} &= x_1\mathbf{w}_1 + x_2\mathbf{w}_2 + \cdots + x_m\mathbf{w}_m \\
&= x_1(a_{11}\mathbf{v}_1 + a_{21}\mathbf{v}_2 + \cdots + a_{n1}\mathbf{v}_n) + x_2(a_{12}\mathbf{v}_1 + a_{22}\mathbf{v}_2 \\
&\quad + \cdots + a_{n2}\mathbf{v}_n) + \cdots + x_m(a_{1m}\mathbf{v}_1 + a_{2m}\mathbf{v}_2 + \cdots + a_{nm}\mathbf{v}_n) \\
&= (x_1a_{11} + x_2a_{12} + x_3a_{13} + \cdots + x_ma_{1m})\mathbf{v}_1 \\
&\quad + (x_1a_{21} + x_2a_{22} + x_3a_{23} + \cdots + x_ma_{2m})\mathbf{v}_2 \\
&\quad + \cdots + (x_1a_{n1} + x_2a_{n2} + x_3a_{n3} + \cdots + x_ma_{nm})\mathbf{v}_n. \tag{32.3}
\end{aligned}$$

Since B_1 is a basis, the vectors $\mathbf{v}_1, \mathbf{v}_2, \ldots, \mathbf{v}_n$ are linearly independent. So each coefficient in (32.3) is 0. Setting each coefficient to 0 produces a homogeneous linear system of n equations in the m unknowns $x_1, x_2, \ldots, x_m$. Since this system has more unknowns than equations, this system has infinitely many solutions. But then B_2 is not linearly independent and is therefore not a basis for V. We conclude that B_2 can have no more than n vectors. But if B_2 has fewer than n vectors, the same argument reversing the roles of B_2 and B_1 would show that B_1 is not a basis for V. We conclude that if a vector space V has a basis of n elements, then any other basis of V also has exactly n elements.

Another way to interpret the result of Theorem 32.6 is that if a vector space V has a basis with a finite number of vectors, then the number of vectors in a basis for that vector space is a well-defined number. In other words, the number of vectors in a basis is an *invariant* of the vector space. This important number is given a name.

Definition 32.7. A **finite-dimensional** vector space is a vector space that can be spanned by a finite number of vectors. The **dimension** of a nontrivial finite-dimensional vector space is the number of vectors in a basis for V. The dimension of the trivial vector space is defined to be 0.

Since $\{1, i\}$ is a basis for $\mathbb{C}$ over R, the dimension of $\mathbb{C}$ over $\mathbb{R}$ is 2. We will call the dimension of an extension the *degree* of the extension.

Definition 32.8. If E is an extension field over a field F with finite dimension as a vector space over F, then the **degree** of E over F is the dimension of E.

If E has a finite dimension as an extension field over F, we say that E is a *finite* extension of F. We will denote the degree of an extension field E over F as $[E : F]$. As discussed later in the investigation on Galois theory, we use the notation $[E : F]$ for the degree of an extension because there is an important connection between extensions and indices $[G : H]$ of subgroup H of a group G.

Extension Fields and the Degree of an Extension

In Investigation 6 we introduced the topic of field extensions through the example of $\mathbb{Q}(\sqrt{2})$. The field $\mathbb{Q}(\sqrt{2})$ is the smallest field that contains both $\mathbb{Q}$ and the real number $\sqrt{2}$. We showed that the set $\{1, \sqrt{2}\}$ is a basis for $\mathbb{Q}(\sqrt{2})$ over $\mathbb{Q}$ and so

$$\mathbb{Q}(\sqrt{2}) = \{a + b\sqrt{2} : a, b \in \mathbb{Q}\}.$$

This example illustrated a more general idea. Theorem 6.10 on page 89 tells us that if r is a root of a quadratic polynomial $p(x) \in F[x]$ that is not in F, then $F(r) = \{u + vr : u, v \in F\}$. That is, the set $\{1, r\}$ is a basis for $F(r)$ over F.

Even more generally (see Definition 6.9 on page 89), if F is a field, E is an extension of F, and S is a subset of E, then the set $F(S)$ (the extension of F *generated by* S) is the smallest subfield of E that contains all the elements of both F and S. In the case that S contains a single element $\alpha \in E$, then $F(\alpha)$ is called a *simple extension*. The field $F(S)$ is a *multiple extension* of F in the case that S contains a finite number (greater than one) of elements. For multiple extensions, if $S = \{r_1, r_2, \ldots, r_n\}$ for some integer n, we also write $F(S)$ as $F(r_1, r_2, \ldots, r_n)$.

If $E = F(r_1, r_2, \ldots r_n)$ is a multiple extension of a field F, then E may contain several subfields between E and F ($F(r_1)$, for example, if r_1 is not in F). Whenever E is an extension of a field F, any subfield K with $F \subseteq K \subseteq E$ is called an *intermediate* field.

Preview Activity 32.9. Consider $E = \mathbb{Q}(\sqrt{2}, \sqrt{3})$ as an extension of $\mathbb{Q}$. Theorem 6.10 on page 89 tells us that a basis for $K = \mathbb{Q}(\sqrt{2})$ over $\mathbb{Q}$ is $\{1, \sqrt{2}\}$. Let $E = K(\sqrt{3})$.

(a) Using the basis $\{1, \sqrt{2}\}$ for K, write the general form of an element in $\mathbb{Q}(\sqrt{2})$ over $\mathbb{Q}$.

(b) Show that $\sqrt{3}$ is not in K.

(c) The fact that $\sqrt{3} \notin K$ implies that K is a subfield of E not equal to E. Since $\sqrt{3}$ satisfies the polynomial $x^2 - 3$ over K, Theorem 6.10 shows that $\{1, \sqrt{3}\}$ is a basis for E over K. Use the basis $\{1, \sqrt{3}\}$ for E over K, write the general form of an element in E.

(d) Combine the results of (a) and (c) to explain why every element in E can be written in the form
$$r + s\sqrt{2} + t\sqrt{3} + u\sqrt{6}$$
for some r, s, t, and u in $\mathbb{Q}$.

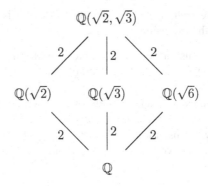

Figure 32.1
A subfield lattice for $\mathbb{Q}(\sqrt{2}, \sqrt{3})$ over $\mathbb{Q}$.

(e) Assume that the set $\{1, \sqrt{2}, \sqrt{3}, \sqrt{6}\}$ is a basis for the extension E over $\mathbb{Q}$. Find all intermediate fields between $\mathbb{Q}$ and E.

We can use the result of Preview Activity 32.9 to construct a subfield lattice, analagous to a subgroup lattice of a group, as shown in Figure 32.1. The numbers represent the dimensions – for example, $[\mathbb{Q}(\sqrt{2}, \sqrt{3}) : \mathbb{Q}(\sqrt{2})] = 2$.

Multiple extensions can be viewed as sequences of simple extensions. The next activity provides an example.

Activity 32.10. Consider the extension $\mathbb{Q}(\sqrt{2}, \sqrt{3})$ of $\mathbb{Q}$. It is reasonable to wonder what the exact connection is between $\mathbb{Q}(\sqrt{2}, \sqrt{3})$ and the simple extension $\mathbb{Q}(\sqrt{2})(\sqrt{3})$ of $\mathbb{Q}(\sqrt{2})$ (or the simple extension $\mathbb{Q}(\sqrt{3})(\sqrt{2})$ of $\mathbb{Q}(\sqrt{3})$). We address that question in this activity.

(a) Recall that if F is a field and S is a set, then $F(S)$ is the smallest field that contains F and S. Use this idea to explain why $\mathbb{Q}(\sqrt{2})(\sqrt{3})$ is a subset of $\mathbb{Q}(\sqrt{2}, \sqrt{3})$.

(b) Explain why $\mathbb{Q}(\sqrt{2}, \sqrt{3})$ is a subset of $\mathbb{Q}(\sqrt{2})(\sqrt{3})$.

(c) What can we say about the relationship between $\mathbb{Q}(\sqrt{2}, \sqrt{3})$ and $\mathbb{Q}(\sqrt{2})(\sqrt{3})$?

The result of Activity 32.10 is true in general, as the next theorem indicates.

Theorem 32.11. *Let F be a field and let r_1, r_2, ..., r_n be elements in some extension of F. Then*

$$F(r_1, r_2, \ldots, r_{n-1})(r_n) = F(r_1, r_2, \ldots, r_{n-1}, r_n).$$

Proof. Let F be a field and let r_1, r_2, ..., r_n be elements in some extension of F. To prove the desired equality, we exhibit containments in both directions. First, we prove that $F(r_1, r_2, \ldots, r_{n-1})(r_n) \subseteq F(r_1, r_2, \ldots, r_n)$. The field $F(r_1, r_2, \ldots, r_n)$ contains F and the elements r_1, r_2, ..., r_{n-1}, and r_n. So $F(r_1, r_2, \ldots, r_n)$ must contain the smallest field that has F and the elements r_1, r_2, $\cdots$, r_{n-1}. Thus, $F(r_1, r_2, \ldots, r_n)$ contains $F(r_1, r_2, \ldots, r_{n-1})$. The fact that $F(r_1, r_2, \ldots, r_n)$ also contains r_n means that $F(r_1, r_2, \ldots, r_n)$ contains the smallest field that has $F(r_1, r_2, \ldots, r_{n-1})$ and r_n, which is $F(r_1, r_2, \ldots, r_{n-1})(r_n)$. Therefore, $F(r_1, r_2, \ldots, r_{n-1})(r_n) \subseteq F(r_1, r_2, \ldots, r_n)$.

To demonstrate that $F(r_1, r_2, \ldots, r_n) \subseteq F(r_1, r_2, \ldots, r_{n-1})(r_n)$ we note that the extension $F(r_1, r_2, \ldots, r_{n-1})(r_n)$ contains F and the elements $r_1, r_2, \ldots, r_n$. So $F(r_1, r_2, \ldots, r_{n-1})(r_n)$ contains the smallest field that contains F and the elements $r_1, r_2, \ldots, r_n$. This smallest field is $F(r_1, r_2, \ldots, r_n)$. Therefore, $F(r_1, r_2, \ldots, r_n) \subseteq F(r_1, r_2, \ldots, r_{n-1})(r_n)$, and we conclude that $F(r_1, r_2, \ldots, r_n) = F(r_1, r_2, \ldots, r_{n-1})(r_n)$. ∎

Theorem 32.11 shows that any multiple extension can be obtained from a sequence of simple extensions. That is, if F is a field and $r_1, r_2, \ldots, r_n$ are elements in some extension of F, then letting $F_1 = F(r_1)$, $F_2 = F_1(r_2)$, and $F_i = F_{i-1}(r_i)$ for $2 \leq i \leq n$, we have

$$F \subseteq F_1 \subseteq F_2 \subseteq \cdots \subseteq F_{n-1} \subseteq F_n = F(r_1, r_2, \ldots, r_n).$$

We can use this idea to find bases for multiple extensions and uncover a result about degrees of extensions.

Activity 32.12. Let us return to the example of $\mathbb{Q}(\sqrt{2}, \sqrt{3}) = \mathbb{Q}(\sqrt{2})(\sqrt{3})$ from Preview Activity 32.9. The set $\{1, \sqrt{2}\}$ is a basis for $\mathbb{Q}(\sqrt{2})$ over $\mathbb{Q}$. Similarly, the set $\{1, \sqrt{3}\}$ is a basis for $\mathbb{Q}(\sqrt{2})(\sqrt{3})$ over $\mathbb{Q}(\sqrt{2})$. So

$$[\mathbb{Q}(\sqrt{2}) : \mathbb{Q}] = 2 \text{ and } [\mathbb{Q}(\sqrt{2}, \sqrt{3}) : \mathbb{Q}(\sqrt{2})] = 2.$$

We might now wonder if this information is enough to tell us what $[\mathbb{Q}(\sqrt{2}, \sqrt{3}) : \mathbb{Q}]$ is.

We should have seen in Preview Activity 32.9 that we can combine these two bases to show that the set $\{1, \sqrt{2}, \sqrt{3}, \sqrt{6}\}$ spans $\mathbb{Q}(\sqrt{2}, \sqrt{3})$ over $\mathbb{Q}$, so $[\mathbb{Q}(\sqrt{2}, \sqrt{3}) : \mathbb{Q}] \leq 4$. The question we want to address in this activity is if $[\mathbb{Q}(\sqrt{2}, \sqrt{3}) : \mathbb{Q}]$ is actually equal to 4. In other words, is the set $\{1, \sqrt{2}, \sqrt{3}, \sqrt{6}\}$ linearly independent? Consider the equation

$$x_1 + x_2\sqrt{2} + x_3\sqrt{3} + x_4\sqrt{6} = 0 \tag{32.4}$$

with x_1, x_2, x_3, and x_4 in $\mathbb{Q}$.

(a) What do we need to know about (32.4) to conclude that the set $\{1, \sqrt{2}, \sqrt{3}, \sqrt{6}\}$ is linearly independent?

(b) Write $\sqrt{6}$ as $\sqrt{2}\sqrt{3}$. Then rewrite the left-hand side of (32.4) as a linear combination of 1 and $\sqrt{3}$ over $\mathbb{Q}(\sqrt{2})$. How can we conclude that

$$x_1 + x_2\sqrt{2} = 0 \text{ and } x_3 + x_4\sqrt{2} = 0?$$

(c) Using the result of (b), how can we conclude that $x_1 = x_2 = x_3 = x_4 = 0$? What does this tell us about the set $\{1, \sqrt{2}, \sqrt{3}, \sqrt{6}\}$ and about $[\mathbb{Q}(\sqrt{2}, \sqrt{3}) : \mathbb{Q}]$?

In the particular situation of Activity 32.12, we see that the degrees multiply – that is $[\mathbb{Q}(\sqrt{2}, \sqrt{3}) : \mathbb{Q}] = [\mathbb{Q}(\sqrt{2}, \sqrt{3}) : \mathbb{Q}(\sqrt{2})] [\mathbb{Q}(\sqrt{2}) : \mathbb{Q}]$. This is no coincidence, as the next theorem shows.

Theorem 32.13. *Let F be a field, K an extension of F, and E an extension of K. If $[K : F]$ and $[E : K]$ are both finite, then $[E : F]$ is finite and*

$$[E : F] = [E : K] [K : F].$$

Proof. Let $\{e_1, e_2, \ldots, e_m\}$ be a basis for E over K and let $\{k_1, k_2, \ldots, k_n\}$ be a basis for K over F. Let $e \in E$. Then

$$e = x_1 e_1 + x_2 e_2 + \cdots + x_m e_m$$

for some $x_1, x_2, \ldots, x_m$ in K. For each x_i with $1 \leq i \leq k$, the fact that $x_i \in K$ implies that there exist $y_{1i}, y_{2i}, \ldots, y_{ni}$ in F such that

$$x_i = y_{1i} k_1 + y_{2i} k_2 + \cdots + y_{ni} k_n.$$

We then have

$$
\begin{aligned}
e &= (y_{11}k_1 + y_{21}k_2 + \cdots + y_{n1}k_n)e_1 + (y_{12}k_1 + y_{22}k_2 + y_{n2}k_n)e_2 + \cdots \\
&\quad + (y_{1m}k_1 + y_{2m}k_2 + \cdots + y_{nm}k_n)e_m \\
&= y_{11}k_1 e_1 + y_{21}k_2 e_1 + \cdots + y_{n1}k_n e_1 + \cdots + y_{1m}k_1 e_m + y_{2m}k_2 e_m + \cdots + y_{nm}k_n e_m \\
&= \sum_{i=1}^{n} \sum_{j=1}^{m} y_{ij} k_i e_j.
\end{aligned}
$$

We conclude that the set $S = \{k_i e_j : 1 \leq i \leq n, 1 \leq j \leq m\}$ spans E over F. Now we need to show that the set S is linearly independent. Suppose

$$\sum_{i=1}^{n} \sum_{j=1}^{m} z_{ij} k_i e_j = 0$$

for some $z_{ij} \in F$. Then

$$
\begin{aligned}
0 &= \sum_{i=1}^{n} \sum_{j=1}^{m} z_{ij} k_i e_j \\
&= z_{11}k_1 e_1 + z_{21}k_2 e_1 + \cdots + z_{n1}k_n e_1 + \cdots + z_{1m}k_1 e_m + z_{2m}k_2 e_m + \cdots + z_{nm}k_n e_m \\
&= (z_{11}k_1 + z_{21}k_2 + \cdots + z_{n1}k_n)e_1 + (z_{12}k_1 + z_{22}k_2 + z_{n2}k_n)e_2 + \cdots \\
&\quad + (z_{1m}k_1 + z_{2m}k_2 + \cdots + z_{nm}k_n)e_m.
\end{aligned}
$$

Since the set $\{e_1, e_2, \ldots, e_m\}$ is linearly independent, it follows that

$$z_{1t}k_1 + z_{2t}k_2 + \cdots + z_{nt}k_n = 0$$

for each t between 1 and m. But $\{k_1, k_2, \ldots, k_n\}$ is also linearly independent, so $z_{st} = 0$ for every s between 1 and n. We conclude that the set S is a basis for E over F and that

$$[E : F] = mn = [E : K]\,[K : F].$$

∎

Simple Extensions

If F is a field and r is an element in some extension of F, then $F(r)$ is the smallest field that contains F and r. Another perspective is that $F(r)$ is the intersection of all fields that contain both F and r. For a simple extension $F(r)$, we call the element r a *generating element* for the extension.

Elements that satisfy polynomials over a field are said to be *algebraic* over that field. If an element is not algebraic, it is said to be *transcendental*. This provides two different possibilities for a simple extension. If the element r is algebraic over F and satisfies a degree n polynomial over F, as we will show below, we can write r^n in terms of the lower powers of r and so $F(r)$ is a finite dimensional extension of F. An example of this is $\mathbb{Q}(\sqrt{2})$, where $\sqrt{2}$ is a root of $p(x) = x^2 - 2$ in $\mathbb{Q}[x]$. But if r does not satisfy any polynomial over F, then $F(r)$ is an infinite dimensional extension of F. This would be the case of $F(\pi)$. It is generally very difficult to show that a number is transcendental. Some examples of transcendental numbers over $\mathbb{Q}$ are π and e. We will not attempt to prove that these numbers are transcendental over $\mathbb{Q}$.

In some cases, every element in a field is algebraic over a base field. For example, if $a + bi$ is a complex number, then $a + bi$ is a root of

$$(x - (a - bi))(x - (a + bi)) = x^2 - 2ax + (a^2 + b^2) \in \mathbb{R}[x].$$

So every element in $\mathbb{C}$ is algebraic over $\mathbb{R}$.

Definition 32.14. An extension field E of a field F is an **algebraic** extension of F if every element in E is algebraic over F.

In our work we will mostly be interested in finite-dimensional extensions of fields, and the next theorem shows that finite-dimensional extensions are always algebraic extensions.

Theorem 32.15. *Let E be a finite-dimensional extension of a field F. Then E is an algebraic extension of F.*

Proof. If E is a finite-dimensional extension over a field F with $[E : F] = n$, and if a is an element in E, then the set $\{1, a, a^2, \ldots, a^n\}$ must be a linearly dependent set because it contains $n + 1$ elements. Therefore, there exist $c_0, c_1, \ldots, c_n$ in F, not all zero, such that

$$c_0(1) + c_1(a) + c_2(a^2) + \cdots + c_n(a^n) = 0.$$

But then a is a root of the polynomial $f(x) = c_0 + c_1 x + c_2 x^2 + \cdots + c_n x^n$ in $F[x]$. We conclude that a is algebraic over F, and E is an algebraic extension of F. ∎

Since $\mathbb{C} = \mathbb{R}(i)$, Theorem 32.15 shows that $\mathbb{C}$ is algebraic over $\mathbb{R}$. In fact, $\mathbb{C} = \mathbb{R}(i)$, where i satisfies the irreducible polynomial $x^2 + 1$ over $\mathbb{R}$. So $[\mathbb{C} : \mathbb{R}] = 2$. (A nonconstant polynomial is irreducible if it can't be factored into a product of smaller degree polynomials.) The next activity expands on this idea.

Activity 32.16. Consider the extension $\mathbb{Q}(\sqrt[3]{2})$ over $\mathbb{Q}$. Note that $a = \sqrt[3]{2}$ satisfies the polynomial $p(x) = x^3 - 2$. In Activity 31.17 and in the proof of Theorem 31.18 on page 454, we defined $\varphi : \mathbb{Q}[x] \to \mathbb{Q}(a)$ by $\varphi(f(x)) = f(a)$, and showed that $\text{Ker}(\varphi) = \langle p(x) \rangle$ and $\text{Im}(\varphi) = \mathbb{Q}(a)$.

(a) Explain why $p(x)$ is irreducible over $\mathbb{Q}$. (Hint: Review Eisenstein's criterion on page 164.)

(b) First, we want to show that the set $\{1, a, a^2\}$ spans $\mathbb{Q}(a)$ over $\mathbb{Q}$. (Notice that $\sqrt[3]{4} = (\sqrt[3]{2})^2$.) The polynomial $p(x)$ will be important in this argument. Let $z \in \mathbb{Q}(a)$. The fact that $\text{Im}(\varphi) = \mathbb{Q}(a)$ means that there is a polynomial $f(x)$ in $\mathbb{Q}[x]$ such that $z = \varphi(f(x)) = f(a)$. The Division Algorithm tells us that there exist polynomials $q(x)$ and $r(x)$ in $\mathbb{Q}[x]$ such that

$$f(x) = q(x)p(x) + r(x) \text{ with } r(x) = 0 \text{ or } 0 \le \deg(r(x)) < \deg(p(x)) = 3.$$

Explain how this can be used to show that $f(x)$ is in the span of the set $\{1, a, a^2\}$.

(c) The final step is to show that the set $\{1, a, a^2\}$ is linearly independent. Suppose there exist b_0, b_1, and b_2 in $\mathbb{Q}$ such that

$$b_0 + b_1 a + b_2 a^2 = 0.$$

Consider the polynomial $g(x) = b_0 + b_1 x + b_2 x^2$.

  (i) What must be true about $g(a)$?

  (ii) The minimal polynomial of an algebraic element a is the monic polynomial of smallest degree that has a as a root (see Definition 31.16 on page 454). The minimal polynomial of a divides all other polynomials that have a as a root (see Exercise (5)). What does the fact that $p(x)$ is the minimal polynomial of a over $\mathbb{Q}$ tell us about the relationship between $p(x)$ and $g(x)$?

  (iii) Use the results of (i) and (ii) to explain why the set $\{1, a, a^2\}$ is linearly independent. (Hint: Compare the degrees of $g(x)$ and $p(x)$.)

Activity 32.16 suggests that if $p(x)$ is the minimal polynomial of an element a over a field F, and $\deg(p(x)) = n$, then $\{1, a, a^2, \ldots, a^{n-1}\}$ is a basis for $F(a)$ over F. The following theorem clarifies this statement.

Theorem 32.17. *Let E be an extension of a field F and let a be an algebraic element over F with minimal polynomial $p(x)$ of degree n in $F[x]$. Then*

(1) $F(a) \cong F[x]/\langle p(x)\rangle$.

(2) $\{1, a, a^2, \ldots, a^{n-1}\}$ is a basis for $F(a)$ over F.

(3) $[F(a) : F] = n$.

Proof. We proved part (1) in Theorem 31.18 on page 455. In the proof of Theorem 31.18, we defined $\varphi : F[x] \to F(a)$ by $\varphi(f(x)) = f(a)$, and we showed that $\mathrm{Ker}(\varphi) = \langle p(x)\rangle$ and $\mathrm{Im}(\varphi) = F(a)$. So the First Isomorphism Theorem (Theorem 13.48 on page 204) shows that

$$F(a) = \mathrm{Im}(\varphi) \cong F[x]/\mathrm{Ker}(\varphi) = F[x]/\langle p(x)\rangle.$$

Part (3) follows directly from part (2), so it remains to prove part (2). First, we show that the set $B = \{1, a, a^2, \ldots, a^{n-1}\}$ spans $F(a)$ over F. Let $y \in F(a)$. Since $\mathrm{Im}(\varphi) = F(a)$, every element in $F(a)$ is the image of some element in $F[x]$ under φ. That is, there is a polynomial $f(x) \in F[x]$ such that $y = \varphi(f(x)) = f(a)$. By the Division Algorithm, there exist polynomials $q(x)$ and $r(x)$ in $F[x]$ such that

$$f(x) = q(x)p(x) + r(x) \text{ with } r(x) = 0 \text{ or } \deg(r(x)) < \deg(p(x)) = n.$$

Let $r(x) = c_0 + c_1 x + c_2 x^2 + \cdots + c_{n-1} x^{n-1}$, with $c_i \in F$ for each i. Since $p(a) = 0$, it follows that

$$y = f(a) = r(a) = c_0 + c_1 a + c_2 a^2 + \cdots + c_{n-1} a^{n-1}.$$

Therefore, y is in the span of B. To conclude our proof, we demonstrate that the set B is linearly independent. Suppose

$$b_0 + b_1 a + b_2 a^2 + \cdots + b_{n-1} a^{n-1} = 0$$

for some elements $b_i \in F$. Let $g(x) = b_0 + b_1 x + b_2 x^2 + \cdots + b_{n-1} x^{n-1}$. Then $g(a) = 0$. Since $p(x)$ is the minimal polynomial of a over F it must be the case that $p(x)$ divides $g(x)$. But $\deg(g(x)) < \deg(p(x))$, so the only possibility is that $g(x) = 0$. But this makes $b_0 = b_1 = \cdots = b_{n-1} = 0$, which shows that B is linearly independent. ∎

A consequence of Theorem 32.17 is the following. Suppose that a and b have the same minimal polynomial $p(x)$ in $F[x]$. Then $F(a) \cong F[x]/\langle p(x)\rangle$ and $F(b) \cong F[x]/\langle p(x)\rangle$. So $F(a)$ is isomorphic to $F(b)$. In fact, given a and b in an extension of F, Theorem 32.17 provides an isomorphism between $F(a)$ and $F(b)$. This consequence can be generalized even more as the next corollary indicates.

Before we discuss the corollary, we need to consider how we can extend an isomorphism of fields to the corresponding polynomial rings. Suppose F and F' are fields and that $\varphi : F \to F'$ is an isomorphism. We can *extend* φ to an isomorphism from $F[x]$ to $F'[x]$ as follows. If $p(x) = c_0 + c_a x + \cdots + c_n x^n \in F[x]$, then $\varphi p(x)$ is the polynomial in $F'[x]$ defined by

$$\varphi p(x) = \varphi(c_0) + \varphi(c_1)x + \cdots + \varphi(c_n)x^n.$$

It is left to Exercise (9) to prove that this mapping is an isomorphism. Generally, there is no confusion between φ mapping F to F' and φ mapping $F[x]$ to $F'[x]$, so we use the same symbol for both.

Corollary 32.18. *Let* $\sigma : F \to F'$ *be an isomorphism of fields and let a be an element that is algebraic over F with minimal polynomial $p(x)$ in $F[x]$. Let b be an algebraic element over F' with minimal polynomial $\sigma p(x)$ in $F'[x]$. Then σ extends to an isomorphism $\overline{\sigma}$ from $F(a)$ to $F'(b)$ such that $\overline{\sigma}(a) = b$ and $\overline{\sigma}(c) = \sigma(c)$ for all $c \in F$.*

Proof. Let $\sigma : F \to F'$ be an isomorphism of fields, and let a be an element that is algebraic over F with minimal polynomial $p(x)$ in $F[x]$. Let b be an algebraic element over F' with minimal polynomial $\sigma p(x)$ in $F'[x]$. We will construct an isomorphism from $F(a)$ to $F'(b)$.

Let $\varphi : F'[x] \to F'(b)$ be defined by $\varphi(f(x)) = f(b)$ as in Activity 31.17 on page 454. Recall that $\mathrm{Ker}(\varphi) = \langle \sigma p(x)\rangle$ and $\mathrm{Im}(\varphi) = F'(b)$. This gives an isomorphism $\overline{\varphi}$ from $F'[x]/\langle \sigma p(x)\rangle$ to $F'(b)$ defined by $\overline{\varphi}(f(x) + \langle \sigma p(x)\rangle) = f(b)$ by the First Isomorphism Theorem (Theorem 13.48 on page 204).

This provides the sequence

$$F[x] \xrightarrow{\sigma} F'[x] \xrightarrow{\pi} F'[x]/\langle \sigma p(x)\rangle \xrightarrow{\overline{\varphi}} F'(b)$$

where

$$\sigma(f(x)) = \sigma f(x)$$
$$\pi(\sigma f(x)) = \sigma f(x) + \langle \sigma p(x)\rangle$$
$$\overline{\varphi}(\sigma f(x) + \langle \sigma p(x)\rangle) = \sigma f(b).$$

Now σ and $\overline{\varphi}$ are both isomorphisms and π is a surjection, so the composite $\overline{\varphi} \circ \pi \circ \sigma$ is a surjection. Now we consider the kernel of this composite. Suppose $(\overline{\varphi} \circ \pi \circ \sigma)(f(x)) = 0$. Then $\sigma f(b) = 0$. Now $\overline{\varphi}$ is an injection, so it must be the case that $\sigma f(x) + \langle \sigma p(x)\rangle$ is the zero element in $F'[x]/\langle \sigma p(x)\rangle$. So $\sigma f(x) \in \langle \sigma p(x)\rangle$ and $\sigma p(x)$ divides $\sigma f(x)$. So $\sigma f(x) = g'(x)\sigma p(x)$ for some $g'(x) \in F'[x]$. Since $\sigma : F[x] \to F'[x]$ is an isomorphism, there is a polynomial $g(x) \in F[x]$ such that $g'(x) = \sigma g(x)$. The fact that σ is an injection implies that $f(x) = g(x)p(x)$. This tells us that $\mathrm{Ker}(\overline{\varphi} \circ \pi \circ \sigma) = \langle p(x)\rangle$ in $F[x]$. The First Isomorphism Theorem then shows that $F[x]/\langle p(x)\rangle \cong F'(b)$. In fact, the isomorphism $\gamma : F[x]/\langle p(x)\rangle \to F'(b)$ is given by

$$\gamma(h(x) + \langle p(x)\rangle) = \sigma h(b).$$

Once again Theorem 31.18 gives us an isomorphism $\bar{\alpha} : F[x]/\langle p(x)\rangle \to F(a)$ defined by $\bar{\alpha}(h(x) + \langle p(x)\rangle) = h(a)$. Then $\bar{\alpha}^{-1}$ is an isomorphism from $F(a)$ to $F[x]/\langle p(x)\rangle$. Putting $\bar{\alpha}^{-1}$ and γ together gives us the isomorphism

$$F(a) \xrightarrow{\bar{\alpha}^{-1}} F[x]/\langle p(x)\rangle \xrightarrow{\gamma} F'(b).$$

What remains to show is that $\bar{\sigma} = \gamma \circ \bar{\alpha}^{-1}$ satisfies $\bar{\sigma}(c) = \sigma(c)$ for all $c \in F$ and $\bar{\sigma}(a) = b$.

Notice that if $k(x) = x$, $k(a) = a$. So $\bar{\alpha}(x + \langle p(x)\rangle) = a$, and $\bar{\alpha}^{-1}(a) = x + \langle p(x)\rangle$. Then

$$\bar{\sigma}(a) = \gamma(\bar{\alpha}^{-1}(a)) = \gamma(x + \langle p(x)\rangle) = b.$$

Let $c \in F$. Then $\bar{\alpha}(c + \langle p(x)\rangle) = c$, so $\bar{\alpha}^{-1}(c) = c + \langle p(x)\rangle$. Since $\sigma c(x) = \sigma(c)$, it follows that

$$\bar{\sigma}(c) = \gamma(\bar{\alpha}^{-1}(c)) = \gamma(c + \langle p(x)\rangle) = \sigma c(b) = \sigma(c).$$

∎

Activity 32.19. Consider the irreducible polynomial $p(x) = x^3 - 2$ in $\mathbb{Q}[x]$. Explain how Corollary 32.18 shows that the fields $\mathbb{Q}(\sqrt[3]{2})$ and $\mathbb{Q}(\sqrt[3]{2}\omega)$ are isomorphic, where ω is a primitive cube root of unity. (What makes ω a primitive cube root of unity is that $\omega^3 = 1$ but $\omega^k \neq 1$ for $1 \leq k < 3$. A background on roots of unity can be found in the online supplemental materials.)

If we let $F' = F$ in Corollary 32.18 with a and b roots of the same minimal polynomial $p(x)$ in $F[x]$, then using the identity mapping as our σ, there is an isomorphism $\bar{\sigma}$ from $F(a)$ to $F(b)$ satisfying $\bar{\sigma}(a) = b$ and $\bar{\sigma}(c) = c$ for every $c \in F$.

We return once again to the example of $\mathbb{Q}(\sqrt{2}, \sqrt{3})$ as an extension of $\mathbb{Q}$. This time we ask if it is possible to recognize $\mathbb{Q}(\sqrt{2}, \sqrt{3})$ as a simple extension of $\mathbb{Q}$. That is, does there exist an element a in $\mathbb{Q}(\sqrt{2}, \sqrt{3})$ such that $\mathbb{Q}(\sqrt{2}, \sqrt{3}) = \mathbb{Q}(a)$? The element $\sqrt{6} = \sqrt{2}\sqrt{3}$ is in $\mathbb{Q}(\sqrt{2}, \sqrt{3})$, so we might be tempted to think that we could use $a = \sqrt{6}$. But we know that

$$[\mathbb{Q}(\sqrt{2}, \sqrt{3}) : \mathbb{Q}] = [\mathbb{Q}(\sqrt{2}, \sqrt{3}) : \mathbb{Q}(\sqrt{2})] \, [\mathbb{Q}(\sqrt{2}) : \mathbb{Q}] = 4,$$

so $\mathbb{Q}(\sqrt{2}, \sqrt{3})$ has dimension 4 over $\mathbb{Q}$. The element $\sqrt{6}$ has minimal polynomial $x^2 - 6$ over $\mathbb{Q}$, so $[\mathbb{Q}(\sqrt{6}) : \mathbb{Q}] = 2$. Thus, $\mathbb{Q}(\sqrt{6})$ is a proper subfield of $\mathbb{Q}(\sqrt{2}, \sqrt{3})$. We make a different attempt in the next activity.

Activity 32.20. Our goal in this activity is to show that $\mathbb{Q}(\sqrt{2}, \sqrt{3})$ is a simple extension of $\mathbb{Q}$. That is, there is an element a in $\mathbb{Q}(\sqrt{2}, \sqrt{3})$ such that $\mathbb{Q}(\sqrt{2}, \sqrt{3}) = \mathbb{Q}(a)$. Let us try $a = \sqrt{2} + \sqrt{3}$. It is certainly the case that $\sqrt{2} + \sqrt{3}$ is in $\mathbb{Q}(\sqrt{2}, \sqrt{3})$, so $\mathbb{Q}(\sqrt{2} + \sqrt{3}) \subseteq \mathbb{Q}(\sqrt{2}, \sqrt{3})$. To demonstrate the equality, we consider the dimension of $\mathbb{Q}(\sqrt{2} + \sqrt{3})$ over $\mathbb{Q}$. To determine this degree, we look for the minimal polynomial of $\sqrt{2} + \sqrt{3}$ over $\mathbb{Q}$. To make the notation a bit easier, let $x = \sqrt{2} + \sqrt{3}$.

(a) Square both sides of $x = \sqrt{2} + \sqrt{3}$ and isolate all radicals that appear in the equation.

(b) Take an appropriate square again to create the polynomial equation $x^4 - 10x^2 + 1 = 0$.

(c) Explain why $x^4 - 10x^2 + 1$ is irreducible over $\mathbb{Q}$. (Hint: Reduce modulo a prime.)

(d) Find all the roots of $x^4 - 10x^2 + 1$. Use any technological tool that is helpful.

(e) Explain why $[\mathbb{Q}(\sqrt{2} + \sqrt{3}) : \mathbb{Q}] = 4$, and then why $\mathbb{Q}(\sqrt{2} + \sqrt{3}) = \mathbb{Q}(\sqrt{2}, \sqrt{3})$.

Activity 32.20 shows that multiple extensions can sometimes be realized as simple extensions. Before getting to the general result, we need two preliminary results.

Theorem 32.21. *Let F be a field of characteristic 0. If $p(x)$ is an irreducible polynomial over F, then $p(x)$ has no repeated roots.*

Proof. Let F be a field of characteristic 0. We proceed by contradiction and assume that $p(x)$ is an irreducible polynomial in $F[x]$ with repeated root r. Then we can write $p(x) = a_0 + a_1x + a_2x^2 + \cdots + a_nx^n = (x - r)^2q(x)$ for some polynomial $q(x) \in F[x]$. Now we apply some calculus and consider the derivative $p'(x)$ of $p(x)$. Note that $p'(x) = a_1x + a_2x + \cdots + a_nx^{n-1} = 2(x - r)q(x) + (x - r)^2q'(x) = (x - r)(2q(x) + (x - r)q'(x))$ is also in $F[x]$. So r is also a root of $p'(x)$. The fact that $p(x)$ is irreducible over F means that $p(x)$ is the minimal polynomial of r over F. Since r is a root of $p'(x)$, we can conclude that $p(x)$ divides $p'(x)$. But the fact that $\deg(p'(x)) < \deg(p(x))$ makes this impossible unless $p'(x)$ is the zero polynomial. But if $p'(x)$ is the zero polynomial, then $a_1 = a_2 = \cdots = a_n = 0$ and $p(x) = a_0$ is a constant polynomial and therefore not irreducible. We conclude that $p(x)$ cannot have a repeated root. ∎

It is a good idea to understand why the condition of F having characteristic 0 is necessary in Theorem 32.21. We leave that discussion for Exercise (7) in Investigation 33.

Lemma 32.22. *Let E be a finite extension of an infinite field F of characteristic 0 with elements u and v in E which are algebraic over F. Then there exists $w \in E$ such that*

$$F(u, v) = F(w).$$

Proof. Let $f(x)$ be the minimal polynomial of u over F and $g(x)$ the minimal polynomial of v over F. As a minimal polynomial, $f(x)$ is irreducible. Let $r_1 = u$ and let $r_2, r_3, \ldots, r_n$ be the other roots of $f(x)$. Theorem 32.21 shows that an irreducible polynomial over F cannot have multiple roots, so the roots of $f(x)$ are all distinct. Let $v = s_1$ and let $s_2, \ldots, s_m$ be the other roots of $g(x)$. Consider the set of equations of the form $r_i + xs_j = r_1 + xs_1$ for $1 \leq i \leq n$ and $2 \leq j \leq m$. There are $n(m - 1)$ equations in this system

$$r_1 + xs_2 = r_1 + xs_1$$
$$r_1 + xs_3 = r_1 + xs_1$$

$$\vdots$$

$$r_2 + xs_m = r_1 + xs_1$$
$$r_2 + xs_1 = r_1 + xs_1$$

$$\vdots$$

$$r_2 + xs_m = r_1 + xs_1$$

$$\vdots$$

$$r_n + xs_1 = r_1 + xs_1$$
$$r_n + xs_2 = r_1 + xs_1$$

$$\vdots$$

$$r_n + xs_m = r_1 + xs_1$$

Let $K = F(r_1, r_2, \ldots, r_n, s_1, s_2, \ldots, s_m)$. The equation $r_i + xs_j = r_1 + xs_1$ has the unique solution $x = \frac{r_1 - r_i}{s_j - s_1}$ in K, and so the equation $r_i + xs_j = r_1 + xs_1$ has at most one solution in F.

Since F is an infinite field, we can choose an element $t \in F$ such that t is different than $\frac{r_1 - r_i}{s_j - s_1}$ for each i and j. It follows that $r_i + ts_j \neq r_1 + ts_1$ for every i and j. Let $w = u + tv$.

To complete our proof, we verify that $F(u,v) = F(w)$. The fact that $t \in F$ means that $w \in F(u,v)$ and $F(w) \subseteq F(u,v)$. So we only have to demonstrate that $F(u,v) \subseteq F(w)$. Note that if $v \in F(w)$, then $u = w - tv$ is also in $F(w)$. Thus, we will prove that $v \in F(w)$. This can be accomplished by demonstrating that v is a root of a linear polynomial in $F(w)[x]$. Consider the polynomials $g(x) \in F[x] \subseteq F(w)[x]$ and $h(x) = f(w - tx)$ in $F(w)[x]$. We know that $g(v) = 0$, and

$$h(v) = f(w - tv) = f(u) = 0.$$

Thus $x - v$ divides both $g(x)$ and $h(x)$, and so $x - v$ divides $\gcd(g(x), h(x))$ in $K[x]$. Since v is a root of $g(x)$ with multiplicity 1, we can conclude that v is a root of $\gcd(g(x), h(x))$ with multiplicity 1. We check to see if $h(x)$ and $g(x)$ have any other roots in common. The remaining roots of $g(x)$ are $s_1, s_2, \ldots, s_m$. In order for one of these roots to also be a root of $h(x)$, it would have to be the case that $w - ts_j = r_i$ for some i and j. But then

$$w = u + tv = r_1 + ts_1 = r_i + ts_j,$$

which is impossible by the choice of t. So $h(x)$ and $g(x)$ have no roots in common in K other than v. So $\gcd(g(x), h(x)) = x - v$ in $K[x]$. The coefficients of $h(x)$ are in $F(w)$ and so $\gcd(g(x), h(x)) \in F(w)[x]$. It follows that v is in $F(w)$. ∎

Example 32.23. The techniques from the proof of Lemma 32.22 allow us to show that $\mathbb{Q}(\sqrt{2}, \sqrt{3}) = \mathbb{Q}(\sqrt{2} + \sqrt{3})$ without finding the minimal polynomial of $\sqrt{2} + \sqrt{3}$ as we did in Activity 32.20. Let $E = \mathbb{Q}(u,v)$ with $u = \sqrt{2}$, $v = \sqrt{3}$ and $F = \mathbb{Q}$. Then $f(x) = x^2 - 2$ and $g(x) = x^2 - 3$. The roots of $f(x)$ are $r_1 = u$ and $r_2 = -u$ and the roots of $g(x)$ are $s_1 = v$, $s_2 = -v$. We have

$$\frac{r_1 - r_1}{s_2 - s_1} = 0$$

$$\frac{r_2 - r_1}{s_2 - s_1} = \frac{\sqrt{2}}{\sqrt{3}}.$$

We can choose 1 in $\mathbb{Q}$ for our value of t. So it must be the case that

$$\mathbb{Q}(\sqrt{2}, \sqrt{3}) = \mathbb{Q}(\sqrt{2} + \sqrt{3}).$$

The result of Lemma 32.22 can be extended to more than just an extension by two elements.

Theorem 32.24. *Let E and F be fields of characteristic 0. The following statements are equivalent:*

  (i) *E is a finite extension of F.*

 (ii) *E is a multiple algebraic extension of F.*

(iii) *E is a simple algebraic extension of F.*

Proof. First, we show that statement (i) implies statement (ii). Assume that $[E : F] = n$. Theorem 32.15 shows that E is algebraic over F. To show that E is a multiple extension of F, let $\{b_1, b_2, \ldots, b_n\}$ be a basis for E over F. We claim that $E = F(b_1, b_2, \ldots, b_n)$. Since b_1, b_2, ..., b_n are all elements of E, closure of E under addition and multiplication shows that $F(b_1, b_2, \ldots, b_n) \subseteq E$. Conversely, suppose $e \in E$. Then there exist scalars f_1, f_2, ..., f_n in F such that

$$e = f_1 b_1 + f_2 b_2 + \cdots + f_n b_n.$$

So $e \in F(b_1, b_2, \ldots, b_n)$ and it follows that $E = F(b_1, b_2, \ldots, b_n)$.

Next we prove that statement (ii) implies statement (iii). We must show that E is a simple extension of F. Since E is a multiple extension of F, there exist elements $a_1, a_2, \ldots, a_n$ in E such that

$$E = F(a_1, a_2, \ldots, a_n).$$

We can alternatively think of E as the culmination of a succession of simple extensions. Start with $F = F_0$ and let $F_1 = F(a_1)$. For each $2 \le i \le n$, let $F_i = F_{i-1}(a_i)$. Then

$$F = F_0 \subseteq F_1 \subseteq F_2 \subseteq \cdots \subseteq F_{n-1} \subseteq F_n = E.$$

Lemma 32.22 shows that there is an element c_2 such that $F(a_1, a_2) = F(c_2)$. Then there is an element c_3 such that $F(a_1, a_2, a_3) = F(c_2, a_3) = F(c_3)$. Continuing inductively, for each i there is an element c_i such that $F(a_1, a_2, \ldots, a_i) = F(c_{i-1}, a_i) = F(c_i)$. So in the end we have

$$F(a_1, a_2, \ldots, a_n) = F(c_{n-1}, a_n) = F(c_n).$$

That statement (iii) implies statement (i) is due to the fact that $E = F(a)$ is algebraic, so a is algebraic over F. ∎

Field Automorphisms

There is an important connection between field theory and group theory that we will explore in more detail in Investigation 33. This connection allows us to translate problems in field theory where they can be more difficult to understand to the alternate setting of group theory where the problems might be easier. The main idea that connects field theory to group theory is field automorphisms.

Preview Activity 32.25. In Preview Activity 6.8 on page 88, we showed that $\mathbb{Q}(\sqrt{2})$ is an extension of the field $\mathbb{Q}$. Now consider the function $\alpha : \mathbb{Q}(\sqrt{2}) \to \mathbb{Q}(\sqrt{2})$ defined by

$$\alpha(a + b\sqrt{2}) = a - b\sqrt{2}.$$

(a) Show that α is a ring isomorphism.

(b) We think of $\mathbb{Q}$ as a subfield of $\mathbb{Q}(\sqrt{2})$ as the set of elements of the form $a + 0\sqrt{2}$, where $a \in \mathbb{Q}$. Show that $\alpha(\mathbb{Q}) = \mathbb{Q}$.

(c) Show that $\alpha^2 = \alpha \circ \alpha$ is the identity mapping id on $\mathbb{Q}(\sqrt{2})$. Conclude that $\langle \alpha \rangle = \{id, \alpha\}$ is a group of order 2 under composition.

We can picture the action of α from Preview Activity 32.25 geometrically if we think of the elements of the form $a + b\sqrt{2}$ as points in the $\mathbb{Q} \times \mathbb{Q}$ plane with the a coordinate along the horizontal axis and the $b\sqrt{2}$ coordinate along the vertical axis. Then α applied to an element $a + b\sqrt{2}$ reflects $a + b\sqrt{2}$ across the horizontal axis as shown in Figure 32.2. Applying α twice reflects $a - b\sqrt{2}$ back across the horizontal axis to $a + b\sqrt{2}$. There is a connection between α and the polynomial $p(x) = x^2 - 2$. The roots of $p(x)$ are $\sqrt{2}$ and $-\sqrt{2}$, and α permutes those roots. That is, $\alpha(\sqrt{2}) = -\sqrt{2}$ and $\alpha(-\sqrt{2}) = \sqrt{2}$. Isomorphisms from a field to itself, especially extensions of $\mathbb{Q}$, are the foundation of Galois theory. These isomorphisms are given a special name.

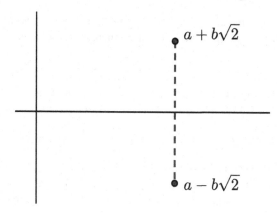

Figure 32.2
The action of α.

Definition 32.26. A field **automorphism** of a field F is an isomorphism from F to F.

For example, the mapping $\alpha : \mathbb{Q}(\sqrt{2}) \to \mathbb{Q}(\sqrt{2})$ defined by $\alpha(a + b\sqrt{2}) = a - b\sqrt{2}$ is an automorphism of $\mathbb{Q}(\sqrt{2})$.

Activity 32.27. Determine which of the following is an automorphism of the given field.

(a) $\varphi : \mathbb{C} \to \mathbb{C}$ defined by $\varphi(a + bi) = a - bi$

(b) $\varphi : \mathbb{C} \to \mathbb{C}$ defined by $\varphi(a + bi) = b + ai$

(c) $\varphi : \mathbb{Q}(\omega) \to \mathbb{Q}(\omega)$ defined by $\varphi(a + b\omega + c\omega^2) = a + b\omega^2 + c\omega$, where $\omega = -\frac{1}{2} + i\frac{\sqrt{3}}{2}$ is a primitive cube root of 1. (Note that $x^2 + x + 1$ is the minimal polynomial of ω over $\mathbb{Q}$, and so $\{1, \omega, \omega^2\}$ is a basis for $\mathbb{Q}(\omega)$ over $\mathbb{Q}$.)

As Activity 32.27 indicates, it can be difficult to determine if a mapping from a field to itself is an automorphism. A good question to ask is if there is a simpler way to determine if a mapping is an automorphism. We will explore that question in this investigation. Our first result is about automorphisms of extensions of $\mathbb{Q}$.

Note that in Activity 32.27, the automorphism φ from $\mathbb{C}$ to $\mathbb{C}$ that sent $a+bi$ to $a-bi$ fixes all the elements of $\mathbb{Q}$. That is, $\varphi(q) = q$ for all $q \in \mathbb{Q}$. Similarly, the automorphism that sends $a + b\omega + c\omega^2$ in $\mathbb{Q}(\omega)$ to $a + b\omega^2 + c\omega$ also fixes all the elements in $\mathbb{Q}$. In general, if φ is an automorphism of a field E and F is a subfield of E such that $\varphi(a) = a$ for all $a \in F$, we say that φ *fixes* the elements of F. Our next activity shows that automorphisms of extensions of $\mathbb{Q}$ always fix the elements of $\mathbb{Q}$.

Activity 32.28. Suppose E is an extension of $\mathbb{Q}$ and $\varphi : E \to E$ is an automorphism. Our goal in this activity is to show that $\varphi(q) = q$ for every rational number $\mathbb{Q}$.

(a) As an isomorphism, we know that $\varphi(0) = 0$. Let's consider what φ does to the rational number 1. Let $y \in E$ such that $\varphi(1) = y$. Use the properties of automorphisms to show that $y^2 = y$. From this, explain why $\varphi(1) = 1$.

(b) Now explain why $\varphi(2) = 2$ and $\varphi(3) = 3$. Then show that $\varphi(n) = n$ for every positive integer n.

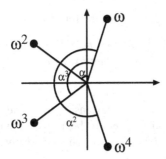

Figure 32.3
The automorphisms α, α^2, and α^3.

(c) Recall that since φ is an automorphism, we know that $\varphi(-x) = -\varphi(x)$ for every $x \in E$. Use this idea to show that $\varphi(n) = n$ for all negative integers n.

(d) Finally, let $q = \frac{r}{s} = rs^{-1}$ be a rational number. Show that $\varphi(q) = q$.

We summarize the result of Activity 32.28 in the following theorem.

Theorem 32.29. *Let φ be an automorphism of an extension E of $\mathbb{Q}$. Then $\varphi(q) = q$ for all $q \in \mathbb{Q}$.*

Based on the result of Theorem 32.29 we might wonder if it is always the case that if field E is an extension of field F and φ is an automorphism of E, then $\varphi(x) = x$ for all $x \in F$. The answer is no, and the problem is left to Exercise (3).

Example 32.30. As another example, consider the polynomial

$$p(x) = x^5 - 1 = (x - 1)(x^4 + x^3 + x^2 + x + 1).$$

The roots of $p(x)$ are $1, \omega, \omega^2, \omega^3$, and ω^4, where $\omega = \cos\left(\frac{2\pi}{5}\right) + i \sin\left(\frac{2\pi}{5}\right)$ is a primitive fifth root of 1. The polynomial $x^4 + x^3 + x^2 + x + 1$ is the minimum polynomial for ω over $\mathbb{Q}$ (See Exercise (6)). So the field $\mathbb{Q}(\omega)$ is a degree 4 extension of $\mathbb{Q}$. Any automorphism of $\mathbb{Q}(\omega)$ has to fix $\mathbb{Q}$, so an automorphism of $\mathbb{Q}(\omega)$ is completely determined by its action on ω. As we will show later, the mapping α defined by $\alpha(\omega) = \omega^2$ is an automorphism of $\mathbb{Q}(\omega)$. It follows that α^2, α^3, α^4, and α^5 are also automorphisms of $\mathbb{Q}(\omega)$. Note that

$$\alpha^2(\omega) = \alpha(\alpha(\omega)) = \alpha(\omega^2) = \omega^4$$
$$\alpha^3(\omega) = \alpha(\alpha^2(\omega)) = \alpha(\omega^4) = \omega^8 = \omega^3$$
$$\alpha^4(\omega) = \alpha(\alpha^3(\omega)) = \alpha(\omega^3) = \omega^6 = \omega.$$

So α^4 is the identity automorphism. Geometrically, these automorphisms cycle the roots of $x^4 + x^3 + x^2 + x + 1$ as illustrated in Figure 32.3.

Note also in Activity 32.27 that the automorphism from $\mathbb{C}$ to $\mathbb{C}$ that sent $a + bi$ to $a - bi$ fixes all the elements of $\mathbb{Q}$ and sends i to $-i$. This is because $\mathbb{C}$ is the splitting field of $x^2 + 1$ over $\mathbb{R}$. Similarly, the automorphism that sends $a + b\omega + c\omega^2$ in $\mathbb{Q}(\omega)$ to $a + b\omega^2 + c\omega$ also fixes all the elements in $\mathbb{Q}$ and sends ω to ω^2. In this case we recognize that ω and ω^2 are roots of $x^2 + x + 1$ over $\mathbb{Q}$, and that $\mathbb{Q}(\omega)$ is the splitting field of $x^2 + x + 1$ over $\mathbb{Q}$. In addition, our earlier example of an automorphism of $\mathbb{Q}(\sqrt{2})$ (the splitting field of $x^2 - 2$ over $\mathbb{Q}$) that sends $\sqrt{2}$ to $-\sqrt{2}$ also permutes the roots of the polynomial $x^2 - 2$ over $\mathbb{Q}$. This behavior is true in general, as the next activity demonstrates.

Activity 32.31. In this activity we investigate the following theorem.

Theorem 32.32. *Let E be an extension of a field F and let $r \in E$ be a root of a polynomial $p(x) \in F[x]$. If φ is an automorphism of E that fixes the elements in F, then $\varphi(r)$ is also a root of $p(x)$.*

First, a bit of terminology. Let E be an extension of a field F. We say that an automorphism φ of E *fixes* F if $\varphi(t) = t$ for all $t \in F$. To begin our activity, we let $r \in E$ be a root of a polynomial $p(x) \in F[x]$ and let φ be an automorphism of E that fixes F. As a reminder, our goal is to demonstrate that $\varphi(r)$ is also a root of $p(x)$.

(a) Suppose $p(x) = c_0 + c_1 x + c_2 x^2 + \cdots + c_n x^n$ for some $c_0, c_1, \ldots, c_n$ in F. Apply φ to $p(r)$ to explain why

$$0 = \varphi(c_0) + \varphi(c_1)\varphi(r) + \varphi(c_2)\varphi(r)^2 + \cdots + \varphi(c_n)\varphi(r)^n.$$

(Hint: What do we know about $\varphi(0)$ and $p(r)$?)

(b) How can we conclude $\varphi(r)$ is a root of $p(x)$?

(c) Let R be the set of roots of $p(x)$. Why is φ a permutation of R?

Theorem 32.32 puts significant limits on what kinds of automorphisms of fields we can have. In particular, automorphisms permute roots of polynomials. Two questions seem reasonable to ask at this point. First, is any permutation of the roots of a polynomial an automorphism, and, second, what happens if a polynomial has a multiple root? We address the first question in the next activity.

Activity 32.33. The field $E = \mathbb{Q}(\sqrt{2}, \sqrt{3})$ is an extension of $\mathbb{Q}$. Since $\sqrt{2}$ is a root of $x^2 - 2$ and $\sqrt{3}$ is a root of $x^2 - 3$, it follows that $\{1, \sqrt{2}, \sqrt{3}, \sqrt{6}\}$ is a basis for E over $\mathbb{Q}$. Note that $\pm\sqrt{2}$ and $\pm\sqrt{3}$ are all roots of $(x^2 - 2)(x^2 - 3)$.

(a) Let $\alpha : E \to E$ be defined by

$$\alpha(a + b\sqrt{2} + c\sqrt{3} + d\sqrt{6}) = a - b\sqrt{2} + c\sqrt{3} + d\sqrt{6}.$$

That is, α permutes the roots $\sqrt{2}$ and $-\sqrt{2}$. Show that α is an automorphism of E. (Hint: If $x = a + b\sqrt{2} + c\sqrt{3} + d\sqrt{6}$ and $y = e + f\sqrt{2} + g\sqrt{3} + h\sqrt{6}$, then

$$xy = (ae + 2bf + 3cg + 6dh) + (af + eb + 3ch + 3dg)\sqrt{2} + (ag + 2bh + ce + 2df)\sqrt{3}$$
$$+ (ae + bg + cf + de)\sqrt{6}.)$$

(b) Let $\beta : E \to E$ be defined by

$$\beta(a + b\sqrt{2} + c\sqrt{3} + d\sqrt{6}) = a + b\sqrt{3} + c\sqrt{2} + d\sqrt{6}.$$

That is, β permutes the roots $\sqrt{2}$ and $\sqrt{3}$. Show that β is not an automorphism of E. (Hint: What does β do to $(1 + \sqrt{2})^2$?)

Activity 32.33 shows that not every permutation of roots of a polynomial will produce an automorphism. Rather, it appears that automorphisms will permute roots of the irreducible factors of a polynomial. We will return to this idea a little later. Now we address the question of multiple roots. Our interest is primarily in the characteristic 0 case, and Theorem 32.21 shows that if $p(x)$ is an irreducible polynomial over a field of characteristic 0, then $p(x)$ has no repeated roots. So we don't have to consider that possibility. The next question we want to address, and this is a more difficult question, is the following: If E is an extension of a field F with $p(x)$ irreducible in $F[x]$, is any mapping from E to E that fixes F and sends one root of $p(x)$ to another an automorphism of E? We will build up a collection of intermediate results about extending isomorphisms that will help us answer this question.

The first result concerns dimensions of isomorphic extensions. The idea is that if E and K are finite extensions of the same field F, and E and K are isomorphic, we might expect that the dimensions of E and K over F are related.

Activity 32.34. Let $E = \mathbb{Q}(\sqrt[4]{3})$ and $K = \mathbb{Q}(\sqrt[4]{3}i)$. In Exercise (8) you are asked to show that E and K are isomorphic as extensions of $\mathbb{Q}$. In this activity we are interested in the degrees of these extensions.

(a) Explain why E and K are different fields.

(b) Find a basis for E over $\mathbb{Q}$. What, then, is $[E : \mathbb{Q}]$?

(c) Find a basis for K over $\mathbb{Q}$. What, then, is $[K : \mathbb{Q}]$?

In Activity 32.34 we had two different but isomorphic field extensions of $\mathbb{Q}$, and the two extensions have the same dimensions as vector spaces over $\mathbb{Q}$. That is, $[E : \mathbb{Q}] = [K : \mathbb{Q}]$. The next lemma shows that this is true in general.

Lemma 32.35. *Let E and K be finite extensions of a field F and suppose that $\varphi : E \to K$ is an isomorphism such that $\varphi(x) = x$ for all $x \in F$. Then $[K : F] = [E : F]$.*

Proof. Suppose $\{u_1, u_2, \ldots, u_n\}$ is a basis for E over F. Let $v_i = \varphi(u_i)$ for each i. We will show that $\{v_1, v_2, \ldots, v_n\}$ is a basis for K over F. Let $k \in K$. There exists $e \in E$ such that $\varphi(e) = k$. Then

$$e = x_1 u_1 + x_2 u_2 + \cdots + x_n u_n$$

for some $x_1, x_2, \ldots, x_n$ in F. Since φ fixes F, it follows that

$$
\begin{aligned}
k &= \varphi(e) \\
&= \varphi(x_1 u_1 + x_2 u_2 + \cdots + x_n u_n) \\
&= \varphi(x_1)\varphi(u_1) + \varphi(x_2)\varphi(u_2) + \cdots + \varphi(x_n)\varphi(u_n) \\
&= x_1 v_1 + x_2 v_2 + \cdots + x_n v_n
\end{aligned}
$$

and the set $\{v_1, v_2, \ldots, v_n\}$ spans K over F. To verify linear independence, suppose that

$$y_1 v_1 + y_2 v_2 + \cdots + y_n v_n = 0$$

for some $y_1, y_2, \ldots, y_n$ in F. Since $v_i = \varphi(u_i)$ for each i, we have

$$
\begin{aligned}
0 &= y_1 v_1 + y_2 v_2 + \cdots + y_n v_n \\
&= y_1 \varphi(u_1) + y_2 \varphi(u_2) + \cdots + y_n \varphi(u_n) \\
&= \varphi(y_1)\varphi(u_1) + \varphi(y_2)\varphi(u_2) + \cdots + \varphi(y_n)\varphi(u_n) \\
&= \varphi(y_1 u_1 + y_2 u_2 + \cdots + y_n u_n).
\end{aligned}
$$

Since φ is an injection, it follows that

$$y_1 u_1 + y_2 u_2 + \cdots + y_n u_n = 0.$$

The fact that $\{u_1, u_2, \ldots, u_n\}$ is linearly independent allows us to conclude that $y_i = 0$ for each i. Thus, the set $\{v_1, v_2, \ldots, v_n\}$ is linearly independent and is a basis for E over F. So $[K : F] = [E : F]$. ∎

$$E \xrightarrow[?\varphi]{\rightarrow} E'$$
$$\cup| \qquad \cup|$$
$$F \xrightarrow[\varphi]{\rightarrow} F'$$

Figure 32.4
Extending an isomorphism.

Suppose that E is the splitting field of some polynomial $p(x)$ over a field F. (Recall that an extension E of a field F is the splitting field of a polynomial $p(x)$ in $F[x]$ if E is the smallest extension of F that contains all the roots of $p(x)$. Refer to Definition 31.14 on page 454.) Also, suppose that $\varphi : F \to F'$ is an isomorphism. The next lemma shows that we can extend φ to an isomorphism from E to E', where E' is the splitting field of $\varphi p(x)$ as illustrated in Figure 32.4.

Lemma 32.36. *Let $\varphi : F \to F'$ be an isomorphism of fields, $p(x) \in F[x]$ a nonconstant polynomial, and $\varphi p(x)$ the image of $p(x)$ in $F'[x]$. Let E be the splitting field of $p(x)$ over F and E' the splitting field of $\varphi p(x)$ over F'. Then φ extends to an isomorphism from E to E'.*

Proof. The proof proceeds by induction on $\deg(p(x))$. For the base case, suppose that $\deg(p(x)) = 1$. Then $p(x) = c(x - r)$ for some c, r in E. In this case we have $E = F(r)$. But $p(x) = cx - cr \in F[x]$ implies that c, cr in F. But then $c^{-1}(cr) = r$ is also in F. So $E = F$. Exercise (9) shows that φ extends to an isomorphism from $F[x]$ to $F'[x]$. Then $\varphi p(x)$ is also a degree 1 polynomial. Using the same argument, we have $E' = F'(\varphi(r)) = F'$. So φ itself is an isomorphism from E to E'.

For the induction step, assume that the lemma is true for all polynomials of degree $n - 1$ and let $\deg(p(x)) = n$ for some integer $n \geq 2$. We can factor $p(x)$ into a product of irreducible polynomials in $F[x]$. So $p(x)$ has a monic irreducible factor $q(x)$ in $F[x]$. Using the extension of φ to an isomorphism from $F[x]$ to $F'[x]$, we have that $\varphi q(x)$ is a monic irreducible factor of $\varphi p(x)$ in $F'[x]$. Since $q(x)$ is a factor of $p(x)$, every root of $q(x)$ is also a root of $p(x)$. So E contains all the roots of $q(x)$. Similarly, E' contains all the roots of $\varphi q(x)$.

Let r be a root of $q(x)$ in E and s a root of $\varphi q(x)$ in E'. Corollary 32.18 shows that φ can be extended to an isomorphism φ from $F(r)$ to $F'(s)$ such that $\varphi(r) = s$. Since r is a root of $p(x)$, we have that $p(x) = (x - r)u(x)$ for some $u(x) \in F(r)[x]$. Then

$$\varphi p(x) = (x - \varphi(r))\varphi u(x) = (x - s)\varphi u(x).$$

Let $r, r_2, r_3, \ldots, r_n$ be the roots of $p(x)$ in E. So

$$p(x) = c(x - r)(x - r2) \cdots (x - r_n) = (x - r)u(x)$$

and so

$$u(x) = (x - r_2) \cdots (x - r_n)$$

has degree $n - 1$. The fact that E is the smallest extension of F that contains all the roots of $p(x)$ implies that E is the smallest extension of $F(r)$ that contains all the roots of $u(x)$. So E is the splitting field of $u(x)$ over $F(r)$. Similarly, E' is the splitting field of $\varphi u(x)$ over $F'(s)$. By the induction hypothesis, the isomorphism φ from $F(r)$ to $F'(s)$ can be extended to an isomorphism φ from E to E'. ∎

Now we can clarify what automorphisms of splitting fields look like.

Theorem 32.37. *Let E be the splitting field of some polynomial $p(x) \in F[x]$, and let a and b be in E. There exists an automorphism α of E that fixes F such that $\alpha(a) = b$ if and only if a and b have the same minimal polynomial over F.*

Proof. Suppose that there is an automorphism α of E that fixes F such that $\alpha(a) = b$, and suppose that $p(x)$ is the minimal polynomial of a over F. Theorem 32.32 shows that $b = \alpha(a)$ is also a root of $p(x)$. Thus, $p(x)$ is the minimal polynomial for b over F.

Now assume that a and b have the same minimal polynomial $p(x)$ in $F[x]$. Theorem 32.17 shows that there is an isomorphism $\sigma : F(a) \to F(b)$ that fixes F with $\sigma(a) = b$. The fact that E is the splitting field of some polynomial over F implies that E is the splitting field of the same polynomial over $F(a)$ and $F(b)$. So by Lemma 32.36, σ extends to an automorphism of E that fixes F. That is, the extension of σ (which we also denote as σ) is an automorphism of E that fixes F with $\sigma(a) = b$. ∎

We conclude this investigation with an example.

Example 32.38. Let $E = \mathbb{Q}(\sqrt[3]{2}, \omega)$, where $\omega = -\frac{1}{2} + \frac{\sqrt{3}}{2}i$ is a primitive cube root of 1. Since $\{1, \omega\}$ is a basis for $\mathbb{Q}(\sqrt[3]{2}, \omega)$ over $\mathbb{Q}(\sqrt[3]{2})$ and $\{1, \sqrt[3]{2}, \sqrt[3]{4}\}$ is a basis for $\mathbb{Q}(\sqrt[3]{2})$ over $\mathbb{Q}$, we can see that a basis for $\mathbb{Q}(\sqrt[3]{2}, \omega)$ over $\mathbb{Q}$ is $\{1, \sqrt[3]{2}, \sqrt[3]{4}, \omega, \sqrt[3]{2}\omega, \sqrt[3]{4}\omega\}$. So the action of any automorphism of $\mathbb{Q}(\sqrt[3]{2}, \omega)$ is completely determined by its action on $\sqrt[3]{2}$ and ω.

The minimal polynomial of $\sqrt[3]{2}$ over $\mathbb{Q}(\omega)$ is $x^3 - 2$, and the roots of $x^3 - 2$ are $\sqrt[3]{2}$, $\sqrt[3]{2}\omega$, and $\sqrt[3]{2}\omega^2$. The nontrivial automorphisms of $\mathbb{Q}(\sqrt[3]{2}, \omega)$ that fix $\mathbb{Q}(\omega)$ are determined by what the automorphisms do to $\sqrt[3]{2}$, and there are two choices. The first, α, satisfies $\alpha(\sqrt[3]{2}) = \sqrt[3]{2}\omega$. The second is α^2 because

$$\alpha^2(\sqrt[3]{2}) = \alpha(\sqrt[3]{2}\omega) = (\sqrt[3]{2}\omega)(\omega) = \sqrt[3]{2}\omega^2.$$

Note that α^3 is the identity.

The minimal polynomial of ω over $\mathbb{Q}(\sqrt[3]{2})$ is $x^2 + x + 1$, whose roots are ω and ω^2. So there is only one nontrivial automorphism β of E that fixes $\mathbb{Q}(\sqrt[3]{2})$, the one with $\beta(\omega) = \omega^2$. Note that β^2 is the identity.

Now

$$(\alpha\beta)(\sqrt[3]{2}) = \alpha(\sqrt[3]{2}) = \sqrt[3]{2}\omega$$
$$(\alpha\beta)(\omega) = \alpha(\omega^2) = \omega^2$$

$$(\alpha^2\beta)(\sqrt[3]{2}) = \alpha^2(\sqrt[3]{2}) = \sqrt[3]{2}\omega^2$$
$$(\alpha^2\beta)(\omega) = \alpha^2(\omega^2) = \omega^2.$$

We leave it to the reader to show that $\beta\alpha = \alpha^2\beta$ and $\beta\alpha^2 = \alpha\beta$.

We summarize the actions of these distinct automorphisms as follows:

x	$\sqrt[3]{2}$	ω
$id(x)$	$\sqrt[3]{2}$	ω
$\alpha(x)$	$\sqrt[3]{2}\omega$	ω
$\alpha^2(x)$	$\sqrt[3]{2}\omega^2$	ω
$\beta(x)$	$\sqrt[3]{2}$	ω^2
$\alpha\beta(x)$	$\sqrt[3]{2}\omega$	ω^2
$\alpha^2\beta(x)$	$\sqrt[3]{2}\omega^2$	ω^2.

So α and β generate a group of order 6 with operation table

	id	α	α^2	β	$\alpha\beta$	$\alpha^2\beta$
id	id	α	α^2	β	$\alpha\beta$	$\alpha^2\beta$
α	α	α^2	id	$\alpha\beta$	$\alpha^2\beta$	β
α^2	α^2	id	α	$\alpha^2\beta$	β	$\alpha\beta$
β	β	$\alpha^2\beta$	$\alpha\beta$	id	α^2	α
$\alpha\beta$	$\alpha\beta$	β	$\alpha^2\beta$	α	id	α^2
$\alpha^2\beta$	$\alpha^2\beta$	$\alpha\beta$	β	α^2	α	id

This group is a non-Abelian group of order 6 and so is isomorphic to S_3.

In the next investigation, we will see that the order of the group of automorphisms of E over $\mathbb{Q}$ is equal to $[E : \mathbb{Q}] = 6$, so we have identified all of these automorphisms. This example indicates a direct connection between automorphisms of E that fix $\mathbb{Q}$ and roots of minimal polynomials. We will explore this connection in more detail in the next investigation.

Concluding Activities

Activity 32.39. As an example of Theorem 32.37, let $E = \mathbb{Q}(\sqrt{2}, \sqrt{3}) = \mathbb{Q}(\sqrt{2})(\sqrt{3})$. Then E is the splitting field of the polynomial $x^4 - 5x^2 + 6 = (x^2 - 2)(x^2 - 3)$ over $\mathbb{Q}$. In this activity we determine the automorphisms of E. Recall that every automorphism of E must fix $\mathbb{Q}$ by Theorem 32.29.

(a) Explain why $x^2 - 3$ is an irreducible polynomial over $\mathbb{Q}(\sqrt{2})$. This makes $x^2 - 3$ the minimal polynomial of $\sqrt{3}$ over $\mathbb{Q}(\sqrt{2})$.

(b) Use Theorem 32.37 to explain why there is only one nontrivial automorphism α_1 of E that fixes $\mathbb{Q}(\sqrt{2})$. Explicitly explain what α_1 does to the elements of E.

(c) Now explain why there is exactly one automorphism, α_2, of E that fixes $\mathbb{Q}(\sqrt{3})$. Explicitly explain what α_2 does to the elements of E. (Hint: Repeat parts (a) and (b) with the polynomial $x^2 - 2$.)

(d) Since α_1 and α_2 are automorphisms of E, so is their composite. Determine the action of $\alpha_1 \circ \alpha_2$ and $\alpha_2 \circ \alpha_1$. The group of automorphisms of E generated by α_1 and α_2 is isomorphic to a familiar group. Which group is this?

Activity 32.40. In Exercise (7) of Investigation 12 on page 181, we introduced the idea of constructible numbers – those numbers that can be geometrically constructed using a collapsible compass and unmarked ruler. We showed that the collection of constructible numbers is an extension of $\mathbb{Q}$ that is contained in $\mathbb{R}$. Three classical problems from ancient Greek mathematics that were influential to the development of geometry and algebra are squaring of the circle, trisecting an angle, and doubling the cube. In this activity we explore the problem of doubling the cube. That is, given a cube of volume V, is it possible to construct, with compass and unmarked ruler, a cube of volume $2V$? (The problems of squaring the circle and trisecting an angle are left for Exercises (20) and (21).) The first step is to classify which numbers are constructible numbers. That is, a number c is constructible if we can geometrically construct a segment of length $|c|$ from a given line segment of length 1 using only a collapsible compass and unmarked straightedge. Using a compass, we can construct a circle whose radius is a constructible number, and we can consider lines whose coefficients are constructible numbers. More formally, if F is a subfield of $\mathbb{R}$ (in this problem think of F as the field of constructible numbers), we define the plane of F to be the set $\{(a, b) \in \mathbb{R}^2 : a, b \in F\}$. Any line in the plane of F has an equation of the form

$$ax + by = c$$

for some a, b, and c in F, and a circle in the plane of F has an equation of the form

$$x^2 + y^2 + ax + by = c$$

for some a, b, and $c \in F$. We can construct new points by intersecting lines and circles.

(a) Our first step is to show that any new constructible number must lie in a quadratic extension of the field of constructible numbers.

  (i) Show that if two lines in the plane of F intersect, then they intersect at a point that is in the plane of F. So no new points that aren't in F can be obtained by intersecting lines in the plane of F.

  (ii) Show that if two circles in the plane of F intersect, then they intersect at a point that is in the plane of F. So no new points that aren't in F can be obtained by intersecting circles in the plane of F.

  (iii) Show that if a line and a circle in the plane of F intersect, then the point of intersection lies in the plane of F or in the plane of $F(\sqrt{\alpha})$ for some positive $\alpha \in F$. That is, the point of intersection is in the plane of a quadratic extension of F.

(b) Use the result of (a) to explain why if c is a constructible number, then there is a sequence

$$\mathbb{Q} \subseteq F_1 \subseteq F_2 \cdots \subseteq F_n$$

where for each k, $F_k = F_{k-1}(\sqrt{\alpha_k})$ for some positive $\alpha_k \in F_{k-1}$.

(c) Explain why if c is a constructible number, then $[\mathbb{Q}(c) : \mathbb{Q}]$ is a power of 2.

(d) Suppose we have a cube of side length 1 and volume 1. To double the cube means to construct, with collapsing compass and unmarked straightedge, a cube whose volume is 2. The length of a side of such a cube must be $\sqrt[3]{2}$. Explain why it is not possible to double the cube.

Exercises

(1) For each field E, find a basis for E over $\mathbb{Q}$ and determine $[E : \mathbb{Q}]$.

  (a) $E = \mathbb{Q}(\sqrt{5})$

  (b) $E = \mathbb{Q}(\sqrt[6]{2})$

  (c) $E = \mathbb{Q}(\sqrt[4]{3}, i)$

(2) Let $r = \sqrt{2} + i$. Show, by finding the minimal polynomials, that the minimal polynomial for r is different over $\mathbb{Q}$ than it is over $\mathbb{R}$.

(3) Show by example that if φ is an automorphism of an extension field E of a field F, that φ does not have to fix the elements of F.

(4) Let m be a square-free integer (that is, no square divides m).

   (a) Show that the mapping $\alpha_m : \mathbb{Q}(\sqrt{m}) \to \mathbb{Q}(\sqrt{m})$ defined by $\alpha(a + b\sqrt{m}) = a - b\sqrt{m}$ is an automorphism.

   (b) Use the automorphism α_m to prove that $\sqrt{m}$ is irrational.

(5) Recall that if F is a field and a is an algebraic element over F, then the minimal polynomial of a over F is the irreducible monic polynomial in $F[x]$ with a as a root. In this exercise we prove the following theorem.

   Theorem 32.41. *Let F be a field and let a be an element that is algebraic over F.*

   1. The minimal polynomial of a over F is unique.

   2. If $p(x)$ is a polynomial in $F[x]$ such that $p(a) = 0$, then the minimal polynomial of a over F divides $p(x)$.

   (a) Our first step is to find a candidate for a minimal polynomial of a over F. To do this, we look for a monic polynomial if smallest degree in $F[x]$ that has a as a root. Let S be the set of all nonzero polynomials in $F[x]$ having a as a root.

      (i) Explain why S is not empty.

      (ii) Explain why S must contain a polynomial of smallest positive degree.

      (iii) Explain why S must contain a monic polynomial $m(x)$ of smallest positive degree.

   (b) Show that $m(x)$ as found in the previous part is an irreducible polynomial.

   (c) Suppose that $p(x)$ is a polynomial in $F[x]$ such that $p(a) = 0$. That is, $p(x) \in S$. Show that $m(x)$ divides $p(x)$. (Hint: Use the Division Algorithm for Polynomials.)

   (d) Finally, show that $m(x)$ is unique.

(6) Use Theorem 11.15 on page 164 with $p = 2$ to show that the polynomial $x^4 + x^3 + x^2 + x + 1$ is irreducible over $\mathbb{Q}$.

(7) (a) Prove that $\mathbb{Q}(\sqrt{2})$ is not isomorphic to $\mathbb{Q}(\sqrt{3})$ as extensions of $\mathbb{Q}$.

   (b) Notice that if a and b are positive integers and $a = k^2 b$ for some positive integer k, then $\mathbb{Q}(\sqrt{b}) = \mathbb{Q}(k\sqrt{a}) = \mathbb{Q}(\sqrt{a})$. So if we are interested in when two extensions of $\mathbb{Q}$ of the form $\mathbb{Q}(\sqrt{a})$ and $\mathbb{Q}(\sqrt{b})$ are isomorphic but not equal, we want to avoid this situation. So, we say that a positive integer is square-free if no perfect square divides the integer. Let a and b be square-free integers. For what values of a and b is $\mathbb{Q}(\sqrt{a})$ isomorphic to $\mathbb{Q}(\sqrt{b})$?

(8) Let $E = \mathbb{Q}(\sqrt[4]{3})$ and $K = \mathbb{Q}(\sqrt[4]{3}i)$. Show that E and K are isomorphic as extensions of $\mathbb{Q}$.

(9) Let F and F' be fields and let $\varphi : F \to F'$ be an isomorphism. Show that the mapping from $F[x]$ to $F'[x]$ that assigns a polynomial $p(x) = a_0 + a_a x + \cdots + a_n x^n$ to the polynomial $\varphi p(x) = \varphi(a_0) + \varphi(a_1)x + \cdots + \varphi(a_n)x^n$ is an isomorphism between $F[x]$ and $F'[x]$.

(10) Let F be a field and E an extension of F. Show that E has exactly one basis over F if and only if $E = F \cong \mathbb{Z}_2$.

(11) Let a and b be algebraic elements over a field F, with $[F(a) : F] = m$ and $[F(b) : F] = n$.

  (a) Show by example that if $\gcd(m, n) > 1$, then $F(a, b)$ need not be of degree mn over F.

  (b) Prove that if $\gcd(m, n) = 1$, then $F(a, b)$ is of degree mn over F.

(12) Let E be an extension of a field F and let $r \in E$.

  (a) Show that $F(r^2) \subseteq F(r)$.

  (b) Find an element r whose minimal polynomial over $\mathbb{Q}$ has degree 3. What is the connection between $\mathbb{Q}(r)$ and $\mathbb{Q}(r^2)$? Be as specific as you can.

  (c) Show that if r is an element whose minimal polynomial over a field F has odd degree, then $F(r^2) = F(r)$.

  (d) Is it true that if $F(r^2) = F(r)$, then the minimal polynomial of r over F has odd degree? Justify your answer.

(13) Let $E_1 = \mathbb{Q}(3 + i)$ and $E_2 = \mathbb{Q}(1 - i)$. Are E_1 and E_2 different fields? If yes, is there any containment relation between them? If no, show that $E_1 = E_2$.

(14) Let a be an algebraic element in an extension E of a field F. Suppose that the minimal polynomial of a has prime degree. Show that if K is a field with $F \subseteq K \subseteq F(a)$, then $K = F$ or $K = F(a)$.

(15) Let φ be an automorphism of $\mathbb{R}$. Show that if a is a positive real number, then $\varphi(a) > 0$.

(16) The field $\mathbb{Q}(\sqrt[3]{2}, \omega)$ is a finite extension of $\mathbb{Q}$, so we can realize $\mathbb{Q}(\sqrt[3]{2}, \omega)$ as a simple extension. Find a single element a such that $\mathbb{Q}(a) = \mathbb{Q}(\sqrt[3]{2}, \omega)$.

(17) Find all automorphisms of the following fields over $\mathbb{Q}$:

  (a) $\mathbb{Q}(\sqrt{5})$

  (b) $\mathbb{Q}(\sqrt[3]{5})$

  (c) $\mathbb{Q}(\sqrt[4]{5})$

(18) Let $E = \mathbb{Q}(\sqrt{2}, i)$.

  (a) Find a basis for E over $\mathbb{Q}$.

  (b) Let K be an extension of a field F with basis $\{k_1, k_2, \ldots k_n\}$ over F. Show that any automorphism φ of K over F that fixes F is completely determined by the values of $\varphi(k_i)$ for $1 \le i \le n$.

  (c) Exhibit all automorphisms of E over $\mathbb{Q}$ using (b) and the basis found in (a).

(19) Let $f(x) = x^2 + [1]$ in $\mathbb{Z}_3[x]$.

    (a) Show that $f(x)$ is irreducible over $\mathbb{Z}_3$.

    (b) Let $E = \mathbb{Z}_3[x]/\langle f(x)\rangle$. Since $f(x)$ is irreducible over $\mathbb{Z}_3$, we know that E is a field by Theorem 12.11 on page 175. Show that E is the splitting field for $f(x)$ over $\mathbb{Z}_3$.

    (c) Find all automorphisms of E that fix $\mathbb{Z}_3$.

(20) In Concluding Activity 32.40, we considered the classical problem of doubling the cube. That is, from a cube of fixed volume V, is it possible to construct, with compass and unmarked ruler, a cube of volume $2V$? We showed that the answer to this question is no. We consider another classical problem, squaring the circle, in this exercise.

    (a) Recall that a number is transcendental if it is not algebraic. There are many familiar numbers that are transcendental, including the famous constants e and π. The number e was proved to be transcendental by Charles Hermite in 1873, and Ferdinand von Lindemann published the first complete proof of the transcendence of π 1882. We will not prove that these numbers are transcendental, but you may assume them to be in this exercise.

    Prove that no transcendental number is constructible.

    (b) The problem of squaring the circle is, given a circle in the plane, if it is possible to construct a square with the same area as the given circle. Show that it is not possible to square the circle. (Hint: Assume that our constructible unit is the radius of the circle, which allows us to work with a circle of radius 1. The result of Exercise (7) on page 181 might be helpful.)

(21) In Concluding Activity 32.40, we showed that it is not possible to double the cube. In Exercise (20) we show that it is not possible to square the circle. We consider the third classical problem, trisecting an angle, in this exercise.

The problem of trisecting an angle is, given an angle of magnitude φ, if it is possible to construct an angle of magnitude $\frac{\varphi}{3}$. (Note that this does not mean that no angle can be trisected, just that there exist angles that can't be trisected.) For example, if it is possible to construct an angle of $60°$, then it would be possible to construct an angle of $20°$. We will show that it is not possible to trisect an angle of measure $20°$.

    (a) Show that if an angle of magnitude θ is constructible, then $\sin(\theta)$ and $\cos(\theta)$ are constructible numbers. (See Exercise (7) on page 181 for what constructions are possible.)

    (b) To show that $\cos(20°)$ is not constructible, we will use the trigonometric identity

$$\cos(3\theta) = 4\cos^3(\theta) - 3\cos(\theta). \qquad (32.5)$$

    Use the angle sum formulas:

$$\cos(\alpha + \beta) = \cos(\alpha)\cos(\beta) - \sin(\alpha)\sin(\beta)$$
$$\sin(\alpha + \beta) = \sin(\alpha)\cos(\beta) + \cos(\alpha)\sin(\beta)$$

    to verify the identity in (32.5).

    (c) Use Equation (32.5) with $\theta = 20°$ and $t = \cos(20°)$ to show that t is a root of the polynomial $p(x) = 8x^3 - 6x - 1$. From this, argue that $\cos(20°)$ is not constructible.

Connections

In this investigation, we used material from both field theory and group theory to study extensions of fields. Of course, any field is a ring, as introduced in Investigation 4, and we saw that facts about polynomials and polynomial rings (Investigation 8), field extensions (Investigation 6), roots of polynomials (Investigation 10), and irreducible polynomials (Investigation 10) came up in our examination of finite extensions of fields. The idea of ring isomorphisms (Investigation 7) was critical to understand automorphisms of fields. All of these ideas will lead us to more fascinating connections between rings, fields, and groups in our subsequent study of Galois theory in Investigation 33.

Investigation 33

Galois Theory

Focus Questions

By the end of this investigation, you should be able to give precise and thorough answers to the questions listed below. You may want to keep these questions in mind to focus your thoughts as you complete the investigation.

- What is a Galois group?

- What is the fixed field of a Galois group?

- What is a normal extension of a field?

- How is a normal extension related to a splitting field in characteristic 0?

- What do we mean by the restriction of a function?

- What is the Galois correspondence? Why is the Galois correspondence important?

- What does the Fundamental Theorem of Galois Theory tell us about the Galois correspondence?

- What does it mean for a polynomial to be solvable by radicals?

- What is a solvable group, and what is the connection between solvable groups and solvability by radicals?

Introduction

Galois theory is a beautiful area of mathematics that provides connections between polynomials, fields, and groups. One of the fundamental applications of Galois theory is the resolution of the question of which polynomials can be solved by radicals. We are familiar with using the quadratic formula

$$x = \frac{-b \pm \sqrt{b^2 - 4ac}}{2a}$$

for solving the quadratic equation $ax^2 + bx + c = 0$ with real coefficients. The basic method for solving quadratics was known to Egyptian mathematicians in the 1500s BCE. A more formal and general solution was produced by the Babylonians around 400 BCE. By the middle of the 16th century similar methods were developed to solve the general cubic. For example, the solution to the cubic $x^3 + px + q = 0$ is

$$z = \sqrt[3]{A} - \frac{p}{3\sqrt[3]{A}}, \text{ where } A = -\frac{q}{2} + \sqrt{\frac{q^2}{4} + \frac{p^3}{27}},$$

and $\sqrt[3]{A}$ is any one of the three cube roots of A (the cubic formula for finding roots of arbitrary cubic equations can be found in the online supplemental materials). There is also a more complicated solution to quartic (fourth degree) polynomial equations. Scipione del Ferro (1465–1526) from Italy is generally credited with the solution to the cubic equation and another Italian, Rafael Bombelli (1526–1572), is regarded as the person who saw how these methods could be applied to solving the general quartic equation. These methods express the solutions to polynomial equations in terms of the elements of the field and radicals (like square roots for quadratics and cube roots for cubics) involving elements of the field or extensions of the field. For this reason, we say that these polynomials are *solvable by radicals*. It wasn't until many years later that it was shown that no such analogous methods exist to solve the general fifth degree or higher polynomial.

In 1824 the Norwegian mathematician Niels Henrik Abel (1802–1829) [*], proved the existence of a fifth degree polynomial that cannot be solved by radicals. Even earlier (1799) the Italian Paolo Ruffini (1765–1822) [†] outlined a proof of the same result, but the argument was incomplete.

It wasn't until later that Évariste Galois (1811–1832), a French mathematician and political activist, completely solved the problem by proving necessary and sufficient conditions for a polynomial to be solvable by radicals. Galois' work was mostly published privately through his friend Auguste Chevalier. Galois died in a duel on May 30, 1832 at the young age of 21.

As we will see, Galois theory is built on characterizing solvable polynomial equations in terms of field extensions. This characterization creates the Galois correspondence between intermediate fields and groups. Through this Galois correspondence, we connect solvability questions from field theory to the group setting. By proving that the symmetric group of degree 5 or higher is not solvable, we can demonstrate the existence of polynomials whose Galois groups are these symmetric groups and are therefore not solvable by radicals. We will work through these items in this investigation, referring often to results in field extensions from Investigation 32. While some of our results hold in the general case, we are mostly interested in polynomials over $\mathbb{Q}$, and so we will focus on the characteristic 0 case.

The Galois Group

A Galois group will provide a connection between subfields of a field and subgroups of a symmetric group. Applying Theorem 32.37 to the splitting field $\mathbb{Q}(\sqrt{2}, \sqrt{3})$ of the polynomial $(x^2 - 2)(x^2 - 3)$ in $\mathbb{Q}[x]$, we can see that there are four automorphisms of $\mathbb{Q}(\sqrt{2}, \sqrt{3})$ that fix $\mathbb{Q}$. These automorphisms permute roots of the irreducible factors of $(x^2 - 2)(x^2 - 3)$. So one automorphism α sends $\sqrt{2}$ to $-\sqrt{2}$ (roots of $x^2 - 2$ over $\mathbb{Q}$) and fixes everything else, while another β sends $\sqrt{3}$ to $-\sqrt{3}$ (roots of $x^2 - 3$ over $\mathbb{Q}$) and fixes everything else. The complete set of automorphisms of $\mathbb{Q}(\sqrt{2}, \sqrt{3})$

---

[*] Niels Henrik Abel, *Mémoire sur les équations algébriques, où on démontre l'impossibilité de la résolution de l'équation générale du cinquième degré* (PDF), in Sylow, Ludwig; Lie, Sophus (eds.), Œvres Complètes de Niels Henrik Abel (in French), vol. I (2nd ed.), Grøndahl & Søn, pp. 28–33.

[†] Paolo Ruffini, *Teoria generale delle equazioni, in cui si dimostra impossibile la soluzione algebraica delle equazioni generali di grado superiore al quarto* (in Italian), Stamperia di S. Tommaso d'Aquino.

that fix $\mathbb{Q}$ is $\{id, \alpha, \beta, \alpha\beta\}$. This set of automorphisms forms a group isomorphic to $\mathbb{Z}_2 \oplus \mathbb{Z}_2$. This group of automorphisms is what we will call the *Galois group* of $\mathbb{Q}(\sqrt{2}, \sqrt{3})$ over $\mathbb{Q}$.

In general, if E is an extension of a field F, we let $\text{Gal}(E/F)$ be the set of isomorphisms from E to E that fix F (the automorphisms of E that fix F). That is, $\varphi \in \text{Gal}(E/F)$ if $\varphi : E \to E$ is an isomorphism of the field E and $\varphi(x) = x$ for all $x \in F$. The next activity demonstrates that $\text{Gal}(E/F)$ is always a group.

Preview Activity 33.1. Let F be a field and let E be an extension of F. Let α and β be in $\text{Gal}(E/F)$.

(a) Explain why $\alpha \circ \beta$ is in $\text{Gal}(E/F)$.

(b) Explain why the identity map from E to E is in $\text{Gal}(E/F)$.

(c) Explain why α^{-1} exists and is in $\text{Gal}(E/F)$.

(d) How can we conclude that $\text{Gal}(E/F)$ is a group?

We formalize the result of Preview Activity 33.1 in the following definition.

Definition 33.2. Let E be an extension of a field F. The **Galois group** of E over F is the set $\text{Gal}(E/F)$ of all automorphisms of E that fix F using the operation of composition. In addition, if E is the splitting field of the polynomial $p(x)$ over F, then $\text{Gal}(E/F)$ is the **Galois group** of $p(x)$.

We will denote the Galois group of a polynomial $p(x)$ as $\text{Gal}(p(x))$. The next activity provides two other examples of Galois groups.

Activity 33.3.

(a) Let $F = \mathbb{Q}$ and $E = F(\sqrt{2})$. The field E is the splitting field of $x^2 - 2$ over $\mathbb{Q}$. Use Theorem 32.37 to determine the elements of $\text{Gal}(E/F)$. To what familiar group is $\text{Gal}(E/F)$ isomorphic?

(b) Let $F = \mathbb{Q}$ and $E = F(\sqrt[3]{2})$. Use Theorem 32.37 to determine the elements of $\text{Gal}(E/F)$. To what familiar group is $\text{Gal}(E/F)$ isomorphic?

A more complicated example is in the next activity.

Activity 33.4. Let $\omega = e^{2\pi i/5}$ be a primitive fifth root of unity. Then ω is a root of the polynomial $p(x) = x^5 - 1 = (x - 1)(x^4 + x^3 + x^2 + x + 1)$ in $\mathbb{Q}[x]$. Let E be the splitting field of $p(x)$ over $\mathbb{Q}$. In this activity we will determine the elements in $\text{Gal}(E/\mathbb{Q})$. (The polynomial $x^4 + x^3 + x^2 + x + 1$ is irreducible in $\mathbb{Q}[x]$ – see Exercise (6) on page 484.)

(a) Explain why $E = \mathbb{Q}(\omega)$. Then explain why any element in $\text{Gal}(E/F)$ is completely determined by what the mapping does to ω.

(b) Explain why the mapping α from E to E that sends ω to ω^2 and fixes the elements of $\mathbb{Q}$ is in $\text{Gal}(E/F)$.

(c) Determine what the mappings α^2, α^3, and α^4 do to ω.

(d) Explain why $\text{Gal}(E/F) = \{id, \alpha, \alpha^2, \alpha^3\}$. To what familiar group is $\text{Gal}(E/F)$ isomorphic?

Activity 33.4 indicates a general result that if E is the splitting field of $x^p - 1$ over $\mathbb{Q}$, where p is a prime, then $\mathrm{Gal}(E/\mathbb{Q}) \cong \mathbb{Z}_{p-1}$. The proof is left for Exercise (11).

Example 33.5. Consider the polynomial $p(x) = x^3 - 2 = (x - \sqrt[3]{2})(x - \sqrt[3]{2}\omega)(x - \sqrt[3]{2}\omega^2)$, where $\omega = -\frac{1}{2} + \frac{\sqrt{3}}{2}i$ is a primitive cube root of unity. The element ω itself has minimal polynomial $x^2 + x + 1$. We have seen that the splitting field for $p(x)$ over $\mathbb{Q}$ is $E = \mathbb{Q}(\sqrt[3]{2}, \omega)$.

We saw in Activity 32.38 that the Galois group $\mathrm{Gal}(\mathbb{Q}(\sqrt[3]{2}, \omega)/\mathbb{Q})$ is generated by the automorphisms α and β, where $\alpha(\sqrt[3]{2}) = \sqrt[3]{2}\omega$ and $\beta(\omega) = \omega^2$. The group $\mathrm{Gal}(\mathbb{Q}(\sqrt[3]{2}, \omega)/\mathbb{Q})$ has the operation table

	id	α	α^2	β	$\alpha\beta$	$\alpha^2\beta$
id	id	α	α^2	β	$\alpha\beta$	$\alpha^2\beta$
α	α	α^2	id	$\alpha\beta$	$\alpha^2\beta$	β
α^2	α^2	id	α	$\alpha^2\beta$	β	$\alpha\beta$
β	β	$\alpha^2\beta$	$\alpha\beta$	id	α^2	α
$\alpha\beta$	$\alpha\beta$	β	$\alpha^2\beta$	α	id	α^2
$\alpha^2\beta$	$\alpha^2\beta$	$\alpha\beta$	β	α^2	α	id

and so $\mathrm{Gal}(\mathbb{Q}(\sqrt[3]{2}, \omega)/\mathbb{Q}) \cong S_3$.

The field $\mathbb{Q}(\sqrt[3]{2}, \omega)$ is also an extension of the field $\mathbb{Q}(\omega)$. Since $\sqrt[3]{2}$ is not in $\mathbb{Q}(\omega)$, it is the case that $x^3 - 2$ is the minimal polynomial of $\sqrt[3]{2}$ over $\mathbb{Q}(\omega)$ and that $\mathbb{Q}(\sqrt[3]{2}, \omega)$ is the splitting field of $x^3 - 2$ over $\mathbb{Q}(\omega)$. In this case, the automorphisms of $\mathbb{Q}(\sqrt[3]{2}, \omega)$ that fix $\mathbb{Q}(\omega)$ are id, α and α^2. So $\mathrm{Gal}(\mathbb{Q}(\sqrt[3]{2}, \omega)/\mathbb{Q}(\omega)) = \langle \alpha \rangle \cong \mathbb{Z}_3$. Notice that $\mathbb{Q}(\omega)$ is exactly the set of elements in $\mathbb{Q}(\sqrt[3]{2}, \omega)$ that are fixed by α. This set of elements is called the *fixed field* of α. As the next activity indicates, the set of elements fixed by a particular automorphism is always a field.

Activity 33.6. Let E be an extension of a field F and $\varphi : E \to E$ an automorphism. Let

$$E_\varphi = \{e \in E : \varphi(e) = e\}.$$

That is, E_φ is the set of all elements in E that are fixed by φ. We explore this set E_φ in this activity.

(a) First, we consider a specific example. Let $E = \mathbb{Q}(\sqrt{2}, \sqrt{3})$ and $F = \mathbb{Q}$. The minimal polynomial of $\sqrt{2}$ over $\mathbb{Q}$ is $x^2 - 2$, and the minimal polynomial of $\sqrt{3}$ over $\mathbb{Q}$ is $x^2 - 3$. Theorem 32.37 tells us that $\varphi : E \to E$ is an automorphism if and only if φ permutes roots of these irreducible polynomials. Suppose that $\varphi(\sqrt{2}) = -\sqrt{2}$ and $\varphi(\sqrt{3}) = -\sqrt{3}$. Determine specifically what the set E_φ is. Explain why E_φ is a subfield of E. (Hint: Recall that $\{1, \sqrt{2}, \sqrt{3}, \sqrt{6}\}$ is a basis for $\mathbb{Q}(\sqrt{2}, \sqrt{3})$ over $\mathbb{Q}$.)

(b) Now we consider the general case where E and F are any fields and φ is an automorphism of E. As in part (a), we will show that E_φ is a subfield of E.

    (i) Is it possible for E_φ to be empty?

    (ii) Is E_φ closed under subtraction?

    (iii) Is E_φ closed under multiplication?

    (iv) Does E_φ contain an inverse of each of its nonzero elements?

    (v) Explain why E_φ is a subfield of E.

The result of Activity 33.6 is that the set of elements fixed by an automorphism is a field. This prompts the following definition.

Definition 33.7. Let φ be an automorphism of a field E. The **fixed field** of φ is the set $E_\varphi = \{e \in E : \varphi(e) = e\}$.

Notice that Theorem 32.29 shows that if E is an extension of $\mathbb{Q}$ and if φ is an automorphism of E, then E_φ contains $\mathbb{Q}$.

Remember that the Galois group of an extension E over a field F is the set of all automorphisms of E that fix F using the operation of composition. If G is a subgroup of the Galois group, it might be the case that the elements in G fix a larger set of elements than just F. For example, if $E = \mathbb{Q}(\sqrt[3]{2}, \omega)$ as in Example 32.38 on page 481, then the set of elements fixed by the entire group $\langle \alpha \rangle$ is $\mathbb{Q}(\omega)$ (recall that $\alpha(\sqrt[3]{2}) = \sqrt[3]{2}\omega$)). In the following activity we see why this set is called the *fixed field of the group* $\langle \alpha \rangle$.

Activity 33.8. Let G be a group of automorphisms of a field E. Let

$$K = \{e \in E : \varphi(e) = e \text{ for all } \varphi \in G\}.$$

In this activity we will demonstrate that K is a subfield of E.

(a) Explain why $0 \in K$.

(b) Show that K is closed under subtraction.

(c) Show that K is closed under multiplication.

(d) Show that K contains an inverse for each of its nonzero elements. Explain why K is a subfield of E.

The result of Activity 33.8 is summarized in the following definition.

Definition 33.9. Let G be a group of automorphisms of a field E. The **fixed field** of G is the set $\{e \in E : \varphi(e) = e \text{ for all } \varphi \in G\}$.

There is a connection between the fixed field of a group of automorphisms and the fixed fields of individual automorphisms. That is, the fixed field of a group G of automorphisms of a field E is the intersection of all the fixed fields E_φ where $\varphi \in G$. The verification of this statement is left to Exercise (5).

Activity 33.10. Consider the splitting field $E = \mathbb{Q}(\sqrt[3]{2}, \omega)$ for $p(x) = x^3 - 2$ over $\mathbb{Q}$. Recall that

$$\{1, \omega, \sqrt[3]{2}, \sqrt[3]{2}\omega, \sqrt[3]{4}, \sqrt[3]{4}\omega\}$$

is a basis for E over $\mathbb{Q}$, where $\omega = -\frac{1}{2} + \frac{\sqrt{3}}{2}i$ is a primitive cube root of unity. We have seen that $\mathrm{Gal}(E/\mathbb{Q}) = \langle \alpha, \beta \rangle$ as in Example 32.38. The subgroups of $\mathrm{Gal}(E/\mathbb{Q})$ are $\langle \alpha \rangle$, $\langle \beta \rangle$, $\langle \alpha\beta \rangle$, and $\alpha^2 \beta \rangle$. As discussed above, the fixed field of $\langle \alpha \rangle$ is $\mathbb{Q}(\omega)$.

Recall that the actions of the elements of $\mathrm{Gal}(E/\mathbb{Q})$ are as follows.

x	$\sqrt[3]{2}$	ω
$id(x)$	$\sqrt[3]{2}$	ω
$\alpha(x)$	$\sqrt[3]{2}\omega$	ω
$\alpha^2(x)$	$\sqrt[3]{2}\omega^2$	ω
$\beta(x)$	$\sqrt[3]{2}$	ω^2
$\alpha\beta(x)$	$\sqrt[3]{2}\omega$	ω^2
$\alpha^2\beta(x)$	$\sqrt[3]{2}\omega^2$	ω^2.

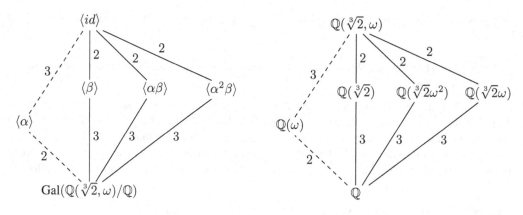

Figure 33.1
Subgroup lattice of $\mathrm{Gal}(\mathbb{Q}(\sqrt[3]{2},\omega)/\mathbb{Q})$ and subfield lattice of $\mathbb{Q}(\sqrt[3]{2},\omega)$ over $\mathbb{Q}$.

Determine the fixed fields of the following groups.

(a) $\langle\beta\rangle$

(b) $\langle\alpha\beta\rangle$

(c) $\langle\alpha^2\beta\rangle$

Note that $(\sqrt[3]{4}\omega)^2 = 2\sqrt[3]{2}\omega^2$, so it follows that $\mathbb{Q}(\sqrt[3]{4}\omega) = \mathbb{Q}(\sqrt[3]{2}\omega^2)$. The result of Activity 33.10 can be summarized by the subgroup lattice of $\mathrm{Gal}(\mathbb{Q}(\sqrt[3]{2},\omega)/\mathbb{Q})$ and the corresponding fixed field lattice that are shown in Figure 33.1. The fixed fields are arranged to correspond to the related subgroup. The numbers in the subgroup lattice correspond to indices of the subgroups. For example, $[\mathrm{Gal}(\mathbb{Q}(\sqrt[3]{2},\omega)/\mathbb{Q}) : \langle\alpha\rangle] = 2$. The numbers in this subfield lattice indicate the dimensions of the subfields. For example, $[\mathbb{Q}(\sqrt[3]{2},\omega) : \mathbb{Q}(\omega)] = 3$.

We can make some observations from this example.

Observation 1: Let K be an extension of $\mathbb{Q}$. Any automorphism of $\mathbb{Q}(\sqrt[3]{2},\omega)$ that fixes K will also fix $\mathbb{Q}$, so $\mathrm{Gal}(\mathbb{Q}(\sqrt[3]{2},\omega)/K)$ is a subgroup of $\mathrm{Gal}(\mathbb{Q}(\sqrt[3]{2},\omega)/\mathbb{Q})$. In fact, as we saw in Activity 33.10, we have the following Galois groups:

- $\mathrm{Gal}(\mathbb{Q}(\sqrt[3]{2},\omega)/\mathbb{Q}(\omega)) = \langle\alpha\rangle$,
- $\mathrm{Gal}(\mathbb{Q}(\sqrt[3]{2},\omega)/\mathbb{Q}(\sqrt[3]{2})) = \langle\beta\rangle$,
- $\mathrm{Gal}(\mathbb{Q}(\sqrt[3]{2},\omega)/\mathbb{Q}(\sqrt[3]{2}\omega)) = \langle\alpha^2\beta\rangle$,
- $\mathrm{Gal}(\mathbb{Q}(\sqrt[3]{2},\omega)/\mathbb{Q}(\sqrt[3]{2}\omega^2)) = \langle\alpha\beta\rangle$.

So there is a one-to-one correspondence between subgroups of $\mathrm{Gal}(\mathbb{Q}(\sqrt[3]{2},\omega)/\mathbb{Q})$ and fields between $\mathbb{Q}(\sqrt[3]{2},\omega)$ and $\mathbb{Q}$ given by associating to each subgroup the corresponding fixed field.

Observation 2: The lattices also seem to imply that if K is an extension of $\mathbb{Q}$, then the order of $\mathrm{Gal}(\mathbb{Q}(\sqrt[3]{2},\omega)/K)$ is equal to the index of K in $\mathbb{Q}(\sqrt[3]{2},\omega)$.

Observation 3: The subgroup $\langle\alpha\rangle$ is the only nontrivial normal subgroup of $\mathrm{Gal}(\mathbb{Q}(\sqrt[3]{2},\omega)/\mathbb{Q})$. This is depicted by the dashed lines in Figure 33.1. The corresponding fixed field is $\mathbb{Q}(\omega)$. This fixed field has a property that is distinct from the other subfields. The polynomial $x^3 - 2$

$$
\begin{array}{ccc}
E & & E(s) \\
\cup| & & \cup| \\
F(r) & \cong & F(s) \\
\cup| & & \cup| \\
F & = & F
\end{array}
$$

Figure 33.2
Field extensions.

is an irreducible polynomial over $\mathbb{Q}$, and each of the fields $\mathbb{Q}(\sqrt[3]{2})$, $\mathbb{Q}(\sqrt[3]{2}\omega)$, and $\mathbb{Q}(\sqrt[3]{2}\omega^2)$, contains only one of the roots of $x^3 - 2$. In contrast, the field $\mathbb{Q}(\omega)$ is the splitting field of the irreducible polynomial $x^2 + x + 1$ over $\mathbb{Q}$. As we will see, splitting fields have the property that every irreducible polynomial that has a root in a splitting field has all of its roots in that field. These subfields correspond to normal subgroups of the Galois group, so we will call these *normal* extensions.

These observations are the key components to Galois theory, and we devote much of this investigation to addressing them. We start with normal extensions.

Definition 33.11. An algebraic extension E of a field F is a **normal** extension of F if wherever E contains a root of an irreducible polynomial $p(x)$ in $F[x]$, then E contains all the roots of $p(x)$.

The Fundamental Theorem of Algebra shows that $\mathbb{C}$ is a normal extension of $\mathbb{R}$. Moreover, $\mathbb{C}$ is the splitting field of $p(x) = x^2 + 1$ over $\mathbb{R}$. It is reasonable to ask if there is a general connection between splitting fields and normal extensions. This question is answered in the next theorem.

Theorem 33.12. *Let F be a field of characteristic 0. The field E is the splitting field over F of a polynomial in $F[x]$ if and only if E is a finite-dimensional normal extension of F.*

Proof. First, assume that E is the splitting field of a polynomial $p(x)$ in $F[x]$. Then $E = F(r_1, r_2, \ldots, r_n)$ where $r_1, r_2, \ldots, r_n$ are the roots of $p(x)$. If $F_i = F(r_1, r_2, \ldots, r_i)$ for i from 1 to n, then

$$[E : F] = [E : F_{n-1}][F_{n-1}, F_{n-2}] \cdots [F_1 : F].$$

Since $F_i = F_{i-1}(r_i)$ and the minimum polynomial of r_i over F_{i-1} must divide $p(x)$, we conclude that $[F_i : F_{i-1}] = d_i$, where d_i is the degree of the minimal polynomial of r_i over F_{i-1} by Theorem 32.17. Thus, E is a finite dimensional extension of F.

To prove that E is a normal extension of F, let $p(x)$ be an irreducible polynomial in $F[x]$ that has a root r in E. We need to show that all the roots of $p(x)$ are in E – that is, that E contains the splitting field of $p(x)$. Let K be the splitting field of $p(x)$ considered as a polynomial in $E[x]$. So we have $F \subseteq E \subseteq K$. We will prove that every root of $p(x)$ in K is actually in E. Let s be a root of $p(x)$ in K different from r. There is an isomorphism $\sigma : F(r)$ to $F(s)$ that extends the identity isomorphism from F to F by Corollary 32.18, giving us the situation depicted in Figure 33.2. Now

$$E(s) = F(r_1, r_2, \ldots, r_n)(s) = F(r_1, r_2, \ldots, r_n, s) = F(s)(r_1, r_2, \ldots, r_n),$$

so $E(s)$ is the splitting field of $p(x)$ over $F(s)$. Since E is also the smallest extension of $F(r)$ that contains all the roots of $p(x)$, E is the splitting field of $p(x)$ over $F(r)$. So the isomorphism from $F(r)$ to $F(s)$ extends to an isomorphism from E to $E(s)$ that maps r to s and fixes the elements of F, again by 32.18. We conclude that $[E : F] = [E(s) : F]$ by Lemma 32.35. Then

$$[E : F] = [E(s) : F] = [E(s) : E][E : F],$$

which implies that $[E(s) : E] = 1$ or that $E(s) = E$. Thus, $s \in E$. So every root of $p(x)$ is in E as desired and E is a normal extension of F.

For the converse, assume that E is a finite-dimensional normal extension of F. So $E = F(a_1, a_2, \ldots, a_n)$ for some elements a_1, a_2, ..., a_n. Since E is finite-dimensional, each a_i is algebraic over F by Theorem 32.24. For each i, let $p_i(x)$ be the minimal polynomial for a_i over F. Now E is a normal extension, so each $p_i(x)$ splits in E. It follows that the polynomial $p(x) = p_1(x)p_2(x), \cdots p_n(x)$ splits in E. To contain the roots of $p(x)$, the splitting field of $p(x)$ must contain each a_i. So E is the splitting field of $p(x)$ over F.   ∎

We can say even more about splitting fields.

Corollary 33.13. *Let E be the splitting field of a polynomial over a field F of characteristic 0. If K is an intermediate field between E and F, then E is a normal extension of K.*

Proof. Let E be the splitting field of a polynomial $f(x) \in F[x]$ for some field F, and let K be an intermediate field between E and F. Then $f(x) \in K[x]$ and E is the splitting field of $f(x)$ over K. Theorem 33.12 then shows that E is a normal extension of K.   ∎

As we discussed in Observation 3 after Activity 33.10, the subfield $\mathbb{Q}(\omega)$ of $\mathbb{Q}(\sqrt[3]{2}, \omega)$, where ω is a primitive cube root of unity, is the splitting field of $x^2 + x + 1$ over $\mathbb{Q}$, and so $\mathbb{Q}(\omega)$ is a normal extension of $\mathbb{Q}$. However, the fields $\mathbb{Q}(\sqrt[3]{2})$, $\mathbb{Q}(\sqrt[3]{2}\omega)$, and $\mathbb{Q}(\sqrt[3]{2}\omega^2)$ are not splitting fields of any polynomials over $\mathbb{Q}$ and so are not normal extensions of $\mathbb{Q}$.

Normal extensions are important in that they correspond to normal subgroups of the Galois group, as we will see. Splitting fields are important in that we can completely factor a polynomial into a product of linear factors in its splitting field. We have just seen that a splitting field E over a field F is a normal extension of any intermediate field between E and F.

Suppose that E is an extension of a field F and K is an intermediate field. If $\varphi \in \mathrm{Gal}(E/F)$, what happens if we restrict the domain of φ to K? Do we obtain an element of $\mathrm{Gal}(K/F)$? The next activity addresses this question.

Activity 33.14. Let $E = Q(\sqrt[3]{2}, \omega)$, where $\omega = -\frac{1}{2} + \frac{\sqrt{3}}{2}i$ is a primitive cube root of unity. Recall that β is the automorphism of E that sends ω to ω^2 and fixes everything else and that α is the automorphism of E that sends $\sqrt[3]{2}$ to $\sqrt[3]{2}\omega$ and fixes everything else.

   (a) Let $K = \mathbb{Q}(\omega)$, and let γ be the mapping with domain K such that $\gamma(k) = \beta(k)$ for every $k \in K$. We call the function γ the restriction of β to K. You may assume that γ is a homomorphism.

      (i) Show that $\gamma(K) = K$. (Hint: What is a basis for K over $\mathbb{Q}$? Recall that $\omega^2 + \omega + 1 = 0$.)
      (ii) Explain why $\gamma \in \mathrm{Gal}(K/F)$.

   (b) Let $J = \mathbb{Q}(\sqrt[3]{2})$, and δ be the mapping with domain J such that $\delta(j) = \alpha(j)$ for every $j \in J$. That is, δ is the restriction of α to J. Explain why δ is not in $\mathrm{Gal}(J/F)$. What difference is there between K and J that might account for this behavior?

Activity 33.14 seems to suggest that restricting an element in a Galois group to a subfield is not always an element of the Galois group of the subfield. Before we proceed with the result, we provide a definition.

Definition 33.15. Let f be a function from a set A to a set B and let C be a subset of A. The **restriction** of f to C is the function $f|_C : C \to B$ satisfying

$$f|_C(c) = f(c) \text{ for all } c \in C.$$

Theorem 33.16. *Let E be a finite-dimensional extension of a field F and let K be an intermediate field between E and F that is a normal extension of F. If $\varphi \in Gal(E/F)$, then $\varphi|_K \in Gal(K/F)$.*

Proof. Let E be a finite-dimensional extension of a field F and let K be an intermediate field between E and F that is a normal extension of F. Let $\varphi \in Gal(E/F)$. We know that φ fixes all of F, so it follows that $\varphi|_K$ also fixes all of F. In order for $\varphi|_K$ to be an automorphism of K, it is necessary that $\varphi|_K$ be a homomorphism.

Let a and b be in K. Then

$$\varphi|_K(a+b) = \varphi(a+b) = \varphi(a) + \varphi(b) = \varphi|_K(a) + \varphi|_K(b)$$

and

$$\varphi|_K(ab) = \varphi(ab) = \varphi(a)\varphi(b) = \varphi|_K(a)\varphi|_K(b).$$

So $\varphi|_K$ is a homomorphism.

Next we demonstrate that the range of $\varphi|_K$ is K. Let $r \in K$. We know that K is a finite-dimensional extension of F, so r is algebraic over F. Let $m(x)$ be the minimal polynomial of r over F. The fact that $\varphi \in Gal(E/F)$ implies that $\varphi(r)$ is a root of $m(x)$, but K is a normal extension of F and so all the roots of $m(x)$ are in K. So $\varphi|_K(r) = \varphi(r)$ is in K and $\varphi(K) \subseteq K$.

Since φ is an automorphism of E, it follows that $K \cong \varphi(K)$. Finally, we show that $\varphi(K) = K$. By Lemma 32.35 we have that $[K : F] = [\varphi(K) : F]$. Then

$$[K : F] = [K : \varphi(K)][\varphi(K) : F] = [K : \varphi(K)][K : F],$$

which implies that $[K : \varphi(K)] = 1$. So $K = \varphi(K)$. We conclude that $\varphi|_K \in Gal(K/F)$. ∎

We can use the idea of restrictions to determine the structure of certain Galois groups. The next activity will set the stage.

Activity 33.17. Let $\omega = \cos\left(\frac{2\pi}{6}\right) + i \sin\left(\frac{2\pi}{6}\right) = \frac{\sqrt{3}}{2} + \frac{1}{2}i$ be a primitive sixth root of unity. In this activity we determine the Galois group of $\mathbb{Q}(\omega)$. As a sixth root of unity, ω is a root of $x^6 - 1$.

(a) Determine the minimal polynomials of $\omega, \omega^2, \omega^3, \omega^4$, and ω^5. (Hint: Factor $x^6 - 1$.)

(b) Use the fact that elements of $Gal(\mathbb{Q}(\omega)/\mathbb{Q})$ permute roots of irreducible polynomials to find two generators of $Gal(\mathbb{Q}(\omega)/\mathbb{Q})$.

(c) Explain why $Gal(\mathbb{Q}(\omega)/\mathbb{Q}) \cong \mathbb{Z}_2 \oplus \mathbb{Z}_2$.

We saw in Activity 33.4 that if ω is a primitive fifth root of unity, then $\mathbb{Q}(\omega)$ is the splitting field of $x^5 - 1$ over $\mathbb{Q}$ and that $Gal(\mathbb{Q}(\omega)/\mathbb{Q})$ is isomorphic to $\mathbb{Z}_4$. Activity 33.17 shows that $Gal(\mathbb{Q}(\omega)/\mathbb{Q})$ is isomorphic to $\mathbb{Z}_2 \oplus \mathbb{Z}_2$ when ω is a primitive sixth root of unity. These examples suggest that when we adjoin a primitive nth root of unity to a field of characteristic 0, we obtain an Abelian group, and if n is prime we obtain a cyclic group. The next lemma addresses these ideas.

Lemma 33.18. *Let F be a field of characteristic 0 and let ω be a primitive nth root of unity for some positive integer n. Then $F(\omega)$ is a normal extension of F and $Gal(F(\omega)/F)$ is Abelian. Moreover, if n is prime, then $Gal(F(\omega)/F)$ is cyclic.*

$$E \longrightarrow \mathrm{Gal}(E/E)$$
$$\cup | \qquad\qquad \cap |$$
$$K \longrightarrow \mathrm{Gal}(E/K)$$
$$\cup | \qquad\qquad \cap |$$
$$F \longrightarrow \mathrm{Gal}(E/F)$$

Figure 33.3
The Galois correspondence.

Proof. Let F be a field, n a positive integer, and let ω be a primitive nth root of unity. Then $F(\omega)$ contains all the nth roots of unity and so $F(\omega)$ is the splitting field of $x^n - 1$ over F. Thus, $F(\omega)$ is a normal extension of F. We know that the roots of $x^n - 1$ are ω^k for k from 1 to $n - 1$. Recall that $\{1, \omega, \omega^2, \ldots, \omega^{n-1}\}$ is a basis for $F(\omega)$ over F, so any automorphism of $F(\omega)$ is completely determined by its action on ω. Let α and β be $\mathrm{Gal}(F(\omega)/F)$. Since automorphisms of $F(\omega)$ must permute roots of polynomials, we know that $\alpha(\omega) = \omega^s$ and $\beta(\omega) = \omega^t$ for some positive integers s and t. Now

$$\alpha\beta(\omega) = \alpha(\beta(\omega)) = \alpha(\omega^t) = (\alpha(\omega))^t = \omega^{s+t} = \omega^{t+s} = (\beta(\omega))^s = \beta(\omega^s) = \beta(\alpha(\omega)) = \beta\alpha(\omega).$$

Thus, $\alpha\beta = \beta\alpha$ and $\mathrm{Gal}(F(\omega)/F)$ is Abelian.

The proof that $\mathrm{Gal}(F(\omega)/F)$ is cyclic when n is prime is left to Exercise (11). ∎

The Galois Correspondence

Let E be a finite-dimensional extension of a field F. If K is an intermediate field between E and F, then we assign the subgroup $\mathrm{Gal}(E/K)$ (the set of automorphisms of E that fix K) of $\mathrm{Gal}(E/F)$ to the field K. This correspondence, called the *Galois correspondence*, defines a mapping from the set of intermediate fields between E and F to the set of subgroups of $\mathrm{Gal}(E/F)$. Note that the containments are reversed in the codomain as indicated in Figure 33.3.

Example 33.19. Return to the example of $E = \mathbb{Q}(\sqrt[3]{2}, \omega)$ with $K = \mathbb{Q}(\omega)$ and $F = \mathbb{Q}$, where ω is a primitive cube root of unity. As we saw in Observation 1 following Activity 33.10, $\mathrm{Gal}(E/F) = \langle \alpha, \beta \rangle$, where $\alpha(\sqrt[3]{2}) = \sqrt[3]{2}\omega$ and $\beta(\omega) = \omega^2$. The result of Activity 33.10 gave us the subgroups of $\mathrm{Gal}(E/F)$ and the corresponding fixed fields as given in Table 33.1.

One question of interest is whether the Galois correspondence is a bijection. The next result answers the surjective part of this question. Note that to prove that the Galois correspondence is

Table 33.1
The Galois Correspondence for $\mathbb{Q}(\sqrt[3]{2}, \omega)$ over $\mathbb{Q}$.

Fixed field	Subgroup of $\mathrm{Gal}(\mathbb{Q}(\sqrt[3]{2}/\mathbb{Q})$
$\mathbb{Q}(\sqrt[3]{2}, \omega)$	$\mathrm{Gal}(\mathbb{Q}(\sqrt[3]{2}, \omega)/\mathbb{Q}(\sqrt[3]{2}, \omega)) = \langle id \rangle$
$\mathbb{Q}(\omega)$	$\mathrm{Gal}(\mathbb{Q}(\sqrt[3]{2}, \omega)/\mathbb{Q}(\omega)) = \langle \alpha \rangle$
$\mathbb{Q}(\sqrt[3]{2})$	$\mathrm{Gal}(\mathbb{Q}(\sqrt[3]{2}, \omega)/\mathbb{Q}(\sqrt[3]{2})) = \langle \beta \rangle$
$\mathbb{Q}(\sqrt[3]{2}\omega)$	$\mathrm{Gal}(\mathbb{Q}(\sqrt[3]{2}, \omega)/\mathbb{Q}(\sqrt[3]{2}\omega)) = \langle \alpha^2\beta \rangle$
$\mathbb{Q}(\sqrt[3]{2}\omega^2)$	$\mathrm{Gal}(\mathbb{Q}(\sqrt[3]{2}, \omega)/\mathbb{Q}(\sqrt[3]{2}\omega^2)) = \langle \alpha\beta \rangle$

surjective, we need to demonstrate that if E is an extension of a field F and if H is a subgroup of $\text{Gal}(E/F)$, then there is an intermediate field K between E and F such that $H = \text{Gal}(E/K)$.

Theorem 33.20. *Let E be a finite-dimensional normal extension over a field F of characteristic 0. Let H be a subgroup of $\text{Gal}(E/F)$ and let K be the fixed field of H. Then $H = \text{Gal}(E/K)$ and $|H| = [E : K]$.*

Proof. Let E be a finite-dimensional normal extension over a field F of characteristic 0. Note that this implies that E is a splitting field over F. Let H be a subgroup of $\text{Gal}(E/F)$ and let K be the fixed field of H. The fact that K is the fixed field of H means that $H \subseteq \text{Gal}(E/K)$. To prove the equality, we will demonstrate that $|H| = |\text{Gal}(E/K)|$.

Since E is a finite-dimensional extension of K, Theorem 32.24 shows that $E = K(r)$ for some element $r \in E$. Let $m(x)$ be the minimal polynomial of r over K and let $n = \deg(m(x))$. Theorem 32.17 shows that $[E : K] = n$. Theorem 32.21 shows that the roots of $m(x)$ are all distinct. We know that every element of $\text{Gal}(E/K)$ maps r to some other root of $m(x)$ by Theorem 32.32. So there are at most n elements in $\text{Gal}(E/K)$. Thus,

$$|H| \leq |\text{Gal}(E/K)| \leq n = [E : K]. \tag{33.1}$$

As elements of $\text{Gal}(E/K)$, every element of H also maps r to one of the roots of $m(x)$. So there are only finitely many distinct images of r under elements from H. Let these distinct images be $r = r_1, r_2, r_3, \ldots, r_k$ in E. Next we show the elements of H permute the elements in the set $\{r_1, r_2, \ldots, r_k\}$.

Let σ be in H. There is an element $\alpha_i \in H$ such that $\alpha_i(r) = r_i$. Then

$$\sigma(\alpha_i(r)) = \sigma(r_i).$$

But $\sigma\alpha_i \in H$, so $\sigma(r_i)$ must equal one of the r_j. Now σ is an injection, so the set $\{\sigma(r_1), \sigma(r_2), \ldots, \sigma(r_k)\}$ is equal to the set $\{r_1, r_2, \ldots, r_k\}$. Thus, the automorphisms in H permute the elements of the set $\{r_1, r_2, \ldots, r_k\}$. Therefore, $|H| \geq k$.

Let

$$f(x) = (x - r_1)(x - r_2) \cdots (x - r_k).$$

We will show that $f(x) \in K[x]$. If $\varphi \in H$, then φ is an automorphism of E. So φ induces an automorphism $\overline{\varphi}$ from $E[x]$ to $E[x]$ such that

$$\overline{\varphi}f(x) = (x - \varphi(r_1))(x - \varphi(r_2)) \cdots (x - \varphi(r_k)).$$

Since elements of H permute the roots $r_1, r_2, \ldots, r_k$ we have that $\overline{\varphi}f(x) = f(x)$. The jth coefficient c_j of $f(x)$ is a polynomial function of the roots r_i, that is $c_j = c_j(r_1, r_2, \ldots, r_k)$. It is also the case that $c_j = c_j(\varphi(r_1), \varphi(r_2), \ldots, \varphi(r_k)) = \varphi(c_j(r_1, r_2, \ldots, r_k)) = \varphi(c_j)$. So every element of H fixes all the coefficients of $f(x)$. Thus, the coefficients of $f(x)$ are in the fixed field of H, which is K. Therefore, $f(x) \in K[x]$.

Now r is a root of $f(x)$ as well as a root of $m(x)$, both as polynomials in $K[x]$. As $m(x)$ is the minimum polynomial of r in $K[x]$, it follows that $m(x)$ divides $f(x)$. So $n = \deg(m(x)) \leq \deg(f(x)) = k$. That makes

$$|H| \geq k = \deg(f(x)) \geq \deg(m(x)) = n. \tag{33.2}$$

Equations (33.1) and (33.2) show that

$$|H| = |\text{Gal}(E/K)| = [E : K].$$

So $H = \text{Gal}(E/K)$ and $|H| = [E : K]$. ∎

Theorem 33.20 proves that the Galois correspondence is a surjection if E is a splitting field over F. Now we address the question of when the Galois correspondence is an injection.

Activity 33.21. Let E be a finite dimensional extension of a field F.

(a) Explain in detail what it means for the Galois correspondence to be an injection.

(b) Consider the example with $E = K = \mathbb{Q}(\sqrt[3]{2})$ and $K' = F = \mathbb{Q}$. Determine the groups $\text{Gal}(E/K)$ and $\text{Gal}(E/K')$. What does this tell us about the Galois correspondence?

Activity 33.21 shows that the Galois correspondence is not in general an injection. An example where the Galois correspondence is an injection is the case where $E = \mathbb{Q}(\sqrt[3]{2}, \omega)$ in Example 33.19.

The previous examples illustrate the additional condition on the fields in order for the Galois correspondence to be an injection. The issue in Activity 33.21 is that $\mathbb{Q}(\sqrt[3]{2})$ is not a splitting field of any polynomial over $\mathbb{Q}$. In comparison, the field $\mathbb{Q}(\sqrt[3]{2}, \omega)$ is the splitting field of $x^3 - 2$ over $\mathbb{Q}$. If we add that condition, we can then prove that the Galois correspondence is injective.

Before we proceed, we make the following note. We are primarily interested in fields of characteristic 0, but some of the results we discuss will be more general. Recall that Theorem 33.12 shows that finite-dimensional normal extensions are splitting fields. In the nonzero characteristic case, an additional condition is needed – separability. In the characteristic 0 case, every irreducible polynomial of degree n has n distinct roots in some extension field by Theorem 32.21. This is not true in the case where the characteristic is not 0, although examples are a bit complicated. When a polynomial of degree n has n distinct roots, we say that the polynomial is *separable*. Then an algebraic element over a field is *separable* if its minimal polynomial is separable. Finally, an extension E of a field F is *separable* if every element in E is separable. It is the case that every extension of a field of characteristic 0 is separable (the verification is left to Exercise (6)). So in characteristic zero, a finite-dimensional, normal, separable extension is a splitting field. This is not true when the characteristic is not zero, so we give these fields a name.

Definition 33.22. A **Galois extension** E of a field F is a finite-dimension, normal, separable extension of F.

If E is a Galois extension of F, we also say that E is *Galois over F*. But remember that in characteristic zero a Galois extension is just a splitting field. We address the question of when we can be sure that the Galois correspondence is an injection in the next theorem.

Theorem 33.23. *If E is a Galois extension of a field F and K is an intermediate extension, then K is the fixed field of $\text{Gal}(E/K)$.*

Proof. Let J be the fixed field of $\text{Gal}(E/K)$. We will prove that $K = J$. Since every automorphism in $\text{Gal}(E/K)$ fixes K, we know that $K \subseteq J$.

We proceed by contradiction and assume that $K \neq J$. Since $K \subseteq J$, there must be an element $j \in J$ that is not in K. The fact that E is a finite-dimensional extension of F means that E is an algebraic extension of K by Theorem 32.15. So j is algebraic over K with some minimal polynomial $m(x) \in K[x]$. The fact that $j \notin K$ implies that $\deg(m(x)) \geq 2$. Since E is separable, the roots of $m(x)$ are all distinct. The roots of $m(x)$ are also all in E because E is a normal extension of K. Let r be a root of $m(x)$ distinct from k. Theorem 32.37 shows that there is an automorphism σ of E with $\sigma(j) = r$. But then j is not in the fixed field of $\text{Gal}(E/K)$, or $j \notin J$. This is our contradiction. We conclude that $K = J$ and K is the fixed field of $\text{Gal}(E/K)$.   ∎

A consequence of Theorem 33.23 is that the Galois correspondence is an injection when E is a Galois extension of F. To understand why, suppose that E is a Galois extension of a field F and assume that K and K' are intermediate fields with $\mathrm{Gal}(E/K) = \mathrm{Gal}(E/K')$. Since $\mathrm{Gal}(E/K) = \mathrm{Gal}(E/K')$, it must be the case that the fixed fields of $\mathrm{Gal}(E/K)$ and $\mathrm{Gal}(E/K')$ are the same. But Theorem 33.23 tells us that K is the fixed field of $\mathrm{Gal}(E/K)$ and K' is the fixed field of $\mathrm{Gal}(E/K)$. So $K = K'$ and the Galois correspondence is injective.

The Fundamental Theorem of Galois Theory

Recall the observations we made about the Galois correspondence in Activity 33.10 and in the remarks following the activity. We now state the fundamental theorem that addresses these observations.

Theorem 33.24 (The Fundamental Theorem of Galois Theory). *Let E be a Galois extension of a field F of characteristic 0.*

(1) The mapping that assigns the fixed field K to the subgroup $\mathrm{Gal}(E/K)$ of $\mathrm{Gal}(E/F)$ is a bijection between the subgroups of the Galois group $\mathrm{Gal}(E/F)$ and the intermediate fields K with $F \subseteq K \subseteq E$. Moreover,

$$[E : K] = |\mathrm{Gal}(E/K)| \text{ and } [K : F] = [\mathrm{Gal}(E/F) : \mathrm{Gal}(E/K)].$$

(2) An intermediate field K between E and F is a normal extension of F if and only if $\mathrm{Gal}(E/K)$ is a normal subgroup of $\mathrm{Gal}(E/F)$. In this case $\mathrm{Gal}(K/F) \cong \mathrm{Gal}(E/F)/\mathrm{Gal}(E/K)$.

Proof. Let E be a Galois extension of a field F.

(1) The fact that the Galois correspondence is a bijection is shown in Theorems 33.20 and 33.23. Let K be an intermediate field between F and E. Theorem 33.23 shows that K is the fixed field of $\mathrm{Gal}(E/K)$ and so $|\mathrm{Gal}(E/K)| = [E : K]$ by Theorem 33.20. This result applies as well when $K = F$, so $|\mathrm{Gal}(E/F)| = [E : F]$. It follows that

$$[E : K]\,[K : F] = [E : F] = |\mathrm{Gal}(E/F)| = |\mathrm{Gal}(E/K)|\frac{|\mathrm{Gal}(E/F)|}{|\mathrm{Gal}(E/K)|}.$$

Recall that $\frac{|\mathrm{Gal}(E/F)|}{|\mathrm{Gal}(E/K)|} = [\mathrm{Gal}(E/F) : \mathrm{Gal}(E/K)]$, so we then have

$$[E : K]\,[K : F] = |\mathrm{Gal}(E/K)|[\mathrm{Gal}(E/F) : \mathrm{Gal}(E/K)] = [E : K][\mathrm{Gal}(E/F) : \mathrm{Gal}(E/K)].$$

We conclude that
$$[K : F] = [\mathrm{Gal}(E/F) : \mathrm{Gal}(E/K)].$$

(2) Assume that K is an intermediate field between E and F and that K is a normal extension of F. Theorem 33.16 tells us that if $\varphi \in \mathrm{Gal}(E/F)$, then $\varphi|_K \in \mathrm{Gal}(K/F)$. Define a function $\Gamma : \mathrm{Gal}(E/F) \to \mathrm{Gal}(K/F)$ by $\Gamma(\varphi) = \varphi|_K$. It is left to Exercise (13) to show that Γ is a homomorphism.

We will next show that Γ is a surjection. Let $\sigma \in \mathrm{Gal}(K/F)$. Since E is a finite-dimensional normal extension of F, E is a splitting field by Theorem 33.12. Also, E is the splitting field of the same polynomial over K. Since σ is an isomorphism from K to K that fixes F, as shown in the figure below, Lemma 32.36 tells us that σ can be extended to an isomorphism Σ from E to E.

$$
\begin{array}{ccc}
E & \xrightarrow{\Sigma} & E \\
\cup| & & \cup| \\
K & \xrightarrow[\sigma]{} & K
\end{array}
$$

Then the restriction $\Sigma|_K$ is σ and γ is an epimorphism.

Finally, if φ is in the kernel of γ, then $\Gamma(\varphi) = \varphi|_K = id$. That is, φ is an automorphism of E that fixes K. So $\mathrm{Ker}(\Gamma) = \mathrm{Gal}(E/K)$. As a kernel, $\mathrm{Gal}(E/K)$ is a normal subgroup of $\mathrm{Gal}(E/F)$. The First Isomorphism Theorem (page 395) then shows that

$$
\mathrm{Gal}(E/F)/\mathrm{Gal}(E/K) \cong \mathrm{Gal}(K/F)
$$

as desired.

For the last piece of the proof, assume that K is an intermediate field between E and F with $\mathrm{Gal}(E/K)$ a normal subgroup of $\mathrm{Gal}(E/F)$. We must show that K is a normal extension of F. Let $p(x)$ be an irreducible polynomial in $F[x]$ that has a root r in K. We will prove that every root of $p(x)$ is in K. Let s be a root of $p(x)$. By Theorem 32.37 we know that there is an automorphism α in $\mathrm{Gal}(E/F)$ such that $\alpha(r) = s$. To show that $s \in K$, we will prove that s is in the fixed field of $\mathrm{Gal}(E/K)$, which is K by Theorem 33.23. Let σ be in $\mathrm{Gal}(E/K)$. Since $\mathrm{Gal}(E/K)$ is normal in $\mathrm{Gal}(E/F)$, there is a $\sigma_1 \in \mathrm{Gal}(E/K)$ such that $\alpha\sigma_1 = \sigma\alpha$. The fact that r is in K means that $\sigma_1(r) = r$ and so

$$
\sigma(s) = \sigma(\alpha(r)) = \alpha(\sigma_1(r)) = \alpha(r) = s.
$$

Thus, every root of $p(x)$ is fixed by every element in $\mathrm{Gal}(E/K)$. We conclude that the roots of $p(x)$ are in the fixed field K of $\mathrm{Gal}(E/K)$. Therefore, K is a normal extension of F.

∎

In the following sections, we will see how to use the Galois correspondence to translate a problem of polynomials and fields to a problem in group theory. We will then be able to solve the problem using results from group theory.

Solvability by Radicals

We are probably familiar with the idea of using the quadratic formula to solve any quadratic equation with real coefficients. That is, if a, b, and c are real numbers with $a \neq 0$, then the solutions to the quadratic

$$
ax^2 + bx + c = 0
$$

are

$$
\frac{-b + \sqrt{b^2 - 4ac}}{2a} \quad \text{and} \quad \frac{-b - \sqrt{b^2 - 4ac}}{2a}.
$$

We can describe the solvability of a quadratic equation $ax^2 + bx + c = 0$ over $\mathbb{Q}$ in terms of field extensions of $\mathbb{Q}$. All the roots of $ax^2 + bx + c$ are in the extension $\mathbb{Q}(\sqrt{b^2 - 4ac})$, and $(b^2 - 4ac)^2 \in \mathbb{Q}$.

In Investigation 11 we exhibited an analogous method for solving the general cubic formula that also involves radicals. There is also a general method for solving an arbitrary quartic (degree 4) polynomial, which we will not discuss. An interesting question is whether or not there is a corresponding formula for finding the solutions to any polynomial equation at all. Mathematicians unsuccessfully searched for a long time to find a general solution to the quintic (degree 5) polynomials until a proof was finally given that no such method exists. The approach to this proof is very elegant and combines elements of field theory and group theory. This is the topic of the remaining sections of this investigation.

The main idea in this approach is deciding when a polynomial equation can be *solved by radicals*. For example, consider the polynomial $x^4 - 3x^2 + 2 = 0$. We can start to solve this equation by substituting y for x^2 to rewrite the equation as $y^2 - 3y - 2 = 0$. The solutions to this equation are

$$y_1 = \frac{3 + \sqrt{9 + 8}}{1} = 3 + \sqrt{17} \text{ and } y_2 = \frac{3 - \sqrt{9 + 28}}{1} = 3 - \sqrt{17}.$$

Since $y = x^2$, we obtain the following four solutions to $x^4 - 3x^2 + 2 = 0$:

$$x_1 = \sqrt{3 + \sqrt{17}}, \ x_2 = \sqrt{3 - \sqrt{17}}, \ x_3 = \sqrt{3 - \sqrt{17}}, \text{ and } x_4 = \sqrt{3 - \sqrt{17}}.$$

These solutions are found by taking radicals of radicals. Informally, we say that a polynomial equation is solvable by radicals if all its roots can be expressed in terms of its coefficients using the standard arithmetic operations of addition, subtraction, multiplication and division, along with extraction of roots. We will provide a formal definition a bit later.

Just as with the quadratic equation, we can describe the solvability by radicals of the equation $x^4 - 3x^2 + 2 = 0$ in terms of field extensions. Note that the roots of the quadratic $y^2 - 3y - 2$ are in the simple extension $F_1 = \mathbb{Q}(\sqrt{17})$ of $\mathbb{Q}$. Then two of the roots of $x^4 - 3x^2 + 2$ can be found in $F_2 = F_1(\sqrt{3 + \sqrt{17}})$ and the remaining two in the extension $F_3 = F_2(\sqrt{3 - \sqrt{17}})$. So we can construct a sequence of simple extensions of $F_0 = \mathbb{Q}$, namely

$$F_0 \subseteq F_1 \subseteq F_2 \subseteq F_3,$$

where $x^4 - 3x^2 + 2$ splits completely in F_3 and each F_i is a simple extension of F_{i-1}. Moreover, if $F_i = F_{i-1}(r_i)$, then $r_i^2 \in F_{i-1}$. Such a sequence of field extensions is called a *radical extension*.

Definition 33.25. A field E is a **radical extension** of a field F if there exist fields $F_1, F_2, \ldots, F_n$ such that

(1) there exists $r_i \in E$ such that $F_i = F_{i-1}(r_i)$ for i from 1 to n,

(2) $r_i^{k_i} \in F_{i-1}$ for some positive integer k_i,

(3) $F = F_0 \subseteq F_1 \subseteq F_2 \subseteq \cdots \subseteq F_n = E$.

Note that the condition $r_i^{k_i} \in F_{i-1}$ means that r_i is a root of $x^{k_i} - a_i$ for some $a_i \in F_{i-1}$. If $n = 1$ in Definition 33.25, that is if $E = F(r)$ with $r^k \in F$ for some r and some positive integer k, then we call E a *simple radical extension* of F.

We can then define what it means for a polynomial to be *solvable by radicals*.

Definition 33.26. Let F be a field. A polynomial $p(x) \in F[x]$ is **solvable by radicals** if the splitting field of $p(x)$ is contained in a radical extension of F.

Our examples above show that all quadratics over $\mathbb{Q}$ are solvable by radicals, as is the polynomial $x^4 - 3x^2 + 2$.

One main consequence of Galois theory is that polynomials exist of degree 5 and higher whose roots cannot be found by radicals. The proof is made by connecting the question of solvability by radicals to group theory. The main idea is to connect radical extensions to group theory using the Galois correspondence.

Example 33.27. In certain cases we can say something about simple radical extensions. For example, let $E = \mathbb{Q}(\omega)(\sqrt[3]{2})$, where ω is a primitive cube root of unity. The field E is the splitting field of $x^3 - 2$ over $\mathbb{Q}(\omega)$ and is therefore a normal extension. We have seen that $\mathrm{Gal}(E/\mathbb{Q}(\omega)) = \langle \alpha \rangle$, where $\alpha(\sqrt[3]{2}) = \sqrt[3]{2}\omega$. So $\mathrm{Gal}(E/\mathbb{Q}(\omega))$ is an Abelian group. However, the field $\mathbb{Q}(\sqrt[3]{2})$ is not a splitting field for any polynomial over $\mathbb{Q}$ and so is not a normal extension.

We want to connect the solvability of a polynomial to group theory using the Galois correspondence. To understand what the correct formulation will be in terms of groups, we examine the idea of solvability by radicals in more detail. In order to have a useful group-theoretic analog of solvability, it will be helpful to be able to characterize polynomials that are solvable by radicals in terms of normal extensions rather than radical extensions. This will allow us to utilize the Fundamental Theorem of Galois Theory.

We begin with a result about normal and radical extensions whose main idea is illustrated in the next activity.

Activity 33.28. Let $F = \mathbb{Q}$, $K = \mathbb{Q}(\sqrt{2})$, and let $L = K(\sqrt[3]{2})$.

(a) Is L a finite-dimensional extension of F? Explain.

(b) Is L a radical extension of K? Explain.

(c) Is L a normal extension of F? Explain.

(d) Let $E = \mathbb{Q}(\sqrt{2}, \sqrt[3]{2}, \sqrt{3}i)$. Note that $L \subset E$. Is E a normal extension of F? Is E a radical extension of K? Explain.

Activity 33.28 provides a situation where we can extend a radical extension to a normal extension. We formalize the idea in the following theorem.

Theorem 33.29. *Let F, K, and L be fields of characteristic 0 with $F \subseteq K \subseteq L$. Assume that K is normal over F, that $L = K(r)$ is a radical extension of K with $r^d \in K$ for some positive integer d, and that L is a finite-dimensional extension of F. Then there exists an extension E of L which is a radical extension of K and a normal extension of F.*

Proof. Let F, K, and L be fields of characteristic 0 with $F \subseteq K \subseteq L$. Assume that K is normal over F, that $L = K(r)$ is a radical extension of K with $r^d \in K$ for some positive integer d, and that L is a finite-dimensional extension of F. A normal extension of F will be a splitting field of some polynomial in $F[x]$, so we proceed to find an appropriate polynomial. Since K is finite-dimensional and normal over F, we know that K is the splitting field of some polynomial $p(x)$ in $F[x]$ by Theorem 33.12. The fact that L is finite-dimensional over F implies that L is algebraic over F by Theorem 32.24. Let $m(x)$ be the minimal polynomial of r in $F[x]$. Now let $f(x) = p(x)m(x)$ in $F[x]$ and let E be the splitting field of $f(x)$. Then E is normal over F. Since K is the splitting field of $p(x)$ and E is the splitting field of $p(x)m(x)$ over F, E contains all the roots of $p(x)$ and so $K \subseteq E$. The field E also contains r (as a root of $m(x)$), so $L = K(r) \subseteq E$. It remains to prove that E is a radical extension of K.

Let $r_1, r_2, \ldots, r_n$ be the roots of $m(x)$ with $r = r_1$. For each i from 1 to n, Theorem 32.37 provides an element $\alpha_i \in \mathrm{Gal}(E/F)$ such that $\alpha_i(r) = r_i$. Theorem 33.16 shows that $\alpha_i|_K \in \mathrm{Gal}(K/F)$. So $\alpha_i(K) \subseteq K$. Now $r^d \in K$ implies that

$$r_i^d = (\alpha_i(r))^d = \alpha_i(r^d)$$

is in K. Letting $E_i = K(r_1, r_2, \ldots, r_i)$ for i from 1 to m we have that $r_{i+1}^m \in E_i$ and so

$$K \subseteq E_1 \subseteq E_2 \subseteq \cdots \subseteq E_m = E$$

is a radical extension of K. ∎

We can use Theorem 33.29 to now characterize solvability in terms of normal extensions.

Theorem 33.30. *Let F be a field of characteristic 0. If $p(x) \in F[x]$ is solvable by radicals, then there is a normal radical extension of F that contains the splitting field of $p(x)$.*

Proof. Let F be a field of characteristic 0 and assume that $p(x)$ in $F[x]$ is solvable by radicals. Let E be the splitting field of $p(x)$ over F. Since $p(x)$ is solvable by radicals, there is a sequence F_0, $F_1, \ldots, F_n$ of fields with

$$F = F_0 \subseteq F_1 \subseteq F_2 \subseteq \cdots \subseteq F_n,$$

where $E \subseteq F_n$, $F_i = F_{i-1}(r_i)$ for i from 1 to n, and $r_i^{k_i} \in F_{i-1}$ for some positive integer k_i.

Let $E_0 = F$. Since F is a normal extension of F, we can use $F = K = F_0$, $L = F_1$, and $r = r_1$ in Theorem 33.29 to give us an extension E_1 of F_1 that is a radical extension of F_0 and is normal over F.

Now that E_1 is a normal extension of F and $r_2^{k_2} \in F_1 \subseteq E_1$, we can apply Theorem 33.29 now to $E_0 = F$, $K = E_1$, and $L = E_1(r_2)$ to produce an extension E_2 of $E_1(r_2)$ that is a radical extension of E_1 and is normal over F. Given that E_1 is a radical extension of F and E_2 is a radical extension of E_1, we can say that E_2 is a radical extension of F as well. Moreover, E_2 contains E_1 and r_2, so E_2 contains $F_2 = F_1(r_2)$.

Proceeding inductively, assume that for some positive integer k we can construct a sequence

$$F = E_0 \subseteq E_1 \subseteq E_2 \subseteq \cdots \subseteq E_k$$

such that each E_i is a normal radical extension of F, E_i is a radical extension of E_{i-1}, and $F_i \subseteq E_i$ for each i. Applying Theorem 33.29 to F, E_k, and $E_k(r_{k+1})$ we obtain an extension E_{k+1} that is a normal extension of F and a radical extension of E_k. The fact that E_{k+1} contains $E_k(r_{k+1})$ means that E_{k+1} also contains F_{k+1}. So in the final step, we have an extension E_n that is a normal radical extension of F that contains F_n, so E_n contains the splitting field of $p(x)$. ∎

Solvable Groups

To relate the question of solvability by radicals to group theory using the Galois correspondence, we need to formulate what the group theory analog of a radical extension is. Recall that E is a radical extension of a field F if there is a sequence

$$F = F_0 \subseteq F_1 \subseteq F_2 \subseteq \cdots \subseteq F_n = E$$

such that $F_i = F_{i-1}(r_i)$ and there exists k_i such that $r_i^{k_i}$ is an element of F_{i-1} for each i.

Theorem 33.30 allows us to reformulate the idea of a radical extension to normal extensions. That is, if $p(x) \in F[x]$ is solvable by radicals, then there is a sequence $E_0, E_1, \ldots, E_n$ of fields such that

$$F = E_0 \subseteq E_1 \subseteq E_2 \subseteq \cdots \subseteq E_n$$

where each E_i is a normal radical extension of F and $F_i \subseteq E_i$ for each i, and E_n contains a splitting field of $p(x)$.

Activity 33.31. Consider the polynomial $p(x) = x^3 - 2 \in \mathbb{Q}[x]$. We can view the splitting field of $p(x)$ as a normal radical extension of $\mathbb{Q}$ as follows:

$$\mathbb{Q} \subset \mathbb{Q}(\omega) \subset \mathbb{Q}(\omega)(\sqrt[3]{2}),$$

where ω is a primitive cube root of unity. Recall that $\mathrm{Gal}(\mathbb{Q}(\omega)(\sqrt[3]{2})/\mathbb{Q})$ is isomorphic to S_3 and is generated by the automorphisms α and β, where $\alpha(\sqrt[3]{2}) = \sqrt[3]{2}\omega$ and $\beta(\omega) = \omega^2$ while $\mathrm{Gal}(\mathbb{Q}(\omega)/\mathbb{Q})$ is cyclic and is generated by β. Let $G = N_0 = \mathrm{Gal}(\mathbb{Q}(\omega)(\sqrt[3]{2})/\mathbb{Q})$, $N_1 = \mathrm{Gal}(\mathbb{Q}(\omega)(\sqrt[3]{2})/\mathbb{Q}(\omega))$, and $N_2 = \mathbb{Q}(\omega)(\sqrt[3]{2})/\mathbb{Q}(\omega)(\sqrt[3]{2}))$.

(a) Identify the groups N_0, N_1, and N_2 in terms of generators.

(b) Explain why N_1 is normal in N_0 and N_2 is normal in N_1.

(c) Explain why N_0/N_1 and N_1/N_2 are Abelian groups.

In Activity 33.31 we see that the Galois correspondence gives us a sequence $N_0 \supset N_1 \supset N_2$ of groups for which each subgroup is normal in the next larger subgroup, and every quotient is Abelian. While Activity 33.31 is only a small example, it provides the basic ideas for us to define a solvable group.

Definition 33.32. A group G is solvable if there exist subgroups $N_0, N_1, \ldots, N_k$ of G such that

(1) $G = N_0 \supset N_1 \supset N_2 \supset \cdots \supset N_k = \{e\}$,

(2) N_{i+1} is normal in N_i for i from 0 to $k - 1$,

(3) N_i/N_{i+1} is Abelian for i from 0 to $k - 1$.

Our goal is now to show that if $p(x)$ is a separable polynomial over a field F (of characteristic 0) with splitting field E, then $\mathrm{Gal}(E/F)$ is solvable. To understand the idea of a solvable group, we look at some examples and results about solvable groups.

Activity 33.33.

(a) Explain why every Abelian group is solvable.

(b) Let $G = S_3$, and let $N_0 = S_3$, $N_1 = \langle (1\ 2\ 3) \rangle$, and $N_2 = \{e\}$. Use these subgroups to explain why S_3 is solvable.

Activity 33.33 shows that S_3 is a solvable group. It is left to Exercise (16) to show that S_4 is also a solvable group. A key idea in the solution to the problem of polynomials not being solvable by radicals is that S_n is not a solvable group for $n \geq 5$. For the proof we will need the following lemma.

Lemma 33.34. *If N is a normal subgroup of a group G and G/N is Abelian, then $aba^{-1}b^{-1}$ is in N for all a, b in G.*

Proof. Suppose N is a normal subgroup of a group G, and let a, b be in G. Then

$$(aba^{-1}b^{-1})N = (aN)(bN)(a^{-1}N)(b^{-1}N) = (aN)(a^{-1}N)(bN)(b^{-1}N) = N,$$

and so $aba^{-1}b^{-1}$ is in N. ∎

The element $aba^{-1}b^{-1}$ is called the *commutator* (Exercise (9) in Investigation 21) of the elements a and b. The *commutator subgroup* G' of a group G is the subgroup of G generated by all the commutators of pairs of elements in G. For example, if G is Abelian, then the commutator subgroup of G is the group consisting of just the identity. If a and b commute, then the commutator of a with b is the identity of G. So if a is in the center of G, then the commutator of a with any element in G is the identity of G. As another example, the commutator subgroup of S_3 is A_3 (the proof is left for Exercise (17)). It is important to note that the set of commutators of a group is itself not a subgroup (see Exercise (18)).

Now we prove that S_n is not solvable if $n \geq 5$.

Theorem 33.35. *If $n \geq 5$, then S_n is not solvable.*

Proof. We proceed by contradiction and assume that S_n is solvable for some $n \geq 5$. Let N_0, N_1, ..., N_k of S_n such that

(1) $S_n = N_0 \supseteq N_1 \supseteq N_2 \supseteq \cdots \supseteq N_k = \{e\}$,

(2) N_{i+1} is normal in N_i for i from 0 to $k-1$,

(3) N_i/N_{i+1} is Abelian for i from 0 to $k-1$.

We will prove that $N_i = S_n$ for each i and so no such sequence of subgroups exists. To do this, we prove that N_i contains all 3-cycles for each i. This will imply that $\{e\}$ contains all 3-cycles, which is impossible. First, we argue that N_1 contains all 3-cycles.

Let $(a\ b\ c)$ be an arbitrary 3-cycle in S_n. The fact that $n \geq 5$ means that we can choose two additional elements d and e, different from a, b, and c. Since N_1 is normal in N_0 and N_0/N_1 is Abelian, Lemma 33.34 shows that

$$(a\ b\ e)(a\ c\ d)(a\ b\ e)^{-1}(a\ c\ d)^{-1} = (a\ b\ e)(a\ c\ d)(a\ e\ b)(a\ d\ c) = (a\ b\ c) \in N_1.$$

So N_1 contains all 3-cycles.

Proceeding by induction, we assume that for some $i \geq 1$ we know that N_i contains all 3-cycles. Since every 3-cycle is in N_i, repeating the argument from above shows that

$$(a\ b\ e)(a\ c\ d)(a\ b\ e)^{-1}(a\ c\ d)^{-1} = (a\ b\ e)(a\ c\ d)(a\ e\ b)(a\ d\ c) = (a\ b\ c) \in N_{i+1}.$$

So we conclude that $N_k = \{e\}$ contains all 3-cycles, which is impossible. We conclude that S_n is not solvable for $n \geq 5$. ∎

There are many other useful results about solvable groups. Among them are the following:

- Any homomorphic image of a solvable group is solvable.

- If G is a solvable group and N is a normal subgroup of G, then G/N is a solvable group.

- Let G be a group and N a normal subgroup of G. If both N and G/N are solvable, then G is solvable.

We will provide a proof of the first of these results, which will be needed later, and leave the others to Exercises (19) and (20).

Theorem 33.36. *Any homomorphic image of a solvable group is solvable.*

Proof. Let G be a solvable group and let H be a group such that there exists a surjection φ from G to H. Since G is solvable, there exist subgroups $N_0, N_1, \ldots, N_k$ of G such that

$$G = N_0 \supset N_1 \supset N_2 \supset \cdots \supset N_k = \{e_G\},$$

where N_{i+1} is normal in N_i and N_i/N_{i+1} is Abelian for i from 0 to $k-1$.

For i from 0 to k let $M_i = \varphi(N_i)$. Then

$$H = M_0 \supset M_1 \supset M_2 \supset \cdots \supset M_k = \{e_H\}.$$

To show that H is solvable, we need to prove that M_{i+1} is normal in M_i and that M_i/M_{i+1} is Abelian for i from 0 to $k-1$.

To demonstrate that M_{i+1} is normal in M_i, let $m \in M_i$ and let $m' \in M_{i+1}$. The fact that φ is a surjection means that there exist $n \in N_i$ and $n' \in N_{i+1}$ such that $\varphi(n) = m$ and $\varphi(n') = m'$. The normality of N_{i+1} in N_i implies that $nn'n^{-1} \in N_{i+1}$. Then $\varphi(nn'n^{-1}) \in M_{i+1}$. So

$$mm'm^{-1} = \varphi(n)\varphi(n')\varphi(n)^{-1} = \varphi(nn'n^{-1}) \in M_{i+1}$$

and M_{i+1} is a normal subgroup of M_i.

To prove that M_i/M_{i+1} is Abelian, we will use Lemma 33.34. Let a and b be in M_i. There exist u and v in N_i such that $\varphi(u) = a$ and $\varphi(v) = b$. The fact that N_i/N_{i+1} is Abelian and Lemma 33.34 show that $uvu^{-1}v^{-1}$ is in N_{i+1}. It follows that

$$aba^{-1}b^{-1} = \varphi(u)\varphi(v)\varphi(u^{-1})\varphi(v^{-1}) = \varphi(uvu^{-1}v^{-1}) \in M_{i+1}.$$

So

$$M_{i+1} = aba^{-1}b^{-1}M_{i+1} = (aM_{i+1})(bM_{i+1})(a^{-1}M_{i+1})(b^{-1}M_{i+1}).$$

Multiplying both sides on the right by $(bM_{i+1})(aM_{i+1})$ gives us

$$(bM_{i+1})(aM_{i+1}) = (aM_{i+1})(bM_{i+1}),$$

and we conclude that M_i/M_{i+1} is an Abelian group. Thus, H is a solvable group. ∎

Polynomials Not Solvable By Radicals

Now we can connect solvability by radicals to solvable groups. The idea is that if $p(x)$ is solvable by radicals, then Theorem 33.30 tells us that there is a normal radical extension of F that contains a splitting field E of $p(x)$. As a splitting field, we know that E is normal over F. The following theorem will show that $\text{Gal}(E/F)$ is a solvable group.

$$
\begin{array}{ccc}
E & & \mathrm{Gal}(E/E) \\
\cup | & & | \cap \\
K & & \mathrm{Gal}(E/K) \\
\text{normal} \ \cup | & & | \cap \quad \text{normal} \\
F & & \mathrm{Gal}(E/F)
\end{array}
$$

Figure 33.4
The Fundamental Theorem.

Theorem 33.37. *Let L be a normal radical extension of a field F and E an intermediate field, with F, L, and E all of characteristic 0. If E is normal over F, then $\mathrm{Gal}(E/F)$ is a solvable group.*

Proof. Let L be a normal radical extension of a field F and E an intermediate field, with F, L, and E all of characteristic 0. Assume that E is normal over F. Since L is a radical extension of F, we know that there exist fields $F_0, F_1, \ldots, F_n$ such that

$$
F = F_0 \subseteq F_1 \subseteq F_2 \subseteq \cdots \subseteq F_n = L
$$

with $F_i = F_{i-1}(r_i)$ and $r_i^{k_i} \in F_i$ for i from 1 to n. Since L is a normal extension of F, we know that L is the splitting field of some polynomial $p(x)$ in $F[x]$. Let k be the least common multiple of $k_1, k_2, \ldots, k_n$ and let ω be a primitive kth root of unity. Note that for each i, some power of ω is a primitive k_ith root of unity. Let $E_i = F_i(\omega)$ for each i from 0 to n. We will show that E_i is a radical extension of E_{i-1}. Note that

$$
E_i = F_i(\omega) = F_{i-1}(r_i)(\omega) = F_{i-1}(r_i, \omega) = F_{i-1}(\omega)(r_i) = E_{i-1}(r_i).
$$

Since $r_i^{k_i} \in F_i \subseteq E_i$, we have

$$
F \subseteq E_0 \subseteq E_1 \subseteq E_2 \subseteq \cdots \subseteq E_n
$$

such that $E_i = E_{i-1}(r_i)$ and $r_i^{k_i} \in E_i$ for i from 1 to n and $L \subseteq E_n$. So E_n is a radical extension of F that contains L, and E as well.

The fact that $E_n = F_n(\omega) = L(\omega)$ means that E_n is the splitting field of the polynomial $p(x)(x^k - 1)$ over F. So E_n is a normal extension of F. We next show that $\mathrm{Gal}(E_n/F)$ is a solvable group.

Recall that the Fundamental Theorem of Galois Theory tells us that if E is a splitting field over a field F and K is an intermediate field, then $\mathrm{Gal}(E/K)$ is a normal subgroup of $\mathrm{Gal}(E/F)$ as illustrated in Figure 33.4 and $\mathrm{Gal}(K/F) \cong \mathrm{Gal}(E/F)/\mathrm{Gal}(E/K)$.

For i from 0 to n, let $G_i = \mathrm{Gal}(E_n/E_i)$. Recall that $\mathrm{Gal}(E_n/E_{i+1})$ is a subgroup of $\mathrm{Gal}(E_n/E_i)$ for each i, and so we have the sequence of groups

$$
\mathrm{Gal}(E_n/F) \supseteq G_0 \supseteq G_1 \supseteq G_2 \supset \cdots \supseteq G_{n-1} \supseteq G_n = \{id\}.
$$

We have to prove that G_i is normal in G_{i-1} and that G_{i-1}/G_i is Abelian for each i. First, we address normality. Recall that $E_i = F_i(\omega)$, so E_i contains $r_i^{k_i}$ and a primitive k_i root of unity. Thus, E_i is a normal extension of E_{i-1} by Lemma 33.18 and $\mathrm{Gal}(E_i/E_{i-1})$ is Abelian. Now E_n is normal over F, so E_n is normal over every E_i with $i \leq n$. By the Fundamental Theorem of Galois Theory, using the extensions $E_{i-1} \subseteq E_i \subseteq E_n$, we have that $G_i = \mathrm{Gal}(E_n/E_i)$ is a normal subgroup of $G_{i-1} = \mathrm{Gal}(E_n/E_{i-1})$.

Similarly, since $E_0 = F(\omega)$ Lemma 33.18 shows that E_0 is a normal extension of F and $\mathrm{Gal}(E_0/F)$ is Abelian. Using the Fundamental Theorem with the extensions $F \subseteq E_0 \subseteq E_n$ we have that $\mathrm{Gal}(E_n/E_0)$ is normal in $\mathrm{Gal}(E_n/F)$ and $\mathrm{Gal}(E_n/F)/\mathrm{Gal}(E_n/E_0) \cong \mathrm{Gal}(E_0/F)$, which is an Abelian group.

Finally, we prove that $\mathrm{Gal}(E/F)$ is solvable by showing that $\mathrm{Gal}(E/F)$ is the homomorphic image of a solvable group. Since E is normal over F, the Fundamental Theorem of Galois Theory applied to the extensions $F \subseteq E \subseteq E_n$ shows that $\mathrm{Gal}(E_n/E)$ is normal in $\mathrm{Gal}(E_n/F)$ and $\mathrm{Gal}(E_n/F)/\mathrm{Gal}(E_n/E) \cong \mathrm{Gal}(E_F)$. So $\mathrm{Gal}(E/F)$ is a homomorphic image of the solvable group $\mathrm{Gal}(E_n/F)$ and so is solvable by Theorem 33.36. ∎

We can show that not every polynomial is solvable by radicals if we can exhibit a polynomial whose Galois group is isomorphic to S_n with $n \geq 5$. The next results will provide conditions under which certain Galois groups are exactly the symmetric groups.

Recall that every permutation in S_n is a product of cycles, and every cycle can be written as a product of transpositions. So S_n is generated by the transpositions. We will also use the result from Exercise (16) in Investigation 22 that if $\beta = (b_1\ b_2\ \ldots\ b_k)$ is a k-cycle in S_n and if σ is in S_n, then

$$\sigma\beta\sigma^{-1} = (\sigma(b_1)\ \sigma(b_2)\ \ldots\ \sigma(b_k)).$$

This result also shows that any conjugate of a k-cycle is a k-cycle.

Activity 33.38. We know that S_3 is generated by the 3-cycle $(1\ 2\ 3)$ and the transposition $(1\ 2)$. In this activity we ask the question if there is a similar result for S_5. That is, can S_5 be generated by the 5-cycle $(1\ 2\ 3\ 4\ 5)$ and the transposition $(1\ 2)$? Let $\alpha = (1\ 2\ 3\ 4\ 5)$ and $\beta = (1\ 2)$, and let $G = \langle \alpha, \beta \rangle$. We will show that G contains all the transpositions, which means that $G = S_5$.

(a) What cycle is $\alpha\beta\alpha^{-1}$?

(b) What cycle is $\alpha^2\beta\alpha^{-2}$?

(c) There is a pattern. Continue the calculations to determine cycles $\alpha^m\beta\alpha^{-m}$ for $m = 3$ and $m = 4$. In general, what cycle is $\alpha^m\beta\alpha^{-m}$?

(d) The previous part shows that G contains all transpositions of the form $(1 + m\ \ 2 + m)$ for any integer m, with elements reduced into the set $\{1, 2, 3, 4, 5\}$ modulo 5. Let s be an integer between 1 and 5.

   (i) Write this product $(s + 1\ \ s + 2)(s\ \ s + 1)(s + 1\ \ s + 2)$ as a single cycle. Why is this product in G?

   (ii) Now calculate the product $(s + 2\ \ s + 3)(s\ \ s + 2)(s + 2\ \ s + 3)$.

   (iii) Continue in this way to show that G contains all the transpositions.

Activity 33.38 contains the basic ideas for the general proof.

Theorem 33.39. *Let p be a prime. If α is a p-cycle and β is a transposition in S_p, then $\langle \alpha, \beta \rangle = S_p$.*

Proof. Let p be a prime and let $\alpha = (a_1\ a_2\ \ldots\ a_p)$ be a p-cycle and β a transposition in S_p. Let $H = \langle \alpha, \beta \rangle$ be the subgroup of S_p generated by α and β.

For any $\gamma \in S_p$ we have $H \cong \gamma H \gamma^{-1}$. Letting γ be the permutation that sends a_i to i gives

$$\gamma\alpha\gamma^{-1} = (\gamma(a_1)\ \gamma(a_2)\ \ldots\ \gamma(a_p)) = (1\ 2\ \ldots\ p).$$

So without loss of generality we can assume that $\alpha = (1\ 2\ \ldots\ p)$.

Let $\beta = (i\ j)$ for some $1 \leq i < j \leq p$. We will prove that H contains all transpositions, from which we can conclude that $H = S_p$. Let $(s\ t)$ be an arbitrary transposition with $1 \leq s < t \leq p$.

First, consider the conjugate $\alpha\beta\alpha^{-1}$. This conjugate is a transposition with

$$\alpha\beta\alpha^{-1}(i+1) = \alpha\beta(i) = \alpha(j) = j+1$$

and

$$\alpha\beta\alpha^{-1}(j+1) = \alpha\beta(j) = \alpha(i) = i+1.$$

So $\alpha\beta\alpha^{-1} = (i+1\ j+1)$. Now we will show that

$$\alpha^m\beta\alpha^{-m} = (i+m\ j+m) \tag{33.3}$$

for every $m \in \mathbb{Z}^+$, where the integers are all considered elements of the set $\{1, 2, 3, \ldots, p\}$ modulo p. We have shown that (33.3) is true when $m = 1$. We proceed by induction and assume that (33.3) is true for some positive integer m. Now

$$\alpha^{m+1}\beta\alpha^{-(m+1)} = \alpha(\alpha^m\beta\alpha^{-m})\alpha^{-1} = \alpha(i+m\ j+m)\alpha^{-1}.$$

Notice that

$$\alpha(i+m\ j+m)\alpha^{-1}(i+m+1) = \alpha(i+m\ j+m)(i+m) = \alpha(j+m) = j+m+1$$

and

$$\alpha(i+m\ j+m)\alpha^{-1}(j+m+1) = \alpha(i+m\ j+m)(j+m) = \alpha(i+m) = i+m+1,$$

completing the induction argument.

Let $k = j - i$. Then $(i+m\ j+m) = (i+m\ i+m+k)$ for every $m \in \mathbb{Z}^+$, with the additions taken modulo p. As m runs through the integers 1 through p, so does $i + m$ (modulo p). In this way we have all transpositions of the form $(u\ u+k)$ in H.

Next we show that H contains the transpositions $(s\ s+nk)$ for any positive integer n, with all additions done modulo p. First, note that $(s+k\ s+2k) = ((s+k)\ (s+k)+k)$ is in H, and so

$$(s+k\ s+2k)(s\ s+k)(s+k\ s+2k) = (s\ s+2k)$$

is also in H. Proceeding by induction we assume that $(s\ s+nk)$ is in H for some positive integer n. Then

$$(s+nk\ s+(n+1)k)(s\ s+nk)(s+nk\ s+(n+1)k) = (s\ s+(n+1)k)$$

is in H, completing the induction proof.

The facts that p is prime and $1 < k < p$ imply that the integers k, $2k$, $\ldots$ take on all values between 1 and p modulo p. So there is a value of n such that $s + nk = t$ modulo p, which means that H contains all transpositions. We conclude that $H = S_p$. ∎

We can use Theorem 33.39 to determine conditions under which a polynomial will have S_p as its Galois group.

Corollary 33.40. *If $p(x)$ is an irreducible polynomial over $\mathbb{Q}$ of prime degree p with exactly $p - 2$ real roots, then the Galois group of $p(x)$ is S_p.*

Proof. Let $p(x)$ be an irreducible polynomial over $\mathbb{Q}$ of prime degree p with roots $r_1, r_2, \ldots, r_p$. Without loss of generality we assume that r_1 and r_2 are nonreal, complex roots and that $r_3, \ldots, r_p$ are real roots. Note that these roots are all distinct because $p(x)$ is irreducible by Theorem 32.21. Let E be the splitting field of $p(x)$ over $\mathbb{Q}$ and let $G = \text{Gal}(E/\mathbb{Q})$ be the Galois group of $p(x)$ over $\mathbb{Q}$.

Since $p(x)$ is irreducible, $p(x)$ is the minimal polynomial of any of its roots. So $[\mathbb{Q}(r_1) : \mathbb{Q}] = p$. Then $|G| = [E : \mathbb{Q}] = [E : \mathbb{Q}(r_1)]\,[\mathbb{Q}(r_1) : \mathbb{Q}]$ and p divides $|G|$.

We know that elements of G permute the roots of $p(x)$, and since p divides $|G|$ it must be the case that G contains an element α of order p, which we identify with a permutation of the roots of $p(x)$. So we consider α as a permutation of order p. We now show that α is a p-cycle. We know that α can be written as a product $\sigma_1 \sigma_2 \cdots \sigma_k$ of disjoint cycles, and that disjoint cycles commute. Recall that the order of a cycle is the length of the cycle. By Exercise (14) in Investigation 22 we have that

$$p = |\alpha| = \text{lcm}(|\sigma_1|, |\sigma_2|, \ldots, |\sigma_k|).$$

So $|\alpha|$ divides the product $|\sigma_1||\sigma_2| \cdots |\sigma_k|$. Thus, p divides $|\sigma_t|$ for some t. If $|\sigma_r| \neq p$ for any r, then $\text{lcm}(|\sigma_1|, |\sigma_2|, \ldots, |\sigma_k|) > p$, so it follows that $|\sigma_i| = p$ for every i and every σ_i is a p-cycle. But there cannot be two disjoint p-cycles in S_p, so α is a p-cycle.

There is also an element β in G that permutes the two complex conjugate roots r_1 and r_2 and leaves the other roots fixed. That is, β is a 2-cycle. Thus, G contains a p-cycle and a permutation, and so G is all of S_p by Theorem 33.39. ∎

We conclude this investigation by exhibiting a polynomial that is not solvable by radicals.

Activity 33.41. We can now exhibit a polynomial whose Galois group is S_5. Consider the polynomial $p(x) = x^5 - 6x + 2$.

(a) Explain why $p(x)$ is irreducible over $\mathbb{Q}$.

(b) Use the derivative of $p(x)$ and some results from calculus to show that $p(x)$ has exactly three real roots.

(c) How do we conclude that $p(x)$ is not solvable by radicals?

One final note. The polynomial $x^k p(x)$ is a polynomial of degree $5 + k$ over $\mathbb{Q}$. Since not *all* roots of this polynomial can be solved by radicals, we have found polynomials of any degree greater than or equal to 5 that are not solvable by radicals.

Concluding Activities

Activity 33.42. There are certain families of polynomials for which we can make general statements about their Galois groups. One such family is the collection of polynomials $x^n - 1$ where n is a positive integer. In this activity we will demonstrate that the Galois group of $x^n - 1$ is an Abelian group. (It is a more complicated argument that we will not make, but the Galois group of $x^n - 1$ is isomorphic to $\mathbb{Z}_n^*$, the group of units of $\mathbb{Z}_n$. This implies that the Galois group of $x^p - 1$ is cyclic when p is a prime, an argument left for the exercises.)

(a) We begin with an example.

(i) Find a primitive fourth root ω of 1. We know that these powers of ω are the roots of $x^4 - 1$. What are the powers of ω?

(ii) We know that any element of $\text{Gal}(x^4 - 1)$ must permute the roots of $x^4 - 1$. That is, an element of $\text{Gal}(x^4 - 1)$ satisfies $\alpha_i(\omega) = \omega^k$ for k from 1 to 4. Explain why not all of these α_k are in $\text{Gal}(x^4 - 1)$. Which ones are?

(iii) Explain why $\text{Gal}(x^4 - 1)$ is isomorphic to $\mathbb{Z}_4^*$.

(b) Now we turn to the general case.

(i) What are the roots of $x^n - 1$? Explain why the splitting field of $x^n - 1$ has the form $\mathbb{Q}(\omega)$ for some element ω.

(ii) We know that any element of $\text{Gal}(x^n - 1)$ permutes the roots of $x^n - 1$. With this in mind, describe an arbitrary element of $\text{Gal}(x^n - 1)$.

(iii) Show that if α and β are in $\text{Gal}(x^n - 1)$, then $\alpha\beta = \beta\alpha$. Conclude that $\text{Gal}(x^n - 1)$ is an Abelian group.

Activity 33.43. We have seen that elements of Galois groups permute roots of certain polynomials. In this way we can view Galois groups as groups of permutations. We make that idea more explicit in this activity and find an upper bound on the size of a Galois group. We will prove the following theorem.

Theorem 33.44. *Let E be the splitting field of a separable polynomial $f(x)$ of degree n over a field F. Then $\text{Gal}(E/F)$ is isomorphic to a subgroup of S_n.*

A consequence of this theorem is that $|\text{Gal}(E/F)|$ divides $n!$.

(a) Explain why $f(x)$ has n distinct roots $r_1, r_2, \ldots, r_n$ in E.

(b) Recall that S_n is the group of permutations of the set $\{1, 2, \ldots, n\}$. But we can also realize S_n as the group of permutations of any set of n elements. So we can consider S_n to be the group of permutations of the set $R = \{r_1, r_2, \ldots, r_n\}$. Let $\alpha \in \text{Gal}(E/F)$.

(i) Explain why $\alpha(r_i)$ is in R for each i.
(ii) Now explain why $\{\alpha(r_1), \alpha(r_2), \ldots \alpha(r_n)\} = R$.

(c) The previous result shows that if we restrict the domain of α to R, denoted $\alpha|_R$, then $\alpha|_R(R) = R$ and $\alpha|_R$ is a permutation of R. Now define $\Phi : \text{Gal}(E/F) \to S_n$ by $\Phi(\alpha) = \alpha|_R$. We will show that Φ is an injective homomorphism. Let α and β be in $\text{Gal}(E/F)$.

(i) Show that $(\alpha\beta)|_R = \alpha|_R\beta|_R$.
(ii) Use the previous result to show that Φ is a homomorphism.
(iii) Now show that Φ is a monomorphism. (Hint: Why is $E = F(r_1, r_2, \ldots, r_n)$?)

(d) Explain why $\text{Gal}(E/F)$ is isomorphic to a subgroup of S_n.

Exercises

(1) (a) Describe, by representing a general element, the smallest extension $\mathbb{Q}(\sqrt{2}, i)$ of $\mathbb{Q}$ that contains $\sqrt{2}$ and i. Verify your description.

(b) Find all subfields of $\mathbb{Q}(\sqrt{2}, i)$ and draw a subfield lattice.

(c) Determine the elements in $\mathrm{Gal}(\mathbb{Q}(\sqrt{2}, i)/\mathbb{Q})$. Identify all the fixed fields of the subgroups of $\mathrm{Gal}(\mathbb{Q}(\sqrt{2}, i)/\mathbb{Q})$. Draw a subgroup lattice to correspond to the subfield lattice from part (b).

(2) Show that $x^2 - 3$ and $x^2 - 2x - 2$ have the same Galois group over $\mathbb{Q}$.

(3) We have seen that the Galois group of $\mathbb{Q}(\sqrt{2})$ over $\mathbb{Q}$ has order 2. Generalize this result to show that the Galois group of $\mathbb{Q}(a + b\sqrt{d})$ over $\mathbb{Q}$ has order 2 if $a, b, d \in \mathbb{Q}$ and $\sqrt{d} \notin \mathbb{Q}$.

(4) Activity 33.42 shows that the Galois groups of $x^n - 1$ are all Abelian. But this result does not necessarily apply to even a slightly different situation. In this exercise we determine the Galois group of $x^4 - 2$.

(a) One root of $x^4 - 2$ is $\sqrt[4]{2}$. Use this root to express all roots of $x^4 - 2$. Then explain why $\mathbb{Q}(\sqrt[4]{2}, i)$ is the splitting field of $x^4 - 2$.

(b) What are $[\mathbb{Q}(\sqrt[4]{2}, i) : \mathbb{Q}(\sqrt[4]{2})]$ and $[\mathbb{Q}(\sqrt[4]{2}) : \mathbb{Q}]$? What does this tell us about $|\mathrm{Gal}(x^4 - 2)|$.

(c) Some elements in $\mathrm{Gal}(x^4 - 2)$ will permute the roots of $x^4 - 2$. Define $\alpha \in \mathrm{Gal}(x^4 - 2)$ to satisfy $\alpha(\sqrt[4]{2}) = \sqrt[4]{2}i$ and $\alpha(i) = i$. Determine the order of α.

(d) Some elements of $\mathrm{Gal}(\mathbb{Q}(\sqrt[4]{2}, i))$ will permute the roots of $x^2 + 1$, the minimal polynomial of i over $\mathbb{Q}$. Define $\beta \in \mathrm{Gal}(\mathbb{Q}(\sqrt[4]{2}, i))$ to satisfy $\beta(i) = -i$ and $\beta(\sqrt[4]{2}) = \sqrt[4]{2}$. Determine the order of β.

(e) Show that α and β do not commute, so that $\mathrm{Gal}(x^4 - 2)$ is a non-Abelian group.

(f) Show that $\mathrm{Gal}(x^4 - 2)$ is isomorphic to D_4.

(5) Let E be an extension of a field F and let G be a group of automorphisms of E. Show that the fixed field of G is the intersection of all the fixed fields $E_\varphi = \{e \in E : \varphi(e) = e\}$ where $\varphi \in G$.

(6) As we discussed in this investigation, we distinguish between those polynomials that have repeated roots, like $x^4 + 2x^2 + 1 \in \mathbb{Q}[x]$ that factors as $(x^2 + 1)^2 = (x + i)^2(x - i)^2$ in $\mathbb{C}[x]$, and polynomials with no repeated roots like $x^2 + 1 \in \mathbb{Q}[x]$ that factors as $(x - i)(x + i)$ in $\mathbb{C}[x]$ and has no multiple roots in $\mathbb{C}$. Recall that if F is a field, then a degree n polynomial $p(x) \in F[x]$ is separable if $p(x)$ has n distinct roots in some extension of F while an algebraic element r in some extension field of F is separable over F if the minimal polynomial of r over F is separable. Then an extension field E of F is a separable extension over F (or is **separable** over F) if every element of E is separable over F.

For example, every element in $\mathbb{C}$ is separable over $\mathbb{Q}$ since the minimal polynomial for $z = a + bi$ with $b \neq 0$ is the quadratic $x^2 - 2ax + (a^2 + b^2)$ with $a + bi$ and $a - bi$ as roots, and if $b = 0$ then the minimal polynomial for z is $x - a$. nonseparable extensions are not that common. In fact, as the next theorem shows, the only possible nonseparable extensions are extensions of fields of characteristic different than 0. We prove this theorem in this exercise.

Theorem 33.45. *Every extension of a field of characteristic zero is separable.*

Let F be a field of characteristic 0 and let E be an extension of F. We proceed by contradiction and assume that there is an element $r \in E$ with minimal polynomial $m(x) = a_n x^n + a_{n-1} x^{n-1} + \cdots + a_2 x^2 + a_1 x + a_0$ over F with $a_n \neq 0$ that does not have n distinct roots. That is, $m(x)$ has a multiple root s.

(a) We define the formal derivative $f'(x)$ of a polynomial $f(x) = c_m x^m + c_{m-1}x^{m-1} + \cdots + c_2 x^2 + c_1 x + c_0$ over a field as

$$f'(x) = mc_m x^{m-1} + (m-1)c_{m-1}x^{m-2} + \cdots + 2c_2 x + c_1.$$

All the standard derivative rules (sum, product, chain) apply to calculating formal derivatives. Show that if t is a multiple root of $f(x)$ (that is, $(x-t)^2$ divides $f(x)$), then t is a root of $f'(x)$.

(b) Show that $m'(x)$ is not the zero polynomial.

(c) Use the fact that $\deg(m'(x)) < \deg(m(x))$ to draw the conclusion that every extension of F is separable.

(7) Exercise (6) shows that every extension of a field of characteristic 0 is separable. In this exercise we explore an example of a nonseparable extension.

Let s be an element that is not in $\mathbb{Z}_2$ but is algebraic over $\mathbb{Z}_2$. Let $F = \mathbb{Z}_2(s)$. Kronecker's Theorem (Theorem 12.12 on page 176) shows that there is an element t that satisfies the polynomial $m(x) = x^2 - s$ in $F[x]$. Let $E = F(t)$. Use the fact that these fields all have characteristic 2 to factor $m(x)$ over E. Explain why $m(x)$ is the minimal polynomial of t over F, and then why E is a nonseparable extension of F.

(8) Show that the polynomial $x^5 - 1$ is solvable by radicals over the rationals.

(9) Let $p(x) = ax^8 + bx^6 + cx^2 + d$ be a polynomial in $\mathbb{Q}[x]$ with $a \neq 0$. Is $p(x)$ necessarily solvable by radicals? Explain.

(10) Is the polynomial $p(x) = 2x^5 - 5x^4 + 5$ solvable by radicals over $\mathbb{Q}$? Justify your answer.

(11) Let p be an odd prime and let $f_p(x) = x^{p-1} + x^{p-2} + \cdots + x + 1$. Our goal in this exercise is to determine the Galois group of $f_p(x)$ over $\mathbb{Q}$. Note that since $x^p - 1 = (x-1)f_p(x)$, the Galois group of $f_p(x)$ is the same as the Galois group of $x^p - 1$. For this exercise we will use the Binomial Theorem (see Exercise (10) in Investigation 5).

(a) One result we will need is that the binomial coefficients $\binom{p}{k}$ are all divisible by p for any integer k with $1 \leq k \leq p-1$. Prove this fact. Note that $\binom{n}{k}$ is an integer whenever $0 \leq k \leq n$.

(b) It can be difficult to show directly that $f_p(x)$ is irreducible over $\mathbb{Q}$. However, we can translate $f_p(x)$ so that we can apply Eisenstein's Criterion (Theorem 11.17 on page 164) to the result. Use this idea to show that $f_p(x+1)$ is irreducible over $\mathbb{Q}$. (Hint: Use the fact that $f_p(x)(x-1) = x^p - 1$.)

(c) Explain why $f_p(x)$ is irreducible if $f_p(x+1)$ is irreducible.

(d) Now that we know that $f_p(x)$ is irreducible over $\mathbb{Q}$, show that the Galois group of $f_p(x)$ over $\mathbb{Q}$ is isomorphic to $\mathbb{Z}_{p-1}$.

(12) Show that D_n is a solvable group for each n.

(13) Let E be a Galois extension of a field F. Assume that K is an intermediate field between E and F and that K is a normal extension of F. Define a function $\Gamma : \text{Gal}(E/F) \rightarrow \text{Gal}(K/F)$ by $\Gamma(\varphi) = \varphi|_K$. Show that Γ is a homomorphism.

(14) Prove that every subgroup of a solvable group is solvable.

(15) Show that if G is a group with subgroups N_1 and N_2 with N_2 normal in N_1, and if H is any subgroup of G, then $N_2 \cap H$ is normal in $N_1 \cap H$.

(16) In this exercise we show that S_4 is solvable. We start by letting $N_0 = S_4$ and $N_1 = A_4$.

   (a) Explain why N_1 is normal in N_0.

   (b) Exercise (16) in Investigation 22 shows that if $\beta = (b_1 \; b_2 \; \ldots \; b_k)$ is a k-cycle in S_n, and $\sigma \in S_n$, then
$$\sigma \beta \sigma^{-1} = (\sigma(b_1) \; \sigma(b_2) \; \cdots \; \sigma(b_k)). \tag{33.4}$$
   Let $N_2 = \{(1), (1\;2)(3\;4), (1\;3)(2\;4), (1\;4)(2\;3)\}$, so that N_2 is a Sylow 2-subgroup of A_4. Show that N_2 is an Abelian group isomorphic to $\mathbb{Z}_2 \oplus \mathbb{Z}_2$.

   (c) Explain why N_1/N_2 is Abelian.

   (d) Explain why N_2 is a normal subgroup of A_4.

   (e) Complete the argument that S_4 is solvable.

(17) Show that the commutator subgroup of S_3 is A_3. (Hint: What kind of permutation is $a^{-1}b^{-1}ab$ if a and b are in S_3?)

(18) If G is a group of order less than 96, then it has been shown that the subset of commutators of G is in fact a group. So, it can be difficult to find examples of groups whose set of commutators is not a subgroup. One example has been given by P.J. Cassidy. Let E be a field and x and y indeterminates over E. Let G be the set of all upper triangular matrices of the form
$$\begin{bmatrix} 1 & f(x) & h(x,y) \\ 0 & 1 & g(y) \\ 0 & 0 & 1 \end{bmatrix}, \tag{33.5}$$
where $f(x) \in E[x]$, $h(x,y) \in E[x,y]$, and $g(y) \in E[y]$. As a shorthand, we denote the matrix (33.5) as (f, g, h).

   (a) Verify that G is a group using the standard matrix product.

   (b) Show that every commutator has the form $(0, 0, k(x, y))$, where $k(x, y) \in E[x, y]$.

   (c) Now we complete the following steps to show that if $k(x, y) \in E[x, y]$, then $(0, 0, k(x, y))$ is in the commutator group of G. Conclude that the commutator group G' of G is $G' = \{(0, 0, k) : k(x, y) \in E[x, y]\}$.
      (i) Show that every monomial of the form $ax^i y^j$ is a commutator.
      (ii) Show that $(0, 0, u)(0, 0, v) = (0, 0, u + v)$ for any $u(x, y), v(x, y)$ in $E[x, y]$.
      (iii) The element $k(x, y)$ is a sum of monomials, say $k(x, y) = \sum_{i,j} a_{ij} x^i y^j$. Use the previous two parts to write $(0, 0, k)$ as a product of commutators.

   (d) The element $(0, 0, x^2 + xy + y^2)$ is in G'. Finally we show that the element $(0, 0, x^2 + xy + y^2)$ is not a commutator. Hence, the set of commutators is not necessarily equal to the commutator subgroup.
      (i) Suppose to the contrary that $(0, 0, x^2 + xy + y^2)$ is a commutator of the form $[(f, g, h), (r, s, t)]$. Assume that $f(x) = \sum_{i=0}^{m} b_i x^i$ and $r(x) = \sum_{j=0}^{n} c_i x^j$. Show that
$$b_1 s(y) + c_1 g(y) = y$$
$$b_2 s(y) + c_2 g(y) = 1$$
$$b_0 s(y) + c_0 g(y) = y^2.$$

(ii) Based on the result of the previous part, what two things can we say about the dimension of Span$\{s(y), g(y)\}$? How does this conclude our exercise?

(19) Prove the following theorem:

Theorem 33.46. *If G is a solvable group and N is a normal subgroup of G, then G/N is a solvable group.*

(20) Prove the following theorem:

Theorem 33.47. *Let G be a group and N a normal subgroup of G. If both N and G/N are solvable, then G is solvable.*

(21) (a) Let $E = \mathbb{Q}(\sqrt{2}, \sqrt{3})$. Determine the elements of $G = \mathrm{Gal}(E/\mathbb{Q})$. Determine all subgroups of G and the corresponding fixed fields.

   (b) Now let p and q be distinct primes and let $E = \mathbb{Q}(\sqrt{p}, \sqrt{q})$. Determine the elements of $G = \mathrm{Gal}(E/\mathbb{Q})$. Determine all subgroups of G and the corresponding fixed fields.

(22) Let $E = \mathbb{Q}(\sqrt{2}, \sqrt{3}, \sqrt{5})$.

   (a) Show that $\mathrm{Gal}(E/\mathbb{Q})$ is isomorphic to $\mathbb{Z}_2 \oplus \mathbb{Z}_2 \oplus \mathbb{Z}_2$.

   (b) For each subgroup of $\mathrm{Gal}(E/\mathbb{Q})$, identify its fixed field.

(23) Determine the Galois groups of each of the following extensions of $\mathbb{Q}$. Is the extension a normal extension? Explain. If an extension is not normal, find a normal extension containing the field.

   (a) $\mathbb{Q}(\sqrt{7})$

   (b) $\mathbb{Q}(\sqrt[4]{5})$

   (c) $\mathbb{Q}(\sqrt{2}, \sqrt[3]{2}, i)$

(24) Determine the Galois groups of $(x^3 - 2)(x^2 + 3)$ and $(x^3 - 2)(x^2 - 3)$ over $\mathbb{Q}$. Are these Galois groups the same? Explain.

(25) Determine the Galois groups of each of the following polynomials in $\mathbb{Q}[x]$. Which polynomials are solvable by radicals? Explain.

   (a) $x^5 - 10x^2 + 2$

   (b) $x^3 + 5$

   (c) $x^4 - x^2 - 6$

   (d) $x^4 + 1$

   (e) $x^8 - 1$

   (f) $x^8 + 1$

Connections

The material in this investigation connected almost all the topics discussed in this text, building bridges to solve problems in ring and field theory problem using topics in groups theory. Galois theory takes information involving rings (Investigation 4), polynomials and polynomial rings (Investigation 8), field extensions (Investigation 6), roots of polynomials (Investigation 10), irreducible polynomials (Investigation 10), and ring isomorphisms (Investigation 7) and connects them to to groups and subgroups (Investigations 17 and 19) to recognize Galois groups as symmetric groups (Investigation 22). Along the way, we encountered Galois groups as cyclic groups (Investigation 20), direct sums of groups (Investigation 25), and dihedral groups (Investigation 21) through group isomorphisms (Investigation 26).

Index

Printed in the United States
by Baker & Taylor Publisher Services